THE ROUTLEDGE HANDBOOK OF ENVIRONMENT AND COMMUNICATION

This Handbook provides a comprehensive statement and reference point for theory, research and practice with regard to environment and communication, and it does this from a perspective that is both international and multidisciplinary in scope. Offering comprehensive critical reviews of the history and state-of-the-art research into key dimensions of environmental communication, the chapters of this Handbook together demonstrate the strengths of multidisciplinary and cross-disciplinary approaches to understanding the centrality of communication and how the environment is constructed, and indeed contested, socially, politically and culturally.

Organized in five thematic parts, *The Routledge Handbook of Environment and Communication* includes contributions from internationally recognized leaders in the field. The first part looks at the history and development of the discipline from a range of theoretical perspectives. Part II considers the sources, communicators and media professionals involved in producing environmental communication. Part III examines research on news, entertainment media and cultural representations of the environment. The fourth part looks at the social and political implications of environmental communication, with the final part discussing likely future trajectories for the field.

The first reference Handbook to offer a state-of-the-art comprehensive overview of the emerging field of environmental communication research, this authoritative text is a must for scholars of environmental communication across a range of disciplines, including environmental studies, media and communication studies, cultural studies and related disciplines.

Anders Hansen is a Senior Lecturer in the Department of Media and Communication, at the University of Leicester, UK. He has published extensively on media and the environment. He is founder and ex-Chair of the IAMCR Environment, Science and Risk Communication Group, founding member and Secretary to the International Environmental Communication Association (IECA), and is an Associate Editor for *Environmental Communication*.

Robert Cox is Professor Emeritus at the University of North Carolina at Chapel Hill, USA. His principal research areas are environmental communication, climate change communication and strategic studies of social movements. He is author of *Environmental Communication and the Public Sphere* and serves on the editorial board of the journal *Environmental Communication*; he also advises US environmental groups on their communication programmes.

'Bringing together the foremost scholars in the field and examining environmental communication from a wide variety of angles and theoretical perspectives, this Handbook is a crucial addition to existing literature. It looks at all the relevant arenas and actors, and accurately maps recent developments in theory and practice. The greatest value of this volume resides however in the thorough and insightful nature of the reviews offered in its chapters: a must for any course on environmental communication and anyone interested in learning about the area.'

Anabela Carvalho, Associate Professor at the Department of
Communication Sciences of the University of Minho, Portugal

'In this timely and comprehensive Handbook, Hansen and Cox expertly organize the growing body of research in the area of environmental communication – growing in volume, importance and impact. The scope of the Handbook is a testament to the maturity of the field; it will serve as an indispensable reference for students and practitioners of communication about the natural world and our role in it.'

Lee Ahern, Associate Professor, College of Communications, Penn State University, USA,
and Chair of the International Environmental Communication Association

'Celebrating achievements whilst acknowledging challenges is a difficult balance to accomplish. Hansen and Cox's Handbook expertly navigates the past, present and future study of environment and communication to offer a highly engaging account of this multidisciplinary field. With passion and authority, the editors and authors of this volume demonstrate the significance of communication to the multiple practices and politics of the environment. In doing so, they remind us of how far the field has come since the 1970s, whilst providing an ethical and critical basis for future research practice.'

Julie Doyle, Reader in Media Studies, University of Brighton, UK

THE ROUTLEDGE HANDBOOK OF ENVIRONMENT AND COMMUNICATION

Edited by Anders Hansen and Robert Cox

LONDON AND NEW YORK

First published 2015
by Routledge
2 Park Square, Milton Park, Abingdon, Oxon OX14 4RN

and by Routledge
52 Vanderbilt Avenue, New York, NY 10017, USA

First issued in paperback 2020

Routledge is an imprint of the Taylor and Francis Group, an informa business

British Library Cataloguing in Publication Data
A catalogue record for this book is available from the British Library

Library of Congress Cataloging in Publication Data
A catalog record has been requested for this book

ISBN 13: 978-0-367-58183-1 (pbk)
ISBN 13: 978-0-415-70435-9 (hbk)

Typeset in Bembo
by GreenGate Publishing Services, Tonbridge, Kent

CONTENTS

PART II
Producing environmental communication: sources, communicators,
media and media professionals   **61**

Sources/communicators

Media and media professionals

FIGURES

TABLES

CONTRIBUTORS

Stuart Allan is Professor of Journalism and Communication, as well as Deputy Head of School (Academic), in the School of Journalism, Media and Cultural Studies (JOMEC) at Cardiff University, UK. He is the author or editor of several books, including most recently *Citizen Witnessing: Revisioning Journalism in Times of Crisis* (Polity Press, 2013). Stuart co-edits the journal *Journalism Education* and serves on several editorial boards, including *Environmental Communication*. His current research interests include examining the interface between citizen science and citizen journalism, where he has been focusing on alternative forms of environmental reporting in evolving online news and social media contexts.

Alison Anderson is Professor in the School of Government and Director of the Centre for Culture, Community and Society, Plymouth University, UK. She is author of numerous publications on media, politics and the environment including *Media, Environment and the Network Society* (Palgrave, 2014).

Maxwell T. Boykoff, Associate Professor, Cooperative Institute for Research in Environmental Sciences, University of Colorado-Boulder, USA.

Curtis Brainard is Blogs Editor for *Scientific American* and contributing editor at *Columbia Journalism Review*. Previously, he spent six years as a staff writer covering science, environment and medical news at CJR. While there, he established The Observatory, focusing on science coverage in the media.

Pat Brereton is currently Head of School of Communications at Dublin City University in Ireland. As a film and media scholar, he has published widely on ecology and environmental communication. His works include *Hollywood Utopia: Ecology in Contemporary American Cinema* (2005) and most recently *Smart Cinema: DVD Add-ons and New Audience Pleasures* (2012).

James G. Cantrill received his PhD from the University of Illinois-Urbana in 1985 and currently is a Professor and the Department Head of Communication and Performance Studies at Northern Michigan University. Dr Cantrill has a lengthy history of research and consulting related to environmental communication and, along with colleagues from around the world, was instrumental in the creation of conservation psychology as a distinct disciplinary focus in the social sciences.

Simon Cottle is Professor of Media and Communications and Head of the School of Journalism, Media and Cultural Studies at Cardiff University. His latest books are *Disasters and the Media* (2012) with Mervi Pantti and Karin Wahl-Jorgensen, *Transnational Protests and the Media* (2011) co-edited with Libby Lester, and *Global Crisis Reporting* (2009). He is Series Editor of the Global Crises and the Media Series for Peter Lang Publishing.

Robert Cox (PhD University of Pittsburgh) is Professor Emeritus in the Department of Communication Studies and Curriculum in the Environment at the University of North Carolina at Chapel Hill. His main research areas are environmental communication, climate communication and strategic studies of social movements. He is author of *Environmental Communication and the Public Sphere* (Sage). One of the leading scholars in the US in environmental communication, he has served as President of the Sierra Club, the largest grassroots environmental non-governmental organization in the U.S. His published work includes critical studies of the discourse of the civil rights, environmental, labour, climate and environmental justice movements. He serves as an advisory editor for *Environmental Communication*. Cox also serves on the Board of Directors for Earth Echo International, dedicated to youth education about oceans and marine ecology.

Steven E. Daniels is a Community Development Specialist in University Extension and Professor in the Department of Sociology, Social Work, and Anthropology and in the Department of Environment and Society at Utah State University, USA.

Mary Beth Deline is a graduate student in the Department of Communication at Cornell University, USA.

Stephen Depoe is Professor and Head, Department of Communication at the University of Cincinnati; he is former editor of *Environmental Communication: A Journal of Nature and Culture*, and past chair of the International Environmental Communication Association.

William Dinan lectures in public relations in the division of Communications, Media and Culture at the University of Stirling. His principal research interests are in the fields of lobbying, public relations and political communication. He is a steering committee member of the Alliance for Lobbying Transparency and Ethics Regulation in Europe (ALTER EU) and a founder of Spinwatch (www.spinwatch.org).

Sharon Dunwoody (PhD 1978) is Evjue-Bascom Professor Emerita of Journalism and Mass Communication at the University of Wisconsin-Madison, where she has taught science and environmental communication and conducted research for more than 30 years. She studies all aspects of the construction and impact of media science/environment messages, including relationships between scientists and the public.

Jens Emborg is an Associate Professor in the Department of Food and Resource Economics, Section for Environment and Natural Resources, University of Copenhagen, Denmark.

Jacqui Ewart is an Associate Professor of Journalism and Media Studies in the School of Humanities at Griffith University, Australia. Jacqui was a journalist and media manager for more than a decade and has worked as an academic for the past 18 years. Her research interests include media coverage of disasters, media representations of terrorism, community and talkback radio. She is the author of *Haneef: A Question of Character* (Halstead Press, 2009) and a co-author of *Media Framing of the Muslim World* (Palgrave Macmillan, 2014). Jacqui is an editorial board member of the journal *Media International Australia* as well as an editorial board member of Anthem Press' *Global Media and Communication* series.

Kimberly Freeman Brown is a social change strategist who has worked for over 20 years in service to national and community-based advocacy organizations including Green For All, American Rights at Work, the Robert F. Kennedy Memorial, the Smithsonian Institution and the Children's Defense Fund. She has developed or managed policy and public relations campaigns on environmental and economic justice, civil and human rights, racial equity, poverty alleviation, public health and the plight of women and disadvantaged youth.

Sharon M. Friedman is Professor and Director of the Science and Environmental Writing Program in the Department of Journalism and Communication at Lehigh University. Her research focuses on risk communication and on how scientific, environmental and health issues are communicated to the public.

Michael K. Goodman, Professor of Environment and Development, Department of Geography and Environmental Science, University of Reading, UK.

Natalie Grecu, Doctoral Candidate, The Edward R. Murrow College of Communication, Washington State University, USA. Natalie Grecu teaches strategic communication and advertising. Her research focuses on issues orientation among stakeholders in contentious environmental policy domains such as individual management of lakeshore property and its impact on water quality. She is especially interested in dialogic engagement of communication campaign audiences.

Anders Hansen is Senior Lecturer in the Department of Media and Communication, University of Leicester, UK. He is Associate Editor of *Environmental Communication*, Founder and immediate-past Chair of the IAMCR Group on Environment, Science and Risk Communication, and Executive Board Member and Secretary to the International Environmental Communication Association (IECA). His main research interests are in environmental, science and health communication. He is editor of the four-volume *Media and the Environment* (Routledge, 2014) and (with David Machin) *Visual Environmental Communication* (Routledge, 2015). He is author of *Environment, Media and Communication* (Routledge, 2010) and editor (with Stephen Depoe) of the *Palgrave Studies in Media and Environmental Communication* book series.

Se-Jin 'Sage' Kim is a doctoral candidate in the Department of Journalism and Technical Communication at Colorado State University. His research interests include media and climate change, information processing and risk communication. His teaching experience includes professional and technical communication, and international mass communication.

Anthony Leiserowitz (PhD University of Oregon, 2003) is a research scientist at the Yale University School of Forestry and Environmental Studies and Director of the Yale Project on Climate Change Communication. An expert on public opinion about climate change and the environment, his research investigates the psychological, cultural and political factors that influence environmental attitudes, policy support and behaviour.

Libby Lester is Professor and Head of Journalism, Media and Communications at the University of Tasmania. Her books include *Media and Environment: Conflict, Politics and the News* (Polity, 2010) and *Environmental Conflict and the Media* (co-edited with Brett Hutchins, 2013). Recent research has appeared in *Media, Culture and Society*, *International Journal of Communication*, *Environmental Communication* and *Media International Australia*.

Pieter Maeseele (PhD Ghent University) is Research Professor at the Department of Communication Studies of the University of Antwerp (Belgium). As a media sociologist, his research and teaching focuses on Science, Media and Democracy, and more specifically on the contribution of public discourse about science and environment to democratic debate and citizenship. The role of ideology and de/politicization, and their influence on pluralism, are central concerns.

Edward Maibach (MPH San Diego State, 1983; PhD Stanford, 1990) is a University Professor and Director of the Center for Climate Change Communication at George Mason University. He co-chairs the Engagement and Communication Working Group of the National Climate Assessment Development and Advisory Committee, and conducts research on a range of topics related to public engagement in climate change.

Katherine McComas is a Professor in the Department of Communication at Cornell University, USA.

Marisa B. McNatt, PhD student, Environmental Studies, University of Colorado-Boulder, USA.

Mark Meister is Professor and Chair of the Department of Communication at North Dakota State University, Fargo, ND, USA.

David Miller is Professor of Sociology in the Department of Social and Policy Sciences at the University of Bath, UK. His research interests mainly revolve around the role of communication in the constitution and reproduction of power relations. He is co-founder of Spinwatch, a website devoted to public interest reporting on spin lobbying and political corruption (on which he maintains an occasional blog), and Editor of Powerbase, a wiki that monitors power networks.

Susanne Moser, PhD, is Director of Susanne Moser Research and Consulting in Santa Cruz, California and a Social Science Research Fellow at Stanford Woods Institute for the Environment. Her work focuses on communication, adaptation to climate change and science–policy interactions. Dr Moser has contributed to IPCC and US national and regional assessments. She has been recognized as a fellow of the Leopold Leadership, Kavli Frontiers of Science, UCAR Leadership, Donella Meadows Leadership, Google Science Communication and Walton Sustainability Solutions Programs.

Todd P. Newman, MS is a doctoral student in the School of Communication at American University in Washington, DC. His research interests focus on the interplay between public opinion and the mass media in debates over science, technology and the environment.

Matthew C. Nisbet, PhD is Associate Professor of Communication Studies, Public Policy and Urban Affairs at Northeastern University, Boston, MA. The author of more than 70 peer-reviewed studies, chapters and reports, his research and teaching focuses on the role of communication, journalism and advocacy in environmental politics. You can follow his research and writing at www.climateshiftproject. org and on Twitter @MCNisbet.

Todd Norton, Association Professor, The Edward R. Murrow College of Communication, and Affiliate Faculty, School of the Environment, Washington State University, Pullman, USA; former Director: Center for Environmental Research, Education, and Outreach. Todd Norton teaches strategic communication and science communication. His research explores public engagement and change within contentious and complex environmental and commons resource issues such as public lands and water management.

Andy Opel, PhD is a Professor in the School of Communication and the Director of the Digital Media Production Program at Florida State where he teaches documentary production and environmental communication. To see more about his work visit www.andyopel.net.

Jennifer Peeples is an Associate Professor of Communication Studies at Utah State University and the past-president of the Environmental Communication Division for the National Communication Association. Her research area is environmental rhetoric with a focus on the construction of identity, place and toxicity.

Susanna Priest has taught journalism and media studies, public opinion and public affairs, science communication and environmental studies since 1987. She is Editor of *Science Communication: Linking Theory and Practice*. Her research work has been published in dozens of reviewed articles, book chapters and books, many focused on public opinion about emerging technologies.

Justin Rolfe-Redding is a doctoral candidate in the Department of Communication at George Mason University. His research at Mason's Center for Climate Change Communication has focused on messaging strategies, the role of cultural values in climate beliefs and the attitude structures of those who identify with the Republican party.

Connie Roser-Renouf, PhD (Stanford, 1986) is an Associate Research Professor at the Center for Climate Change Communication at George Mason University. Her research focuses on identifying effective communication strategies for building audience engagement with climate change.

Charlotte Ryan is Associate Professor of Sociology at University of Massachusetts Lowell and a member of its Climate Change Initiative, www.uml.edu/Research/Climate-Change. She co-directs the Movement and Media Research Action Project www.mrap.info.

David B. Sachsman is West Chair of Excellence in Communication and Public Affairs at the University of Tennessee at Chattanooga. He is the author of numerous works, including *Environment Reporters in the 21st Century*, *The Reporter's Environmental Handbook* and *Environmental Risk and the Press*.

Steve Schwarze is Associate Professor and Chair, Department of Communication at the University of Montana, USA.

James Shanahan is a Professor in the College of Communication at Boston University, USA.

Neil Stenhouse is a doctoral student in communication and graduate research assistant for the Center for Climate Change Communication at George Mason University. His interests include communicating the scale and nature of the risks involved with climate change, and building understanding of (and support for) the transformation of global energy systems.

Anne Marie Todd is Professor of Communication Studies at San José State University. Her book *Communicating Environmental Patriotism* (Routledge, 2013) suggests patriotism as a persuasive framework for sustainability. Her research has appeared in *Public Understanding of Science*, *Environmental Communication*, *Ethics and the Environment* and *Critical Studies in Media Communication*.

Craig Trumbo is a Professor in the Department of Journalism and Technical Communication at Colorado State University. His research addresses a range of interests involving health, risk and the environment. His areas of teaching include mass media effects, communication of science and technology, and risk communication.

JoAnn Myer Valenti, Emerita Professor of Communications, is co-author of *Environment Reporters in the 21st Century* and *Developing Protocol for Ethical Communication in Environmental News Coverage*. She was the founding SEJ Academic Board member and is an AAAS Fellow.

Gregg B. Walker is a Professor of Communication and an Adjunct Professor in the Environmental Sciences, Forest Ecosystems and Society, Public Policy and Water Resource Management programmes at Oregon State University, USA.

Lorraine Whitmarsh is an environmental psychologist, specializing in perceptions and behaviour in relation to climate change, energy and transport. She is a Senior Lecturer in the School of Psychology at Cardiff University, UK; and partner coordinator for the Tyndall Centre for Climate Change Research.

Andy Williams is a lecturer and researches about news and journalism at Cardiff School of Journalism, Media and Cultural Studies. He was previously the RCUK Research Fellow in Risk, Health and Science Communication. His current major research interests relate to news sources and the influence of public relations on the UK media, especially in the area of environment, science and health news. He also researches the decline of local newspapers and the rise of hyperlocal community news, and citizen participation in news production more generally.

ACKNOWLEDGEMENTS

We are grateful to Andrew Mould, our editor at Routledge, for suggesting that we put together this Handbook and for his guidance and support throughout the process. Thank you also to Andrew's colleagues at Routledge, who have been involved at various stages, and in particular to Sarah Gilkes for assisting us through the final stages. In realizing this project, we are fortunate and privileged to have been able to draw on some of the foremost scholars in the field of environmental communication and beyond, and we wish to express our gratitude to all our contributors – it's been a great pleasure working with you. Anders Hansen gratefully acknowledges support from the University of Leicester in the form of academic study leave during semester 2 of 2013/14, and he thanks his colleagues at the University of Leicester and elsewhere for their support.

ABBREVIATIONS

AC	appropriate collaboration
BCSD	Business Council for Sustainable Development
BELC	Business Environmental Leadership Council
CDA	Critical Discourse Analysis
CEQ	Council of Environmental Quality
CIRES	Cooperative Institute for Research in Environmental Sciences
CMA	conflict management assessment
COCE	Conference on Communication and Environment
COPUS	Committee on the Public Understanding of Science
CP	collaborative potential
CSTPR	Center for Science and Techology Policy Research
DSP	Dominant Social Paradigm
E&E	Energy & Environment Publishing
ECC	environmental communication campaign
ECREA	European Communication Education and Research Association
EHS	Environmental Health Sciences
EJ	environmental justice
EJN	Earth Journalism Network
ELM	Elaboration Likelihood Model
E-ELM	Extended-Elaboration Likelihood Model
EMA	Environmental Media Association
EPA	Environmental Protection Agency
FDA	Food and Drug Administration
FERN	Food & Environment Reporting Network
FONSI	finding of no significant impact
GCC	Global Climate Coalition
GFA	Green For All
GIS	Geographic Information System
GSS	General Social Survey
HBCUs	Historically Black Colleges and Universities
HSM	Heuristic Systematic Model

IAMCR	International Association for Media and Communication Research
IAP2	International Association of Public Participation
ICA	International Communication Association
IECA	International Environmental Communication Association
IPCC	Intergovernmental Panel on Climate Change
JATAN	Japan Tropical Forest Action Network
MAO	Motivation, Ability and Opportunity
NCA	National Communication Association
NEPA	National Environmental Policy Act
NFMA	National Forest Management Act
NRC	National Research Council
PCB	polychlorinated biphenyl
OPAL	Open Air Laboratories
SARF	social amplification of risk framework
SFL	systemic-function linguistics
SMC	Science Media Centre
SOC	Stage of Change
STS	Science and Technology Studies
Tech-reg	technical-regulatory
TEK	traditional ecological knowledge
TMDL	determining total maximum daily load
TMoC	Transtheoretical Model of Change
TSL	The Sainsbury Laboratory
UNF	Unifying Negotiation Framework
WBCSD	World Business Council on Sustainable Development

INTRODUCTION

Environment and communication

Anders Hansen and Robert Cox

The 'environment' has become one of the key public and political concerns of our time. The very concept of 'the environment' as well as the public discourse on the environment, which now seems so familiar, are however relatively recent phenomena dating back only to the 1960s. This is not to say that some of the concerns now expressed about the environment were not present prior to the 1960s, but rather that they have coalesced and become focused in the last half century into what we might call an environmental discourse, replete with its own distinctive vocabulary, themes, frames and perspectives.[1]

The seeming familiarity of the environmental discourse and its vocabulary is due to a large extent to the concurrent growth witnessed in the last half century in (mass) mediated forms of communication, including perhaps particularly the rise and increasing ubiquity of visual media. Traditional broadcast and print media and newer forms of digital communication have been instrumental in defining 'the environment' as a concept and domain, and in bringing environmental issues and problems to public and political attention. Although the role of neither personal experience nor education should be underestimated, it is likely the case – not least because of the long timescales and low visibility of many environmental problems – that we as publics get most of our information and understanding about the challenges facing our environment in mediated form.

Given the central role of media and communication in defining 'the environment' and in defining it as an issue or problem for public and political concern, it is thus not surprising that the last three to four decades have also witnessed a very considerable expansion in research interest in media and communication processes regarding the environment, and in the field of environmental communication scholarship.

This Handbook brings together international scholars and multidisciplinary perspectives to offer state-of-the-art reviews charting the history and development of environmental communication scholarship, and examining core concepts, theories and research in the study of environment and communication. It is our belief that such examination can help, not only to understand the centrality of communication processes and communications media in the public sphere, but political definition, elaboration and contestation of environmental issues and problems. Starting with overviews that chart the emergence and development of the field of environmental communication research, we proceed with research examining the three major domains of the communication process: the sources and production of communication about the environment; the study of media and cultural representations of the environment; and the study of how communication about the environment impacts on and interacts with public and political beliefs about the environment, as well as political action regarding the environment. In the concluding part, future trajectories for the field of environmental communication research are proposed, mapped out and discussed. In the following, we provide an overview of what is covered under each of the five parts of the Handbook.

Part I: Environment, communication and environmental communication: emergence and development of a field

Research on media, communication and the environment dates back to the 1970s. Early pioneering studies that have been influential in shaping the development include Anthony Downs' (1972) examination of the public careers of social issues, including the environment as a social problem, David Sachsman's (1976) study of source-influence on environmental reporting, and Harvey Molotch and Marilyn Lester's (1975) study of news reporting of a major oil spill. At the end of the 1970s came one of the first studies to offer a comprehensive perspective on the key role of the news media in the public construction of the environment as a social problem, namely the study by Schoenfeld, Meier and Griffin (1979). The 1980s saw the publication of important work that in several ways was directly relevant to the rise of media/communication research on the environment: this included work on media and nuclear power (Mazur, 1984; Rubin, 1987; Gamson and Modigliani, 1989), crises/disasters (Nimmo and Combs, 1985; Walters *et al.*, 1989), environmental news journalism (Lowe and Morrison, 1984), and media and science/technology communication (Friedman *et al.*, 1986; Nelkin, 1987).

While the 1970s and 1980s thus produced a steady trickle of media and communication studies relevant to or directly touching on the environment and environmental issues, the 1990s can be characterized as the decade where these trends first coalesced into a distinctive focus on 'media and the environment'. A special issue entitled 'Covering the Environment' of the journal *Media, Culture and Society* in 1991 provided an early thematic focus on media and environment, and this was consolidated further in one of the first book-length academic collections of its kind on *The Mass Media and Environmental Issues* (Hansen (ed.), 1993). This was followed by several book-length introductions with a core focus on media and the environment (Hannigan, 1995; Anderson, 1997; Chapman *et al.*, 1997; Lacey and Longman, 1997; Shanahan and McComas, 1999) as well as others touching directly or indirectly on the key roles of discourse, rhetoric and communication in relation to the environment and nature (Cronon, 1995; Hajer, 1995; Cantrill and Oravec, 1993 and 1996; Macnaghten and Urry, 1998).

The 1990s consolidation of environmental communication research has continued and become significantly more pronounced during the start of the present century. This is evident not just in a marked increase in scholarly research on media, communication and the environment, but in the embedding of environmental communication research within university-level curricula and in sections and groups within national and international communication associations. Sustaining this trend and its consolidation is the growing body of books on environmental communication and closely related fields (e.g. Allan *et al.*, 2000; Allan, 2002; Senecah, 2004 and 2005; Cox, 2006 (3rd edition 2013); Corbett, 2006; Hannigan, 2006; Boyce and Lewis, 2009; Cottle, 2009; Hansen, 2010; Lester, 2010; Hendry, 2010; Doyle, 2011; Boykoff, 2011; Lester and Hutchins, 2013; Merry, 2014; Todd, 2014), and the rapid growth in environment-focused journal articles across a range of science/environment/health and communications journals, including the establishment of academic journals specifically focused on environmental communication.

In Chapter 1, Robert Cox and Stephen Depoe set the scene by tracing the emergence and growth of the field of environmental communication. They examine the field's institutional bases and delineate some of its key assumptions and research questions. Focusing on four emerging issues – climate change communication, sustainability science, visual communication or the 'imaging' of nature, and the problematizing of the human/nature binary – they demonstrate the rapid expansion and diversification that the field has experienced in the recent decades.

What is clear, and evident in the chapters of the Handbook, is that the broad field of research, that we can now label as 'environmental communication' research, owes much of its innovativeness, dynamic development and diversity to the fact that it draws from a wide range of theoretical and disciplinary traditions. Research has developed in and from such varied fields as sociology, geography, political science, historical studies, psychology, social psychology, media studies, cultural studies and literary, linguistic and rhetorical studies.

Theoretically, key inspiration has come from a similarly diverse range, although several dominant trends – often cutting across traditional disciplinary boundaries – can be identified. Social constructionist theory in the tradition of Blumer (1971) and Spector and Kitsuse (1977) has featured prominently. In Chapter 2, Anders Hansen traces the application of social constructionist theory in environmental communication research, showing, *inter alia*, how social constructionist perspectives merged, in the 1970s and 1980s, with other developments in the sociology of news to help propel journalism and news research out of theoretically limited and circular concerns about bias, balance and objectivity in news reporting. Given the social constructionist perspective's central emphasis on communication in the form of 'claims-making', its attractiveness to communication researchers, sociologists and others interested in the role of media and communication processes in relation to defining the environment as an issue for public and political concern can hardly come as a surprise.

A common feature across much analysis of media, communication and the environment is the close attention often paid to language, rhetoric and discourse in public communication about the environment. There is thus a clear recognition that lexical choice, narrative and discursive practices are central components of how issues are rhetorically constructed and 'framed' and how in turn particular messages/ meanings are conveyed and boundaries set for public understanding and public interpretation/opinion regarding environmental issues. Jennifer Peeples (Chapter 3) traces the influence of rhetorical and discourse analytical approaches in environmental communication research, and she shows how the insights afforded by these approaches have proved particularly productive in uncovering the meanings of media content and other public communication about the environment.

Much study in the field of environmental communication, but also cutting across the closely adjacent fields of science and risk communication, has drawn on German sociologist Ulrich Beck's work on 'risk society' (1992, 1995, 1999, 2009) and to a lesser extent on prominent globalization scholar Manuel Castells (1997, 2009). Work in psychology, including cognitive psychology, has been an important inspiration particularly for research on environmental risk communication and associated public perception. The work of Slovic, Fischhoff, Renn and associates (e.g. Slovic and Fischhoff, 1982; Slovic *et al.*, 1985; Renn *et al.* 1992; Morgan *et al.*, 2002) has provided key models for understanding public perception of risk and much work in recent decades on climate change communication and public perception similarly draws on psychology and social psychology for understanding some of the complexities of how publics perceive and make sense of media and public communication about climate change and other environmental risks (Leiserowitz, 2006; Butler and Pidgeon, 2009; Pidgeon and Fischhoff, 2011).

In Chapter 4, James Cantrill – demonstrating the importance of interdisciplinary approaches in environmental communication – charts the contribution to environmental communication from six central disciplines in the social sciences: economics, history, human geography, political science, sociology and psychology. Providing a synopsis of the development of an environmental focus in each of the disciplines and their particular conceptual and methodological approaches, he offers illustrative examples – and a critique – of how research in these core social science disciplines can inform our current understanding of the relationship between media, communication and the environment at large.

A sign of the increasing maturity of 'environmental communication' is the increasing diversification and broadening of scope, both in terms of theoretical and disciplinary traditions, and significantly in terms of the types of media, communications genres and communications processes studied. This widening of scope is reflected throughout the Handbook in the range of media, genres, cultural representations and communication forms that are considered.

Part II: Producing environmental communication: sources, communicators, media and media professionals

Research on the 'production of environmental communication' focuses traditionally on the sources, who make claims in the public sphere and/or try to influence what is publicly communicated (including through the media), and the media and media professionals, whose task it is to report on or cover the environment and 'environmental issues'. Much of the research on the relationship between sources and journalists has focused on specialist reporters and on three types of sources: scientists/ experts, environmental pressure groups and government/big business. As the media and communications landscape, broadly speaking, has changed enormously in the last few decades – whether looked at in terms of technology, communication flows, organizational arrangements, or ownership and control – so too have the nature of communication, professional roles and types of actors involved in public communication and discourse about the environment. The chapters in this section chart these changes and show how research focused on the key actors involved in communicating about the environment can help in understanding the dynamics and politics of how public discourse on the environment is shaped and contested.

The first six chapters in this section focus predominantly on key types of actors among the variety of primary sources involved in communicating about the environment. The second group of chapters focuses primarily on the changing nature of news organizations and implications for environmental journalism and journalists.

In Chapter 5, Sharon Dunwoody explores the historical evolution of scientists as sources of information for science and environmental journalism. Charting the historically changing emphases on public communication of science, she argues that 'scientists generally – and environmental scientists specifically – continue to hew to a model of science communication that emphasizes educating the public rather than engaging them'.

From scientists in Chapter 5, the focus turns in Chapter 6 to Robert Cox and Steve Schwarze's discussion of environmental pressure groups and NGOs as key sources/actors/claims-makers in environmental debate. Cox and Schwarze examine the communication strategies and media/new media uses of these groups with particular emphasis on their cultivation of news media, rhetorical appeals, issue frames, and alignment of media and audiences.

In Chapter 7, David Miller and William Dinan examine the role of elite planning groups, think tanks, institutes and other lobbying organizations in public relations and campaigning on climate change. Noting the tension between dominant proactive responses to climate change and climate change scepticism/ denial, they argue for the need to go well beyond examinations of the relative success of sources/claims in mainstream media reporting and to attend also to these sources' claims-making and influence in a wider range of scientific and political arenas.

Libby Lester and Simon Cottle (Chapter 8) move the examination of key sources and claims-makers in environmental protest and debate beyond the specific types/groups of sources/actors to focus more broadly on the changing opportunities and strategies that are emerging with new communications technologies and increasing global connectedness of media and communications. They explore the implications for environmental publics and citizenship and related debates on transnational public spheres and cosmopolitanism.

The two next chapters offer a unique combination of theory-based scholarly insight and experience of campaigning and public/stakeholder engagement on environmental issues. In Chapter 9, Gregg Walker, Steven Daniels and Jens Emborg draw on over two decades of field experience with collaborative learning training and field projects in the United States and Scandinavia. Pointing to the significance of the United States' National Environmental Policy Act as an archetype for public participation policies throughout the world, they show how collaborative stakeholder engagement has evolved in recent decades. They conclude by offering six key lessons about communication, participation and stakeholder engagement in the environmental and natural resource policy arenas.

Combining critical scholarly insight with extensive experience of campaign communication, Charlotte Ryan and Kimberly Freeman Brown (Chapter 10) address the question of how marginalized communities who are directly affected by power inequities build communication power as a component of strategic movement building. Critically reviewing and assessing two current models of communication and social change – social marketing and media advocacy – and drawing on recent environmental campaign examples, they make a compelling case for a 'movement-building communication model' and conclude with a discussion of the implications of the different models for marginalized communities pursuing structural change.

Moving the focus from the sources discussed in the first six chapters of this section to the media and media professionals involved in representing/mediating the environmental claims, Chapters 11 and 12 chart the changes in environmental journalism and the roles of environmental journalists in the US in recent decades. Sharon Friedman (Chapter 11) discusses the implications for environmental journalism of three key factors: the Internet, downsizing and mainstreaming. She argues that the Internet has had both negative (downsizing, mainstreaming – resulting in the reduction or outright elimination of environmental beats and reporters) and positive (enhancing the range and nature of information resources available) implications for environmental journalism.

Drawing on their extensive surveys of American environmental journalism, David Sachsman and JoAnn Myer Valenti (Chapter 12) chart the rise and decline of environmental journalism in the newspaper and television industries, providing valuable and detailed statistical evidence that confirms the trends also delineated by Friedman in the previous chapter. Leading science and environmental journalist Curtis Brainard (Chapter 13) complements the previous two chapters with a first-hand account and professional insight into the significant upheavals and transformations in environmental journalism of recent times. Drawing on key examples from changes in environmental journalism in the last two decades, he argues that US news organizations largely failed to adapt to the rapidly changing news environment, with severe implications for the environment beat, environmental correspondents and environmental journalism.

Alison Anderson (Chapter 14) connects the trends observed in the three previous chapters to the significant body of literature that has emerged from the sociology of news production generally and the production of environmental news particularly. She shows how organizational arrangements, journalistic practices, cultural assumptions and source relations combine to influence the nature of environmental news, while also arguing that the significant changes witnessed in these areas and highlighted in the previous two chapters call for a rethink of traditional approaches to the study of news production.

The rapidly changing nature of environmental journalism and the need to rethink approaches to the study of this are explored by Stuart Allan and Jacqui Ewart in Chapter 15 through their focus on citizen scientists and citizen journalists. Drawing on recent examples from major environmental disaster scenarios, they show how citizen journalists have played a key role in putting environmental science on the public agenda, and they conclude by outlining key issues for further study of citizen science, citizen journalism and the production of public knowledge about environmental problems.

Pursuing the key questions outlined in several of the previous chapters regarding the changing balance of power between sources and media professionals, Andy Williams in the final chapter (Chapter 16) of this section examines the growing role of environment public relations practices and their impact on the quality and independence of environment news. He argues that the enhanced use of PR by sources, combined with increasing institutional and economic constraints on journalism, lead to increased dependence on authority sources.

Part III: Covering the environment: news media, entertainment media and cultural representations of the environment

Of the three major domains of the communication process – production, content and audiences – it is the media representations (content) of environment and environmental issues that have attracted the bulk of communication research interest. As with media and communication research generally and wider topics, the main focus of environmental communication research interest has been on news reporting and coverage with regard to environmental problems, disasters, crises and policies. Recognizing, however, that the symbolic environment through which images, ideas and messages are communicated and circulated in society goes far beyond the news media, the chapters in this section broaden the scope to examine research on a much more diverse range of entertainment genres/media and cultural representations of the environment.

The first three chapters in this section examine the research evidence on news coverage of the environment. Anders Hansen (Chapter 17) shows how longitudinal studies are better equipped than short-term snapshots of news reporting to help us understand what drives the often observed 'ups' and 'downs' in news attention to the environment. Max Boykoff, Marisa McNatt and Michael Goodman (Chapter 18) review and discuss the evidence from large-scale longitudinal and cross-cultural comparative research on news reporting of climate change. They demonstrate the importance of context – cultural, social, political, economic and environmental – in analysing the exercise of power and media roles in the public and political construction of climate change. These two chapters are complemented by Libby Lester's (Chapter 19) examination of how online media and new media are not only adding to but in many ways significantly changing the roles of traditional news media. She shows how the changing ecology of environmental communication holds opportunities and has implications for the communications strategies of key stakeholders (environmental groups, government, businesses – and the media organizations themselves), with ultimate implications for the balances of political power associated with these.

The next six chapters (Chapters 20 through 25) in this section focus on non-news media and cultural representations of the environment. James Shanahan, Katherine McComas and Mary Beth Deline (Chapter 20) survey the comprehensive body of research built up over the last couple of decades on television entertainment representations of the environment. Coupled, within a Cultural Indicators framework, with research on television audiences, the evidence shows, they argue, relatively consistent trends of television entertainment viewing being associated with lower levels of concern about the environment.

In Chapter 21, Anne Marie Todd discusses the flexibility and scope for a more critical view offered by cartoons and animation, compared with other genres and media forms, and she argues that animated 'environmental messages construct and contest the environment through apocalyptic visions, risk communication and crisis response'. In the following chapter (Chapter 22), Pat Brereton surveys the emerging body of scholarly analysis of the representation of environment and ecology in film, increasingly referred to as 'eco-film criticism'. Exemplified through mass audience 'eco-films' of the recent decade, Brereton – like Todd in the previous chapter – shows the narrative flexibility, and associated enhanced scope for a more nuanced view of the environment, afforded by film compared with more restrictive media or genres.

In Chapter 23, the focus turns to the representation and uses of nature and environment in commercial advertising. Reviewing a growing body of research on this topic, Anders Hansen shows how advertising articulates and reworks deep-seated cultural categories and understandings of nature, the natural and the environment, and in doing so, communicates important boundaries and public definitions of appropriate consumption and 'uses' of the natural environment.

In the following chapter (Chapter 24), Mark Meister examines the rising role of celebrities and celebrity culture in environmental campaigning and in popular/public environmental discourse. The chapter surveys key categories of celebrity-endorsed environmental activism and discusses the relationship between celebrity environmental activism and corporate social responsibility.

In the final chapter (Chapter 25) of this section, Andy Opel shows how environmental language and images are deeply embedded in, what he refers to as, 'the backdrop of our consumer culture' and the 'wallpaper of contemporary life', exemplified through such media and genres as greeting cards, board games, computer screens and theme parks.

Part IV: Social and political implications of environmental communication

Environmental communication research is concerned, ultimately, with mapping and understanding how media and communication processes impact on and shape public understanding/opinion and political decision making in society. Like research on the production and content of media representations of environmental issues, studies of the wider social implications of such coverage have been characterized by increasing sophistication and appreciation of the highly dynamic and complex ways in which environmental messages, images and beliefs are promoted, contested and circulated in society. The rapidly changing nature of the media and communications landscape, combined with increasingly differentiated models of how we as individuals interact with the media and communications environments, have helped move the emphasis in the study of 'publics' for environmental communication away from notions of a largely passive mass audience towards notions of a more active and highly differentiated audience (see particularly Priest, Chapter 26, Whitmarsh, Chapter 29, Norton and Grecu, Chapter 30, and Roser-Renouf *et al.*, Chapter 31).

Much of what we now know about social and political implications of mediated environmental communication has drawn on prominent media and communication research models such as cultural indicators/cultivation analysis (see Shanahan *et al.* in Chapter 20), agenda-setting research (Trumbo and Kim, Chapter 27) and framing research (Nisbet and Newman, Chapter 28), while research from the disciplines of psychology and social psychology has drawn on, for example, the social-amplification-of-risk model to provide increasingly differentiated insights into how different publics interpret environmental issues (see Whitmarsh, Chapter 29 and Roser-Renouf *et al.*, Chapter 31).

Susanna Priest (Chapter 26) charts the changes in approaches to the study of how publics acquire, interact with and interpret science, environment and risk information and communication. Noting the rise of the Internet-dominated world, and the ability of individuals to readily seek out information compatible with their existing views, she argues that public views are shaped more by ideology and trust than by the nature and quality of the underlying science. 'Critical science literacy', she argues, is therefore now more important than traditional concerns about public understanding of science.

Noting the longstanding prominence of 'environment' in agenda-setting research, Trumbo and Kim provide, in Chapter 27, an exemplary overview and review of how this type of research has helped throw light on the role of media and communication processes in shaping public and political perception, views, behaviour and decisions. This is complemented in Chapter 28 by Nisbet and Newman's review of framing research (often seen as closely associated, or indeed integrated, with the agenda-setting perspective) and its contribution to understanding public and political opinion and debate regarding the environment. Drawing on examples from public engagement campaigns on climate change, they outline new directions for framing research in environmental communication.

Lorraine Whitmarsh (Chapter 29) extends the previous chapters' discussions by examining in particular the contributions from psychology and closely related disciplines to our understanding of how public perceptions and images of environmental change are formed. Drawing on studies – deploying a range of approaches and research methodologies – from across several countries, she shows that while public concern for environmental issues changes over time in ways that are influenced by information/communication sources, processes and context, deeper environmental values tend to endure. Building on exciting, emerging evidence from experimental research, she concludes by delineating how environmental change may be communicated more effectively.

Effective communication of environmental change is very much the focus of the following chapter (Chapter 30) in which Todd Norton and Natalie Grecu offer a comprehensive review of communication campaigns and persuasive communication in environmental communication. They confirm the need – also voiced in other chapters in this section – for a differentiated/segmented view of publics and an associated tailoring of communication strategies. They note how the increasing complexity of the science of environmental issues and the interdisciplinary nature of environmental campaigning call for an assessment of traditional strategies and tools, while at the same time opening up new and exciting opportunities for interdisciplinary scholarship and application.

The theme of audience segmentation and associated tailoring of message strategies is also at the core of the final chapter (Chapter 31) in this section. Connie Roser-Renouf and her colleagues draw on their extensive research, identifying six major perspectives of American publics on global climate change, and they argue that understanding the sources and cultural/political underpinnings of people's views is key to effective communication aimed at engaging with and changing public understanding and action with regard to climate change.

Part V: Conclusions: future trajectories of environment and communication

Environmental communication research has come a long way in the last few decades, on the one hand consolidating itself as a distinctive subfield of media and communication research, while at the same time healthily diversifying in terms of theoretical frameworks, analytical approaches and types of media and communications processes examined.

The main achievement is perhaps the considerable advances in the last two decades towards an increasingly sophisticated understanding of the complex processes involved in the social 'construction' of the environment as an issue for public and political concern. Not least, as the chapters in this Handbook demonstrate, environmental communication research has made great strides towards showing the complex and highly unequally distributed resources and power-relations involved in public communication and definitions regarding the environment.

As argued elsewhere (Hansen, 2011), and while recognizing the significant achievements of environmental communication research as collectively evidenced by the chapters in this Handbook, there is scope for further strengthening the integration of research across the three core domains – production, content, publics – of environmental communication to enhance our understanding of the complex (communication) processes and unequal power relations involved in the social construction and politics of the environment.

The final part concludes with two different, yet complementary, accounts of the field of environmental communication and two different, yet complementary, visions of the future trajectories of the field. In Chapter 32, Pieter Maeseele argues for a reorientation of research aims in environmental communication 'towards social roles of media in liberal democratic societies and the relationship between media(ted) discourses, power and democratic politics'. Drawing from the literature on radical democracy and post-politics, he argues that a politicization of academic discourse is needed to facilitate research designs aimed at revealing processes of de/politicization in public discourse about environmental risk and issues.

Finally, in Chapter 33, Susanne C. Moser asks whether the environmental communication field may be losing touch with the very 'heart' of communication at a time of perpetual environmental crises. Such crises and the pervasiveness of technology-based communication, she argues, 'open a gap, a profound need', that calls for a more humanistic view of the field. Adding a layer to the notion of environmental communication as a 'crisis discipline', Moser urges a humanistic turn that 'is about communicating meaningfully and supportively to those *living through crisis*', one that is oriented towards compassion, truth telling, grieving, envisioning alternative futures and hope. Moser closes with an appeal to environmental communication researchers and practitioners: communicating in a time of crisis requires 'not just warnings and clarion calls to action but to partake in the restoration of our relationships to each other and between ourselves and the more-than-human world'.

Collectively, the following chapters explore the core concepts, theories and findings that characterize the complex and diverse terrain of the field of environmental communication. In the twenty-first century's second decade, scholars and practitioners in environmental communication have much to guide us in our understanding of the sources and production, the media and cultural representations, and the public and political impacts of communication about the environment, and more. These reviews occur, as several authors emphasize, at a time when political, economic and ideological interests, and the uneven distribution of communicative resources are increasingly entwined with environmental crises, placing both 'nature' and human communities at risk. The contributions in this volume, therefore, address not simply the occurrence of crises, but the prospects for a field of environmental communication to fashion a response, one attuned to the environmental well-being of both human communities and the 'more-than-human world'.

Note

1   Parts of this chapter draw on Hansen, 2011.

References

Allan, S. (2002). *Media, Risk and Science*. Maidenhead: Open University Press.

Allan, S., Adam, B. and Carter, C. (eds) (2000). *Environmental Risks and the Media*. London: Routledge.

Anderson, A. (1997). *Media, Culture and the Environment*. London: UCL Press.

Beck, U. (1992). *Risk Society: Towards a New Modernity* (M. Ritter, Trans.). London: Sage.

Beck, U. (1995). *Ecological Politics in an Age of Risk*. Cambridge: Polity Press.

Beck, U. (1999). *World Risk Society*. Cambridge: Polity Press.

Beck, U. (2009). *World at Risk*. Cambridge: Polity Press.

Blumer, H. (1971). Social problems as collective behavior. *Social Problems*, 18(3), 298–306.

Boyce, T. and Lewis, J. (eds) (2009). *Climate Change and the Media*. Oxford: Peter Lang.

Boykoff, M. (2011). *Who Speaks for the Climate? Making Sense of Media Reporting on Climate Change*. Cambridge: Cambridge University Press.

Butler, C. and Pidgeon, N. (2009). Media communications and public understanding of climate change: reporting scientific consensus on anthropogenic warming. In T. Boyce and J. Lewis (eds), *Media and Climate Change* (pp. 43–58). Oxford: Peter Lang.

Cantrill, J. and Oravec, C. (eds) (1993). *Environmental Discourse: Perspectives on Communication and Advocacy*. Louisville, KY: University Press of Kentucky.

Cantrill, J. G. and Oravec, C. L. (eds) (1996). *The Symbolic Earth: Discourse and Our Creation of the Environment*. Lexington, KY: University of Kentucky Press.

Castells, M. (1997). *The Power of Identity* (Vol. II). Oxford: Blackwell.

Castells, M. (2009). *Communication Power*. Oxford: Oxford University Press.

Chapman, G., Kumar, K., Fraser, C. and Gaber, I. (1997). *Environmentalism and the Mass Media: The North–South Divide*. London: Routledge.

Corbett, J. B. (2006). *Communicating Nature: How We Create and Understand Environmental Messages*. Washington, DC: Island Press.

Cottle, S. (2009). *Global Crisis Reporting: Journalism in the Global Age*. Maidenhead: Open University Press.

Cox, R. (2006). *Environmental Communication and the Public Sphere*. London: Sage.

Cronon, W. (ed.). (1995). *Uncommon Ground: Toward Reinventing Nature*. New York: Norton.

Downs, A. (1972). Up and down with ecology – the issue attention cycle. *The Public Interest*, 28(3), 38–50.

Doyle, J. (2011). *Mediating Climate Change*. Aldershot: Ashgate.

Friedman, S. M., Dunwoody, S. and Rogers, C. L. (eds). (1986). *Scientists and Journalists: Reporting Science as News*. New York: The Free Press.

Gamson, W. A. and Modigliani, A. (1989). Media discourse and public opinion on nuclear power: a constructionist approach. *American Journal of Sociology*, 95(1), 1–37.

Hajer, M. (1995). *The Politics of Environmental Discourse: Ecological Modernization and the Policy Process.* Oxford: Clarendon Press.

Hannigan, J. A. (1995). *Environmental Sociology.* London: Routledge.

Hannigan, J. A. (2006). *Environmental Sociology* (2nd edn). London: Routledge.

Hansen, A. (ed.). (1993). *The Mass Media and Environmental Issues.* Leicester: Leicester University Press.

Hansen, A. (2010). *Environment, Media and Communication.* London: Routledge.

Hansen, A. (2011). Communication, media and environment: towards reconnecting research on the production, content and social implications of environmental communication. *International Communication Gazette, 73*(1–2), 7–25.

Hendry, J. (2010). *Communication and the Natural World.* State College, PA: Strata Publishing.

Lacey, C., and Longman, D. (1997). *The Press as Public Educator: Cultures of Understanding, Cultures of Ignorance.* Luton: University of Luton Press.

Leiserowitz, A. (2006). Climate change risk perception and policy preferences: the role of affect, imagery, and values. *Climatic Change, 77*(1–2), 45–72.

Lester, L. (2010). *Media and Environment.* Cambridge: Polity.

Lester, L. and Hutchins, B. (eds). (2013). *Environmental Conflict and the Media.* New York: Peter Lang.

Lowe, P. and Morrison, D. (1984). Bad news or good news: environmental politics and the mass media. *The Sociological Review, 32*(1), 75–90.

Macnaghten, P., and Urry, J. (1998). *Contested Natures.* London: Sage.

Mazur, A. (1984). The journalists and technology: reporting about Love Canal and Three Mile Island. *Minerva, 22*(Spring), 45–66.

Merry, M. K. (2014). *Framing Environmental Disaster: Environmental Advocacy and the Deepwater Horizon Oil Spill.* New York: Routledge.

Molotch, H. and Lester, M. (1975). Accidental news: the great oil spill. *American Journal of Sociology, 81*(2), 235–260.

Morgan, M. G., Fischhoff, B., Bostrom, A. and Atman, C. J. (2002). *Risk Communication: A Mental Models Approach:* Cambridge University Press.

Nelkin, D. (1987). *Selling Science: How the Press Covers Science and Technology.* New York: W. H. Freeman & Company.

Nimmo, D. and Combs, J. E. (1985). *Nightly Horrors: Crisis Coverage in Television Network News.* Knoxville, TN: University of Tennessee Press.

Pidgeon, N. and Fischhoff, B. (2011). The role of social and decision sciences in communicating uncertain climate risks. *Nature Climate Change, 1*(1), 35–41.

Renn, O., Burns, W. J., Kasperson, J. X., Kasperson, R. E., and Slovic, P. (1992). The social amplification of risk: theoretical foundations and empirical applications. *Journal of Social Issues, 48*(4), 137–160.

Rubin, D. M. (1987). How the news media reported on Three Mile Island and Chernobyl. *Journal of Communication, 37*(3), 42–57.

Sachsman, D. B. (1976). Public relations influence on coverage of environment in San Francisco Area. *Journalism Quarterly, 53*(1), 54–60.

Schoenfeld, A. C., Meier, R. F. and Griffin, R. J. (1979). Constructing a social problem – the press and the environment. *Social Problems, 27*(1), 38–61.

Senecah, S. (ed.). (2004). *The Environmental Communication Yearbook* (Vol. 1). London: Lawrence Erlbaum Associates.

Senecah, S. (ed.). (2005). *The Environmental Communication Yearbook* (Vol. 2). London: Lawrence Erlbaum Associates.

Shanahan, J. and McComas, K. (1999). *Nature Stories: Depictions of the Environment and their Effects.* Cresskill, NJ: Hampton Press.

Slovic, P. and Fischhoff, B. (1982). Determinants of perceived and acceptable risk. In L. C. Gould and C. A. Walker (eds), *Too Hot to Handle: The Management of Nuclear Wastes* (pp. 112–150). New Haven, CT: Yale University Press.

Slovic, P., Fischhoff, B. and Lichtenstein, S. (1985). Characterizing perceived risk. In R. W. Kates, C. Hohenemser and J. Kasperson (eds), *Perilous Progress: Managing the Hazards of Technology* (pp. 91–123). Boulder, CO: Westview Press.

Spector, M. and Kitsuse, J. I. (1977). *Constructing Social Problems.* Menlo Park, CA: Cummings.

Todd, A. M. (2014). *Communicating Environmental Patriotism: A Rhetorical History of the American Environmental Movement.* London: Routledge.

Walters, L. M., Wilkins, L. and Walters, T. (eds). (1989). *Bad Tidings: Communication and Catastrophe.* Hillsdale, NJ: Lawrence Erlbaum Associates.

PART I

Environment, communication and environmental communication

Emergence and development of a field

1

EMERGENCE AND GROWTH OF THE "FIELD" OF ENVIRONMENTAL COMMUNICATION

Robert Cox and Stephen Depoe

Introduction

The first decade of the twenty-first century may be considered an inflection point for scholarship in environment and communication. In asking whether such scholarship was a "crisis discipline," the inaugural issue of *Environmental Communication: A Journal of Nature and Culture* (2007) posed explicitly, for the first time, the question of why, and in what ways, studies investigating the nexus between environment and human communication constituted an academic "field" as such. By 2011, scholars and practitioners had also launched the International Environmental Communication Association, "to advance the practice, study, and teaching of Environmental Communication in civic, political, educational, business, and cultural contexts" (IECA, 2013, para. 2).

This chapter traces the growth of research and institutional support that, by the early twenty-first century, purported to define a field of inquiry in environment and communication, as well as where this field has gone since. Following a summary of the early history, we identify major areas of research, key assumptions and questions defining this field, and emerging issues and controversies in the field.

History and growth of field

Environmental communication, as a definable area within, and beyond, the communication discipline, emerged in North America, Europe, and elsewhere over the past three decades. Initial work from North American scholars came out of the rhetorical tradition, including Oravec's critical study of naturalist John Muir's writings on the sublime response (1981) and the battles between conservation and preservation advocates over wilderness (1984); Farrell and Goodnight's examination of the rhetorical failures at Three Mile Island (1981); Cox's essay on the irreparable (1982); and Peterson's analyses of Dust Bowl-era rhetoric (1986). Other early scholarship in the U.S. and Europe focused on the content, cycles of attention, and production of environmental meaning in news media (Downs, 1972; Nimmo and Combs, 1982; Ostman and Parker, 1987; Anderson, 1990, 1991; Burgess, 1990). Related work also appeared on the relationship of environmental pressure groups and the news media (Anderson, 1991; Hansen, 1993). Separately, scholars in risk analysis had begun to question objectivist assumptions of risk (Slovic, 1987) and lay the basis for a communicative approach to environmental risk studies (Plough and Krimsky, 1987, 1988).

The 1990s also saw other book-length studies on a variety of subjects, including communication and environmental politics (Killingsworth and Palmer, 1992); emerging patterns of environmental discourse (Myerson and Rydin, 1996; Dryzek, 1997); media, culture, and the environment (Hansen, 1993; Anderson, 1997); the rhetoric of sustainable development (Peterson, 1997), media coverage of environmental issues (Neuzil and Kovarik, 1996; Shanahan and McComas, 1999); and image strategies of environmental activists (DeLuca, 1999). Further evidence of an emerging field included several important edited volumes (Hansen, 1993; Cantrill and Oravec, 1996; Herndl and Brown, 1996; Muir and Veenendall, 1996). By the end of the 1990s, the first significant literature review of environmental communication (Cantrill, 1993; see also Pleasant *et al.*, 2002) and a "landmark" anthology of environmental essays (Waddell, 1998) had been published.

Scholarship in environmental communication in the 1980s and 1990s was both accompanied and advanced by professional organizing work in the U.S., Europe, and elsewhere (Milstein, 2012). The International Association for Media and Communication Research (IAMCR) formed a working group on environmental/science/risk communication in 1988. Two years later, the Society of Environmental Journalists, including researchers and working journalists, was created. At the same time, in 1990 the Society for Risk Analysis formed a Risk Communication Specialty Group for coverage of a range— including environmental—of risk-related research. Meanwhile, in 1991 a group of scholars in the United States held the first of what would become a biennial Conference on Communication and Environment (COCE) (Oravec and Cantril, 1992). Interest from these conferences led, in 1996, to the founding of an Environmental Communication interest group, later a division, within the National Communication Association (NCA). Fifteen years later, a similar group had formed in the International Communication Association (ICA). Finally, 2008 saw the creation of a Science and Environment Communication section within the European Communication Education and Research Association (ECREA).

These and other developments led to three formative achievements: 1) a peer-reviewed periodical publication, the *Environmental Communication Yearbook* (2004–06), followed by*: Environmental Communication* (2007–present); 2) publication of a number of college-level textbooks (Flor, 2004; Parker, 2005; Corbett, 2006; Hansen, 2010; Hendry, 2010; Jurin, Roush and Danter, 2010; Cox, 2013); and 3) the launch, in 2011, of the International Environmental Communication Association (IECA). With members from thirty countries and contributions to *Environmental Communication* from Sweden, Turkey, Portugal, China, Australia, New Zealand, the U.K., Canada, the U.S., and other nations, the IECA and the field of environmental communication by 2014 was becoming global in scope.

Emerging assumptions and questions

In its early decades, the field of environmental communication witnessed differing subjects, approaches, and even conceptions of communication, sometimes constituting distinct "discourse communities" (Coppola and Karis, 2000, p. xviii). By the first decade of the twenty-first century, however, scholars had begun, explicitly, to identify some of the underlying assumptions of this earlier scholarship. The inaugural issue of *Environmental Communication*, in 2007, featured an exchange of views regarding the "broad agreements or working hypotheses of a field that is defined by the articulation of 'environment' and 'communication'" (Cox, 2007, p. 12). In subsequent years, other scholars extended this conversation, elaborating basic assumptions of scholarship in the emerging field (Milstein, 2009, 2012). This section describes these "working hypotheses" or assumptions and some of the major questions that have characterized research in environmental communication in recent decades.

Emerging assumptions

Although addressing a range of topics and processes, much of the scholarship in early decades of the field appeared to proceed from a set of agreements or assumptions about the relationships among communication, environment, and the social/cultural/ideological contexts in which such communication occurs. Among these are:

1. social/symbolic and environmental processes that are mutually implicated. That is, environmental problems are both materially produced—through interactions between human actions and bio-physical processes—and are also socially or discursively constructed. Such constructionist assumptions invite our understanding of "environment," "nature," and environmental "problems" as inextricably implicated with meaning, that is, the social/discursive investment of significance in our representations of the natural world (Gamson and Modigliani, 1989; Hansen, 1991, 2010; Depoe, 2007; Cox, 2007; Milstein, 2009, 2012);
2. representations of nature or the environment that embody interested and/or consequential orientations (Oravec, 2004; Cox, 2007, Milstein, 2009). Such representations both reflect and influence our social, economic, or ideological interests;
3. further social, cultural, economic, and other contexts that may enable, sustain, and/or frustrate the production (sense-making) of interested representations of "environment." Such productions, in turn, help to maintain, question, or challenge dominant discourses, representations, and other regimes of power, including those having deleterious consequences for biological systems, human and non-human (DeLuca and Peeples, 2002; Milstein, 2009; Takahashi and Meisner, 2012).

As a consequence, many environmental communication scholars have proceeded from a normative assumption, that understanding of communicative processes about "environment" may serve both pedagogically and in socially beneficial ways to (a) improve our understanding of the cultures or locations in which such communication is produced and (b) strengthen the capacity of societies to deliberate and respond to conditions relevant to the well-being of both society and natural biological communities.

Major research questions

The emerging assumptions underlying scholars' interest in the nexus between communication and the environment typically invite a recurring—albeit quite general—set of questions regarding such communication and its effects. Among such questions have been:

1. *How do human agents represent nature/environment? That is, how are specific environmental phenomena, conditions, or processes discursively or symbolically constituted as subjects for human understanding and/or action?*

Such questions motivated some of the earliest—particularly rhetorical and constructionist—scholarship in the field. More recently, others have examined the elemental acts of pointing and/or naming as "the basic entry to socially discerning and categorizing parts of nature" (Milstein, 2011, p. 4) and as "an orientation to the world" (Oravec, 2004, p. 3). Specific studies of representation of nature/environment have ranged broadly, for example, from depictions of the "pristine" in nineteenth-century photographs of the American West (DeLuca and Demo, 2000) to scientists' warnings of "tipping points" in global climate change (Russill and Nyssa, 2009).

2. *What accounts for the development and reproduction of dominant systems of representation or discourses of "environment," and what communicative practices contribute to the interruption, dilution, or transformation of such discourses?*

Often drawing upon critical social theories, scholars have traced the development, influence, and/or alteration of discourses sustaining dominant social, political, and ideological formations, particularly as these rationalize unsustainable practices of the natural world. Relatedly, others have examined the recuperative role that various communicative practices play in reaffirming or reproducing such discourses. Rogers' (2008) study of television advertisements for the consumption of meat, for example, found these ads "articulate the eating of meat with primitive masculinities as a response to perceived threats to hegemonic masculinity" (p. 281). Conversely, Hodgins and Thompson (2011) identified strategies of citation and parody used by Canadian artists and photographers to subvert the extractive and romantic "gazes" of the Canadian landscape.

3. *What effects do different environmental sources (e.g., media) as well as specific communicative practices have on audiences?*

Studies of the effects of environmental sources (e.g., media, "green" advertising, marketing) constitute some of the earliest and most influential areas of research in environmental communication. Media practices of framing and agenda-setting, for example, have been staples in such scholarship (e.g., Ader, 1995; Anderson, 1997; Shanahan and McComas, 1999; Hansen, 2010). Similarly, studies of the effects of "green" product claims on consumer perceptions and behavior have drawn considerable attention (for a review of this research, see Spack, Board, Crighton, Kostka and Ivory, 2012, pp. 442–446).

4. *What are the relationships between or among communication, individuals' values and beliefs, and their environmental behaviors?*

Closely related to the study of effects, scholars have paid considerable attention to the antecedents of individuals' environmental attitude and/or behavior change. Some, for example, have begun to specify the role of values in light of findings of an "attitude–behavior gap," i.e., the weakness of individuals' attitudes in explaining their environmental behaviors (Kollmuss and Agyeman, 2002; Schultz and Zelezny, 2003; Compton, 2008).

5. *In what ways do different modes of production, dissemination, and reception of scientific or technical information contribute to the understanding of, or constitute "knowledge" of nature or environmental phenomena?*

Building on earlier research on media and environmental risk (Allan, Adam and Carter, 2000), studies of the role of media in the production, dissemination, and reception of environmental knowledge range from technology diffusion (Skjølsvold, 2012) to the uses and influence of new media on public understanding of climate change (O'Neil and Boykoff, 2011; Nielsen and Kjaergaad, 2011).

6. *How do humans discursively or symbolically constitute "space" or places, and how does a sense of one's "self-in-place" influence one's understanding and/or behaviors in relation to such environments?*

Extending early scholarship on the environmental influences of the "self-in-place" (Cantrill, 1998; Cantrill and Senecah, 2001; Cantrill, Thompson, Garrett and Rochester, 2007), scholars have begun to investigate the discursive transformation of environmental "space" (Thompson and Cantrill, 2013), as well as the influence of place-based self-understandings on environmental behaviors. Shellabarger, Peterson, Sills, and Cubbage (2012), for example, traced the impacts of land managers and humanitarian aid volunteers' differing perceptions of U.S. south-west borderlands with Mexico on conservation and human rights. And, studies of the discursive transformation of space have ranged from perceptions of nature online and in socially networked spaces (Uggla and Olausson, 2013; Adams and Gynnild, 2013) to how the esthetic of tourism texts "renders Africa invisible through anthropocentric distance" (Todd, 2010, p. 206).

7. *How do local or indigenous cultures understand "nature" or "environment," and how do such cultures form or convey these understandings in everyday life?*

Along with place, scholars have explored ways in which particular cultures and communities shape human understandings of and responses to their physical surroundings. A leading scholar in this area, Carbaugh (1996a, 1996b, 2007a; 2007b; Carbaugh and Boromisza-Habashi, 2011) has explored ways of seeing, listening to/with, and talking about natural scenes in a variety of local case studies, including observing conversations about local development choices in Massachusetts (Carbaugh, 1992, 1996b), engaging in practices of health and leisure in Finland (Carbaugh, 1996a), and dwelling with Blackfeet tribal members in Montana (Carbaugh, 1999). The goal of this and related research has been "to introduce a way to think through communication to cultural places, and further to link our understanding … to deeply seated notions of identity, and to the affective dimension of belonging which place-based communication often brings with it" (Carbaugh and Cerulli, 2013, p. 8).

Building upon this work, other scholars have examined how environmental communication shapes and is shaped by local cultural understandings and practices. Examples include Scollo Sawyer's (2004) comparison of and discourse about walking and other non-verbal behavior in various cultural settings; Rautio's study of nature writing in Finland (2011); and Salvador and Clarke's articulation of the *weyekin* principle, an embodied rhetorical performance among Nez Perce people (2011).

Emerging issues and controversies

Apart from shared assumptions, the vitality of a field may also be measured by the emergence of new inquiries and by the vigor of debates about consequential claims that a field invites. By this measure, forums within environmental communication are witnessing new, innovative lines of inquiry as well as robust exchange, not only about specific claims, but about the nature of the field itself.

Emerging issues

While early scholarship often focused upon rhetorical texts, media practices (agenda-setting, etc.) and other discrete communicative practices, recent scholarship has significantly widened the scope of environmental communication studies. These inquiries range from discourses of/about food (Retzinger, 2010; Opel, Johnston, and Wilk, 2010), international environmental justice (Sowards, 2012), visual rhetorics of environmental film and photography, sustainability science, and climate change communication, to the problematizing of the culture/nature binary, to name only a few areas.

Four of these areas—climate change communication, sustainability science, visual communication or the "imaging" of nature, and the problematizing of the human/nature binary—are illustrative of the greatly expanded range and sophistication of recent scholarship.

1. Climate change communication

Climate change (and climate science) communication have been subjects of robust investigations by environmental communication scholars recently, virtually becoming a field of their own. Initially, this interest centered around the difficulties in explaining climate change to popular audiences—distant, unobtrusive, and outside of audiences' personal experiences (Moser and Dilling, 2007; Moser, 2010). As a consequence, science communication, often premised on an "information deficit" model (Dickson, 2005, para. 9), appeared to be inadequate to the challenge (Leiserowitz and Smith, 2010). Scholars, therefore, began to explore the different modes of appeal and message framing that were designed not only to heighten understanding of climate phenomena, but also to enhance the public's willingness to act.

While environmental groups and news media, for example, often use fear-inducing or "apocalyptic" rhetoric in their climate communication (Carvalho and Burgess, 2005; Foust and Murphy, 2009), scholars generally have found little evidence for the effectiveness of such appeals (Moser and Dilling, 2007; O'Neil and Nicholson-Cole, 2009). More promising has been the study of the different frames, including "energy security," "morality and ethics," and "public health" (Carvalho and Peterson, 2009; Nisbet, 2009; Lakoff, 2010; Maibach, Nisbet, Baldwin, Akerlof and Diao, 2010; Myers, Nisbet, Maibach and Leiserowitz, 2012).

Finally, some scholars have examined the influence of different sources and communication media on public opinion about climate change, including the communication of climate change deniers (Jacques, Dunlap, and Freeman, 2008; Moore, 2009; Oreskes and Conway, 2010). Others have investigated the decline of science journalism and the rise of partisan or self-interested information providers in new media in nurturing skepticism about climate change (Cox, 2013).

2. Sustainability science and environmental communication

Since the Brundtland Report was issued in 1987, the concept and agenda of "sustainable development" has drawn considerable attention from scholars, policy makers, and critics (Dixon and Fallon, 1989; Aguirre, 2002; Peterson, 1997). In recent decades, the interdisciplinary field of sustainability science has emerged as a response to the conversation/controversy over "sustainable development" (NRC, 1999; Kates et al., 2001; Kates, Parris and Leiserowitz, 2005). Sustainability science has been defined as applied research that spans and integrates multiple physical and social science disciplines and is directed toward the management of human-environment systems in ways that meet needs for human livelihoods while protecting ecosystem and environmental integrity (Chapin et al., 2010).

Environmental communication scholarship has much to contribute to this line of inquiry (Heath, Palenchar, Proutheau, and Hocke, 2007; Lindenfeld, Hall, McGreavy, Silka, and Hart, 2012). Related work is occurring in diverse areas such as risk and science communication (Kurath and Gisler, 2009; Nisbet and Scheufele, 2009), public science argument and participation processes (Kinsella, 2004; Endres, 2009), and media coverage of sustainability issues (Carvalho and Burgess, 2005). We recognize and applaud the call by Lindenfeld et al. (2012) and others for a closer integration between environmental communication and sustainability science, while also continuing to assess sustainability claims by scientific and industrial sources (Smerecnik and Renegar, 2010) as well as the critique of sustainability science itself (Nelson and Vucetich, 2012).

3. Visual representations or the "imaging" of the environment

By the second decade of the twenty-first century, studies of media representations of the environment had moved substantially beyond rhetorical and linguistic analyses to studies of "the visual" in its own right. Two impulses prefigured this interest. In the U.S., DeLuca's and Demo's (2000) "Imaging nature: Watkins, Yosemite, and the birth of environmentalism" illustrated the rhetorical and ideological work of visually mediated cultures in depicting the environment. And, in the U.K. and Europe, studies of televised news coverage of climate change and other environmental risks (e.g., Cottle, 2000; Hansen and Machin, 2008; Lester and Cottle, 2009), documented both the challenge in visually depicting many environmental problems and the ways in which news organizations "lean toward well-trodden frames of reference to make issues recognizable to audiences" (Hansen and Machin, 2013, p. 157).

In more recent years, studies of visual representations or "imaging" of nature have proliferated across a range of subjects, including photographic images of polluted landscapes and peoples (Cammaer, 2009; Peeples, 2011, 2013); images linking "wildness" and "nature as primitive" in sports advertising (Ferrari, 2013); and effects of the "Romantic gaze" and "extractive gaze" in distancing people from nature (Hodgins and Thompson, 2011; Takach, 2013). Beyond such subjects, Hansen and Machin (2013) have called for environmental communication scholars to provide a more comprehensive account of the semiotics, composition, and "sites of meaning-making" characterizing such visual representations (p. 163).

4. Problematizing the human/nature binary

Finally, the assumption, particularly in Western societies and in much of the environmental communication field itself, of a culture/nature or human/nature binary has been extensively questioned in recent work by some environmental communication scholars (Carbaugh, 2007b; Peterson, Peterson, and Peterson, 2007; Milstein and Kroløkke, 2012). Influenced by Rogers' (1998) call for a materialist theory of communication that goes beyond "constitutive" theories, such scholars have sought to appreciate the ways in which nature or non-human species "speak," engage, or influence humans and/or their environments. Milstein (2012), for example, calls for scholars to interrogate the ways nature's communication *mediates* human–nature relations. The emphasis on mediation considers "nature as co-present, active, and [a] dynamic force in human–nature relationships" (pp. 167, 171).

Efforts to avoid reifying the natural world also open the study of environmental communication to the possibility of articulating "agency beyond the human world," or nature as "an active subject" (Milstein, 2012, p. 167). Such a prospect echoes Rogers' (1998) call for a "rehearsal of ways of listening to nondominant voices and extra-human agents and their inclusion in the production of meaning" (p. 268). Further, some have argued that environmental communication practitioners have "a responsibility to amplify and translate the voices of nonspeaking human and extrahuman subjects" as members of a more sustainable community (Peterson et al., 2007, p. 78).

Controversies in the field

While emerging studies push the boundaries of scholarship in environmental communication, they do not necessarily imply disagreement or controversy. In some instances, new studies have refined or corrected earlier research. This has been true particularly with research claims about the effects of media practices, such as agenda-setting and the cultivation theory of media (Dahstrom and Schuefele, 2010).

Other lines of inquiry, however, have invited considerable debate or controversy. One recent controversy has been over the concept and uses of "framing" in environmental campaigns. While the phenomenon of *media frames* is a well-accepted theory (Entman, 1993; Pan and Kosicki, 1993), the role and value of framing in public understanding of subjects such as climate change has been more contentious (Brulle, 2010).

In an exchange among scholars with differing views of framing in *Environmental Communication*, in 2010, cognitive linguist George Lakoff (2010) argued that the idea of frames had been misunderstood. Although frames are expressed in words, he contended, they are not likely to have long-lasting impact unless they draw upon deep, conceptual structures built over a long period in the "neural circuits in the brain" (p. 71). Others took issue, arguing that those influenced by Lakoff's cognitive linguistics are likely to engage in a "mechanistic strategy" that remains "a 'shallow' method of environmental communications" (López, 2010, p. 99). Brulle (2010) similarly argued that, while Lakoff's theory may have short-term advantages, campaigns derived from such theory "are most likely incapable of developing the large-scale mobilization necessary to enact the massive social and economic changes necessary to address global warming" (p. 83).

As certain assumptions of a "field" became clearer, a second—and somewhat controversial—thesis arose. In the inaugural issue of *Environmental Communication*, Cox (2007) proposed that these tenets carried ethical implications. As in other "crisis disciplines," he contended, the basic tenets of a field of environment and communication implied normative premises, including the belief that scholarship should "enhance the ability of society to respond appropriately to environmental signals relevant to the well-being of both human communities and natural biological systems" (p. 15).

The thesis of "crisis discipline" has generated debate and some demurral among environmental communication scholars. Senecah (2007), for example, contended that the ethical premise of a crisis discipline has "multiple danger points," including her concern that it binds scholars to an "adversarial/advocacy" stance (pp. 25, 28). Peterson et al. (2007), on the other hand, would modify Cox's normative claim by broadening its scope, urging scholars to "represent nonspeaking or otherwise silenced members of the community" (p. 81). Still others found that the "aspiration to 'enhance the ability of society to respond appropriately to environmental signals' … underlies much contemporary environmental communication scholarship" (Salvador and Clarke, 2011, p. 244).

By the twenty-first century's second decade, scholars appear to have embraced many of the "working hypotheses" or tenets for a field of environment and communication as well as the value of professional associations and other forums for the support, collaboration, and dissemination of their work. In many ways, Goshorn's (2001) vision for a new field had become real, as environmental communication scholars have come to engage discourses about environment as a "constitutive force, not just a topical cluster of issues, events, or campaigns" (p. 321). The following chapters survey the scholarship defining the principal contributions of this vibrant field.

Further reading

The International Environmental Communication Association [online], at: https://theieca.org/

Milstein, T., 2012. Greening communication. In S. D. Fassbinder, A. J. Nocella II, and R. Kahn, eds. *Greening the academy: Ecopedagogy through the liberal arts*. Rotterdam: Sense Publishers, pp. 161–174.

References

Adams, P. C. and Gynnild, A., 2013. Environmental messages in online media: The role of place. *Environmental Communication: A Journal of Nature and Culture*, 7(1), pp. 113–130.

Ader, C. R., 1995. A longitudinal study of agenda setting for the issue of environmental pollution. *Journalism and Mass Communication Quarterly*, 72(2), pp. 300–311.

Allan, S., Adam, B. and Carter, C., 2000. eds. *Environmental risks and the media*. London: Routledge.

Aguirre, B. E., 2002. "Sustainable development" as collective surge. *Social Science Quarterly*, 83(1), pp. 101–18.

Anderson, A., 1990. The production of environmental meanings in the media: A new era. *Media Education Journal*, 10, pp. 18–20.

Anderson, A., 1991. Source strategies and the communication of environmental affairs. *Media, Culture and Society*, 13(4), pp. 459–476.

Anderson, A., 1997. *Media, culture and the environment*. New Brunswick, NJ: Rutgers University Press.

Brulle, R. J., 2010. From environmental campaigns to advancing the public dialog: Environmental communication for civic engagement. *Environmental Communication: A Journal of Nature and Culture*, 4(1), pp. 82–98.

Brundtland Report, 1987. *Our common future*. World Commission on Environment and Development. Oxford: Oxford University Press.

Burgess, J. (1990). The production and consumption of environmental meaning in the mass media: A research agenda for the 1990s. *Transactions for the Institute of British Geographers*, 15(2), pp. 139–161.

Cammaer, G., 2009. Edward Burtynsky's *Manufactured Landscapes*: The ethics and aesthetics of creating moving still images and stilling moving images of ecological disasters. *Environmental Communication: A Journal of Nature and Culture*, 3(1), pp. 121–130.

Cantrill, J. G., 1993. Communication and our environment: Categorizing research in environmental advocacy. *Journal of Applied Communication Research*, 21, pp. 66–95.

Cantrill, J. G., 1998. The environmental self and a sense of place: Communication foundations for regional ecosystem management. *Journal of Applied Communication Research*, 26, pp. 301–318.

Cantrill, J. G. and Oravec, C. L., eds, 1996. *The symbolic earth: Discourse and our creation of the environment*. Lexington, KY: University Press of Kentucky.

Cantrill, J. G. and Senecah, S. L., 2001. Using the "sense of self-in-place" construct in the context of environmental policy-making and landscape planning. *Environmental Science and Policy*, 4, pp. 185–203.

Cantrill, J. G., Thompson, J. L., Garrett, E., and Rochester, G., 2007. Exploring a sense of self-in-place to explain the impulse for urban sprawl. *Environmental Communication: A Journal of Nature and Culture*, 1(2), pp. 123–145.

Carbaugh, D., 1992. "The mountain" and "the project": Dueling depictions of a natural environment. In C. L. Oravec and J. G. Cantrill, eds, *The conference on the discourse of environmental advocacy*. Salt Lake City, UT: University of Utah Humanities Center, pp. 360–376.

Carbaugh, D., 1996a. Naturalizing communication and culture. In J. G. Cantrill and C. L. Oravec, eds, *The symbolic earth: Discourse and our creation of the environment*. Lexington, KY: University Press of Kentucky, pp. 38–57.

Carbaugh, D., 1996b. *Situating selves: The communication of social identities in American scenes*. Albany, NY: SUNY Press.

Carbaugh, D., 1999. "Just listen": "Listening" and landscape among the Blackfleet. *Western Journal of Communication (includes Communication Reports)*, *63*(3), pp. 250–270.

Carbaugh, D., 2007a. Cultural discourse analysis: Communication practices and intercultural encounters. *Journal of Intercultural Communication Research*, *36*, pp. 167–182.

Carbaugh, D., 2007b. Quoting "the environment": Touchstones on Earth. *Environmental Communication: A Journal of Nature and Culture*, *1*(1), pp. 64–73.

Carbaugh, D. and Boromisza-Habashi, D., 2011. Discourse beyond language: Cultural rhetoric, revelatory insight, and nature. In C. Meyer and F. Girke, eds, *The interplay of rhetoric and culture*. Oxford and New York: Berghahn Books Studies in Rhetoric and Culture III, pp. 101–118.

Carbaugh, D. and Cerulli, T., 2013. Cultural discourses of dwelling: Investigating environmental communication as a place-based practice. *Environmental Communication: A Journal of Nature and Culture*, *7*(1), pp. 4–23.

Carvalho, A. and Burgess, J., 2005. Cultural circuits of climate change in U.K. broadsheet newspapers, 1985–2003. *Risk Analysis: An International Journal*, *25*(6), pp. 1457–1469.

Carvalho, A. and Peterson, T. R., 2009. Discursive constructions of climate change: Practices of encoding and decoding. *Environmental Communication: A Journal of Nature and Culture*, *3*(2), pp. 131–133.

Chapin III, F. S., Crumley, C. L., Gomes, C. P., Graedel, T. E., Levin, J., Matson, P. A., Matus, K., Myers, S., and Smith, V. K., 2010. Working group IV: Managing human-environment systems for sustainability. In S. A. Levin and W. C. Clark, eds, *Toward a science of sustainability*, Report from a Science of SustaProcinability Conference, pp. 56–71. Available at: www.nsf.gov/mps/dms/documents/SustainabilityWorkshop2009Report.pdf [Accessed 13 August 2013].

Compton, T., 2008. Weathercocks and signposts: The environment movement at a crossroads. World Wildlife Fund-UK. Available at: www.wwf.org.uk/wwf_articles.cfm?unewsid=2224 [Accessed 11 May, 2013].

Coppola, N. W. and Karis, B., eds, 2000. Technical communication, deliberative rhetoric, and environmental discourse: Corrections and directions. *Contemporary studies in technical communication*, pp. xi–xxviii. Greenwich, CT: Ablex Publishing.

Corbett, J. B., 2006. *Communicating nature: How we create and understand environmental messages*. Washington DC: Island Press.

Cottle, S., 2000. TV news, lay voices and the visualisation of environmental risks. In S. Allan, B. Adam, and C. Carter, eds, *Environmental risks and the media*. London: Routledge, pp. 29–44.

Cox, R., 1982. The die is cast: Topical and ontological dimensions of the *locus* of the irreparable. *Quarterly Journal of Speech*, *68*(2), pp. 227–239.

Cox, R., 2007. Nature's "crisis disciplines": Does environmental communication have an ethical duty? *Environmental Communication: A Journal of Nature and Culture*, *1*(1), pp. 5–20.

Cox, R., 2013. *Environmental communication and the public sphere*. 3rd ed. Los Angeles, CA: Sage.

Dahstrom, M. F. and Schuefele, D. A., 2010. Diversity of television exposure and its association with the cultivation of concern for environmental risks. *Environmental Communication: A Journal of Nature and Culture*, *4*(1), pp. 54–65.

DeLuca, K. M., 1999. *Image politics*. New York: Guildford Press.

DeLuca, K. M. and Demo, A. T., 2000. Imaging nature: Watkins, Yosemite, and the birth of environmentalism. *Critical Studies in Media Communication*, *17*, pp. 241–261.

DeLuca, M. and Peeples, J., 2002. From public sphere to public screen: Democracy, activism, and the "violence" of Seattle. *Critical Studies in Media Communication*, *19*(2), pp. 125–151.

Depoe, S., 2007. Environmental communication as nexus. *Environmental Communication: A Journal of Nature and Culture*, *1*(1), pp. 1–4.

Dickson, D., 2005. The case for a "deficit model" of science communication. *Science and Development Network*. Available at: www.scidev.net/global/communication/editorial-blog/the-case-for-a-deficit-model-of-science-communic.html [Accessed 30 August, 2013].

Dixon, J. A. and Fallon, L. A., 1989. The concept of sustainability: Origin, extensions, and usefulness for policy. *Society and Natural Resources*, *2*(2), pp. 73–84.

Downs, A., 1972. Up and down with ecology-the attention-issue cycle. *Public Interest*, *28*, pp. 38–50.

Dryzek, J. A., 1997. *The politics of the earth: Environmental discourses*. Oxford: Oxford University Press.

Endres, D., 2009. Science and public participation: An analysis of public scientific argument in the Yucca Mountain controversy. *Environmental Communication: A Journal of Nature and Culture, 3*(1), pp. 49–75.

Entman, R. M., 1993. Framing: Toward clarification of a fractured paradigm. *Journal of Communication, 43*(4), pp. 51–58.

Farrell, T. B. and Goodnight, G. T., 1981. Accidental rhetoric: Root metaphors of Three Mile Island. *Communication Monographs, 48*(4), pp. 271–300.

Ferrari, M. P., 2013. Sporting nature(s): Wildness, the primitive, and naturalizing imagery in MMA and sports advertising. *Environmental Communication: A Journal of Nature and Culture, 7*(2), 277–296.

Flor, A., 2004. *Environmental communication: Principles, approaches and strategies of communication applied to environmental management.* Diliman, Quezon City, Philippines: University of the Philippines-Open University.

Foust, C. R. and Murphy, W. O., 2009. Revealing and reframing apocalyptic tragedy in global warming discourse. *Environmental Communication: A Journal of Nature and Culture, 3*(2), 151–167.

Gamson, W. A. and Modigliani, A., 1989. Media discourse and public opinion on nuclear power: A constructionist approach. *American Journal of Sociology, 95*(1), pp. 1–37.

Goshorn, K., 2001. From beyond our borders: Other readings on environmentalism and communicative action. *Quarterly Journal of Speech, 87*(3), pp. 321–339.

Hansen, A., 1991. The media and the social construction of the environment. *Media, Culture, and Society, 13*(4), pp. 443–458.

Hansen, A., ed., 1993. *The mass media and environmental issues.* Leicester: Leicester University Press.

Hansen, A., 2010. *Environment, media, and communication.* London and New York: Routledge.

Hansen, A. and Machin, D., 2008. Visually branding the environment: Climate change as a marketing opportunity. *Discourse Studies, 10*(6), pp. 777–794.

Hansen, A. and Machin, D., 2013. Researching visual environmental communication. *Environmental Communication: A Journal of Nature and Culture, 7*(2), pp. 151–168.

Heath, R. L., Palenchar, M. J., Proutheau, S., and Hocke, T. M., 2007. Nature, crisis, risk, science, and society: What is our ethical responsibility? *Environmental Communication: A Journal of Nature and Culture, 1*(1), pp. 34–48.

Hendry, J., 2010. *Communication and the natural world.* State College, PA: Strata.

Herndl, C. G. and Brown, S. C., eds, 1996. *Green culture: Environmental rhetoric in contemporary America.* Madison, WI: University of Wisconsin Press.

Hodgins, P. and Thompson, P., 2011. Taking the romance out of extraction: Contemporary Canadian artists and the subversion of the romantic/extractive gaze. *Environmental Communication: A Journal of Nature and Culture, 5*(4), pp. 393–410.

International Environmental Communication Association [IECA]. 2013. About the IECA. [online] Available at: http://theieca.org/about-ieca [Accessed 29 August, 2013].

Jacques, P. J., Dunlap, R. E. and Freeman, M., 2008. The organization of denial: Conservative think tanks and environmental scepticism. *Environmental Politics, 17,* 349–385.

Jurin, R. R., Roush, D., and Danter, K. J., 2010. *Environmental communication: Skills and principles for natural resource managers, scientists, and engineers.* 2nd ed. New York: Springer.

Kates, R. W., Clark, W. C., Corell, R., Hall, J. M., Jaeger, C. C., Lowe, I., McCarthy, J. J., Schellnhuber, H. J., Bolin, B., Dickson, N. M., Faucheux, S., Gallopin, G. C., Grübler, A., Huntley, B., Jäger, J., Jodha, N. S., Kasperson, R. E., Mabogunje, A., Matson, P., Mooney, H., Moore III, B., O'Riordan, T., and Svedin, U., 2001. Sustainability science. *Science, 27*(5517), pp. 641–642.

Kates, R. W., Parris, T. M., and Leiserowitz, A. A. (2005). What is sustainable development? Goals, indicators, values, and practice. *Environment: Science and Policy for Sustainable Development, 47*(3), 8–21.

Killingsworth, M. J. and Palmer, J. S., 1992. *Ecospeak: Rhetoric and environmental politics in America.* Carbondale, IL: Southern Illinois University Press.

Kinsella, W. J., 2004. Public expertise: A foundation for citizen participation in energy and environmental decisions. In S. P. Depoe, J. W. Delicath, and M. Aepli Elsenbeer, eds, *Communication and public participation in environmental decision-making.* Albany, NY: SUNY Press, pp. 83–98.

Kollmuss, A. and Agyeman, J., 2002. Mind the gap: Why do people act environmentally and what are the barriers to pro-environmental behavior? *Environmental Education Research, 8*(3), pp. 96–119.

Kurath, M. and Gisler, P., 2009. Informing, involving, or engaging? Science communication in the ages of atom-, bio- and nanotechnology. *Public Understanding of Science, 18*(5), 559–573.

Lakoff, G., 2010. Why it matters how we frame the environment. *Environmental Communication: A Journal of Nature and Culture, 4*(1), pp. 70–81.

Leiserowitz, A. and Smith, N., 2010. *Knowledge of climate change across global warming's six Americas.* New Haven, CT: Yale Project on Climate Change Communication.

Lester, L. and Cottle, S., 2009. Visualizing climate change: Television news and ecological citizenship. *International Journal of Communication*, 3, pp. 920–936.

Lindenfeld, L. A., Hall, D. M., McGreavy, B., Silka, L., and Hart, D., 2012. Creating a place for environmental communication research in sustainability science. *Environmental Communication: A Journal of Nature and Culture*, 6(1), pp. 23–43.

López, A., 2010. Defusing the cannon/canon: An organic media approach to environmental communication. *Environmental Communication: A Journal of Nature and Culture*, 4(1), pp. 99–108.

Maibach, E. W., Nisbet, M., Baldwin, P., Akerlof, K., and Diao, G., 2010. Reframing climate change as a public health issue: an exploratory study of public reactions. [online] *BMC Pub Health*, 10: 299. www.ncbi.nlm.nih.gov/pubmed/20515503 [Accessed 30 August, 2013].

Milstein, T., 2009. Environmental communication theories. In S. Littlejohn and K. Foss, eds, *Encyclopedia of communication theory*. Thousand Oaks, CA: Sage, pp. 344–349.

Milstein, T., 2011. Nature identification: The power of pointing and naming. *Environmental Communication: A Journal of Nature and Culture*, 5(1), pp. 3–24.

Milstein, T., 2012. Greening communication. In S. D. Fassbinder, A. J. Nocella II, and R. Kahn, eds, *Greening the academy: Ecopedagogy through the liberal arts*. Rotterdam: Sense Publishers, pp. 161–174.

Milstein, T. and Krøløkke, C., 2012. Transcorporeal tourism: Whales, fetuses, and the rupturing and reinscribing of cultural constrains. *Environmental Communication: A Journal of Nature and Culture*, 6(1), pp. 82–100.

Moore, M. P., 2009. The Union of Concerned Scientists on the uncertainty of climate change: A study of synecdochic form. *Environmental Communication: A Journal of Nature and Culture*, 3(2), pp. 191–205.

Moser, S. and Dilling, L., eds, 2007. *Creating a climate for change: Communicating climate change and facilitating social change*. Cambridge: Cambridge University Press.

Moser, S. C., 2010. Communicating climate change: History, challenges, process and future directions. *Wiley Interdisciplinary Reviews: Climate Change*, 1(1), pp. 31–53.

Muir, S. A. and Veenendall, T. L., eds, 1996. *Earthtalk: Communication empowerment for environmental action*. Westport, CT: Praeger.

Myers, T. A., Nisbet, M. C., Maibach, E. W., and Leiserowitz, A. A., 2012. A public health frame arouses hopeful emotions about climate change. *Climatic Change*, 113(3–4), pp. 1105–1112.

Myerson, G. and Rydin, Y., 1996. *The language of environment: A new rhetoric*. London: Routledge.

National Research Council, 1999. *Our common journey: A transition toward sustainability*. Washington DC: National Academy Press.

Nelson, M. P. and Vucetich, J. A., 2012. Sustainability science: Ethical foundations and emerging challenges. *Nature Education Knowledge*, 3(10), p. 12.

Neuzil, M. and Kovark, W., 1996. *Mass media and environmental conflict: America's green crusades*. Thousand Oaks, CA: Sage.

Nielsen, K. H. and Kjaergaad, R. S., 2011. News coverage of climate change in *Nature News* and *ScienceNOW* during 2007. *Environmental Communication: A Journal of Nature and Culture*, 5(1), pp. 25–44.

Nimmo, D. and Combs, J. E., 1982. Fantasies and melodramas in television network news: The case of Three Mile Island. *Western Journal of Speech Communication*, 46, pp. 45–55.

Nisbet M. C., 2009. Communicating climate change: Why frames matter for public engagement. *Environment*, 51(2), 12–23.

Nisbet, M. C. and Scheufele, D. A., 2009. What's next for science communication? Promising directions and lingering distractions. *American Journal of Botany*, 96, pp. 1767–1778.

O'Neill, S. and Nicholson-Cole, S., 2009. "Fear won't do it": Promoting positive engagement with climate change through visual and iconic representations. *Science Communication*, 30(3), 355–379.

O'Neill, S. J. and Boykoff, M., 2011. The role of new media in engaging the public with climate change. In L. Whitmarsh, S. J. O'Neill, and I. Lorenzoni, eds, *Engaging the public with climate change: Behaviour change and Communication*. London: Earthscan, pp. 233–251.

Opel, A., Johnston, J., and Wilk, R., 2010. Food, culture, and the environment: Communicating about what we eat. *Environmental Communication: A Journal of Nature and Culture*, 4(3), pp. 251–254.

Oravec, C., 1981. John Muir, Yosemite, and the sublime response: A study in the rhetoric of preservationism. *Quarterly Journal of Speech*, 67(3), pp. 245–258.

Oravec, C., 1984. Conservationism vs. preservationism: The "public interest" in the Hetch Hetchy controversy. *Quarterly Journal of Speech*, 70(4), pp. 444–458.

Oravec, C. L., 2004. Naming, interpretation, policy, and poetry. In S. L. Senecah, ed. *Environmental communication yearbook, vol. 1*. Mahwah, NJ: Lawrence Erlbaum Associates, pp. 1–14.

Oravec, C. L. and Cantrill, J. G. eds, 1992. *The conference on the discourse of environmental advocacy*. Salt Lake City, UT: University of Utah Humanities Center.

Oreskes, N. and Conway, E, 2010. *Merchants of doubt: How a handful of scientists obscured the truth on issues from tobacco smoke to global warming*. New York: Bloomsbury Press.

Ostman, R. E. and Parker, J. L., 1987. A public's environmental information sources and evaluations of mass media. *Journal of Environmental Education*, *18*, pp. 9–17.

Pan, Z. and Kosicki, G. M., 1993. Framing analysis: An approach to news discourse. *Political Communication*, *10*, pp. 55–76.

Parker, L. J., 2005. *Environmental communication: Messages, media, and methods*. Dubuque, IA: Kendall Hunt.

Peeples, P., 2011. Toxic sublime: Imaging contemporary landscapes. *Environmental Communication: A Journal of Nature and Culture*, *5*(4), pp. 373–392.

Peeples, J., 2013. Imagining toxics. *Environmental Communication: A Journal of Nature and Culture*, 7(2), 191–210.

Peterson, M. N., Peterson, M. J., and Peterson, T. R., 2007. Environmental communication: Why this discipline should facilitate environmental democracy. *Environmental Communication: A Journal of Nature and Culture*, *1*(1), 74–86.

Peterson, T. R., 1986. The will to conservation: A Burkeian analysis of Dust Bowl rhetoric and American farming motives. *Southern Speech Communication Journal*, *52*, pp. 1–21.

Peterson, T. R., 1997. *Sharing the earth: The rhetoric of sustainable development*. Columbia, SC: University of South Carolina Press, pp. 86–118.

Plough, A. and Krimsky, S., 1987. The emergence of risk communication studies: Social and political context. *Science, Technology & Human Values*, *12*(3/4), pp. 4–10.

Plough, A. and Krimsky, S., 1988. *Environmental hazards: Communicating risk as a social process*. Dover, MA: Auburn.

Pleasant, A., Shanahan, J., Cohen, B., and Good, J., 2002. The literature of environmental communication. *Public Understanding of Science*, *11*(2), pp. 197–205.

Rautio, P., 2011. Writing about everyday beauty: Anthropomorphizing and distancing as literary practices. *Environmental Communication: A Journal of Nature and Culture*, *5*(1), pp. 104–129.

Retzinger, J., 2010. Spectacles of labor: Viewing food production through a television screen. *Environmental Communication: A Journal of Nature and Culture*, *4*(4), pp. 441–460.

Rogers, R. A., 1998. Overcoming the objectification of nature in constitutive theories: Toward a transhuman, materialist theory of communication. *Western Journal of Communication*, *62*, pp. 244–272.

Rogers, R. A., 2008. Beats, burgers, and hummers: Meat and the crisis of masculinity in contemporary television advertisements. *Environmental Communication: A Journal of Nature and Culture*, *2*(3), pp. 281–301.

Russill, C. and Nyssa, Z., 2009. The tipping point trend in climate change communication. *Global Environmental Change*, *19*(3), pp. 336–344.

Salvador, M. and Clarke, T., 2011. The Weyekin principle: Toward an embodied critical rhetoric. *Environmental Communication: A Journal of Nature and Culture*, *5*(3), pp. 243–260.

Schultz, P. W. and Zelezny, L., 2003. Reframing environmental messages to be congruent with American values. *Research in Human Ecology*, *10*, pp. 126–136.

Scollo Sawyer, M., 2004. Nonverbal ways of communicating with nature: A cross-case study. In S. L. Senecah, ed., *Environmental communication yearbook, Volume 1*. Mahwah NJ: Erlbaum, pp. 227–249.

Senecah, S. L., 2007. Impetus, mission and future of the environmental communication commission/division. *Environmental Communication: A Journal of Nature and Culture*, *1*(1), pp. 21–33.

Shanahan, J. and McComas, K., 1999. *Nature stories: Depictions of the environment and their effects*. Cresskill, NJ: Hampton Press.

Shellabarger, R., Peterson, M. N., Sills, E., and Cubbage, F., 2012. The influence of place meanings on conservation and human rights in the Arizona Sonora borderlands. *Environmental Communication: A Journal of Nature and Culture*, *6*(3), pp. 383–402.

Skjølsvold, T., 2012. Curb your enthusiasm: On media communication of bioenergy and the role of the news media in technology diffusion. *Environmental Communication: A Journal of Nature and Culture*, *6*(4), pp. 512–531.

Slovic, P., 1987. Perception of risk. *Science*, *236*(4799), pp. 280–285.

Smerecnik, K. R. and Renegar, V. R., 2010. Capitalistic agency: The rhetoric of BP's Helios Power campaign. *Environmental Communication: A Journal of Nature and Culture*, *4*(2), pp. 152–171.

Sowards, S., 2012. Introduction: Environmental justice in international contexts: Understanding intersections for social justice in the twenty-first century. *Environmental Communication: A Journal of Nature and Culture*, *6*(3), pp. 285–289.

Spack, J. A., Board, V. E., Crighton, L. M., Kostka, P. M., and Ivory, J. D., 2012. It's easy being green: The effects of argument and imagery on consumer responses to green product advertising. *Environmental Communication: A Journal of Nature and Culture*, 6(4), pp. 441–458.

Takach, G., 2013. Selling nature in a resource-based economy: Romantic/extractive gazes and Alberta's bituminous sands. *Environmental Communication: A Journal of Nature and Culture*, 7(2), pp. 211–230.

Takahashi, B. and Meisner, M., 2012. Environmental discourses and discourse coalitions in the reconfiguration of Peru's environmental governance. *Environmental Communication: A Journal of Nature and Culture*, 6(3), pp. 346–364.

Thompson, J. L. and Cantrill, J. G. 2013. The symbolic transformation of space. *Environmental Communication: A Journal of Nature and Culture*, 7(1), pp. 1–3.

Todd, A. M., 2010. Anthropocentric distance in *National Geographic*'s environmental aesthetic. *Environmental Communication: A Journal of Nature and Culture*, 4(2), pp. 206–224.

Uggla, Y. and Olausson, U., 2013. The enrollment of nature in tourist information: Framing urban Nature as "the Other." *Environmental Communication: A Journal of Nature and Culture*, 7(1), pp. 97–112.

Waddell, C., ed., 1998. *Landmark essays on rhetoric and the environment*. Mahwah, NJ: Erlbaum.

2

COMMUNICATION, MEDIA AND THE SOCIAL CONSTRUCTION OF THE ENVIRONMENT[1]

Anders Hansen

Introduction

This chapter explores how constructionism and closely associated perspectives have provided a productive framework for studying and understanding the roles of media and communication in relation to the environment and environmental issues. It traces some of the parallels in the emergence in the 1960s and 1970s of public/political interest in and concern about our relationship with the environment, and the emergence of the constructionist perspective (as defined by Herbert Blumer, Spector and Kitsuse, and others) in the sociology of social problems generally and in media and communication research more particularly. It proceeds to review the contribution of research on media and communication roles in the elaboration and contestation of the environment as a subject for public and political concern and action. It examines how research across a range of media and communication forms has provided an increasingly robust body of evidence on the centrality of communication to understanding the social and political 'careers' of environmental problems and issues, and it explores the contribution that environmental communication research can make in the politics of the environment.

Constructing social problems/constructing environmental issues

Why and how should we study media and communication in relation to environmental issues? Perhaps the answer to this emerges from the simple observation that not all environmental problems are publicly recognised as such – as problems requiring some kind of social/political/legislative attention and action – and from the equally puzzling observation that environmental issues or problems – over time – fade in and out of public focus in cycles that often seem to have little to do with whether they have been addressed, resolved, averted or ameliorated. Both observations suggest that communication – what is being said about environmental phenomena – is important, and they suggest that a public forum or arena, e.g. the media, is necessary for environmental phenomena to be recognised as issues for public or political concern.

A key breakthrough in sociology, and one that points directly to the centrality of 'media and communication', was the emergence in the 1960s and early 1970s of what became known as the constructionist perspective on social problems. The fundamental argument of this perspective was the argument that 'social problems' are not some objective condition in society that can be identified and studied independently of what is being 'said' about them. Problems and issues of various kinds only become recognised as such through talk, communication or discourse that defines or 'constructs' them as problems or issues for public and political concern.

One of the first to articulate this perspective was American sociologist Herbert Blumer, who took issue with the way that sociologists had traditionally identified social problems on the basis of public concern. Blumer argued that this was problematic, as many 'ostensibly harmful conditions are not recognised as such by the public, and thus are ignored by sociologists'. Instead, Blumer called for a definition of social problems that recognises these as 'products of a process of collective definition' rather than 'objective conditions and social arrangements' (1971: 298). The key task for research then, according to Blumer, is 'to study the process by which a society comes to recognize its social problems' (1971: 300).

The focus on process and communication evident in the early work of Blumer and fellow American sociologists Malcolm Spector and John Kitsuse received its full articulation in an early article by Spector and Kitsuse (1973) and again in their pioneering 1977 book *Constructing Social Problems*. In this, Spector and Kitsuse define social problems as:

> the activities of individuals or groups making assertions of grievances and claims with respect to some putative conditions. [...] The central problem for a theory of social problems is to account for the emergence, nature, and maintenance of claims-making and responding activities.
>
> (2000: 75–76; emphasis in original)

This approach/framework then suggests (a) that problems/issues only become 'social problems/issues' when someone communicates or makes claims (in public) about them, and (b) that the important dimension to study and understand is the *process* through which claims emerge, are publicised, elaborated and contested. As prominent constructionist sociologist Joel Best (2013: 262) summarises in a recent overview of constructionist social problems theory: 'social problems need to be viewed as a social process, rather than as social conditions'.

The key achievement of the constructionist perspective on social problems lies in the recognition that problems do not become recognised or defined by society as problems by some simple objective existence, but only when someone makes claims in public about them. The construction of a problem as a 'social problem' is then largely a rhetorical or discursive achievement, the enactment of which is perpetrated by claims-makers, takes place in certain settings or public arenas, and proceeds through a number of phases.

Joel Best (1995) suggests that analysis of the construction of social problems needs to focus on 1) the claims themselves, 2) the claims-makers, and 3) the claims-making process. To these we might add the public arenas (Hilgartner and Bosk, 1988) or settings (Ibarra and Kitsuse, 1993) in and through which claims are articulated and made public, as arenas or settings (including the media, the courts, parliamentary politics, the scientific community) set their own boundaries for or impose their own constraints on what can and can not be said. And we should add a further focus, namely what constructionists refer to broadly as the 'cultural context' (Lowney, 2008), and communications researchers have referred to as a cultural or cultural resonances perspective in the analysis particularly (but not exclusively) of the sociology of news (Schudson, 1989; Gamson and Modigliani, 1989).

Constructionism and media/communication

If we accept the constructionist argument that environmental problems – and social problems generally – do not 'objectively' announce themselves, but only become recognised as such through the process of public claims-making, then it is also immediately clear that media, communication and discourse have a central role and should be a central focus for study. In light of this, it is perhaps surprising that the early formulations of the social constructionist perspective offered relatively little comment on media and communications. The development of a social constructionist perspective in mass media and communications research was left to sociologists with a particular interest in communications, notably such prominent American sociologists as Harvey Molotch, Herbert Gans and Gaye Tuchman.

Work by these and many other scholars with a particular interest in media and communications helped move communication research on environmental problems out of journalism studies trapped in circular concerns with balance, bias and objectivity, and proved a productive inspiration for attempts at grappling with sociological interpretations of the media's role in public and political controversy about the environment (Schoenfeld *et al.*, 1979, Mazur, 1981; Lowe and Morrison, 1984; Hilgartner and Bosk, 1988; Gamson and Modigliani, 1989; Burgess, 1990). Within mainstream media and communication research, organisational and culturalist perspectives (Gans, 1979; Schudson, 1989) on news production, agenda-setting research and, in the last two decades or so, the concepts of 'framing', 'interpretive packages' and 'cultural resonances' have provided productive – and often overlapping – frameworks for analysing environmental communication.

The constructionist argument has implications for understanding media roles both in relation to how claims are promoted/produced through the public arena of the media and for understanding how the media are a central, possibly *the* central, forum through which we, as audiences and publics, make sense of our environment, society and politics. This boils down to the argument that most of what we as individuals know, we know, not from direct experience (experiential knowledge), but from the symbolic reality constructed for us through what we have been told (by friends, family, teachers and other 'officials' of a host of social institutions: schools, churches, government departments or agencies) or have read about or have heard/seen *re*-presented to us through media of various kinds. The centrality of the media in this context is further emphasised by the fact that much of the symbolic construction of reality by a host of social institutions is now itself principally encountered through their representation in and through the media.

Public agendas and power

The social constructionist perspective's emphasis on 'claims-making' in public arenas as the constitutive component in the creation of 'social problems' usefully draws our attention to the importance and centrality of getting issues of concern onto the public and, more significantly, the political agenda.

The social constructionist perspective's emphasis on 'claims-making' in public arenas as the constitutive component in the creation of 'social problems' has interesting similarities to the traditions of research in political science and in communication research known as 'agenda-building' and 'agenda-setting', which in turn link with key traditions in the study of 'power' in society. An early and often-quoted formulation from political science that inspired the agenda-setting tradition in media and communication research was Bernard Cohen's statement that 'The press may not be successful much of the time in telling people what to think, but it is stunningly successful in telling its readers what to think *about*' (Cohen, 1963: 13). In other words, the 'power' of the media to influence public and political processes resides principally in signalling what society and the polity should be concerned about and in setting the framework for definition and discussion of such issues.

In their work on agenda-building, political scientists Roger Cobb and Charles Elder (1971) in their discussion of the political process likewise pointed to the centrality of *agendas* recognised by the polity and as forums for the public/political definition of *issues* of conflict 'between two or more identifiable groups over procedural or substantive matters relating to the distribution of positions or resources' (p. 32). In contrast to earlier pluralist perspectives on 'power' in society, which had focused on 'decision making' as a central component of the exercise of power, these perspectives recognised that the ability to control or influence what issues get onto the public/political agenda in the first place was itself a core part of exercising power in society (Lukes, 1974).

But while the tendency in media research has very predominantly been to focus on the issues that *do* make it onto the media and public agenda – perhaps not least because these are conspicuous and lend themselves most easily to being studied – media research has contributed rather less and had significantly less to say about the type of claims-making or publicity management that is aimed principally at keeping issues off or away from the public and/or media agenda. As sociologists critical of power perspectives focused on 'decision making' were pointing out around the same time as the social constructionist perspective emerged, the ability to keep issues from appearing on the political agenda and thus to ensure that they don't become issues for decision making, that they remain 'non-decisions' in other words, is as much an exercise of power as making decisions about issues that *are* on the agenda is. Edelman (1988: 14) takes this argument a step further by hinting that the placing of certain issues on the public agenda simultaneously achieves the granting of 'immunity' from notice and criticism to those issues that are not on the public agenda.

While the constructionist perspective then generally focuses our attention on the importance of propelling *claims* into/onto the public arena, it is also clear that an important aspect of the claims-making process may be to keep issues from emerging in particular (public) forums, and thereby influencing the degree to which issues become recognised, or not as the case may be, as candidates for public concern, discussion or political decision making.

Widespread publicity may, in other words, have a negative influence on a group's or claims-maker's objectives, particularly if the main effect is to galvanise opposition. As Ibarra and Kitsuse (1993) indicate, every claim generates a counter-rhetoric or counter-claim. It is, however, the 'power-framework' delineated by Lukes (1974), Edelman (1988) and others that provides the main corrective or qualification to the constructionist perspective's emphasis on claims-making, namely by stressing that the ability to keep claims and issues *off* the public agenda is just as significant, if not more, an exercise of power as the ability to successfully place claims and issues on the public agenda or in public view.

A related 'side effect' of successful claims-making is the extent to which it prompts a sharpening of opponents' publicity practices. Cox (2013) provides a succinct overview of some of these key publicity practices under the headings of corporate green marketing, green-washing and corporate advocacy campaigns. Signitzer and Prexl (2007), in their analysis of green-washing, likewise point out that activism pressure often tends to stimulate in corporate organisations what public relations theorists call 'an excellent public relations function' (Grunig *et al.*, 2002), a redoubling of corporate public relations efforts to preempt, counter, engage with, accommodate, undermine, frame, etc. the arguments and claims of pressure groups (see also Harlow *et al.*, 2011; Baum, 2012; and Plec and Pettenger, 2012).

A perhaps more sinister extension is the type of counter-claims making that is specifically designed to sow doubt and undermine public sentiment by exploiting the kind of genuine political and scientific uncertainties that in most cases exist, at least in the early stages of attention to new issues or problems. Reference is often made to the long and well-documented history of controversy and action in relation to tobacco promotion and action against smoking (e.g. Michaels and Monforton, 2005; Proctor, 2012), while more recently much research has focused on the claims-making activities geared towards impeding political action on climate change (e.g. Beder, 2002; Proctor and Schiebinger, 2008; Stocking and Holstein, 2009; Oreskes and Conway, 2010; Dunlap and McCright, 2010).

The claims-making process and issue careers

The emphasis on the *process* of claims-making also led early constructionists to identify the distinctive stages/phases that social problems pass through as they emerge, are elaborated, addressed, contested and perhaps resolved, in other words the 'career' path of social problems. Spector and Kitsuse (1977) thus suggest a four-stage natural history model to describe the career of social issues. Downs (1972), in a much quoted article (not least in studies of environmental, science and risk issues), similarly proposed what he called an 'issue-attention cycle' to explain the cyclical manner in which various social problems suddenly emerge on the public stage, remain there for a time and 'then – though still largely unresolved – gradually [fade] from the centre of public attention' (Downs, 1972: 38). Downs suggests – and he happens to use 'ecology' or environmental issues as his example – that the career of public issues takes the shape of five distinctive stages: (1) a pre-problem stage; (2) alarmed discovery and euphoric enthusiasm; (3) realising the cost of significant progress and the sacrifices required to solve the problem; (4) gradual decline of intense public interest; and (5) the post-problem stage, where the issue has been replaced at the centre of public concern and 'moves into a prolonged limbo – a twilight realm of lesser attention or spasmodic recurrences of interest' (Downs, 1972: 40).

Criticism of natural history models has focused on the notion that they offer a much too simplistic and linear model of the evolution and progression of social problems. As Schneider (1985: 225) points out, Wiener (1981) for example 'argues that the sequential aspect of natural history models probably misleads us about the definitional process. She believes a more accurate view is one of "overlapping", simultaneous, continuously ricocheting interaction (Wiener 1981: 7)'. Wiener's evocative metaphor of 'continuously ricocheting interaction' is a particularly prescient and apt formulation relevant to communication research, as it directly counters the long-dominant – in communication research – linear view of communication, and begins to capture the highly dynamic and interactive nature of social communication processes.

Hilgartner and Bosk (1988) similarly criticise natural history models for their 'crude' suggestion of 'an orderly succession of stages' and for inadequate recognition that 'Many problems exist simultaneously in several "stages" of development, and patterns of progression from one stage to the next vary sufficiently to question the claim that a typical career exists' (p. 54). While these are valid points, they may not in fact be entirely fair criticisms of what was actually proposed by Spector and Kitsuse and others. Spector and Kitsuse thus never suggested a simple linear progression, but indeed emphasised the *heuristic* nature of the model and, more importantly, the open-ended nature of the processes described. Downs similarly, particularly with his description of the fifth stage of issue careers and of course through the deliberate use of the word 'cycle', implied a recursive, cyclical process replete with loops.

It is perhaps also relevant, as I have argued elsewhere (Hansen, 2010), for communication researchers in particular to recognise that Spector and Kitsuse, and indeed Downs, discussed the general social career of issues (part of which may relate to the media) but never proposed that the media career was synonymous with the social career or indeed that the social career can be 'read off' or deduced from mapping of the media career of a social issue.

While heeding then the advice from Wiener, Hilgartner and Bosk and others, that the career of social problems rarely follows a simple linear trajectory, the natural history models remain useful as heuristic models for the simple reason that they focus our attention on the notion of a *career*, the idea that issues evolve over time, and the idea that there are distinctive phases or stages in this process. They further alert us to the notion – and this is perhaps the most significant dimension – that issues don't simply evolve in some vague general or abstract location called 'society', but rather that they develop and evolve in particular social arenas (Hilgartner and Bosk, 1988) or forums, including political forums and the media, that (a) interact with each other in important ways that determine how issues evolve; (b) *host* the stages of problem definition and (c) themselves influence or frame these stages in important ways.

The analytical benefit from looking at the claims-making process as a series of distinct phases was also recognised early on in relation to identifying the key tasks or challenges for claims-makers. In a seminal article, published in 1976, Solesbury thus made the crucial distinction between, on the one hand, gaining media coverage for a particular cause or issue, and on the other, ensuring that the tone or framing of the coverage is such that the 'message' conveyed to the larger public is one of legitimacy and support for the claims being made. To these two tasks, Solesbury adds a further crucial task in the claims-making process, namely that of 'invoking action'. The three key tasks for claims-makers making claims about a putative problem then are: 1) commanding attention, 2) claiming legitimacy and 3) invoking action.

Resonant of Downs's (1972) issue–career stages, Solesbury's three tasks critically disentangle – what is often ignored or confused in the hyped-up rhetoric of claims-makers themselves and indeed in some media analyses – these three separate achievements to emphasise that media coverage cannot be assumed to be 'the right kind of media coverage'. Coverage may be positive and be conferring legitimacy, lending credibility to the claims being made, or it may be critical and negative, undermining the legitimacy of both claims and claims-makers, or worse still marginalising and branding the claims-makers as insincere or 'extremist'. Nor does media coverage itself guarantee social or political action, i.e. that something is actually done to address or resolve the problem about which claims are being made. As well as providing a sobering yardstick for claims-makers themselves for the planning and evaluation of claims-making strategies, Solesbury's three tasks serve as a useful reminder and tool for media-analysts seeking to assess media and communication roles in claims-making and campaigning processes. Most significantly perhaps, it provides an important corrective to the comparatively low attention to the mass media in early constructionist theory by identifying the media themselves as not only a key public arena for claims-makers, but as a key claims-maker themselves.

Where the constructionist notion of issue careers has been particularly productive and useful in communication research has been in providing a framework for the longitudinal analysis of how claims about the environment emerge, develop, fare in the media arena and eventually subside or fade from public prominence again (see Hansen, Chapter 17 in this Handbook, for a fuller discussion of longitudinal research). This has, *inter alia*, helped in drawing attention to the complex interactions between different public arenas (media, political, legal, science), to notions of trigger-events and thresholds in bringing issues to public and media attention (Mazur, 1984; Ungar, 1992; Mazur and Lee, 1993) and to the interaction of issues. The latter involves both consideration of how environmental issues fare in competition with other non-environmental issues, and of how different environmental issues fare in competition with each other for the limited space available in the media and other public arenas (see e.g. Djerf-Pierre, 2012; Schäfer et al. 2014; Ungar, 2014). Cross-cultural comparisons drawing on a constructionist issue careers framework have likewise helped advance our understanding of how political, cultural and media-systems differences across countries impact on how environmental issues are construed for public and political concern (see e.g. Hansen and Linné, 1994; Brossard *et al.*, 2004; Shehata and Hopmann, 2012).

Rhetoric/discourse/framing and the claims-making process

Given the centrality of claims-making in constructionist theory, and accepting that claims-making is essentially an act of communication, it is hardly surprising that much attention in constructionist theory – and of course in communication research – has focused on language, rhetorical strategies and discourse.

In their chapter on 'vernacular resources' in claims-making and the construction of social problems, Ibarra and Kitsuse (1993: 34) thus argue that the linguistics focus of constructionist studies could usefully examine *rhetorical idioms* (definitional packages that invoke particular views or narratives: e.g. a rhetoric of loss, a rhetoric of calamity, etc.), *counter-rhetorics* (discursive strategies for countering characterisations made by claimants), *motifs* (recurrent thematic elements, metaphors and figures of speech that encapsulate, highlight or offer a shorthand to some aspect of a social problem, 1993: 47) and *claims-making styles* (e.g. claims can be made in *legalistic* fashion or *comic* fashion, in a *scientific* way or a *theatrical* way, in a *journalistic* ('objective') manner or an 'involved citizen' (or '*civic*') manner – 1993: 35).

Although Ibarra and Kitsuse (1993) do not explicitly position themselves within the context of 'framing', much of what they say resonates very well with the concept of framing as it has grown to be used in media and communication research. Reese (2001; emphasis in original) for example defines frames as '*organizing principles* that are socially *shared* and *persistent* over time, that work *symbolically* to meaningfully *structure* the social world', while Gitlin (1980) sees frames as 'principles of selection, emphasis, and presentation composed of little tacit theories about what exists, what happens, and what matters'. Frames in other words draw attention to particular dimensions or perspectives and they set the boundaries for how we should interpret or perceive what is presented to us. Gamson (1985), with particular reference to the media, suggests that 'news frames are almost entirely implicit and taken for granted. [...] News frames make the world look natural. They determine what is selected, what is excluded, what is emphasised. In short, news presents a packaged world' (Gamson, 1985: 618).

Miller and Riechert (2000: 46) make an important further addition to the definition of framing by suggesting the need to examine whose interests are served. They thus suggest that framing is usually thought of as 'driven by unifying ideologies that shape all content on a topic into a specific, dominant interpretation consistent with the interests of social elites'.

In a synoptic – and much quoted – overview of the 'framing' concept and its various disciplinary origins, Entman (1993) defines framing as follows:

> To frame is to *select some aspects of a perceived reality and make them more salient in a communicating text, in such a way as to promote a particular problem definition, causal interpretation, moral evaluation, and/or treatment recommendation* for the item described.
>
> (Entman, 1993: 53; emphasis in original)

Claims-makers, including media and news professionals, then draw attention to particular interpretations through *selection* (e.g. our attention is drawn to some aspects while others, not selected, are kept out of view) and *salience or emphasis*, which promotes particular definitions/interpretations/understandings rather than others. Perhaps the most significant point about how a problem is defined is that the definition invariably carries with it the allocation of responsibility or blame as well as – implicitly or explicitly – directions for the problem's solution.

This is succinctly captured by Charlotte Ryan (1991: 57), who argues that the notion of framing directs the analysis of claims-making and the construction of social problems to ask three core questions: 1) What is the issue? (definition), 2) Who is responsible? (identification of actors/stakeholders) and 3) What is the solution? (suggested action/remedies). And to answer these questions we can usefully draw on the analytical framework offered by Gamson and Modigliani (1989) which helpfully sets out the notion of a 'signature matrix' to indicate the constituent parts of frames – in other words, enables us to answer the question, by which rhetorical/linguistic and other devices is a frame constituted and sustained?

Gamson and Modigliani's (1989) discussion of framing is particularly useful because it draws attention to the two core meanings of 'framing' in media research: on the one hand, framing as a term for the stories/discourses/ideologies/packages available to us for making sense of our environment, and, on the other hand, framing as the workings or operation of the constituent parts that together contribute to a particular frame. Gamson and Modigliani thus suggest:

> [M]edia discourse can be conceived of as a set of interpretive packages that give meaning to an issue. A package has an internal structure. At its core is a central organizing idea, or frame, for making sense of relevant events, suggesting what is at issue [and] a package offers a number of different condensing symbols that suggest the core frame and positions in shorthand, making it possible to display the package as a whole with a deft metaphor, catchphrase, or other symbolic devices.
>
> (1989: 3)

The constituent parts that together contribute to or build the 'frame' can, according to Gamson and Modigliani (1989), be identified as five *framing devices* that suggest how to think about the issue (metaphors; exemplars (i.e., historical examples from which lessons are drawn); catchphrases; depictions; visual images (e.g., icons)) and three *reasoning devices* that justify what should be done about it (roots (i.e., a causal analysis); consequences (i.e., a particular type of effect); and appeals to principle (i.e., a set of moral claims)).

What is offered then is an analytical framework for characterising and un-packing the interpretive packages in 'a signature matrix that states the frame, the range of positions, and the eight different types of signature elements that suggest this core in a condensed manner' (Gamson and Modigliani, 1989: 4).

The analytical frameworks suggested by Gamson and Modigliani (1989) and Ibarra and Kitsuse (1993) draw attention to the centrality of language and rhetorical/discursive strategies in claims-making and the social construction of issues for public concern. And they offer suggestive lists of key questions/tools for identifying and analysing the devices and frames that come into play in the claims-making process and for identifying and analysing the rhetorical means by which some claims-making is more successful than other types of claims-making.

Constructionist analyses of environmental communication that draw productively on concepts such as agenda-building/agenda-setting and cultural resonance include the work of Birkland and Lawrence (2001, 2002). Most recently, Melissa Merry (2014) offers an exemplary constructionist study of the rhetorical and campaigning strategies – particularly through 'blame casting' – pursued by environmental groups in relation to the 2010 Deepwater Horizon oil spill in the Gulf of Mexico.

Historically much less prominent in constructionist analysis of environmental communication than the focus on language, discourse and rhetoric, analysis – from a constructionist framework – of how environmental communication is constructed *visually* has started to gain momentum only within the last decade or so (Hansen and Machin, 2013). Yet, and not surprisingly, it is clear that the constructionist framework has much to offer also in the analysis of visual environmental communication (e.g. Doyle, 2007; Peeples, 2013).

The claims-makers/actors

As Best (1995 and 2013) and many others have indicated, constructionist theory directs analytical attention crucially to the people who as individuals or representatives of movements, organisations, institutions, government etc. make claims or counter-claims or are otherwise involved in influencing and shaping the claims-making process. These have also long been a central focus for media and communication research, coming broadly under the study of the *production* of news and other media reporting (see the chapters in Part II of this Handbook).

There are close parallels between the constructionist notion of claims-maker and, in media and communication research, the notion of primary definers (Hall *et al.*, 1978). But, where the original notion of primary definers carried with it relatively narrow assumptions about the hegemonic and uniform nature of primary definition through the main media, the constructionist notion of claims-maker has contributed a more open and less predetermined notion of who can be a claims-maker and what interests or ideologies claims-makers might represent (see Schlesinger, 1990, for an insightful critique).

The focus, in media and communication research, on claims-makers has centred on examining: a) the sources who seek to make claims through the media and are quoted or referenced in news and other media reporting; b) the journalists and other media professionals involved in producing such reporting; and c) the nature of the relationship between the two.

There is now a well-developed research literature on the nature and relative prominence of different sources used in news media reporting on the environment and closely associated fields (including science, technology, health, etc.). This has traditionally confirmed a considerable 'authority orientation' in news reporting with a heavy reliance on scientists and other 'expert' professionals (Albaek *et al.*, 2003), politicians and figures in authority within key institutions, with much less prominence given to environmental pressure groups, movements or protestors (Lewis *et al.*, 2008; Andrews and Caren, 2010; and see the chapters in this Handbook by Dunwoody (5), Cox and Schwarze (6) and Lester and Cottle (8)). There is also now a growing body of research aimed at examining the often highly sophisticated source publicity practices – aimed at influencing media coverage – of governments, business and large corporations (Beder, 2002; and see particularly Miller and Dinan, Chapter 7 in this Handbook).

Studies of journalistic practices and values have similarly provided much insight into the ways in which these shape and impinge on the selection of sources, the selection of issues and the framing of these in media reporting. Studies in this mould have for example shown how classic journalistic requirements of objectivity, balance and accuracy can be seen as being behind the 'authority orientation' in the choice of sources (the need for the journalists to cover their backs, so to speak, when reporting on topics characterised by great scientific complexity, uncertainty and – frequently – controversy) as well as the tendency to give equal weight to claims and counter-claims, regardless of the strength of evidence or degree of agreement within for example the scientific community and literature (Boykoff and Boykoff, 2004; and see chapters by Friedman (11), Sachsman and Valenti (12), Anderson (14) and Williams (16) in this Handbook).

Studies of the relationship between journalists and their sources have noted the often symbiotic nature of this relationship, and have sought to establish who has the upper hand: sources, with their specialised knowledge and sometimes powerful publicity resources to draw on, or journalists, with their power to decide on what sources to select from, whom to give access to the media arena and how to frame these sources positively or negatively in their coverage. Of particular interest in this body of research is the question of how economic pressures on news organisations and rapid technological change in the news media environment change not only the organisation of news work and the news-gathering practices of journalists, but also the balance of power in the relationship between journalists and their sources (Trench, 2009; Friedman, Chapter 11 in this Handbook). Increasing economic pressures on news organisations and the rise of new media platforms for claims-making thus would appear to have enhanced the power of sources and reduced the influence of journalists over what gets reported and how. In addition, new media technologies have facilitated the rise of new types of journalism, including so-called citizen journalism (see Allan and Ewart, Chapter 15, in this Handbook), that in turn can be seen to dissipate the power of professional journalists.

Public arenas/forums/settings for claims-making and issue construction

While much attention in constructionist communication research has focused on rhetorical/discursive strategies, on issue careers and on the claims-makers themselves, less analysis has been devoted to examining the key public arenas (Hilgartner and Bosk, 1988), forums (Hansen, 1991) or settings (Ibarra and Kitsuse, 1993) where such claims are made, challenged, modified and developed. Yet, analysis of the characteristics of the public arenas of claims-making can help explain the framing and timing of the claims themselves, as well as the rhetorical and other strategies of key claims-makers. In media and communication research this focus has found its main articulation in what Schudson, Gans and others term the Organisational Perspective on news production and through the still wider framework of the Political Economy perspective, concerned with how ownership and economic factors impact on the (political/ ideological) nature of news and other media output.

Schoenfeld *et al.* (1979) had already drawn attention at an early stage in the development of environmental communication research to the significant implications for environmental news and claims-making of temporal cycles in the organisation of news work and of the organisational arrangements of news beats. They noted that it was not until the establishment within many major news organisations of an environmental news beat and the assignment of specialist environmental correspondents that claims-making on the environment and the rise of environmental reporting took off. In this Handbook, Friedman (Chapter 11) and Sachsman and Valenti (Chapter 12), building on their pioneering and long-standing research on environmental journalism, provide succinct overviews of the important influences of organisational arrangements in the ups and downs of environmental news and journalism.

Different public arenas set different boundaries for what can be said, by whom, how it can be said (e.g. the type of language/discourse that can be used, legalistic, scientific, journalistic, etc. – see also Ibarra and

Kitsuse, 1993) and for how the authority and credibility of claims-makers are constructed. Hansen and Linné (1994) thus note how environmental pressure groups often gain media attention through the forum of public protest or demonstrations, which potentially convey less of an authoritative and credible image than that of scientific experts interviewed against the background of impressive scientific laboratories, graphs or numbers on screens. Work on the narrative, format and stylistic conventions (constraints) of different media genres, such as news and documentaries, likewise demonstrates the importance of attending to the setting of claims-making (e.g. Cottle, 1993 and 2000; Hornig, 1990). However, the characteristics of settings/forums/arenas and their significance for claims-making practices and the ways in which the credibility of claims-makers is constructed (or undermined) is an area ripe for further development in environmental communication research.

Conclusion

The constructionist perspective provides a framework for analysing and understanding why some environmental issues come to be recognised as issues for public and political concern, while others never make it into the public eye, and thus fail to command the political attention and resources required for their resolution. The constructionist perspective focuses our attention on the claims-making *process*, on the role of claims-makers and on the public definition of social problems as essentially a rhetorical/discursive achievement.

One of the key benefits of a constructionist perspective in media and communication research is the departure that it offers from traditional narrow concerns, particularly in journalism studies, with such slippery concepts as objectivity, bias/balance and accuracy. Another major benefit is the move away from traditionally dominant linear models of communication towards an appreciation of the highly complex, dynamic and interactive nature of communication, claims-making, types of claims-maker and claims-making arenas.

The constructionist perspective further shows that social problems do not simply appear in some vague location called society, but that they are actively constructed, defined and contested in identifiable public arenas – notably the media – and that the careers of social problems are characterised by distinctive stages or phases. While claims-making and definition takes place in a number of arenas, the media are a particularly important arena, not least because it is through the media that we as publics predominantly learn about what goes on in other key arenas (such as parliament, science or the courts).

But the media are not simply an open stage; as a public arena they are governed by their own organisational, professional, narrative and format constraints that influence the what, by whom and how of claims-making and communication about environmental issues.

While the constructionist perspective has thus provided a valuable framework for the analysis and understanding of media and communications roles and processes in the construction of the environment as an issue/problem for public and political concern, there is still much work to be done, not least in the analysis of rhetorical/discursive practices and in the analysis of settings/arenas of claims-making. As Lowney (2008) has indicated, we need to know much more about what distinguishes unsuccessful from successful claims and claims-making. In many respects then it may be tempting to agree with prominent constructionist sociologist Joel Best's conclusion in his recent overview/review of constructionist social problems theory: 'The study of the sociology of social problems has just begun' (Best, 2013: 262).

Note

1   This chapter builds on and extends my chapters 2 (Communication and the construction of environmental issues) and 3 (Making claims and managing news about the environment) in Hansen (2010).

Further reading

Schneider, J. W. (1985). Social problems theory: the constructionist view. *Annual Review of Sociology, 11*, 209–229.
An exemplary early summary and stock-taking of the constructionist perspective, with much referencing of media and communication research.

Best, J. (2013). Constructionist social problems theory. In C. T. Salmon (ed.), *Communication Yearbook 36* (Vol. 36, pp. 236–269). London: Routledge.
A succinct overview from one of the foremost constructionist sociologists of our time.

Hannigan, J. A. (2006). *Environmental Sociology* (2nd edn). London: Routledge.
See particularly Chapter 5: Social construction of environmental issues and problems.

Cox, R. (2013). *Environmental Communication and the Public Sphere* (3rd edn). London: Sage.
See particularly Chapter 3: Social-symbolic constructions of environment.

Merry, M. K. (2014). *Framing Environmental Disaster: Environmental Advocacy and the Deepwater Horizon Oil Spill*. New York: Routledge.
Focusing on the concept of 'blame casting' as a rhetorical and campaigning strategy, Melissa Merry provides an exemplary constructionist analysis of environmental pressure groups' claims-making strategies and the associated agenda-building processes in the politics of oil spills.

References

Albaek, E., Christiansen, P. M. and Togeby, L. (2003). Experts in the mass media: researchers as sources in Danish daily newspapers, 1961–2001. *Journalism and Mass Communication Quarterly, 80*(4), 937–948.

Andrews, K. T. and Caren, N. (2010). Making the news: Movement organizations, media attention, and the public agenda. *American Sociological Review, 75*(6), 841–866.

Baum, L. M. (2012). It's not easy being green … or is it? A content analysis of environmental claims in magazine advertisements from the United States and United Kingdom. *Environmental Communication: A Journal of Nature and Culture, 6*(4), 423–440.

Beder, S. (2002). *Global Spin: The Corporate Assault on Environmentalism* (Revised edn). Totnes, Devon: Green Books.

Best, J. (ed.). (1995). *Images of Issues: Typifying Contemporary Social Problems* (2nd edn). New York: Aldine de Gruyter.

Best, J. (2013). Constructionist social problems theory. In C. T. Salmon (ed.), *Communication Yearbook 36* (Vol. 36, pp. 236–269). London: Routledge.

Birkland, T. A. and Lawrence, R. G. (2001). The *Exxon Valdez* and Alaska in the American imagination. In S. Biel (ed.), *American Disasters* (pp. 382–402). New York: New York University Press.

Birkland, T. A. and Lawrence, R. G. (2002). The social and political meaning of the Exxon Valdez oil spill. *Spill Science and Technology Bulletin, 7*(1–2), 17–22.

Blumer, H. (1971). Social problems as collective behavior. *Social Problems, 18*(3), 298–306.

Boykoff, M. T. and Boykoff, J. M. (2004). Balance as bias: global warming and the US prestige press. *Global Environmental Change: Human and Policy Dimensions, 14*(2), 125–136.

Brossard, D., Shanahan, J. and McComas, K. (2004). Are issue-cycles culturally constructed? A comparison of French and American coverage of global climate change. *Mass Communication and Society, 7*(3), 359–377.

Burgess, J. (1990). The production and consumption of environmental meanings in the mass media: a research agenda for the 1990s. *Transactions of the Institute of British Geographers, NS15*, 139–161.

Cobb, R. W. and Elder, C. D. (1971). The politics of agenda building: an alternative perspective for modern democratic theory. *Journal of Politics, 33*, 892–915.

Cohen, B. C. (1963). *The Press and Foreign Policy*. Princeton, NJ: Princeton University Press.

Cottle, S. (1993). Mediating the environment: modalities of TV news. In A. Hansen (ed.), *The Mass Media and Environmental Issues* (pp. 107–133). Leicester: Leicester University Press.

Cottle, S. (2000). TV news, lay voices and the visualisation of environmental risks. In S. Allan, B. Adam and C. Carter (eds), *Environmental Risks and the Media* (pp. 29–44). London: Routledge.

Cox, R. (2013). *Environmental Communication and the Public Sphere* (3rd edn). London: Sage.

Djerf-Pierre, M. (2012). The crowding-out effect: issue dynamics and attention to environmental issues in television news reporting over 30 years. *Journalism Studies*, *13*(4), 499–516.

Doyle, J. (2007). Picturing the clima(c)tic: Greenpeace and the representational politics of climate change communication. *Science as Culture*, *16*(2), 129–150.

Downs, A. (1972). Up and down with ecology: the issue attention cycle. *The Public Interest*, *28*(3), 38–50.

Dunlap, R. E., and McCright, A. M. (2010). Climate change denial: sources, actors and strategies. In C. Lever-Tracy (ed.), *Routledge Handbook of Climate Change and Society* (pp. 240–259). London: Routledge.

Edelman, M. (1988). *Constructing the Political Spectacle*. Chicago, IL: University of Chicago Press.

Entman, R. M. (1993). Framing: toward clarification of a fractured paradigm. *Journal of Communication*, *43*(4), 51–58.

Gamson, W. (1985). Goffman's legacy to political sociology. *Theory and Society*, *14*(5), 605–622.

Gamson, W. A. and Modigliani, A. (1989). Media discourse and public opinion on nuclear power: a constructionist approach. *American Journal of Sociology*, *95*(1), 1–37.

Gans, H. J. (1979). *Deciding What's News*. New York: Vintage.

Gitlin, T. (1980). *The Whole World is Watching: Mass Media in the Making and Unmaking of the New Left*. Berkeley, CA: University of California Press.

Grunig, L. A., Grunig, J. E. and Dozier, D. M. (2002). *Excellent Public Relations and Effective Organizations: A Study of Communication Management in Three Countries*. Mahwah, NJ: Lawrence Erlbaum Associates.

Hall, S., Critcher, C., Jefferson, T., Clarke, J. and Roberts, B. (1978). *Policing the Crisis*. London: Macmillan.

Hannigan, J. A. (2006). *Environmental Sociology* (2nd edn). London: Routledge.

Hansen, A. (1991). The media and the social construction of the environment. *Media Culture and Society*, *13*(4), 443–458.

Hansen, A. (2010). *Environment, Media and Communication*. London: Routledge.

Hansen, A. and Linné, O. (1994). Journalistic practices and television coverage of the environment: an international comparison. In C. Hamelink and O. Linné (eds), *Mass Communication Research: On Problems and Policies* (pp. 369–383). Norwood, NJ: Ablex.

Hansen, A. and Machin, D. (2013). Researching visual environmental communication. *Environmental Communication: A Journal of Nature and Culture*, *7*(2), 151–168.

Harlow, W. F., Brantley, B. C. and Harlow, R. M. (2011). BP initial image repair strategies after the Deepwater Horizon spill. *Public Relations Review*, *37*(1), 80–83.

Hilgartner, S. and Bosk, C. L. (1988). The rise and fall of social problems: a public arenas model. *American Journal of Sociology*, *94*(1), 53–78.

Hornig, S. (1990). Television's NOVA and the construction of scientific truth. *Critical Studies in Mass Communication*, *7*(1), 11–23.

Ibarra, P. R. and Kitsuse, J. I. (1993). Vernacular constituents of moral discourse: an interactionist proposal for the study of social problems. In J. A. Holstein and G. Miller (eds), *Reconsidering Social Constructionism: Debates in Social Problems Theory* (pp. 25–58). Hawthorne, NY: Aldine de Gruyter.

Lewis, J., Williams, A. and Franklin, B. (2008). A compromised fourth estate? UK news journalism, public relations and news sources. *Journalism Studies*, *9*(1), 1–20.

Lowe, P. and Morrison, D. (1984). Bad news or good news: environmental politics and the mass media. *The Sociological Review*, *32*(1), 75–90.

Lowney, K. (2008). Claimsmaking, culture, and the media in the social construction process. In J. A. Holstein and J. F. Gubrium (eds), *Handbook of Constructionist Research* (pp. 331–353). London: The Guildford Press.

Lukes, S. (1974). *Power: A Radical View*. London and Basingstoke: Macmillan.

Mazur, A. (1981). Media coverage and public opinion on scientific controversies. *Journal of Communication*, *31*(2), 106–115.

Mazur, A. (1984). The journalists and technology: reporting about Love Canal and Three Mile Island. *Minerva*, *22*(Spring), 45–66.

Mazur, A. and Lee, J. (1993). Sounding the global alarm: environmental issues in the United States national news. *Social Studies of Science*, *23*(4), 681–720.

Merry, M. K. (2014). *Framing Environmental Disaster: Environmental Advocacy and the Deepwater Horizon Oil Spill*. New York: Routledge.

Michaels, D. and Monforton, C. (2005). Manufacturing uncertainty: contested science and the protection of the public's health and environment. *American Journal of Public Health*, *95*, S39–S48.

Miller, M. M. and Riechert, B. P. (2000). Interest group strategies and journalistic norms: news media framing of environmental issues. In S. Allan, B. Adam and C. Carter (eds), *Environmental Risks and the Media* (pp. 45–54). London: Routledge.

Oreskes, N. and Conway, E. M. (2010). *Merchants of Doubt: How a Handful of Scientists Obscured the Truth on Issues from Tobacco Smoke to Global Warming.* London: Bloomsbury.

Peeples, J. (2013). Imaging toxins. *Environmental Communication: A Journal of Nature and Culture,* 7(2), 191–210.

Plec, E. and Pettenger, M. (2012). Greenwashing consumption: the didactic framing of ExxonMobil's energy solutions. *Environmental Communication: A Journal of Nature and Culture,* 6(4), 459–476.

Proctor, R. (2012). *Golden Holocaust: Origins of the Cigarette Catastrophe and the Case for Abolition.* Berkeley, CA: University of California Press.

Proctor, R. and Schiebinger, L. (eds). (2008). *Agnotology: The Making and Unmaking of Ignorance.* Palo Alto, CA: Stanford University Press.

Reese, S. D. (2001). Prologue – framing public life: a bridging model for media research. In S. D. Reese, O. H. Gandy and A. E. Grant (eds), *Framing Public Life: Perspectives on Media and Our Understanding of the Social World* (pp. 7–31). Mahwah, NJ: Lawrence Erlbaum Associates.

Ryan, C. (1991). *Prime Time Activism: Media Strategies for Grassroots Organizing.* Boston, MA: South End Press.

Schäfer, M. S., Ivanova, A. and Schmidt, A. (2014). What drives media attention for climate change? Explaining issue attention in Australian, German and Indian print media from 1996 to 2010. *International Communication Gazette,* 76(2), 152–176.

Schlesinger, P. (1990). Rethinking the sociology of journalism: source strategies and the limits of media centrism. In M. Ferguson (ed.), *Public Communication: The New Imperatives* (pp. 61–83). London: Sage.

Schneider, J. W. (1985). Social problems theory: The constructionist view. *Annual Review of Sociology,* 11, 209–229.

Schoenfeld, A. C., Meier, R. F., and Griffin, R. J. (1979). Constructing a social problem: the press and the environment. *Social Problems,* 27(1), 38–61.

Schudson, M. (1989). The sociology of news production. *Media, Culture and Society,* 11(3), 263–282.

Shehata, A. and Hopmann, D. N. (2012). Framing climate change: a study of US and Swedish press coverage of global warming. *Journalism Studies,* 13(2), 175–192.

Signitzer, B. and Prexl, A. (2007, 23–25 July). *Communication strategies of 'greenwash trackers': How activist groups attempt to hold companies accountable and to promote sustainable development.* Paper presented at the IAMCR Annual Conference, Paris.

Solesbury, W. (1976). The environmental agenda: an illustration of how situations may become political issues and issues may demand responses from government; or how they may not. *Public Administration,* 54(4), 379–397.

Spector, M. and Kitsuse, J. I. (1973). Social problems: a reformulation. *Social Problems,* 21(2), 145–159.

Spector, M. and Kitsuse, J. I. (1977). *Constructing Social Problems.* Menlo Park, CA: Cummings.

Spector, M. and Kitsuse, J. I. (2000). *Constructing Social Problems* (New edition). New Brunswick, NJ: Transaction Publishers.

Stocking, S. H. and Holstein, L. W. (2009). Manufacturing doubt: journalists' roles and the construction of ignorance in a scientific controversy. *Public Understanding of Science,* 18(1), 23–42.

Trench, B. (2009). Science reporting in the electronic embrace of the internet. In R. Holliman, E. Whitelegg, E. Scanlon, S. Smidt and J. Thomas (eds), *Investigating Science Communication in the Information Age: Implications for Public Engagement and Popular Media* (pp. 166–180). Milton Keynes: Oxford University Press and The Open University.

Ungar, S. (1992). The rise and (relative) decline of global warming as a social-problem. *Sociological Quarterly,* 33(4), 483–501.

Ungar, S. (2014). Media context and reporting opportunities on climate change: 2012 versus 1988. *Environmental Communication,* 8(2), 233–248.

Wiener, C. (1981). *The Politics of Alcoholism: Building an Arena around a Social Problem.* New Brunswick, NJ: Transaction.

3

DISCOURSE/RHETORICAL ANALYSIS APPROACHES TO ENVIRONMENT, MEDIA, AND COMMUNICATION

Jennifer Peeples

Introduction

The complexity of environmental problems, with their intricate connection to the material, cultural, social, economic, political aspects of our lives, invites scholars to bring diverse conceptual, theoretical, and methodological frameworks to the task of understanding and mitigating these expansive concerns. The more perspectives, the more dynamic our understanding of these problems and the greater options available to us for dealing with the most pressing issue of our time.

This chapter explores approaches for analyzing communication that have as their primary goal gaining insight into how communication constructs and influences people's awareness, orientation to, and sense-making of the environment. These humanist and critical approaches are less concerned with unearthing environmental "facts," but rather focus on understanding how communication functions pragmatically and constitutively. The rigor for these approaches comes both in the close, critical engagement of texts, informed by specific conceptual and/or theoretical insights, and, at other times, in the systemic application of methodology and the use of textual evidence to support claims made by researchers.

Rhetorical criticism, the first approach introduced in this chapter, is primarily used to explain how communication functions through the analysis of symbolic acts and artifacts, broadly referred to as the "texts." A broad, multifaceted approach, rhetorical criticism encompasses both a hermeneutic orientation as well as more defined methods of critical interpretation when contemplating a text or set of discursive practices, including what have been termed Neo-Aristotelian, genre, feminist, metaphor, cluster, pentadic, narrative, and ideological (Foss, 2009). Each has its perceptual strengths and limitations, leaving it to the discretion of the critic to formulate an interpretive approach that provides the most insight into the subject being analyzed.

Discourse analysis, the second approach examined in this chapter, has an even more extensive reach, spanning humanist, critical/cultural, and social scientific perspectives. Trappes-Lomax (2004) lists the various discourse analysis approaches: pragmatics (including speech act theory and politeness theory), conversation analysis, ethnography of communication, interactional sociolinguistics, systemic-function linguistics (SFL), Birmingham school discourse analysis, text-linguistics, pragmatic and social linguistic approaches to power in language, and critical discourse analysis (2004: 136). In this chapter, I focus generally on the discourse analysis approaches found in environmental communication that come from humanist or critical perspectives.

The rhetorical and discursive approaches discussed here are "sister disciplines" (Milstein 2009). They explore similar subject matter (e.g. discourse and media) and often present complementary findings. They do, however, have different theoretical groundings that are manifested in their various applications. This stems in part from rhetorical criticism being primarily a U.S. approach with its origins in public oratory. Discourse analysis comes principally from a European tradition with its grounding in linguistics. While somewhat unconventional to address these two approaches together, I will focus this discussion on over-lapping elements of the two approaches that are the most productive for understanding environmental communication scholarship.

Both approaches are used to investigate texts, symbolic acts, discursive practices, and/or language-in-use that reference or constitute the "environment" or the natural world. The materials for analysis include everyday conversations, corporate and scientific documents, public addresses, print and digital media, political actions, protests, books, websites, images and films, and acts of tourism and consumption. The goal of such analyses are broadly to understand the relationships between the material and the symbolic and the textual and the ideological as they are reflected and constructed in a variety of venues and contexts. Rhetorical critics and discourse analysts investigate how discourses socially construct an invested, partial and always subjective understanding of the environment. Scholars also analyze, extend and test existing theories by applying them to environmental case studies. Finally, many scholars intend their research to aid or influence changes in attitudes and practices affecting the environment.

The various modes of analysis within each approach range from a focus on written or spoken words to explorations of the visual, of performance, of hegemonic discourses, or textual fragments. Partially in response to changes in the subject matter and partially in response to a broadening knowledge of the social and political in environmental controversies, practitioners of both methodologies have made a critical—and visual—turn, incorporating European social theorists, such Michel Foucault, Jürgen Habermas, and Roland Barthes. The incorporation of these critical theories have enriched and deepened an appreciation of how power influences the choice of symbols and the construction of meaning.

In the following sections, I define key terms for discourse analysis and rhetorical criticism, offer an introduction to each approach, provide snapshots of the types of environmental communication research being performed, and explain the perspectives and insights revealed in the analyses. While the two approaches function similarly, it is in the application of the various modes of analyses where one can witness their capacity to explain how environmental communication functions.

Rhetorical criticism

Rhetoric

The study of rhetoric is usually said to have begun in early Greece with the writings of philosophers Plato, Isocrates, and Aristotle (fourth century BCE), and continues to the present. Aristotle defined rhetoric as, "the faculty of observing in any given case the available means of persuasion" (Herrick, 2009: 77). Other definitions also emphasize rhetoric's ability to influence audiences, for example, Francis Bacon's eloquent statement of rhetoric as the application of reason "for the better moving of the will" (1828). Contemporarily, Hauser explains that rhetoric is an "*instrumental* use of language." "Its goal is to influence human choices on specific matters that require immediate attention" (1986: 11). Foss expands the notion of rhetoric, arguing that to view it solely as a means of persuasion limits its capacity to invite interaction and understanding. She chooses to define it more broadly as the "human use of symbols to communicate" (2009: 3).

Often citing Kenneth Burke (1966), rhetoric scholars also agree that, in the same act of inviting, informing or persuading, rhetoric also socially constructs meaning based on linguistic and other symbolic choices. Burke, in his theory of terministic screens, maintains that language always reflects, selects, and deflects a particular reality. But as the use of certain symbols becomes normalized and naturalized, the inherent partiality of the rhetorical construction (in shaping an audience's perception) may be overlooked, necessitating a critic to explain and clarify rhetoric's influence.

Critical approaches

Rhetorical criticism is broadly considered to be "the business of identifying the complications of rhetoric and then unpacking or explaining them in a comprehensive and efficient manner" (Hart and Daughton, 2005: 22). In accounting for such "complications," critics have engaged a range of texts and discursive practices from an interpretative, or hermeneutic, fashion (Ricoeur, 1974; Hyde and Smith, 1979), as a "critical rhetoric" (McKerrow, 1989) and as a "[q]ualitative research method that is designed for the systematic investigation and explanation of symbolic acts and artifacts for the purpose of understanding rhetorical processes" (Foss, 2009: 6). With the origins of rhetoric grounded in the speeches of ancient Greece, the first texts chosen by early critics explored the spoken argument. Just as more contemporary critics have expanded to encompass many forms of symbolic representations, environmental communication scholars have also extended the range of texts considered, including advertisements, government and corporate documents, websites, news reports, protests, photographs, literature, films, and books.

Noting concerns that a critic applying a predetermined approach potentially limits findings, Foss (2009) maintained that most scholars use a generative approach to rhetorical criticism. Instead of starting with a particular method, such critics generate "units of analysis or an explanation from [the] artifact rather than from previously developed, formal methods of criticism" (2009: 387). This also allows critics to adapt and adopt forms of analysis that are capable of explaining the symbolic workings of nontraditional rhetorical texts, such as image events (discussed below) that would be difficult to analyze if restricted to language-focused methods.

A more recent approach to rhetorical criticism has been termed "critical rhetoric," explained by McKerrow (1989) as a method for analysis influenced by social theorist Michel Foucault (1926–1984) that has as its primary purpose the critique of domination and freedom: an analysis of the discourse of power. McKerrow argues that critical rhetoric's purpose is to "understand the integration of power/knowledge in society—what possibilities for change the integration invites or inhibits and what intervention strategies might be considered appropriate to effect social change" (1989: 91). Coming to environmental communication through engagement with environmental issues and/or with an awareness that many existing systems and organizations function to the detriment of the environment, scholars of environmental rhetoric often bring such a critical rhetoric perspective to their research.

In the following section, I provide examples of some of the analyses taking place in environmental communication, paying special attention to the research questions the authors are addressing and how the methodological choices they make aid in their analyses.

Rhetorical criticism in practice

One of the first examples of environmental rhetorical criticism in the U.S. was Christine Oravec's (1981) study of the persuasive strategies used by California preservationist John Muir in his natural history essays. Oravec began her analysis by asking how a literary writer so profoundly altered people's understanding of nature that legislators who had never set foot in the region voted to set aside Yosemite Valley for preservation. Closely analyzing the essays written by Muir in the years surrounding the campaign for Yosemite as a National Park, she found two rhetorical strategies at play.

First, Muir incorporated elements of the sublime response in his writing, a religious feeling akin to seeing the face of God. It required "the immediate apprehension of a sublime object; a sense of overwhelming personal insignificance akin to awe; and ultimately a kind of spiritual exaltation" (1981: 248).

Second, Oravec argued that Muir needed to create a literary persona that would "identify the readers' more or less passive literary experience with the activity of the figure," in many cases that of a "true mountaineer" (248). The reader was encouraged to see, hear, smell, and experience that which Muir was describing, at times speaking directly to "you."

These rhetorical strategies turned the audience from inert readers to dynamic participants, feeling the overwhelming awe and wonder that comes when one is in physical contact with the natural world. Oravec's explanation of the sublime response allowed her to investigate and explain the rhetorical strategies used by Muir as well as provide insights into the reasons for his literature's success in changing public and political opinion.

Rhetorical criticism also allows scholars to move beyond the texts or impact of one speaker, author, or "rhetor." Killingsworth and Palmer (1995), for example, began by asking what it is in the messages and style of environmental writers and thinkers that leads to calls of "hysteria" from their opposition. Instead of beginning with a close examination of the text, they embarked on a genealogical analysis (Foucault, 1978) of hysteria in order to comprehend its current application to environmentalists. They argued that their goal was not to "create an accurate, one-to-one correspondence in the analogy between the discourse of environmental crisis and the discourse of hysteria … but rather to use the psychoanalytic model to display and examine the motions at work in the environmental debate" (Killingsworth and Palmer, 1995: 3).

Their analysis found similarities in environmental discourses that had been labeled hysterical: they address human health; they publicly challenge "the value and direction of technological development"; and they use apocalyptic rhetoric in their appeals (1995: 2). Killingsworth and Palmer argued that just as "hysterics" before Freud were often diagnosed as faking their physical symptoms because of the complexity of the mind–body connections, environmental rhetors, such as Rachel Carson and Paul Erlich, are deemed hysterical because of the complexity of proving the relationship between human actions and environmental outcomes. They warn that as hysterics have lashed out at individuals in an unconscious attempt to overwhelm, overpower, and subdue their bodies, rhetorical acts, such as Carson's castigation of "Neanderthal" science and scientists who do not recognize toxins at work in the environment, may come from the same unconscious desire to control outcomes, but may in fact be detrimental to the movement because it does not attempt to "rehabilitate or convert the other side" (1995: 15). As with Oravec's use of the sublime, Killingsworth and Palmer's work illustrates the flexibility of rhetorical criticism as a method of critique as it allows for the use of concepts outside the communication discipline to function as conceptual frameworks for analysis.

In the next two examples, the authors examine the strengths and weaknesses of traditional rhetorical concepts and approaches by applying them to contemporary environmental texts. In these projects, both the rhetorical concept and the text were laid open for critique. In the first, Foust and Murphy (2009) conducted a frame analysis of newspaper articles that featured discussions of climate change in order to understand how apocalyptic claims function in global warming discourse. When performing a frame analysis, critics examine how a rhetor's choice of symbols (words, images, layout, etc.), shape the audiences' understanding of the content, often precluding other equally viable readings. A close analysis of the frame further "permits critics to identify constitutive structures in a discourse, but also to consider the structures' possible impacts in terms of agency, public opinion, policy, and democracy" (2009: 152). This approach is one that is commonly seen in both rhetorical criticism and discourse analysis research, especially in media studies.

Examining the news content that used the terms "catastrophe" in the text and "global warming" or "climate change" in the title, Foust and Murphy found two variations in the apocalyptic rhetoric. They argued that a tragic apocalyptic frame leads the reader to believe that human fate is sealed, while the comic frame argues that, while flawed, humans still have the capacity to influence the outcome of global warming.[1] In conclusion, they contended that the comic frame allows for continued engagement and action from the audience and advocate its use in global warming discourse.

Instead of beginning with a compelling text or a pressing social problem, Schwarze (2006) asked critics to reconsider a rhetorical form that had previously been denounced for its polarizing affect, its simplification of problems, and its emotional appeals. Melodrama, Schwarze argued, "generates stark, polarizing distinctions between social actors and infuses those distinctions with moral gravity and pathos" (2006: 239). Noting its recurrent appearance in environmental controversies, he contended that melodrama "offers environmental advocates a powerful resource for rhetorical invention" (2006: 239).

Using the rhetoric surrounding the asbestos contamination in Libby, Montana as his case study, Schwarze countered that melodrama can, in certain circumstances, "transform ambiguous and unrecognized environmental conditions into public problems; it can call attention to how distorted notions of the public interest conceal environmental degradation; and, it can overcome public indifference to environmental problems by amplifying their moral and emotional dimensions" (2006: 240). He concluded with a call for critics to avoid discounting rhetorical approaches without evaluating them within a particular context. This is especially true in environmental controversies where the rhetorical tactics of the marginalized may be judged differently when wielded by those in power.

Finally, in *Image Politics: The New Rhetoric of Environmental Activism* (1999), Kevin Michael DeLuca also questions disciplinary practice when he asked what are appropriate texts for rhetorical criticism scholarship. He argued that rhetorical critics have not taken images or radical political actions seriously enough in their research. Both have been characterized as illustrations or spectacles that are meant to draw attention to the "real" rhetoric (rational, reasoned, written, or verbal discourse), and that even when critics ostensibly maintain that they are analyzing images or acts of protests, they often center on the words that surround them more than images/acts themselves. In response, he offered an analysis of "image events" in which people engage in public acts, "employing the consequent publicity as a social medium through which to hold corporations and states accountable, help form public opinions, and constitute their own identities as subaltern counterparts" (1999: 22).

Most importantly to DeLuca, image events are social and political "critique through spectacle, not critique versus spectacle" (22). He offered for analysis Greenpeace's anti-whaling campaign in which people on rubber dinghies place themselves between the whaling ship and the whales and EarthFirsts! Redwood Summer, where activists in California sat in trees slated for logging, often putting their lives at risk for the forest. He contended that image events are "the primary rhetorical activity of environmental groups that are radically changing public consciousness" (1999: 14) and are therefore worthy of critical attention as to the ways they function as "mind bombs" that transform the way people see the world. Allowing the image events to motivate the analysis, DeLuca employs a fusion of postmodern theory, media theory, and cultural studies for his analysis.

As illustrated through these examples of rhetorical criticism, the approach has the flexibility and breadth to allow for close analyses of particular texts and can also track and reveal environmental discourses as they appear in multiple venues. Traditional rhetorical concepts (tragic frame), theories, and ideas from other disciplines (hysteria) can be brought to bear in order to better understand how environmental communication functions. The approach allows scholars to use their findings to critique the efficacy of environmental public discourse as well as use them to reflexively analyze how critics work. Rhetorical criticism analyses tie the specific use of symbols to larger environmental and disciplinary understandings.

Discourse analysis

Discourse

James Paul Gee, in his foundational work in discourse analysis, differentiated between what he referred to as "little d" and "big D" discourse. "Little d" discourse is language-in-use—"connected stretches of language that make sense, like conversations, stories, reports, arguments, essays" (Gee, 1990: 142). Capital D

"Discourses" are symbolic guides that attempt to simplify and explain the world around us. They are a "set of meanings, metaphors, representations, images, stories, statements, and so on that in some way together produce a particular version of events" (Gee, 1999: 201). As scripts, they provide order and insight for the vast amounts of data that perpetually engage us. A person who subscribes to a particular discourse is able to take fragments of information and create meaning through their placement within the discourse's narrative (Dryzek, 1997: 9). Dryzek added, "Each discourse rests on assumptions, judgments, and contentions that provide the basic terms for analysis, debates, agreements, and disagreements" (1997: 9).

Discourses can also tell us who we are, how we should act, and define our place of existence and our purpose. "These discourses project certain social values and ideas and in turn contribute to the (re) production of social life" (Hansen and Machin, 2008: 779). Discourses are manifest in language, and, therefore, analyses of language use can reveal the discourses that inform their construction. This revelation is important as discourses, such as neoliberalism, become normalized, naturalized, or "common sense," they maintain power over our sense of "reality," but fail to call attention to themselves as human constructs. Analysis is then essential for unpacking and revealing the powerful and at times elusive discourses that underlie and motivate human thought and action.

Method

Discourse analysis is the "study of language in use" (Gee and Handford, 2012: 1). It shares many of the same assumptions as those found in rhetorical criticism: that "language is ambiguous"; it is always "in the world"; the way we use language is inseparable from who we are and the different social groups to which we belong (Jones, 2012). Combining its roots in linguistics, philosophy, and sociology, discourse analysis examines the socially constructed meanings and relationships found in language. As Gee and Handford explain, "'Discourse analysis' covers both pragmatics (the study of contextually specific meanings of language in use) and the study of 'texts' (the study of how sentences and utterances pattern together to create meaning across multiple sentences or utterances)" (2012: 1). Additionally, discourse analysis unearths the discursive logic often hidden within a text, takes it apart, shows the reader how it works, and then puts it back together again in a way that makes what was previously invisible, visible (Carbaugh and Cerulli, 2013: 11).

Gee and Handford divide discourse analysis into four categories. The first continues the close association with linguistics, with a particular emphasis on grammar and structure. The second looks at themes or images within an oral or written text, with the third focusing on the description and explanation of discourse's function. The fourth type, often called *critical* discourse analysis, is "interested in tying language to politically, socially, or culturally contentious issues and in intervening in these issues in some way" (Gee and Handford, 2012: 5).

Critical Discourse Analysis (CDA) is a wedding of linguistic and social approaches that, according to Stamou and Paraskevopoulos (2004), works to rectify the shortcomings of both. They argue that linguistic discourse analysis, with its focus on the linguistic text, "has failed to account for its social nature." Social discourse analysis, on the other hand, has "neglected the role of language" in shaping symbolically constructed reality (2004: 107). CDA then has "the goal of looking beyond texts and taking into account institutional and sociocultural contexts." It involves "questioning the role of discourse in the production and transformation of social representations of reality, as well as social relations" (Carvalho, 2008). As we will see in the following examples, the reach of discourse analysis approaches allows for multiple perspectives and insights when analyzing environmental communication.

Discourse analysis in practice

Seeking to understand how visitors negotiate the "duality" and potential contradiction of ecotourism (tourism *and* environmentalism), Stamou and Paraskevopoulos (2004) employed a lexical analysis to examine language found within guest books from two visitor centers at a Greek nature reserve. With a lexical analysis, the researcher focuses on the specific word choice of the text producers. Stamou and Paraskevopoulos maintained that it "is through vocabulary that text producers and consumers identify different discourses, whereas speakers summarize their representational image via some key words" (2004: 111–112). After charting the terms used in the guestbook on a tourism–environmental continuum, with the middle representing both activities, they found that a few of the "images" constructed through the guests' language were in the middle of the continuum, leaving the researchers to conclude that the discourse constructed "a dualism rather than a duality of ecotourism, and thus favoring their competition rather than their reconciliation" (2004: 124). The lexical analysis (similar in some ways to rhetorical criticism's close textual and cluster analyses) allows the researcher to make claims to the perspectives and intentionality of the text producers based on the interdependent relationship between individuals' worldview and their word choice.

Moving from a focused investigation of word choice to one that looks more broadly at the social construction of an environmental issue, Carvalho (2005) used CDA to answer the question: "How is political governance of climate change constructed in and by the media?" (2005: 3). Carvalho employed an analysis of discursive strategies that, like the emphasis on persuasion found in much of the rhetorical criticism research, are understood as "forms of (discursive) manipulation of 'reality' by social actors in order to achieve a certain goal" (2005: 3). Her texts for analysis were three British "quality" newspapers' coverage of climate change. Carvalho focused on the newspapers' responses to "critical discourse moments" which are particular events that "challenge existing discursive positions and constructs or, in contrast, may contribute to their further sedimentation" (2005: 6). She complemented the specificity of that analysis with a longitudinal study looking at the evolution of climate change coverage over the period of her case study. This allowed Carvalho to make claims about the government's influence that she would have been unable to do with a smaller set of texts or a limited timeframe. She found that the government's discursive framework directly impacts the language used by the three newspapers, both in being shaped by and departing from the prevailing government frame.

Carbaugh and Cerulli (2013) differ substantially in their means of analysis from the previous examples. They examine discourse collected through ethnographic interviews (as opposed to public, written or spoken texts) and use a *cultural* discourse analysis approach. The goal of their work is to explore "human relations with nature, while embracing cultural and linguistic variability in these processes" (2013: 8). They do this through their study of discourses of dwelling or "place," specifically for this project, nontraditional hunters' descriptions of New England hunting grounds.

Carbaugh and Cerulli maintain that communication is "doubly placed" in that communication happens in and is affected by place and that communication is used to construct people's understanding of it. They explore the cultural knowledge of place created through the interaction of five "discursive hubs," namely, identity, action, feeling, relating, and dwelling, all made explicit in people's communication practices. As with the analyses discussed previously, the authors are attempting to understand and interpret the underlying discourses that influence humans' attitudes and actions. In this case, the authors are focusing on the "cultural logic in the discourse" and making its "radiants of belief and of value, more readily visible for consideration" (2013: 16). The approach outlined by Carbaugh and Cerulli differs most greatly from rhetorical criticism and the other examples of discourse analysis in that they focus on the cultural codes at play within an individual's language as revealed through interviews and critical evaluation.

Hansen and Machin (2008) show us yet another means to analyzing discourse, this one concentrating on visual images as a subject for examination. Following the lead of Kress and van Leeuwen (2006) who examine the language of visual design, the authors contended that "just as we can describe the way that discourses are signified in texts through lexical and grammatical choices so we can look at the visual semiotic choices that realize these in images" (2008: 777). They used a visual approach to explain how discourses presented in the media "might shape public perceptions of the environment and green issues" (2008: 777). Taking a multimodal CDA approach, which allows the researchers to look at "modes" beyond written and spoken language, they examined how the Getty's "Green Collection" of images construct a particular environmental discourse, one, they argued, that is more in harmony with marketing than raising concern for environmental degradation (2008: 778). They contended that a multimodal CDA approach allowed them to "analyse the way that Getty images and the terms available for searching the images, convey particular kinds of scripts, values and identities and what kinds of social relations these favour" (2008: 781). As with DeLuca's work in rhetorical criticism, Hansen and Machin are expanding the parameters of discourse analysis from language to images in order to maintain the relevancy and applicability of their chosen method for the most current iterations of environmental communication.

This review of environmental critical and discursive scholarship is in no way attempting to be encyclopedic. Instead, it provides examples that can be used to argue for the breadth, depth, and significance of these methods as a means for investigating environmental communication.

Conclusion

In her analysis of global warming discourse, Livesey (2002) demonstrated how rhetorical and discursive methods can reveal distinct aspects of the same set of "texts." She applied Kenneth Burke's dramatistic approach for her rhetorical critique and Michel Foucault's theories as the framework for her discourse analysis. Livesey (2002) concluded that rhetorical criticism provides insight into the use of symbols for persuasion, noting specifically how the "rhetor" uses language to influence the "audience's" actions and attitudes through tactics such as identification. The focus of rhetorical criticism, she argued, is on how symbols function in a specific controversy—a micro level of analysis.

Using Foucault's theories to analyze the same set of texts, she maintained that discourse analysis reveals how language reflects and reproduces "taken-for-granted realities that govern practice in the wider social arena" (2002: 140). Discourse analysis, she contended, explains how larger social systems manifest themselves in a particular text—a macro level of analysis.

While her work provides an interesting point for comparison, it restricts the breadth and complexity of both approaches by making methodological-level claims using only two theories/theorists. Livesey recognized this limitation. She observed that many rhetorical critics have begun to embrace the practices and theories of the social, cultural, and critical scholars, contending that "the lines between rhetorical analysis influenced by Burke and Foucauldian-inspired discourse analysis have started to blur" (2002: 140).

Livesey's analysis does however provide a useful framework for understanding the significance and impact of rhetorical and discursive approaches to researching environmental communication. They both explore the use of symbols and explain how those symbols function within the particular context, often illustrated through a case study. They reveal how the language used within that context influences, and is influenced, by larger cultural, political, economic, and/or social systems in play. Because they are not restricted by the need to limit variables and replicate findings, these methods are well suited to investigate and reveal the complex relationship between the symbolic and the material environment.

Note

1   Kenneth Burke explains the comic and tragic frame in *Attitudes Towards History* (1984).

Further reading

Waddell, C. (ed.) (1998). *Landmark Essays on Rhetoric and the Environment*. Mahwah, NJ: Hermagoras Press.
This edited book contains many of the foundational essays in the field of environmental rhetoric.

Meister, M. and Japp, P. M. (eds) (2002). *Enviropop: Studies in Environmental Rhetoric and Popular Culture*. Westport, CT: Praeger Publishers.

Enviropop provides excellent examples of rhetorical criticism's use for analyzing nontraditional texts: Hallmark cards, television shows, and SUV advertisements, to name a few.

Harré, R., Brockmeier, J., and Mühlhäusler (1999). *Greenspeak: A Study of Environmental Discourse*. Thousand Oaks, CA: Sage Publications.

The authors use a linguistic foundation to explore the rhetoric of science, environmental narratives, metaphor, temporal dimensions, and ethno-ecology.

Dryzek, J. S. (1997). *The Politics of the Earth: Environmental Discourses*. Oxford, Oxford University Press.

Dryzek's book provides a big-picture explanation of environmental discourses by charting them into categories for analysis and comparison: problem solver, survivalist, sustainability, and green radical orientations.

References

Bacon, F. (1828). *Of the Proficience and Advancement of Learning: Divine and Human*, London: J. F. Dove.

Burke, K. (1966). *Language as Symbolic Action: Essays on Life, Literature, and Method*, Berkeley, CA: University of California Press.

Burke, K. (1984). *Attitudes Towards History*, 3rd ed. Berkeley, CA: University of California Press.

Carbaugh, D. and Cerulli, T. (2013). Cultural discourses of dwelling: Investigating environmental communication as a place-based practice. *Environmental Communication*, 7(1), 4–23.

Carvalho, A. (2005). Representing the politics of the greenhouse effect. *Critical Discourse Studies*, 2(1), 1–29.

Carvalho, A. (2008). Media(ted) discourse and society. *Journalism Studies*, 9(2), 161–177.

Deluca, K. M. (1999). *Image Politics: The New Rhetoric of Environmental Activism*. New York: Guilford Press.

Dryzek, J. S. (1997). *The Politics of the Earth: Environmental Discourses*. Oxford: Oxford University Press.

Foss, S. K. (2009). *Rhetorical Criticism: Exploration & Practice*, 4th ed. Prospect Heights, IL: Waveland Press.

Foucault, M. (1978). *The History of Sexuality: Volume 1: An Introduction*. Trans. R. Hurley. New York: Pantheon.

Foust, C. and Murphy, W. (2009). Revealing and reframing apocalyptic tragedy in global warming discourse. *Environmental Communication*, 3(2), 151–167.

Gee, J. P. (1990). *Social Linguistics and Literacies: Ideology in Discourses, Critical Perspectives on Literacy and Education*. London: Routledge.

Gee, J. P. (1999). *An Introduction to Discourse Analysis: Theory and Method*. London: Routledge.

Gee, J. P. and Handford, M. (2012). *The Routledge Handbook of Discourse Analysis* [electronic resource]. London: Routledge. Available online at http://www.eblib.com (accessed March 17, 2014).

Hansen, A. and Machin, D. (2008). Visually branding the environment: Climate change as a marketing opportunity. *Discourse Studies*, 10(6), 777–794.

Hart, R. P. and Daughton, S. M. (2005). *Modern Rhetorical Criticism*. Boston, MA: Pearson/Allyn & Bacon.

Hauser, G. A. (1986). *Introduction to Rhetorical Theory*. Cambridge, Harper & Row.

Herrick, J. A. (2009). *The History and Theory of Rhetoric: An Introduction*. Boston, MA: Pearson.

Hyde, M. J. and Smith, C. R. (1979). Hermeneutics and rhetoric: A seen but unobserved relationship. *Quarterly Journal of Speech*, 65(4), 347–363.

Jones, R. H. (2012). *Discourse Analysis: A Resource Book for Students*. Milton Park, Abingdon: Routledge.

Killingsworth, M. and Palmer, J. (1995). The discourse of "environmentalist hysteria." *Quarterly Journal of Speech*, 81(1), 1–19.

Kress, G. and van Leeuwen, T. (2006). *Reading Images: The Grammar of Visual Design*. 2nd ed. London: Routledge.

Livesey, S. M. (2002). Global warming wars: Rhetorical and discourse analytic approaches to ExxonMobil's corporate public discourse. *Journal of Business Communication*, 39(1), 117–148.

Mckerrow, R. E. (1989). Critical rhetoric: Theory and praxis. *Communication Monographs*, 56(2), 91–112.

Milstein, T. (2009). Environmental communication theories. In S. Littlejohn, and K. Foss (eds) *Encyclopedia of Communication Theory* (pp. 344–349). Thousand Oaks, CA: Sage.

Oravec, C. (1981). John Muir, Yosemite, and the sublime response: A study in the rhetoric of preservationism. *Quarterly Journal of Speech*, 67(3), 245–258.

Ricoeur, P. (1974). *The Conflict of Interpretations: Essays in Hermeneutics*. Evanston, IL: Northwestern University Press.

Schwarze, S. (2006). Environmental melodrama. *Quarterly Journal of Speech*, 92(3), 239–261.

Stamou, A. and Paraskevopoulos, S. (2004). Images of nature by tourism and environmentalist discourses in visitors books: A critical discourse analysis of ecotourism. *Discourse & Society*, *15*(1), 105–129.

Trappes-Lomax, H. (2004). Discourse analysis. In A. Davies and C. Elder (eds) *The Handbook of Applied Linguistics* (pp. 133–164). Malden, MA: Blackwell Publishing.

4

SOCIAL SCIENCE APPROACHES TO ENVIRONMENT, MEDIA, AND COMMUNICATION

James Cantrill

Without a doubt, whenever we reference *social science* scholarship regarding environmental issues or behavior, we do well to conceptualize the subject as a large quilt of distinct yet kindred fields. In contrast to various physical sciences such as biology and chemistry, each of which certainly inform our understanding of the environment around us, the social sciences focus on the ways in which humans interact in that environment. As a consortium of many disciplinary fibers—in particular the fields of economics, history, human geography, political science, psychology, and sociology—the social sciences also represent lenses brought to bear upon how and why it is that the natural world seems to be unraveling before our very eyes, especially regarding the onslaught of changing climatic conditions. And, arguably, the thread that holds together the practical implications of this grand academic blanket may be the study of media and human communication; as Latour (2005) observed, attempts to explain the social scientific bases of human behavior bereft of a focus upon what people say and how they act are, at best, shortsighted if not overly reductionistic.

Twenty years ago, I wanted to introduce my peers to a range of literature in the social sciences seldom referenced in the rhetorical tradition that mostly characterized the practice of environmental communication research at the time (Cantrill, 1993). As Cox and Depoe note in their introductory remarks, the field has grown immensely over a mere two decades and, in this chapter, I hope to accomplish a number of objectives that update and expand my earlier efforts. I begin by very briefly reviewing the origin and nature of the social sciences, so as to provide a bit of context. Next, I examine six environmental specializations, referencing allied environmental communication studies and other research examples that epitomize various disciplines' contributions to the study of humans facing a warming planet. I conclude with a brief commentary suggesting how existing theory and research in the social sciences can inform our further understanding of the relationship between human communication and the environment writ large. Overall, my goal is to provide a rather general orientation so that readers of *The Routledge Handbook of Environment and Communication* can better appreciate a myriad of factors that influence the warp and woof of human interaction with and about the dynamic environment around us.

The social science tradition

Most historical treatments (e.g., Flyvbjerg, 2001) point to the rise of positivism in the mid-nineteenth century as the wellspring of the social sciences. Simply put, positivism and its philosophical descendant, logical empiricism, holds that society operates according to general laws in much the same way as does the physical world wherein verifiable empirical evidence is the coin of the realm and more introspective or intuitive knowledge is suspect. As such, the actions of individuals, groups, and institutions can be reduced to a logic-system that captures quantitative variances among people and organizations facing different kinds of social and physical forces.

By the middle of the twentieth century, however, many scholars (e.g., Giddens, 1974; Masterman, 1970) came to argue that a variety of cultural and social processes were not amenable to purely objective empiricism, that the relative superiority of quantitative analysis was illusory, and that positivism failed to account for the fact that value-laden social norms and processes of symbolic exchange among people rendered moot most law-like generalizations. Hence, a range of qualitative methods and research designs began to flourish and what we find today is a plethora of theoretical and methodological commitments—positivist and constructionist, quantitative and qualitative—pointing to the disciplinary richness of the social science academy.

Though likely just a coincidence, the 1970s also saw the social sciences begin to turn their attention toward the environment. In an era of environmental awakenings, as it became increasingly apparent that society was significantly altering the natural world in an unsustainable onslaught of environmental carnage, scholars carved out a number of sub-disciplines to use social science in hopes of altering human behavior (for a general review, see: Moran, 2010). Here, again, the approaches taken to make sense of environmental issues were quite varied and generally reflected the tradeoff between seeing causal relationships at play versus approaching the richness of a person's lived experience. As Brewer and Stern (2005) observe, the social sciences are today seen as adding pragmatic value to the work of those in the hard sciences when it comes down to deciding how humans can and should live on Earth. And even a cursory review of the major environmental specializations in those fields reveals a number of relatively unique vantage points for understanding the role played by communication processes in "The Drama of the Commons" (Dietz, Dolšak, Ostrom and Stern, 2000).

Environmental specialties in the social sciences

Environmental economics

As a distinct specialization within a field often characterized as "the dismal science," environmental economics examines the nexus between environmental quality and the market behavior of individuals or institutions. Arising out of the post-industrialist advent of modern environmentalism in the 1970s, it is a rapidly maturing area of study (e.g., Maréchal, 2012; Stavins, Wagner and Wagner, 2003) typically grounded in well-established economic concepts and somewhat esoteric formulae using survey methods or critical analysis to explore monetary policy and practice related to the environment. A number of excellent texts have been published to introduce the subject to general audiences (e.g., Brown, 2002; List and Price, 2013).

Insofar as financial security abets social and ecological integrity in producing a sustainable society (e.g. Hawken, Lovins and Lovins, 1999), rhetorical scholars have found in the field of environmental economics a treasure trove of concepts and interactions related to natural resource management and pollution prevention (e.g., Kendall, 2008; Smerecnik and Renegar, 2010). It is clear that issues such as the failure of market forces and private externalities to forestall environmental degradation (Porter and Van der Linde, 1995), the pooling of common resources under conditions of scarcity and rivalry (Ostrom, 1990), alternate mechanisms for economic valuation (e.g., Garrod and Willis, 1999), or impacts of globalization on international resource management (Newell, 2013) offer a sundry of avenues for exploring the link between communication practice and environmental conditions. In one way or the other, researchers who index economic interests in their environmental studies invariably encounter the tension that exists between the maintenance of "natural capital" such as ecosystem services attending biologically diverse landscapes and more tangible aspects such as resource extraction (e.g., Kareiva, Tallis, Ricketts, Daily and Polasky, 2011).

For a specific example of the relationship between environmental economics and environmental communication, consider the analysis of Dietz and Morton (2011). In this study, the role of economic considerations and decision making structure was examined in the creation of two policy reviews launched by the United Kingdom. *The Stern Review on the Economics of Climate Change* was, in large part, based upon a purely cost–benefit analysis determining that significant and immediate reductions in carbon emissions were warranted; it relied upon input from a small group of governmental economists who focused almost exclusively upon the long-term financial consequences of incrementally adapting to climate change trends. In contrast, the Committee on Radioactive Waste Management employed a much more inclusive process of public participation that incorporated economic and scientific evidence as well as broader stakeholder values. The Dietz and Morton analysis of these two projects points to the differential impact of problem framing and deliberation strategies when dealing with such wicked problems and concludes that an over-reliance upon expert economic modeling using multiple simulations of economic loss and gain under increasingly uncertain scenarios of climate effects is ill-advised, even if it is politically expedient in the public sphere. Grounded as it is in the field of environmental economics, the Dietz and Morton study thus offers communication scholars an example of how a better understanding of interaction dynamics and high-profile decision making results in more sustainable public policy.

Environmental history

To the extent historians, in general, use critical and interpretive methods to reflect upon and deconstruct the past, it is not surprising that environmental history has a significant niche in the literature of the field. As a specialization, it focuses upon the relationship between nature and human agency, adopts both positivist and subjectivist orientations (i.e., extracting key events versus discursive exchanges as progenitors of environmental problems and solutions) and emerged full-force in the tumultuous 1960s (Bird, 1987). About that time, a handful of texts dealing with agricultural development (e.g., Hays, 1959) and conservation issues (e.g., Nash, 1967) influenced a new generation of scholars (e.g., Cronon, 1991; White, 2004). A number of comprehensive overviews of environmental history exist (e.g., Hughes, 2001), each of which demonstrates the wide scope of subjects approached through the lens of this field. Ultimately, however, most environmental historians figuratively adopt as their creed the words of George Santayana (1905): "Those who cannot remember the past, are condemned to repeat it" (p. 284).

Scholars of environmental communication have certainly found in furrows of environmental history fertile ground upon which to explore the nature and impact of conservation advocacy, science and technology, movement dynamics, and policy deliberations (e.g., Endres, 2011; Opie and Elliot, 1996). In much the same way, historians have often used temporally situated discourse to illuminate the ways in which time unfolds vis-á-vis the environment (e.g., Brinkley, 2009; Flippen, 2008; McCormick, 1989; Merchant, 1980). Clearly, some topics such as the history of wilderness preservation (e.g., Warecki, 1998) and deforestation (e.g., Williams, 2003) appear to be more amenable to communication analysis, but several other subjects suggest themselves as well (e.g., Worster, 1988). Nonetheless, those wishing to study the intertwining threads of historical events and symbolic action would do well to appreciate both the natural and social science underpinnings of any subject thus approached, including a variety of methods for doing so.

A timely example of the goodness-of-fit between environmental history and communication studies can be seen in Langston's (2009) analysis of climate-induced changes threatening the boreal forests of the northern hemisphere. Reminiscent of Rachel Carson's prose in *Silent Spring*, Langston adeptly traces the synergistic and potentially pernicious effects of combating budworm infestations with DDT in the 1950s, logging in the boreal north, and developmental pressures following global warming in an underappreciated yet vital ecosystem. In this case, history is preamble to tomorrow wherein sequestered carbon and bioaccumulative toxins are released as the permafrost warms, landscapes are deforested, and grasslands follow climatic changes creeping toward the Arctic Circle. Her work concludes with a nod toward Santayana that environmental communication scholars could certainly further interrogate:

> Without reference to an ecological past that may no longer resemble our ecological futures, how will we learn to live responsibly in place? Global warming challenges us to re-examine what history means to us when we are changing the earth so quickly that our shared environmental histories are vanishing, possibly never to be witnessed again.
>
> (p. 648)

Environmental human geography

Perhaps no other foci in the social sciences captures the essential integrative quality of modern environmental communication studies as does environmental human geography (a.k.a., "environmental studies," albeit largely distinct from natural science-based geography and the more anthropologically oriented cultural geography or human ecology). Although not as distinct a specialization as others examined in this chapter, environmental human geography stresses the relationships between people and natural resources; it, too, grew out of the critical geography movement of the 1970s (Soja, 1989). As such, the field avails itself to a wide array of interdisciplinary topics ranging from millennial archeology (e.g., Fisher, Hill and Feinman, 2009) to current land use planning (e.g., Hillier, 2002) to persistent problems of environmental justice (e.g., Harvey, 1996) to ongoing interaction between humans and the places they inhabit (Tuan, 1977). A great deal of literature here is speculative and conceptual, though more tightly empirical approaches abound, and environmental human geography represents some of the most qualitatively oriented avenues we have to studying environmental issues. College students are often exposed to this range of subjects and methods by way of various well-received introductory texts related to human environmental geography or environmental studies (e.g., Moseley, Perramond, Hapke and Laris, 2013).

Considering the quite eclectic range of disciplines drawn upon by both communication scholars and environmental human geographers, one would expect a substantial overlap in their areas of research interests. For example, both domains use quantitative and qualitative methods to explore the relationship between environmental conditions and culture (cf. Carbaugh, 1999 and Eden, 2001), habitation patterns (cf. Milstein, Anguiano, Sandoval, Chen and Dickenson, 2011; and Borén and Young, 2013), and adaptive behaviors in times of uncertainty and risk (cf. Shellabarger, Peterson, Sills and Cubbage, 2012; and Pelling and Dill, 2010). And, without a doubt, environmental geographers and communication scholars have shared the limelight along with conservation psychologists in pursuing studies of place identity and attachment (for a review, see: Bott, Cantrill and Meyers, 2003).

Given the sheer expansiveness of environmental human geography, it should not come as a surprise that a great many projects in this scholarly area have focused on the issue of climate change, each of which more-or-less index the role of environmental communication and the media as suggested by Cottle and Lester in this volume (e.g., Birkenholtz, 2011; Cote and Nightingale, 2012; O'Brien, 2010). And much of this body of geographic research references the impact one's perception of place has in communicating and making decisions that can mitigate global warming, such as the relationship between place attachment and the acceptance of alternative energy production siting disputes. For example, various research teams (e.g., Devine-Wright, 2011; Vorkinn and Riese, 2001) have demonstrated that the stronger one's affective and cognitive attachment to a location, the less she or he is likely to embrace local energy developments such as wind farms, hydroelectric projects, and large-scale solar arrays in lieu of continuing the use of existing carbon-based utilities. Thus, perception of relatively local geographic spaces mediates attitudes toward climate change abatement technologies and, certainly, those beliefs may be amenable to persuasion and influence.

Environmental politics

As with other specialties in the social sciences, environmental politics is a political science variant. Grounded in an intellectual tradition going back to at least the writings of Plato, scholars in this area attempt to understand relationships between political behavior and public policy making related to the natural world (Carter, 2001). As such, it encompasses the wide range of theories and methodological orientations the discipline of political science embraces (e.g., Atkinson, 1997; Gollain, 2011; Skjærseth, Stokke and Wettestad, 2006). And, of all the environmental social sciences, this field likely has the most significant impact on the content of environmental journalism and media (Hays, 2000). A number of useful introductions to environmental politics have been crafted by scholars who, following more general trends in the late twentieth century, also galvanized their interests around the study of environmental problems and natural resource issues (e.g., Dauvergne, 2012; Kraft, 2001).

Just as environmental communication scholars (e.g., Besel, 2012; Peterson, 2004) have focused on various aspects of political life so, too, have those in environmental politics studied the same, and often with a nod toward the symbolic dimensions of their subject matter (e.g., Potoski, 2002; Hajer and Versteeg, 2005). As Bell (2013) observes, over the past two decades political scientists have developed a rich compendium of studies exploring several topics including environmental worldviews embodied in different professions (e.g., Dryzek, 1997), the politics of sustainability (e.g., Blühdorn, 2011), changes in the nature of party politics (e.g., Jensen and Spoon, 2011), and even the link between place attachment and public opinion formation (e.g., Dowling, 2010). The breadth of such research and theorizing points to any number of ways in which the study of communication and media related to environmental issues might be fruitfully advanced.

To the extent the foreseeable effects of global climate change pose a significant challenge to political order on Earth, the high degree to which scholars of environmental politics have invested theory and research into this subject is to be expected (e.g., Falkner, 2013). Numerous political scientists have investigated relevant issues such as the framing of climate change communications (e.g., Morton, Rabinovich, Marshall and Bretschneider, 2011) and the role of public deliberation therein (e.g., Bulkeley and Betsill, 2013). And, given the highly partisan nature of reactions to the threat of global climate change, the work of Guber (2013) offers an excellent bridge to environmental communication scholarship. Her analysis of survey data spanning three decades reveals that party affiliation is playing an increasingly significant role in regards to climate change and the mitigation thereof but only insofar as citizens gain their knowledge from and are prompted by partisan elites via the mass media; those that become polarized in their attitudes toward the subject are the same individuals who regularly attend to increasingly virulent framing of commentators and opinion leaders with a vocal political niche.

Environmental and conservation psychology

For many outside the general field of psychology, the idea of a distinctive *environmental* focus to the discipline would likely be associated with studying relationships between cognition and human responses to, say, pollution. However, that characterization has only recently begun to supplant the manner in which environmental psychologists traditionally explored the connection between relatively molar physical settings (e.g., a cityscape) and behavior (Pol, 2006). It was not until after the advent of the modern environmental milieu that the pioneering work of scholars such as George Simmel (1903) was transcended by a new crop of psychologists intent upon using social science to restore or preserve the Earth (e.g., Novic and Sandman, 1974). More recently, we have seen the emergence of *conservation* psychology which brings together researchers and practitioners interested in human sustenance of nature (e.g., Clayton, 2012). As a whole, the fields of environmental and conservation psychology probably represent the most expansive body of literature related to the protection of our planet found anywhere in the social sciences.

Setting aside theory and research associated with behavioral settings approached via environmental psychology (for a review, see: Bechtel and Churchman, 2002), the field of conservation psychology is also quite eclectic. In this academic tradition, scholars attend to a surfeit of subjects: the role of attitudes and values (e.g., McKenzie-Mohr, 2001), the nature of risk perception (e.g., Adeola, 2007), identity and place (e.g., Brehm, Eisenhauer and Stedman, 2012), human–animal relationships (e.g., Meyers, Saunders and Bexell, 2009), and the preservation of biodiversity (e.g., Castro and Mouro, 2011) to name just a few. Clearly, topics such as these have been and continue to be of interest to many in the field of environmental communication studies, especially as they relate to the psychology of climate change communications (e.g., Good, 2009; Schweizer, Davis and Thompson, 2013).

Similar to the work of Whitmarsh in this volume, a specific study regarding climate change that may capture a range of linkages between environmental communication and conservation psychology is the work of Slimak and Dietz (2006). Their research surveyed both the general public and risk professionals, asking respondents to rank a wide range of potential threats to the environment and human well-being. Analysis revealed that laypersons are more concerned about low-probability, high-consequence risks (e.g., radiation exposure) while environmental experts are more concerned about situations posing long-term risks to local and global ecosystems (e.g., global warming). Ancillary analysis indicated that the most consistent predictors of the different rankings were adherence to ecological worldviews and altruistic orientations toward social conduct in the biosphere, far more than other sociocultural variables such as religion, partisanship, or ethnicity. Theorizing that the integration of values, beliefs, and norms differentially amplify perceptions of risk events, the authors conclude that the framing of risk analyses ought to be based upon robust public sphere deliberations. Thus, a tightly knit study in the tradition of conservation psychology corroborates many of the same public engagement strategies advocated by environmental communication scholars (e.g., Daniels and Walker, 2001; Senecah, 2004).

Environmental sociology

The sociological study of interactions between humans and the larger environment has traditionally focused upon processes related to the social construction, causes, and impacts of environmental problems as well as attempts by society to remedy such (cf. Field, Luloff and Krannich, 2013). Somewhat interdisciplinary in nature, this field grew out of anthropological and structuralist critiques of environmentalism in the late 1960s (Dunlap, 1997) and now offers a coherent, group-oriented complement to more individual-oriented approaches ensconced in fields such as conservation psychology. And, as with any mature disciplinary focus in the social sciences, the substance of environmental sociology can be examined using varied methods highlighted in texts that range from introductory overviews (e.g., Bell, 2012) to more highly focused scholarly anthologies (e.g., Buttel, Dickens, Dunlap and Gijswijt, 2002).

Generally interested in collective behavior, environmental sociologists have focused on a wide variety of topics including population and materialism (e.g., Ostrom, 1990), religion and philosophy (e.g., Shibley and Wiggins, 1997), and the relationship between group size and conservation practices (e.g., Yang, Liu, Viña, Tuanmu, He, Dietz and Liu, 2013). In this vein, scholars have also devoted sizable resources to uncover the bases for human concerns regarding the environment (for a review, see: Franzen and Vogl, 2013). And, without question, the environmental sociologist casting the largest shadow in the field must be Riley Dunlap who has spent more than three decades using survey research methods to disentangle the roots of environmental attitudes and the tension between the Dominant Social or "Exemptionalist" Paradigm (Catton and Dunlap, 1978) and the New Environmental Paradigm (e.g., Dunlap and Van Liere, 2008). Indeed, it may be the insights and data provided by Dunlap that have had the greatest impact on those who have invested in developing valid psychometric scales with which to measure environmental communication (e.g., Kassing, Johnson, Kloeber and Wentzel, 2010).

Any significant mitigation in global climate change trends will most certainly pivot on the actions of social collectives more than those of individuals and, consequently, one would expect environmental sociologists to explore the subject in detail. In turn, several have investigated such issues as the manner in which groups deny the existence of global warming (e.g., Dunlap, 2013), the dysfunctional relationship between political necessities and capitalistic imperatives (Giddens, 2009), the role of religious institutions (e.g., Samson, 2013), and attempts to coordinate international responses to a warming planet (e.g., Lane, 2013). A good example of this burgeoning body of research may be Smith, Anderson, and Moore's (2012) study of social resilience in the face of changing climactic conditions. Their findings demonstrate that the communal ability to withstand the impacts of climate change can be predicted by a number of communication-related variables such as the extent to which local social networks serve to dampen individual risk appraisals while, at the same time, increase the potential for information seeking behaviors (cf. Norton, 2007).

Our collective back pages

I began this chapter with reference back to the 1990s and our nascent attempts to integrate social science research into the emerging body of environmental communication scholarship. The field has clearly advanced in some amazing ways, but I am not sure we have traveled far enough to truly affect the health of our home world. In addition, the amount of variance we see in reactions to media and discourse regarding environmental protection and natural resource management may not be, in the end, sufficiently robust to initiate communication campaigns that effectively ameliorate vexing problems such as sustainability and climate change. Most of the social and environmental forces that influence reactions to environmental advocacy in the Anthropocene still must be processed by human minds that vary widely in terms of their cognitive complexity regarding the environment. And, to be certain, a great deal of how lived experiences transform themselves into the mental top soil out of which productive solutions grow is rooted in what people encounter in daily conversations and the media they consume (Jones, 2014; Shanahan, Jones, McBeth and Lane, 2013). Such narrative frames and content are explored in the following chapters and it is left to the reader to integrate and fertilize the material found in this Handbook so as to further advance the use of social science in the service of sustainable environmental lifestyles.

Despite their obvious value to environmental communication scholarship, the social sciences have not been without their detractors. For example, vaunted conservative George Will (2012) links the institutionalization of the social sciences to the values of "progressivism" in opining about their academic origins:

> [The social sciences represented] specialization, expertise, and scientific management [and] would reconcile the public to the transformation of universities, especially public universities, into something progressivism desired but the public found alien. Replicating industrialism's division of labor, universities introduced the fragmentation of the old curriculum of moral instruction into increasingly specialized and arcane disciplines ...—economics, sociology, political science—that were supposed to supply progressive governments with the expertise to manage the complexities of the modern economy and the simplicities of the unstructured masses ... the progressive virtue of subordinating the individual to the collectivity.
>
> (p. A4)

Although we might question the ideological virtues of his brand of conservatism, Will may well be right about the silos we have created in the environmental social sciences. One of the potential pitfalls associated with the fields reviewed here is to be found in the relatively parochial approach each discipline is tempted to take in examining the origins and solutions to various environmental problems; few scholars reference work across the arts and sciences, let alone studies of the physical plane we depend upon for survival. As may be evident from the research examples used in this chapter to exemplify the relationship between any given field and the practice of communication, there is sometimes a tendency to view scholarly puzzles in something of a trans-disciplinary vacuum. That caveat also applies to environmental communication researchers as well despite the integrative nature of our subject matter. Just as Latour (2005) urged social scientists to avoid trying to account for attitudes, values, and behaviors without considering the interaction context in which beliefs and actions arise, so too should scholars in communication heed the way in which the environmental social sciences provide an aesthetic and pragmatic grounding for our own studies. Perhaps the following pages will give each of us an increased ability to conduct just such an integrative synthesis.

Further reading

Buttel, F. H. (1976). Social science and the environment: Competing theories. *Social Science Quarterly, 57*, 307–387. An excellent review of conceptual origins related to the environmental social sciences that was published when the environment was first becoming a distinct focus beyond the physical sciences.

Clayton, S. D. and Meyers, G. (2009). *Conservation psychology: Understanding and promoting human care for nature.* Oxford: Wiley-Blackwell.
A well-developed introduction to not only the emerging field of conservation psychology but also related scholarship in sociology and anthropology.

Flyvbjerg, B. (2001). *Making social science matter: Why social inquiry fails and how it can succeed again.* Cambridge: Cambridge University Press.
A review and critique of the link between social science research and theorizing and practical conduct in society, with more than a passing nod to environmental issues.

Moran, E. (2010). *Environmental Social Science: Human–environment interactions and sustainability.* Hoboken, NJ: John Wiley & Sons.
A contemporary analysis that provides an integrative approach to using social science research and theory to tackle the wicked problems of fomenting a sustainable society.

References

Adeola, F. O. (2007). Nativity and environmental risk perception: An empirical study of native-born and foreign-born residents of the USA. *Human Ecology Review, 14*, 13–25.

Atkinson, A. (1997). Taking sides: Anthropology, environmentalism and the academy. *Environmental Politics, 6*, 183–189.

Bechtel, R. and Churchman, A. (2002). *Handbook of environmental psychology.* New York: Wiley & Sons.

Bell, D. (2013). Coming of age? Environmental politics at 21. *Environmental Politics, 22*, 1–15.

Bell, M. M. (2012). *An invitation to environmental sociology.* Los Angeles, CA: Sage.

Besel, R. (2012). Prolepsis and the environmental rhetoric of congressional politics: Defeating the Climate Stewardship Act of 2003. *Environmental Communication: A Journal of Nature and Culture, 6*, 233-249.

Bird, E. A. (1987). The social construction of nature: Theoretical approaches to the history of environmental problems. *Environmental Review, 11*, 255–264.

Birkenholtz, T. (2011). Network political ecology: Method and theory in climate change vulnerability and adaptation research. *Progress in Human Geography, 36*, 295–315.

Blühdorn, I. (2011). The politics of unsustainability: Symbolic politics and the politics of simulation. *Organization and Environment, 24*, 34–53.

Borén, T. and Young, C. (2013). The migration dynamics of the "creative class": Evidence from a study of artists in Stockholm, Sweden. *Annals of the Association of American Geographers, 103*, 195–210.

Bott, S., Cantrill, J. G., and Myers, O. E. (2003). Place and the promise of conservation psychology. *Human Ecology Review, 10*, 100–112.

Brehm, J. M., Eisenhauer, B. W., and Stedman, R. C. (2012). Environmental concern: Examining the role of place meaning and place attachment. *Society and Natural Resources, 26*, 1–17.

Brewer, G. D. and Stern, P. C. (2005). *Decision making for the environment: Social and behavioral science research priorities.* Washington, DC: National Research Council.

Brinkley, D. G. (2009). *The wilderness warrior: Theodore Roosevelt and the crusade for America.* New York: Harper Collins.

Brown, L. R. (2002). *Eco-economy: Building an economy for the Earth.* Hyderabad, India: Orient Blackswan.

Bulkeley, H. and Betsill, M. M. (2013). Revisiting the urban politics of climate change. *Environmental Politics, 22*, 136–154.

Buttel, F. H. (1976). Social science and the environment: Competing theories. *Social Science Quarterly, 57*, 307–387.

Buttel, F. H., Dickens, P., Dunlap, R. E., and Gijswijt, A. (2002). Sociological theory and the environment: An overview and introduction. In R. E. Dunlap, F. H. Buttel, P. Dickens, and A. Gijswijt (eds), *Sociological theory and the environment: Classical foundations, contemporary insights* (pp. 3–32). Lanham, MD: Rowan & Littlefield.

Cantrill, J. G. (1993). Communication and our environment: Categorizing research in environmental advocacy. *Journal of Applied Communication Research, 21*, 66–95.

Carbaugh, D. (1999). "Just listen": Listening and landscape among the Blackfeet. *Western Journal of Communication, 63*, 250–270.

Carter, N. (2001). *The politics of the environment: Ideas, activism, policy.* London: Cambridge University Press.

Castro, P. and Mouro, C. (2011). Psycho-social processes in dealing with legal innovation in the community: Insights from biodiversity conservation. *American Journal of Community Psychology, 47*, 362–373.

Catton, W. R. and Dunlap, R. E. (1978). Environmental sociology: A new paradigm. *The American Sociologist, 13*, 41–49.

Clayton, S. D. (2012). *Handbook of environmental and conservation psychology.* Oxford: Oxford University Press.

Clayton, S. D. and Meyers, G. (2009). *Conservation psychology: Understanding and promoting human care for nature.* Oxford: Wiley-Blackwell.

Cote, M., and Nightingale, A. J. (2012). Resilience thinking meets social theory: Situating social change in socio-ecological systems (SES) research. *Progress in Human Geography, 36*, 475–489.

Cronon, W. (1991). *Nature's metropolis: Chicago and the Great West.* New York: Norton & Company.

Daniels, S. E. and Walker, G. B. (2001). *Working through environmental conflict: The collaborative learning approach.* Westport, CT: Praeger.

Dauvergne, P. (ed.). (2012). *Handbook of global environmental politics.* Northampton, MA: Edward Elgar Publishing.

Devine-Wright, P. (2011). Place attachment and public acceptance of renewable energy: A tidal energy case study. *Journal of Environmental Psychology, 31*, 336–343.

Dietz, S. and Morton, A. (2011). Strategic appraisal of environmental risks: A contrast between the United Kingdom's Stern Review on the Economics of Climate Change and its Committee on Radioactive Waste Management. *Risk Analysis, 31*, 129–142.

Dietz, T., Dolšak, N., Ostrom, E., and Stern, P. C. (2000). The drama of the commons. In T. Dietz, N. Dolšak, E. Ostrom, and P. C. Stern (eds), *The drama of the commons* (pp. 3–35). Washington, DC: National Academy Press.

Dowling, R. (2010). Geographies of identity: Climate change, governmentality, and activism. *Progress in Human Geography, 34*, 488–495.

Dryzek, J. S. (1997). *The politics of the Earth: Environmental discourses.* Oxford: Oxford University Press.

Dunlap, R. (1997). The evolution of environmental sociology. In M. Redclift and G. Woodgate (eds), *The international handbook of environmental sociology* (pp. 21–39). Northhampton, MA: Edward Elgar Publishing.

Dunlap, R. E. (2013). Climate change skepticism and denial: An introduction. *American Behavioral Scientist, 57*, 691–698.

Dunlap, R. E. and Van Liere, K. D. (2008). The "New Environmental Paradigm." *The Journal of Environmental Education, 40*, 19–28.

Eden, S. (2001). Environmental issues: Nature versus the environment? *Progress in Human Geography, 25*, 79–85.

Endres, D. (2011). Environmental oral history. *Environmental Communication: A Journal of Nature and Culture, 5*, 485–498.

Falkner, R. (2013). *The handbook of global climate and environment policy.* Oxford: Wiley-Blackwell.

Field, D. R., Luloff, A. E., and Krannich, R. S. (2013). Revisiting the origins of and distinctions between natural resource sociology and environmental sociology. *Society and Natural Resources, 26*, 211–225.

Fisher, C. T., Hill, J. B., and Feinman, G. M. (2009). *The archaeology of environmental change: Socionatural legacies of degradation and resilience.* Tucson, AZ: University of Arizona Press.

Flippen, J. B. (2008). Richard Nixon, Russell Train, and the birth of modern American environmental diplomacy. *Diplomatic History, 32,* 613–638.

Flyvbjerg, B. (2001). *Making social science matter: Why social inquiry fails and how it can succeed again.* London: Cambridge University Press.

Franzen, A. and Vogl, D. (2013). Two decades of measuring environmental attitudes: A comparative analysis of 33 countries. *Global Environmental Change, 23,* 1001–1008.

Garrod, G. and Willis, K. G. (1999). *Economic valuation of the environment: Methods and case studies.* Cheltenham: Edward Elgar Publications.

Giddens, A. (1974). *Positivism and sociology.* London: Heinemann.

Giddens, A. (2009). *The politics of climate change.* Cambridge: Polity Press.

Gollain, F. (2011). Economic growth, the environment and international relations. The growth paradigm. *Environmental Politics, 20,* 948–949.

Good, J. E. (2009). The cultivation, mainstreaming, and cognitive processing of environmentalists watching television. *Environmental Communication: A Journal of Nature and Culture, 3,* 279–297.

Guber, D. L. (2013). A cooling climate for change? Party polarization and the politics of global warming. *American Behavioral Scientist, 57,* 93–115.

Hajer, M. and Versteeg, W. (2005). A decade of discourse analysis of environmental politics: achievements, challenges, perspectives. *Journal of Environmental Policy and Planning, 7*(3), 175–184.

Harvey, D. D. (1996). *Justice, nature and the geography of difference.* Oxford: Blackwell.

Hawken, P., Lovins, A., and Lovins, L. H. (1999). *Natural capitalism: Creating the next industrial revolution.* New York: Little, Brown & Company.

Hays, S. P. (1959). *Conservation and the gospel of efficiency: The Progressive conservation movement, 1890–1920.* Cambridge, MA: Harvard University Press.

Hays, S. P. (2000). *A history of environmental politics since 1945.* Pittsburgh, PA: University of Pittsburgh Press.

Hillier, J. (2002). *Shadows of power: An allegory of prudence in land-use planning* (Vol. 5). New York: Psychology Press.

Hughes, J. D. (2001). *An environmental history of the world: Humankind's changing role in the community of life.* London: Routledge.

Jensen, C. and Spoon, J. J. (2011). Testing the "party matters" thesis: Explaining progress towards Kyoto Protocol targets. *Political Studies, 59,* 99–115.

Jones, M. D. (2014). Cultural characters and climate change: How heroes shape our perception of climate science. *Social Science Quarterly, 95,* 1–39.

Kareiva, P., Tallis, H., Ricketts, T. H., Daily, G. C., and Polasky, S. (eds). (2011). *Natural capital: theory and practice of mapping ecosystem services.* Oxford: Oxford University Press.

Kassing, J. W., Johnson, H. S., Kloeber, D. N., and Wentzel, B. R. (2010). Development and validation of the environmental communication scale. *Environmental Communication: A Journal of Nature and Culture, 4,* 1–21.

Kendall, B. E. (2008). Personae and natural capitalism: Negotiating politics and constituencies in a rhetoric of sustainability. *Environmental Communication: A Journal of Nature and Culture, 2,* 59–77.

Kraft, M. E. (2001). *Environmental policy and politics.* New York: Longman.

Lane, J. E. (2013). Global environmental coordination: How to overcome the double collective action problematic? *Sociology, 3,* 83–88.

Langston, N. (2009). Paradise lost: Climate change, boreal forests, and environmental history. *Environmental History, 14,* 641–650.

Latour, B. (2005). *Reassembling the social.* New York: Oxford University Press.

List, J. A. and Price, M. K. (eds) (2013). *Handbook of experimental economics and the environment.* Northampton, MA: Edward Elgar Publishing.

McCormick, J. (1989). *Reclaiming paradise: The global environmental movement.* Bloomington, IN: Indiana University Press.

McKenzie-Mohr, D. (2001). Promoting sustainable behavior: An introduction to community-based social marketing. *Journal of Social Issues, 56,* 543–554.

Maréchal, K. (2012). *The economics of climate change and the change of climate in economics.* London: Routledge.

Masterman, M. (1970). The nature of a paradigm. In I. Lakatos and A. Musgrave (eds), *Criticism and the growth of knowledge* (pp. 59–90). London: Cambridge University Press.

Merchant, C. (1980). *The death of nature: Women, ecology and the scientific revolution.* New York: Harper & Row.

Meyers, O. E., Saunders, C., and Bexell, S. (2009). Fostering empathy with wildlife: Factors affecting free-choice learning for conservation concern and behavior. In J. H. Falk, J. E. Heimlich, and S. Foutz (eds), *Free-choice learning and the environment* (pp. 39–56). Lanham, MD: AltaMira Press.

Milstein, T., Anguiano, C., Sandoval, J., Chen, Y-W., and Dickenson, E. (2011). Communicating a "new" environmental vernacular: A sense of relations-in-place. *Communication Monographs, 78*, 486–510.

Moran, E. (2010). *Environmental social science: Human–environment interactions and sustainability.* Hoboken, NJ: John Wiley & Sons.

Morton, T. A., Rabinovich, A., Marshall, D., and Bretschneider, P. (2011). The future may (or may not) come: How framing changes responses to uncertainty in climate change communications. *Global Environmental Change, 21*, 103–109.

Moseley, W. G., Perramond, E., Hapke, H. M., and Laris, P. (2012). *An introduction to human-environment geography.* Oxford: Wiley-Blackwell.

Nash, R. (1967). *Wilderness and the American mind.* New Haven, CT: Yale University Press.

Newell, P. (2013). *Globalization and the environment: Capitalism, ecology and power.* Cambridge: Polity Press.

Norton, T. (2007). The structuration of public participation: Organizing environmental control. *Environmental Communication: A Journal of Nature and Culture, 1*, 146–170.

Novic, K. and Sandman, P. M. (1974). How use of mass media affects views on solutions to environmental problems. *Journalism and Mass Communication Quarterly, 51*, 448–452.

O'Brien, K. (2010). Responding to environmental change: A new age for human geography? *Progress in Human Geography, 35*, 542–549.

Opie, J. and Elliot, N. (1996). Tracking the elusive jeremiad: The rhetorical character of American environmental discourse. In J. Cantrill and C. Oravec (eds), *The symbolic Earth: Discourse and our creation of the environment* (pp. 9–37). Lexington, KY: University Press of Kentucky.

Ostrom, E. (1990). *Governing the commons: The evolution of institutions for collective action.* New York: Cambridge University Press.

Pelling, M. and Dill, K. (2010). Disaster politics: Tipping points for change in the adaptation of sociopolitical regimes. *Progress in Human Geography, 34*, 21–37.

Peterson, T. R. (2004). *Green talk in the White House: The rhetorical presidency encounters ecology.* College Station, TX: Texas A&M University Press.

Pol, E. (2006). Blueprints for a history of environmental psychology (I): From first birth to American transition. *Medio Ambiente y Comportamiento Humano, 7*, 95–113.

Porter, M. E. and Van der Linde, C. (1995). Toward a new conception of the environment–competitiveness relationship. *The Journal of Economic Perspectives, 9*(4), 97–118.

Potoski, M. (2002). Designing bureaucratic responsiveness: Administrative procedures and agency choice in state environmental policy. *State Politics and Policy Quarterly, 2*, 1–23.

Samson, W. A. (2013). Between God and green: How evangelicals are cultivating a middle ground on climate change. *Sociology of Religion, 74*, 137–138.

Santayana, G. (1905). *The life of reason.* New York: Scribner's.

Schweizer, S., Davis, S., and Thompson, J. L. (2013). Changing the conversation about climate change: A theoretical framework for place-based climate change engagement. *Environmental Communication: A Journal of Nature and Culture, 7*, 42–62.

Senecah, S. (2004). The trinity of voice: The role of practical theory in planning and evaluating the effectiveness of environmental participatory processes. In S. P. Depoe, J. W. Delicath, and A. M. Aepli-Elsenbeer (eds), *Communication and public participation in environmental decision making* (pp. 13–33). Albany, NY: State University of New York Press.

Shanahan, E. A., Jones, M. D., McBeth, M. K., and Lane, R. R. (2013), An angel on the wind: How heroic policy narratives shape policy realities. *Policy Studies Journal, 41*, 453–483.

Shellabarger, R., Peterson, M. N., Sills, E., and Cubbage, F. (2012). The influence of place meanings on conservation and human rights in the Arizona Sonora borderlands. *Environmental Communication: A Journal of Nature and Culture, 6*, 383–402.

Shibley, M. A., and Wiggins, J. L. (1997). The greening of mainline American religion: A sociological analysis of the environmental ethics of the national religious partnership for the environment. *Social Compass, 44*, 333–348.

Skjærseth, J. B., Stokke, O. S., and Wettestad, J. (2006). Soft law, hard law, and effective implementation of international environmental norms. *Global Environmental Politics, 6*, 104–120.

Simmel, G. (1903). *The metropolis and mental life.* Dresden: Petermann.

Slimak, M. W. and Dietz, T. (2006). Personal values, beliefs, and ecological risk perception. *Risk Analysis, 26*, 1689–1705.

Smerecnik, K. R. and Renegar, V. R. (2010). Capitalistic agency: The rhetoric of BP's Helios Power campaign. *Environmental Communication: A Journal of Nature and Culture, 4*, 152–171.

Smith, J. W., Anderson, D. H., and Moore, R. L. (2012). Social capital, place meanings, and perceived resilience to climate change. *Rural Sociology, 77*, 380–407.

Soja, E. (1989). *Postmodern geographies: The reassertion of space in critical social theory.* London: Verso.

Stavins, R. N., Wagner, A. F., and Wagner, G. (2003). Interpreting sustainability in economic terms: Dynamic efficiency plus intergenerational equity. *Economics Letters, 79*, pp. 339–343.

Tuan, Y. F. (1977). *Space and place: The perspective of experience.* Minneapolis, MN: University of Minnesota Press.

Vorkinn, M. and Riese, H. (2001). Environmental concerns in a local context: The significance of place attachment. *Environment and Behavior, 33*, 249–263.

Warecki, G. M. (1998). *Protecting Ontario's wilderness: A history of changing ideas and preservation politics, 1927–1973.* New York: Peter Lang.

White, R. (2004). From wilderness to hybrid landscapes: The cultural turn in environmental history. *The Historian, 66*, 557–564.

Will, G. (2012). College athletics is pigskin progressivism on display. *The Mining Journal.* Marquette, MI: Sunday, September 9, A4.

Williams, M. (2003). *Deforesting the Earth: From prehistory to global crisis.* Chicago, IL: University of Chicago Press.

Worster, D. (1988). *The ends of the Earth: Perspectives on modern environmental history.* New York: Cambridge University Press.

Yang, W., Liu, W., Viña, A., Tuanmu, M. N., He, G., Dietz, T., and Liu, J. (2013). Nonlinear effects of group size on collective action and resource outcomes. *Proceedings of the National Academy of Sciences, 110*, 10916–10921.

## PART II

Producing environmental communication

Sources, communicators, media and media professionals

5

ENVIRONMENTAL SCIENTISTS AND PUBLIC COMMUNICATION

Sharon Dunwoody

When wildlife ecologist Stan Temple decided to reconstruct the morning sounds of birdlife captured in the notes of Aldo Leopold some 70 years ago, at the ecologist's iconic shack in south central Wisconsin, he knew he was onto something interesting but had no idea how dramatic the public response would be.

On many mornings in the 1930s and 1940s, Leopold would rise before dawn, settle onto a bench near the shack—a former chicken coop converted into a rustic but beloved get-away cabin for his family of six—and spend a few minutes meticulously identifying the trills and squawks of awakening birds. Temple, a University of Wisconsin-Madison emeritus professor of forest and wildlife ecology and now a senior fellow at the Aldo Leopold Foundation, enlisted the help of an acoustic ecologist to recreate a "soundscape" from those handwritten notes.

The resulting five minutes of sound (www.news.wisc.edu/21058) is the first historical soundscape to be derived from a written account rather than a tape recording, noted Temple. The mix of species that Leopold heard back in the 1940 differs from those heard today thanks to many factors, including climate change. But most striking in the comparison is the human presence that overwhelms the twenty-first century soundscape. A recording taken at the shack today is dominated not by birdsong but by the rumble of cars speeding along a nearby freeway.

Temple's soundscape captured national attention, and he was inundated with requests for interviews. Although he admits the experience was time consuming, Temple also feels that interactions with journalists and the public are an important part of his work *as a scientist*.

Is Stan Temple unusual in this respect, or does his embrace of public communication and engagement typify scientists generally or, perhaps, environmental scientists more specifically? This chapter will try to answer those questions, first with a brief look at the history of the relationship between scientists and the public and then, through the lens of available literature, by drawing a twenty-first century portrait of that relationship.

But first, let me give away the denouement. While the historical record paints a picture of a sometimes tense and volatile relationship between scientists and both journalists and the public, data gathered over several decades suggest that the relationship is cyclical and that today's scientists are engaging in increasingly productive interactions with reporters and lay audiences. Further, environmental scientists may serve as the poster children for this rapprochement with the public, thanks to the explosion of interest in environmental issues among publics and policy makers over the past 50 years. That said, though, a relationship that—way back in the nineteenth century—might have characterized the scientist as one among equals—remains strongly hierarchical today, with scientists repeatedly affirming their status as "experts" who believe in a mandate to "educate" the public.

A brief history of scientist/public interactions

British journalist-turned-academic Peter Broks captures the historical thread of this relationship in his 2006 volume *Understanding Popular Science*. It is a saga that has caught the attention of others as well, among them U.S. historian John Burnham, who described the American experience in his book *How Superstition Won and Science Lost: Popularizing Science and Health in the United States* (1987).

Both authors track a large cycle that begins in the nineteenth century with efforts by scientists and lay individuals to embed science in the warp and woof of daily life. Scientific discovery was recognized early on by lay people as a source of both practical information and wonder, and nineteenth-century scientists additionally understood that, as members of a newly minted occupation, they needed all the help they could get.

Secord (1994), for example, examined the ways in which educated botanists and "working-class naturalists" together constructed the practice of "scientific botany" in early nineteenth century England. Plant knowledge was important to artisans and farmers, and early botanical societies met regularly to view plant specimens and to borrow or return books purchased through monthly dues. Both artisans and more highly educated scientists benefited from these meetings, noted Secord. The sessions "not only fulfilled a didactic purpose," she explained, "but also allowed the more expert botanists to accumulate information rapidly" (p. 283), for example, through the discovery of rare specimens.

But by the end of the nineteenth century, in both Europe and the United States, science had become the domain of the "expert." Broks explains that public audiences were no longer viewed as participants in the construction of science but, instead, as passive information receptacles and, if properly "educated," as potential cheerleaders for science. Professions of all kinds were developing during this era, and scientists embraced professionalization with great energy, creating scientific societies, specialized educational requirements, and in-house systems of rewards and punishments. Professions also require a distinction between "us" and "them," and that delineation meant that scientists withdrew from the pub and from the author listings of popular science magazines and, instead, embraced the notion of "expertise" as a condition that set them apart from others.

By the twentieth century, the chasm between "us" and "them" was so wide that popularization had become a dirty word for many scientists. The scientific culture not only failed to provide rewards for interacting with publics but also actively punished scientists for doing so. I recall when, as a writer for a U.S. medical center in the 1970s, I learned of a productive researcher at the hospital who had run afoul of norms restricting scientific popularization. His research had been the subject of a story in the local newspaper at one point. Some months later, his application for membership in an honorific scientific society was rejected. The society noted that the newspaper story, while accurate, had identified the researcher by name and that such identification constituted "unethical advertising." The society advised the scientist to avoid such lapses in future, making a subsequent application more likely to succeed.

Scientists who nonetheless braved the waters of popularization found negotiating the resulting public visibility with their colleagues to be rough going. Scholar Rae Goodell interviewed a number of these individuals for her book *The Visible Scientists* (1977) and found them to be committed to public understanding of science but subject to withering criticism from the scientific culture. Those critics lambasted the "visible scientists" for such things as spending too much time talking to the public at the expense of their research and for speaking out about issues beyond the focus of their research expertise. Among her "lab rats" were a number of environmental scientists, including Paul Ehrlich and Barry Commoner.

By the start of the twenty-first century, however, the cycle had come full circle. Scientists have emerged from their professional cocoons and are once again actively engaging audiences of all kinds, from children to adults to industry CEOs. They are investing in communication training and are devoting considerable time and effort to products—documentaries, trade books, blogs, congressional testimony—that historically have not "counted" toward scientists' climb up the professional ladder. I will share some of the factors that prompted this change below. But first, I would ask whether these trends have affected environmental scientists in the same way as scientists in other disciplines.

Environmental scientists and publics

Wildlife ecologist Stan Temple, with whom I opened this chapter, acquired his training when scientists' avoidance of public activity was at its height in the United States. Yet even as a graduate student, he recognized the potential value of his work to publics outside science, and he did not hesitate to wade into the public arena to share what he learned.

Was Temple the young ecologist choosing a risky path, or did circumstances in the mid to late twentieth century actually encourage environmental scientists to engage with publics? Was Temple jumping into the public arena at a preternaturally early stage of his career, or was he part of a cohort of environmental scientists who may have pioneered public engagement and visibility decades before mainstream "science" came back to the table?

Sociologist Dorothy Nelkin would perhaps have opted for the latter explanation. In an analysis of experiences of American ecologists, Nelkin fingered the environmental movement of the 1960s and 1970s as a critical catalyst for public engagement. "As the expertise of ecologists was perceived as a social resource, relevant to a major problem," she noted, "scientists were thrust into the political arena, forced to face many of the issues and implications of social responsibility" (Nelkin, 1977: 75). During this same period, Nelkin explained, universities established environmental courses and programs, and many leading scientists known primarily as scholars began writing popular trade books and giving public lectures.

The passage into U.S. law, in 1970, of the National Environmental Policy Act and, in 1973, of the Endangered Species Act mandated increased scientific involvement in policy making. As a result, ecologists were called on more and more frequently to apply their research skills to policy decisions.

There were costs, of course, both actual and perceived. Ecologists worried in the 1970s that such intense involvement in public problems would erode their autonomy as researchers and would make applied research more valuable (read: more likely to be funded) than basic scholarship. Nelkin also explained that, since societal problems typically play out at an ecosystem level rather than at more micro, biological levels where much scientific research takes place, ecologists feared that the research on large-scale processes needed for policy decisions was premature and would fail.

Despite such ambivalence, many ecologists stepped into roles of increased social responsibility during the period and, in doing so, got a jump on many in the scientific culture. It is possible that, even now, ecologists may be called on more frequently than scientists in other disciplines to help solve problems.

Thus, it appears that Stan Temple was not unusual. As a graduate student at Cornell University, Temple signed on with an adviser, Tom Cade, whose awareness of the dire straits of one of his favorite birds led the senior researcher to establish an organization called "The Peregrine Fund" to raise money to support his conservation work (peregrines are smaller, aerodynamic falcons whose reproduction crashed in the 1960s when accumulation of DDT in their tissues caused eggshells to thin and eggs to break). Temple devoted himself to studies of peregrine recovery strategies and soon found his work attracting the interest of major American media and even the likes of the iconic CBS news anchor Walter Cronkite, who on one evening program featured Temple and his work reintroducing juvenile peregrines to the wild.

Although current data on the frequency with which environmental scientists connect to the public appear to be sparse, the few available studies suggest that these scientists remain popular sources for the public. For example, Jensen (2011), in a study of the popularization practices of French scientists, found great variation in level of activity by discipline, with environmental scientists second only to social scientists in the number of study participants who had engaged in at least one popularization activity in the previous five years. Specifically, three-quarters of the environmental scientists indicated they had been so engaged. In contrast, 61.3 percent of physicists and 45.2 percent of biologists claimed similar levels of activity. Similarly, in a study of Spanish scientists, Torres-Albero et al. (2011) found that those focusing on natural resources had more journalistic encounters and more "open door events" (p. 19) than did other types of scientists.

Still, many environmental scientists eschewed public communication during the heady period of the 1960s and 1970s, and it is likely that many still prefer the anonymity of the lab or the forest depths even today. The extremely complex nature of environmental "truths" continues to lead scholars to recommend that scientists in this field proceed cautiously when communicating with the public and with policy makers (Kriebel et al., 2001; Janse, 2008). But why do some environmental scientists step into the public arena while others hang back? Below, I offer some hypothesized explanations, and indicate when empirical evidence is available.

Status is important

One recurring pattern across time and scientific disciplines is that senior scientists and those who have moved into leadership positions are more likely to interact with publics than are more junior colleagues (see, for example, Boltanski and Maldidier, 1970; Peters et al., 2008a; Bauer and Jensen, 2011). Bentley and Kyvik (2011), for example, surveyed scientists in 13 countries and found that those researchers with popular publications on their CVs also have higher academic rank and have published more in peer-reviewed journals than do researchers who don't popularize.

Although the supposition is that seniority precedes public activity, that scientists tend to wait until they have gained enough in-house credibility to make the rigors of popularizing survivable, it is also possible that this status/popular engagement relationship is recursive, that public visibility can, in turn, enhance a scientist's scientific reputation. I offer some evidence for that pattern below.

Training and socialization

Formal scientific training typically excludes skill-building in a variety of ancillary fields. For example, scientists are increasingly expected to behave like entrepreneurs but rarely receive formal preparation to do so. The collaborative nature of the scientific process makes interpersonal skills critical, but scientists-in-training are rarely exposed to that domain in a systematic way. Similarly, while scientists write for a living (your exciting finding is deemed "science" only when it has been cast as a narrative, vetted by your peers, and published in a journal), they often must settle for a kind of experiential learning by doing. While such training sometimes (but not always!) provides the skills needed for clear, facile scientific writing, the construction of popular narratives is almost never a part of that experiential diet.

The interdisciplinary nature of ecological training—which privileges problem-solving across disciplines and scientific languages—may give environmental scientists better access to communication skill-building during their student years than is available to other science students. But the environmental science community itself acknowledges a serious training deficit. One of the most prominent outreach training programs for scientists in the United States, the Aldo Leopold Leadership Program (www.leopoldleadership.stanford.edu/), was created in 1998 specifically to offer communication training to senior and mid-career environmental scientists.

We don't know how many environmental scientists are taking advantage of communication training opportunities, but we do know that such training is related to scientists' popularization attitudes and activities. For example, in a survey of epidemiological and stem cell researchers in the United States, formal training in communication—via such mechanisms as workshops, internships, formal courses—was one of the strongest predictors of frequency of media contacts (Dunwoody et al., 2009). And regardless of whether a scientist has engaged in formal communication training, feeling efficacious about the process has been identified as a predictor of activity. A recent analysis of a sample of scientists who are members of the Royal Society in the U.K., for example, found that scientists' perceptions of their abilities to handle engagement activities were related to their willingness to do so, as well as to their levels of such activity (Besley et al., 2013).

Rewards and punishment

The fallout from public communication activities has unavoidable impacts on scientists' attitudes and behaviors. Scholars have long assumed that the most powerful influences are negative, that is, that scientists calibrate their popularization activities more in response to potential negative feedback than to positive feedback. And indeed, scientists' recollections of impacts have long highlighted the negative: the scientist who felt "burned" by an inaccurate media account that generated criticism from her peers, the perception that public communication activities have little to no impact on one's merit evaluations, the political pushback that can stem from public dissemination of research related to controversial scientific and environmental issues.

Wildlife ecologist Stan Temple can serve as our poster boy once again. For more than 25 years, he has been studying and speaking out about the toll that outdoor cats take on wildlife, particularly birds. Although he makes clear that he does not shrink from sensitive topics, he admits that the fallout from his cat work has been his "hardest experience." Ad hominem attacks sometimes escalated into death threats; one threat was sinister enough to warrant police action.

But recent research suggests that today's scientists also believe that public communication can bring rewards. And they are right.

Most of these benefits stem from the social legitimacy that public visibility can confer. Widely disseminated information about an environmental scientist's work can be a powerful signaling mechanism. At the least, it suggests to audiences that *this* research is important, worth one's attention. Although one cannot guarantee who will pick up the message, empirical work suggests that public visibility can positively influence potential funders and can make a scientist's research seem more salient to publics and policy makers.

Even more interesting, however, is that this legitimizing function seems to work in similar ways for the scientific community itself. Although only a few of studies have examined this phenomenon, they find that research picked up by the mass media is accorded more weight within the scientific culture. Here are a couple of patterns found by these studies:

- *Folks seeking to follow up on a media story about your research are predominantly experts with an interest in the topic.* Many scientists acknowledge that public dissemination of their research can produce large numbers of requests for interviews, additional information, and copies of the study. But while many scientists would blame journalists and the public for that labor-intensive aftermath, a few studies have found, instead, that contacts come primarily from other scientists. I used myself as a guinea pig at one point when a study of public perceptions of AIDS—then a new and scary phenomenon—got a lot of media attention. I kept a log of subsequent contacts and found that, some two weeks after the story broke, contacts from social scientists and health policy experts seeking further information about the study began to dominate. Ultimately, those more specialized audience members generated 63 percent of the information requests I received (Dunwoody, 1993).
- *More visible studies get cited more often in the scientific literature.* A study that compared two sets of articles published in the *New England Journal of Medicine*—one set that received coverage by *The New York Times* and another, matched set that did not—found that the former were cited 72.8 percent more often in the scientific literature than were the latter (Phillips et al., 1991). A subsequent study that included a broader array of journals, articles, and media outlets found that publicized research was cited 22 percent more frequently by other scientists (Kiernan, 2003).

Today's scientists certainly understand that public visibility can have both good and bad outcomes. But recent surveys find that perceptions of the value of these experiences now trump perceptions of their debits. For example, a five-country study of biomedical research scientists found that three-quarters of the respondents chose the response "mainly good" when summarizing their experiences with the media in the past three years, and more than half (57 percent) indicated they were "mostly pleased" with "their latest appearance in the media" while only 6 percent said they were "mostly dissatisfied" (Peters et al., 2008a; Peters et al., 2008b). Similarly, more than half (62 percent) of the scientists in a recent Danish study acknowledged that media encounters were good for their careers (Wien, 2013).

Additionally, although the apparent value of visibility to scientists' careers could easily become scientists' primary *raison d'être* for seeking public attention, the scientists in several recent studies instead seemed to embrace a more "intrinsic" motive: the enjoyment of sharing their work with others. A recent analysis of U.S. scientists, for example, found that the only "reward" that accounted for significant variance in scientists' frequency of interactions with the media was a measure of their level of enjoyment in explaining their research and its implications to the public (Dunwoody et al., 2009).

Again, data examining environmental scientists' attitudes about public communication is hard to find, but the modest empirical material found is consonant with the studies mentioned above. For example, a survey of members of two large scientific societies with considerable relevance to the environment— the American Meteorological Society and the American Geophysical Union—found that the bulk of respondents evaluated coverage of global warming in national press outlets such as *The New York Times* and the *Wall Street Journal* as "at least somewhat reliable" (Farnsworth and Lichter, 2012: 451), suggesting—albeit indirectly—that such channels offer reasonable visibility venues.

Expertise and environmental sources

Regardless of environmental scientists' motives when it comes to public visibility, attributes of the normative environment of journalism and characteristics of environmental issues themselves play important roles in judgments about what is news and who constitutes a credible source for a particular story. Hansen (1991), in his seminal essay on the role of the media in the "social construction" of the environment, captures the multivariate landscape from which sources are pulled in order to construct "environmental news." Scientists are certainly considered experts by environmental journalists, but it is clear that the fabric of an environmental issue contains many other dimensions, among them political, aesthetic, and citizen-based. As Hansen notes, the presence of scientists in environmental news stories is often trumped by journalists' use of governmental officials and other policy makers.

Trumbo (1996) and, more recently, Nisbet and Huge (2007) argue that environmental issues are characterized by cycles that prompt journalists to operationalize source credibility differently by cycle stage. Early stages—such as efforts to diagnose a problem and determine its causes—lead to a preference for scientists as sources, while later stages—characterized by a search for remedies (often policy-based)— encourage the use of governmental and interest groups. Thus, journalists' judgments about expertise may vary in ways aligned with such issue cycles.

However, some scholars have noted that coverage of contested science and environmental issues can also set experts from different domains against one another. This tendency was given early attention by sociologist Allan Mazur. In his 1981 book *The Dynamics of Technical Controversy*, Mazur conducted four case studies of highly contested issues in the United States—fluoridation of water, nuclear power, siting high-voltage transmission lines, and the use of anti-ballistic missiles for defense—and found a common rhetorical strategy that he characterized as "arguing past one another" (Mazur, 1981: 18). Many debates, he noted, were filled with "conflicting contentions based on differing premises" (p. 19). For example, he noted that while proponents of nuclear power might emphasize the unlikely nature of the prospect of someone receiving a high enough dose of radiation to cause acute poisoning, opponents would focus instead on the long-term risks of chronic poisoning from lower radiation doses.

More recently, Boyce (2006), in a study of coverage by the British media of the vaccine/autism controversy, found that those stories that sought to present differing viewpoints exercised an extreme form of "arguing past one another." These accounts rarely presented dueling scientists; rather, the stories contrasted scientists' voices with those of politicians and parents.

Thus, environmental news presents a complex environment in which judgments of expertise and credibility take place. Exacerbating this pattern is the possibility that many environmental stories fail to include scientists because journalists do not define the issue at hand as "scientific." While this may be related to the concept of issue cycle, discussed above, it can also stem from the ways in which journalists make basic decisions about what's news. This issue was brought home some years ago in a large study of media coverage of social sciences. Weiss and Singer (1988) paired a massive content analysis of stories carried in major print and broadcast news outlets in the United States with interviews of journalists involved in the stories. While the researchers found many social science stories in these media outlets, they were surprised to learn that the reporters did not always define them as "social science" stories. That's a problem, they noted, as that initial definition will govern the reporter's search for sources. A story about community attitudes toward recycling, for example, may eschew scientific expertise if the reporter initially defines the topic not as "scientific" (i.e., given meaning through empirical studies of perceptions and attitudes) but, rather, as a "feature story" that can be captured through anecdotal interviews with families who do or do not recycle.

The medialization of environmental scientists

In this early part of the twenty-first century, the study of scientists as sources has become dominated by the concept of medialization, a process described as the increasingly sophisticated incorporation, by scientists, of public visibility into scientific judgments and decision making. The "medialization" of science requires an understanding of public visibility processes—journalistic deadlines, normative practices of reporters—as well as of the risks and benefits that can accompany such processes. The result, maintain scholars, is a growing tendency for scientists to adapt to journalistic work patterns (Bucchi, 2013) and even to modify their own behaviors to encourage (or discourage) coverage of their work (Rödder, 2009).

While I would argue that the process of medialization was of interest long before the term itself came into vogue—see, for example, Blumler and Gurevitch's development of the concept of a "shared culture" between journalists and sources (Blumler and Gurevitch, 1981)—studies indeed seem to corroborate the idea that scientists are increasingly open to the incorporation of public visibility into decisions they make about research processes (Dudo, 2012) and that their beliefs in the influence of media increase their motivation and efforts to obtain such visibility (Tsfati et al., 2011).

But much of this work also cautions that scientists remain committed to a level of societal interaction that maintains a clear distinction between "them" and "us." Interviews with Dutch plant scientists indicated that, while the scientists displayed tolerance of lay views, they also were quick to criticize the scientific value of those views (Mogendorf et al., 2012). Noted the authors:

> Displays of tolerance in our data seem to indicate here that scientists are concerned about the societal image of their research to some extent; they show with their displays that they do care about lay views but that they do not necessarily need to involve laypeople or their views in their research practices.

(p. 744)

In a recent study of German climate scientists, Ivanova, Schäfer, Schlichting, and Schmidt (2013) found the extent of medialization to differ significantly among subgroups in their sample. It is the younger, less experienced scientists, they found, who are adapting to media norms.

In conclusion

In *Flight Behavior* (2012), a novel by Barbara Kingsolver that exemplifies a new subgenre called "climate fiction"—clifi for short—interactions between environmental scientists and journalists do not go well. In the book, millions of Monarch butterflies are attempting to winter over in the trees of Tennessee rather than in their historic forest roosts in Mexico, and entomologist Ovid Byron has set up a small laboratory to document factors that might be behind the troubling shift. When television reporter Tina Ultner tries to interview him about the "beautiful phenomenon" (p. 364) of tree branches draped in vivid, living orange, the interaction rapidly deteriorates. "Tina," retorts Byron, "to see only beauty here is very superficial. Certainly in terms of news coverage, I would say it's off message" (p. 365). Moments later, Tina's refusal to accept global warming as a factor leads to an abrupt end to the interview, with Byron claiming: "You have no interest in real inquiry" (p. 369).

Entomologist Ovid Byron displays all the characteristics of the classic, wary scientist who believes that only bad things will come from his journalistic encounter. And although this episode encourages an interpretation of the scientist as heroic, as a truth-seeker unwilling to "dumb down" his work for a reporter only interested in sensationalizing the science, I think many modern-day ecologists would eschew that role. Like wildlife ecologist Stan Temple, they instead view interactions with journalists and with the public as important means of leveraging both public understanding and their own legitimacy as scientists.

This doesn't mean that environmental scientists have embarked on an unfettered dash to become household names. History suggests that they have long worked to make outreach and public communication an outcome of high quality science rather than a driver per se. As Temple noted, when asked how he had achieved success as a public figure without putting his scientific career at risk: "The credibility of my research among my peers was always paramount. If my research resulted in findings the public wanted and needed to hear, I wasn't shy about getting the word out, but that was always a secondary goal."

Yet as strategic and sophisticated as scientists have become as public communicators, research suggests that most continue to hew to a fairly hierarchical notion of their relationship to the public. Peters, in a recent overview of research on scientists' relationships with the media and with publics, concludes that today's scientists continue to embrace "the dominant view" of these relationships, a view that assumes *scientific* communication and *science* communication are two separate domains and that, while scientists can be effective actors in science communication, the public is not a legitimate player on the scientific side of the ledger (Peters, 2013). Put another way, despite the popularity of terms such as "public involvement" and "engagement" on this cusp of the twenty-first century, it is a safe bet that many environmental scientists still believe strongly in their traditional mandate: to educate the public.

Acknowledgment

I acknowledge, with thanks, the efforts of UW-Madison graduate student Soo Yun Kim in helping compile relevant literature for this chapter.

Further reading

Baron, N. (2010) *Escape from the Ivory Tower: A Guide to Making Your Science Matter*. Washington, DC: Island Press. This practical guide stems from years of working with ecologists in the Aldo Leopold Leadership Program and, thus, contains strategies germane to environmental issues.

Broks, P. (2006) *Understanding Popular Science*. Maidenhead: Open University Press.
This slim volume offers a rich overview of the role of scientists in popular science, with a particular focus on Great Britain and Europe.

Friedman, S. M., Dunwoody, S., and Rogers, C. L. (eds) (1999) *Communicating Uncertainty: Media Coverage of New and Controversial Science*. Mahwah, NJ: Laurence Erlbaum Associates.
Via a variety of authors, the book illuminates the scientist–media–public arena across disciplines, including environmental science. Several chapters concentrate on the behavior of scientists themselves in this complicated landscape.

Schneider, S. H. (2009) *Science as a Contact Sport: Inside the Battle to Save Earth's Climate*. Washington, DC: National Geographic Books.
Described by many as the consummate climate change scientist, Stanford professor Stephen Schneider wrote this reflection on his years in the global warming trenches just before his own untimely death.

References

Baron, N. (2010) *Escape from the Ivory Tower: A Guide to Making Your Science Matter*. Washington, DC: Island Press.

Bauer, M. W. and Jensen, P. (2011) "The mobilization of scientists for public engagement," *Public Understanding of Science*, 20: 3.

Bentley, P. and Kyvik, S. (2011) "Academic staff and public communication: A survey of popular science publishing across 13 countries," *Public Understanding of Science*, 20: 48–63.

Besley, J. C., Oh, S. H., and Nisbet, M. (2013) "Predicting scientists' participation in public life," *Public Understanding of Science*, 22: 971–987.

Blumler, J. G. and Gurevitch, M. (1981) "Politicians and the press: An essay on role relationships," in D. D. Nimmo, and K. R. Sanders (eds) *Handbook of Political Communication*. Beverly Hills, CA: Sage Publications, 467–493.

Boltanski, L. and Maldidier, P. (1970) "Carrière scientifique, morale scientifique et vulgarization," *Social Science Information*, 9: 99–118.

Boyce, T. (2006) "Journalism and expertise," *Journalism Studies*, 7: 889–906.

Broks, P. (2006) *Understanding Popular Science*. Maidenhead: Open University Press.

Bucchi, M. (2013) *Science and the Media: Alternative Routes to Scientific Communications*. London: Routledge.

Burnham, J.C. (1987) *How Superstition Won and Science Lost: Popularizing Science and Health in the United States*. New Brunswick, NJ: Rutgers University Press.

Dudo, A. (2012) "Toward a model of scientists' public communication activity: the case of biomedical researchers," *Science Communication*, 35: 476–501.

Dunwoody, S. (1993) *Reconstructing Science for Public Consumption*. Geelong, Australia: Deakin University Press.

Dunwoody, S., Brossard, D., and Dudo, A. (2009) "Socialization or rewards? Predicting US scientist–media interactions," *Journalism and Mass Communication Quarterly*, 86: 299–314.

Farnsworth, S. J. and Lichter, S. R. (2012) "Scientific assessments of climate change information in news and entertainment media," *Science Communication*, 34: 435–459.

Friedman, S. M., Dunwoody, S., and Rogers, C. L. (eds) (1999) *Communicating Uncertainty: Media Coverage of New and Controversial Science*. Mahwah, NJ: Laurence Erlbaum Associates.

Goodell, R. (1977) *The Visible Scientists*. Boston, MA: Little, Brown & Company.

Hansen, A. (1991) "The media and the social construction of the environment," *Media, Culture and Society*, 13: 443–458.

Ivanova, A., Schäfer, M. S., Schlichting, I., and Schmidt, A. (2013) "Is there a medialization of climate science? Results from a survey of German climate scientists," *Science Communication*, 35: 626–653.

Janse, G. (2008) "Communication between forest scientists and forest policy-makers in Europe—A survey on both sides of the science/policy interface," *Forest Policy and Economics*, 10: 183–194.

Jensen, P. (2011) "A statistical picture of popularization activities and their evolutions in France," *Public Understanding of Science*, 20: 26–36.

Kiernan, V. (2003) "Diffusion of news about research," *Science Communication*, 25: 3–13.

Kingsolver, B. (2012) *Flight Behavior*. New York: Harper Perennial.

Kriebel, D., Tickner, J., Epstein, P., Lemons, J, Levins, R., Loechler, E.L., Quinn, M., Rudel, R., Schettler, T., and Stoto, M. (2001) "The precautionary principle in environmental science," *Environmental Health Perspective*, 109: 871–876.

Mazur, A. (1981) *The Dynamics of Technical Controversy*. Washington, DC: Communications Press.

Mogendorff, K., te Molder, H., Gremmen, B., and Van Woerkum, C. (2012) "'Everyone may think whatever they like, but scientists …' Or how and to what end plant scientists manage the science–society relationship," *Science Communication*, 34: 727–751.

Nelkin, D. (1977) "Scientists and professional responsibility: The experience of American ecologists," *Social Studies of Science*, 7: 75–95.

Nisbet, M. C. and Huge, M. (2007) "Where do science debates come from? Understanding attention cycles and framing," in D. Brossard, J. Shanahan, and T. C. Nesbitt (eds) *The Media, the Public and Agricultural Biotechnology*. Wallingford: CAB International, 193–230.

Peters, H. P. (2013) "Gap between science and media revisited: Scientists as public communicators," *Proceedings of the National Academy of Sciences*, 110(Supplement 3): 14102–14109.

Peters, H. P., Brossard, D., De Cheveigné, S., Dunwoody, S., Kallfass, M., Miller, S., and Tsuchida, S. (2008a) "Interactions with the mass media," *Science*, 321: 204.

Peters, H. P., Brossard, D., De Cheveigné, S., Dunwoody, S., Kallfass, M., Miller, S., and Tsuchida, S. (2008b). "Science–media interface. It's time to reconsider," *Science Communication*, 30: 266–276.

Phillips, D. P., Kanter, E. J., Bednarczyk, B., and Tastad, P. L. (1991) "Importance of the lay press in the transmission of medical knowledge to the scientific community," *The New England Journal of Medicine*, 325: 1180–1183.

Rödder, S. (2009) "Reassessing the concept of a medialization of science: A story from the 'book of life,'" *Public Understanding of Science*, 18: 452–463.

Schneider, S. H. (2009) *Science as a Contact Sport: Inside the Battle to Save Earth's Climate*. Washington, DC: National Geographic Books.

Secord, A. (1994) "Science in the pub: Artisan botanists in early nineteenth-century Lancashire," *History of Science*, 32: 269–315.

Torres-Albero, C., Fernández-Esquinas, M., Rey-Rocha, J., and Martín-Sempere, M. J. (2011) "Dissemination practices in the Spanish research system: scientists trapped in a golden cage," *Public Understanding of Science*, 20: 12–25.

Trumbo, C. (1996) "Constructing climate change: Claims and frames in US news coverage of an environmental issue," *Public Understanding of Science*, 5: 269–283.

Tsfati, Y., Cohen, J., and Gunther, A. C. (2011). "The influence of presumed media influence on news about science and scientists," *Science Communication*, 33: 143–166.

Weiss, C. H. and Singer, E. (1988) *Reporting of Social Science in the National Media*. New York: Russell Sage Foundation.

Wien, C. (2013) "Commentators on daily news or communicators of scholarly achievements? The role of researchers in Danish news media," *Journalism*, 15: 427–445.

6

THE MEDIA/COMMUNICATION STRATEGIES OF ENVIRONMENTAL PRESSURE GROUPS AND NGOs

Robert Cox and Steve Schwarze

> *The critical element of an effective media advocacy effort is that it is strategic …*
> *(Wallack et al., 1999, p. 9)*

Scholarship on environmental change assumes that actors or entities articulate their demands via a process of public, communicative actions and/or use of media. As Hutchins and Lester (2006) observed:

> It is the media that serves as the primary and hotly contested communicative interface – the structuring intermediary – between environmentalists and [others] as they compete for public awareness and approval. The media is more than a site for environmental action; it plays a significant role in shaping debate and influencing outcomes.
>
> (p. 438)

While we agree that actors' access to media (including new media) and other modes of attention-gaining communication are essential for successful change efforts, we would quickly add other factors as well—struggles over interpretation of messages or frames, the capacity to influence key audiences, and more. The efforts of environmental pressure groups and NGOs, in particular, to respond to such challenges also raise important questions about the *strategies*, or ability of such groups to align their messages, choices of media, and audiences with desired outcomes, as well as adapt to contingent events affecting such outcomes.

In this chapter, we summarize the research on environmental groups' uses of media, including digital media, as well as other modes of communication that are intended to generate publicity. Overall, we surveyed 44 journals in 13 fields, as well as related book-length studies whose findings were related, in some respect, to environmental groups' strategic uses of media or communication.

Section one surveys the role of strategy, generally, and the suite of decisions about content, audiences, and modes of dissemination (media) that constitute the strategic terrain of environmental groups' communicative practices. Section two takes a closer look at environmental NGOs' actual media/communicative practices, including the cultivation of news media, rhetorical appeals, issue frames, and alignment of media and audiences.

In section three, we examine some of the obstacles and contingent events that environmental NGOs sometimes face that affect the outcomes of their strategic uses of media or other attention-gaining communication. This section concludes by placing NGOs' media/communication efforts in a wider landscape of hypermediacy and remediations of groups' communicative actions. Finally, in the conclusion, we identify areas for further study of environmental groups' media/communicative strategies.

Strategic assumptions and decisions

Like other social movement groups, environmental NGOs often lack direct or non-mediated channels for conveying their demands to governmental and other decision makers. As a result, Hansen (2010) notes, "outsider groups" depend on media for two reasons: "(1) Public publicity—and, crucially, legitimacy—to help recruit members and financial support for the group's campaigning activities, and (2) as the main channel for achieving public and political attention and action regarding the issues on which [they] campaign" (p. 52). In this section, we inquire further into environmental groups' strategic assumptions, generally, and the decisions about content, audiences, and modes of dissemination that constitute the strategic terrain for such groups' media or communicative practices.

Although there are numerous studies of environmental groups, these often are less explicitly focused on the strategic dimensions of such groups' uses of media or communication. By "strategy," social movement and media scholars usually mean the use of media "in relation to and in support of, rather than instead of or isolated from," the other elements of a campaign (Wallack, Woodruff, Dorfman and Diaz, 1999, pp. 9, 10). An account of an NGO's strategic uses of media/communication, then, is a description of the relationships among specific communicative efforts, relevant audiences, and the outcomes or effects within a given system (a mobilized public, policy change, etc.) (Cox, 2010). For this reason, scholars such as Ryan (1991), Anderson (1997), Lester (2010), and Hansen (2010) have stressed the importance of attending to an environmental group's media/communicative alignment or the ways in which choices about content, media, and audience, are likely to influence a desired outcome.

In studies of environmental NGOs' media/communication practices, scholars have identified a number of recurring challenges that characterize the strategic terrain in which such groups operate.

1. Access to, and attention of, media

In documenting environmental NGOs' attempts to gain media access and attention, scholars have identified a range of strategic efforts and modes of coverage, particularly what is termed "earned media" or favorable news coverage of "newsworthy" actions or events, (Anderson, 1997; Hansen, 2010; Lester, 2010).

Other modes of dissemination in both "old" and new media include public service announcements; paid announcements, independent media, and an array of websites and other online platforms. Additionally, environmental protests and large demonstrations have taken advantage of the "public screen" (DeLuca and Peeples, 2002, p. 132) as well as the "horizontal media" of social networks, micro-blogging (e.g., Twitter), and other social media ("The people," 2011: 10; DeLuca, Lawson and Sun, 2012).

2. Favorable media framing of a group's issues and identity

Environmental scholars have noted that an NGO not only must gain access to, and attention from, media but also influence media framing of its issue in favorable terms (Nisbet and Huge, 2006). As we shall see in section two, such media efforts face many difficulties. Several studies have reported an incongruence between the message or frame initiated by an environmental NGO, on the one hand, and its eventual representation by media sources (DeLuca et al., 2012; Hansen, 2010).

3. Exposure to relevant audiences or constituencies

Favorable framing of an environmental group's issue presupposes that such media coverage is able to convey a particular interpretation of an issue and consequently influence audiences that are relevant to that group's objectives. For this reason, Wallack et al. (1999) emphasized the importance of a group's "media monitoring," that is, ascertaining whether and how the group's issue "is being reported in the news sources to which … [the] target audience is most likely to respond" (p. 28). The strategic assumption is the group's desire to reach "the person or organization with the power to make the change" that the group seeks (p. 28). Importantly, such pivotal audiences may be reachable only through specific media, requiring NGOs to "adopt a highly targeted approach, carefully 'packaging' their information … to suit the particular needs of selected media" (Hansen, 2010, p. 56).

4. Responses to counter-communication and other contingent events

An environmental group's access to media, the framing of its issues, and audience responses occur in a wider context of contingent events, constraints, and reactions from opposing forces. News cycles, concerted campaigns by opponents, unfavorable media framing, low salience of environmental news, and other obstacles often inhibit the effectiveness of the group's communication. As a result, Hansen (2010) has cautioned that environmental NGOs relying mainly upon "theatrical stunts and visually daring protest actions" are not likely, by themselves, to gain sustained media access or coverage (p. 53).

5. Alignment of media/communication and outcomes

As noted above, few studies of environmental NGOs' use of media involve an analysis of such groups' strategic alignments, that is, an account of the relationships among the group's decisions about content, media framing, audiences, and subsequent outcomes. Lester (2011) illustrates one of the few accounts in her analysis of Sea Shepherd Conservation Society's highly mediated campaigns against the harpooning of whales. In this case, the relationships among the campaign's strategic decisions—the use of celebrity to gain media access, exploitation of media frames of conflict, etc.—enabled greater visibility for the campaign's key figure Paul Watson and his ability "to participate in public debate" over whaling (p. 124).

The reasons so few studies provide accounts of environmental groups' assumptions regarding strategic alignment are not hard to understand. They are likely due to the contingent or over-determined nature of media effects themselves, as well as the difficulties of access to the internal deliberations of environmental NGOs (Fox and Frye, 2010).

NGOs' strategic uses of media/communication

While scholars have devoted little explicit attention to the internal assumptions of environmental NGOs, studies of such groups' actual practices and uses of media are numerous and growing. This research—usually in the form of case studies—documents a series of interrelated decisions by NGOs, reflecting the strategic terrain in which their advocacy operates. In this section, we survey studies addressing these groups' cultivation of media sources, the rhetorical modes used to gain media attention and/or disseminate their message, attempts to frame issues favorably, and, relatedly, NGOs' efforts to align their uses of media and exposure to relevant audiences.

Cultivation of media sources

Until very recently, most studies of outsider groups' relationships with media presupposed the role and influence of traditional news organizations—newspapers, radio, TV—and, hence, emphasized the need for strategies of access and attention from such sources. As Nisbet and Huge (2006) noted, "Power in policy making revolves in part around the ability to control media attention to an issue"; thus, "the media lobbying activities of strategic actors" become an important determination of media coverage that reflects and shapes "where an issue is decided, by whom, and with what outcomes" (p. 3).

Environmental group's access and attention from media occur within a matrix of competing forces and constraints: Shifting news cycles, need for narrative structure, type of reporter assigned to a story, and "competition from other issues for attention" (Nisbet and Huge, 2006, p. 3). Further, Anderson (1991) found that environmental NGOs faced a fundamental difficulty: "while environmental issues tend to be drawn-out processes ... the media feed upon short, sharp, highly visible events" (p. 465). In adapting to these constraints, scholars have identified a range of strategic efforts by environmental groups to ensure access and coverage from traditional media.

Prominent among environmental NGOs' efforts to attract attention from news organizations has been the use of earned media. Such media is earned as a group's public actions fulfill traditional norms of "newsworthiness" (Cox, 2013). Analyses of this strategy have included a range of activities: Protests, demonstrations, and other "image events" (DeLuca, 1999a); the use of celebrities as spokespersons (Brockington, 2008); "toxic tours," (Pezzullo, 2003b); sit-ins, blockades, and other forms of civil disobedience (Short, 1991; Wall, 1999; Hayes, 2006); street theater (DeLuca, 1999b; Pezzullo, 2003a;); and various forms of "ecotage" (Lange, 1990; Wagner, 2008). Resource-poor groups, particularly, have relied upon such strategies. Waisbord and Peruzzotti (2009), for example, document the uses of rallies, parades, street theatre, and blockades of roads and bridges as a way of adapting to media's willingness to cover "conflictive and dramatic events" in the *asambleísmo* movement in Argentina (p. 696).

While environmental NGOs' use of staged activities may attract media coverage, research generally has shown diminishing effectiveness. Complicating such methods has been a recurring "conflict between the need to command attention and the need to claim legitimacy" (Cracknell, 1993, p. 8; Greenberg, 1985). DeLuca (1999a) and Wagner (2008), for example, document environmental groups' dramatic protests or "stunts," which—while generating news coverage—evoked mainstream media frames of controversy that portrayed the groups as radical or extreme. Furthermore, environmental media scholars report that "the newsworthiness of environmental pressure groups would soon wear off if they had to rely solely on their creation of spectacular protest 'performances'" (Hansen, 2010, p. 53). Such staged events may initially be newsworthy or "visually striking," but are not always "sufficient for remaining on the media agenda or for maintaining media visibility for the long term" (p. 53).

Indeed, media scholars documented a change in the relationship between some environmental groups and news media as early as the 1980s. Greenberg (1985), for example, found that the UK group Friends of the Earth had decided by this time to combine "a strong research commitment with its attention-getting tactics ... in order to have credibility" (p. 356). Similarly, Anderson (1997) reported that such groups had become more established news sources.

A research role may be one related factor in explaining the greater effectiveness of environmental NGOs in recent decades. Hansen (2010) has argued that "successful" environmental groups have become "skilled in providing what Gandy (1982) refers to as 'information subsidies' to media news professionals and organisations short of both time and resources" (p. 59). Provision of such information by others (e.g., an environmental group), therefore, reduces a news organization's price or "market cost" of obtaining similar information through its own resources (Gandy, 1982, pp. 8, 30–31).

Relatedly, environmental NGOs have been able to sustain greater coverage by a combination of timing of their information subsidies and differential selection of media outlets. This occurs when such groups provide information about, and draw the media's attention to, "environmental issues which are already being discussed in the forums which the media regularly report on" (Hansen, 2010, p. 53). Particularly relevant, Hansen notes, is the "careful timing of press releases and publication of reports" (pp. 53–54), and the ability to relate information to developments, events, and persons in "existing established and legitimate news forums" (p. 53).

Rhetorical resources and modes of appeals

The resources and modes of appeals used by environmental groups to gain media attention or disseminate messages have been the focus, particularly in the U.S., of environmental rhetoric scholarship. Here, we identify several categories of rhetorical appeals and highlight those studies that explicitly connect those appeals to media.

Tropes

Rhetorical scholarship has paid considerable attention to the strategic use of rhetorical figures or tropes by environmental activists. Sometimes identified simply as figures of speech, tropes also function constitutively, for example, as they transfer meaning from one domain to another (metaphor) or re-figure some aspect of reality in other terms (synecdoche or metonymy).

While some studies highlight specific metaphors to show how they function constitutively on an environmental issue (de Onís, 2012; Patterson and Lee, 1997), others give attention to metaphors' relations to media coverage. Väliverronen and Hellsten (2002), for example, described the pivotal use of metaphors about biodiversity loss in generating media storylines, including apocalyptic narratives of extinction. Such metaphors were pivotal to media coverage because they "evoke[d] powerful images and emotions" and "help[ed] create continuity from previous … frames to current ones" (p. 231). Relatedly, Nerlich and Koteyko (2009) reported that, in press coverage in the UK, "metaphors of finance, religion, and diet frame[d] what [carbon rationing groups] are doing in relatively moralistic terms" (p. 220).

Others have examined tropes in the circulation and appropriation of key phrases by environmental activists and media—Russill's (2008) study of "tipping point" forewarnings in climate change communication, for example, and Peeples' (2011) tracing of the "downwinder" ideograph in a waste incineration dispute.

Narratives

Prominent among the recurring narrative forms used by environmental groups are apocalyptic narratives (Foust and Murphy 2009), the jeremiad (Buehler 1998; Opie and Elliot, 1996; Singer, 2010; Wolfe, 2008); and melodrama (Kinsella, 2008; Schwarze, 2006). Strategically, such narratives align with journalistic norms, that is, stories with cultural resonance. Foust and Murphy (2009), for example, documented the prominence of apocalyptic narratives in media coverage of climate change, and Schwarze (2006) identified the pervasive use of melodrama in media accounts of environmental issues. Narratives also provide important structural elements of mediated representations of environmental issues in documentary and fictional film (Rosteck and Frentz, 2009; Salvador and Norton, 2011).

Scientific evidence

A growing area of inquiry is the use of science as a rhetorical resource. Scholars, particularly, have developed the concepts of "lay" or public expertise (Endres, 2009; Zoller, 2012) and scientific uncertainty (Moore, 2009; Schwarze, 2003) to explain how ordinary citizens and activists generate knowledge about environmental risks. Other studies have found, however, that such appeals can lead to protracted conflict. Ceccarelli's (2011) identification of "manufactured scientific controversies," for example, documented a propensity of news media to funnel competing scientific appeals into a conflict frame.

Images

Analyses of images and image events in environmental media are a growing area of research. DeLuca (1999a) has argued that image events are more than attention-getting devices; they aid in reconstituting the meaning of nature and industrialism in a media system that is dominated by visual images.

As with scientific evidence, however, some have noted the limitations of images for activists seeking to use image-driven media. Doyle (2007), for example, in studying the "representational politics" of Greenpeace's photographic images of climate change, found that the group has struggled with temporality; as impacts of climate change become visible, and thus available for photographic representation, the time for preventive action has passed. The temporal nature of climate change (and other long-term environmental challenges), Doyle noted, presents a constraint.

NGOs and media framing

Another broad domain of scholarship focuses on the concept of framing to understand how environmental issues are discursively packaged for media and public understanding. Rather than offer an account of the extensive framing literature, we attempt to clarify different approaches for studies of framing based on how scholars have engaged its strategic uses. (See Chapter 1 for a discussion of tensions among scholars over the different approaches to framing in environmental campaigns.)

Thematic or audience focus

One main approach to framing might be considered thematic, that is, studies of the tailoring of messages by environmental NGOs or media in relation to audience interests, values, and responsiveness. Recent interest in the public's responsiveness to NGOs' climate change communication, for example, has spurred analyses of frames that foreground "public health" (Maibach, Nisbet, Baldwin, Akerlof and Diao, 2010). Similarly, Van Gorp and van der Goot (2012) examined different stakeholders' use of frames such as "progress" and the archetype of the "Good Mother" in debates over sustainable agriculture (p. 127).

In some cases, environmentalists miss opportunities for strategically adapting to a relevant audience. Haluza-DeLay and Fernhout (2011), for example, concluded that English-speaking Canadian environmental groups often remained locked into an "environmentalist" frame, ignoring the prevalent cultural master-frame of "social inclusion" (e.g., multiculturalism).

Critical interventions

In this approach, framing is used by rhetorical scholars to investigate environmental groups' efforts either to reconstitute or interrupt dominant discourses or frames. This is particularly apparent in studies of environmental justice organizations (Pal and Dutta, 2012) and protest groups (Boudet 2011; Dauvergne and Neville 2011). Brady and Monani (2012), for example, deploy a "just sustainability" frame to "interrogate not only ... economic development but also the use of popular American Indian archetypes like 'the Ecological Indian' in the marketing of sustainable energy" (p. 147).

Other scholars have examined the critical use of NGOs' frames by sympathetic, "alternative" media. In a study of conflict over metallic mining in El Salvador, for example, Hopke (2012) found that alternative media were able to "break a cycle of environmental inequity" by reframing the Salvadorian government's "dominant narratives of economic progress toward community rights and environmental justice" (p. 365).

News media

Building on work by Entman (1993), scholars have identified storylines (frames) used by journalists to characterize environmental issues. Doyle (2011), for example, explored how the UK news media "contributed to the reframing of nuclear power as low carbon," and implications of this for public support of nuclear power (p. 107). Similarly, Hansen (2006) documented media frames of "nature" and "natural" in coverage of genetic research and biotechnology.

Importantly, several studies have reported unfavorable media framing of a group's identity, for example, Greenpeace as "terrorists" (Hansen, 2010, p. 57) or incongruence between the message or frame initiated by an environmental NGO and its representation by media sources. For example, in their study of debate over a Norwegian oil company's drilling, Ihlen and Nitz (2008), found that media tended to use a "typical media frame, the 'horse race' frame," ignoring frames of environmental groups (p. 1).

"Behind the scenes"

A small number of scholars have begun to gain access to the internal, strategic thinking of environmental NGOs and their decisions about frames. Della Porta and Rucht (2002), particularly, have urged scholars to observe more closely the ways environmental campaigns are "organized, orchestrated and framed" (p. 5), while Chesters and Welsh (2004) succeeded in observing the "deliberative processes that were undertaken by activists involved in framing the protests" at World Bank meetings in Prague in 2000 (p. 314).

Alignment of media and audiences

Scholars studying the strategic uses of media have been particularly concerned to understand a group's *selective choice* of media, that is, its attempt to align a particular mode of dissemination and a relevant audience. Hansen (2010), for example, has argued that successful environmental groups are knowledgeable not only about the different requirements of news organizations but, importantly, about "target audiences of different media," who "adopt a highly targeted approach, carefully 'packaging' their information ... to suit the particular needs of selected media" (p. 56).

An early attempt to trace the alignment of media and outcomes was Eyerman and Jamison's (1989) study of Greenpeace's strategic use of media. They concluded that it was the relationship between Greenpeace's "selective gathering of campaign-related facts, the selective dissemination of arguments to the media and other public fora ... that gives Greenpeace its enormous influence" (p. 113, quoted in Hansen, 2010, pp. 54–55). Similarly, DeHanas (2009), in a study of Muslim women's use of radio, concluded that an alignment between the group's message (imbuing environmental ethics with religious meaning) and its selective use of media (Muslim Community Radio) accounted for the program's appeal to female listeners in London's Muslim community.

With the advent of the internet and other digital media, environmental groups have gained new tools for their media/communication efforts. Indeed, "Growth in networked digital communications technology innovation and use since the 1990s has helped to change the conditions for visibility in environmental politics" (Lester and Hutchins, 2012, p. 848). As a result, environmental communication scholars have begun to investigate the role of digital media in altering the relationship between NGOs and traditional news media, as well as the contributions of online and other digital media in raising awareness, organizing resistance, and creating spaces for exchanges (Lester and Hutchins, 2009; Lin, 2012; Pal and Dutta, 2012; DeLuca et al., 2012).

Beyond their function as modes of dissemination, new media have emerged also as significant organizing resources for environmental NGOs. Findings of such uses range from Frederick's early (1992) study of the role of computer networks in environmental NGOs' opposition to the North American

Free Trade Association, and the International Campaign for Justice's use of a website for mobilization in Bhopal, India (Pal and Dutta, 2012), to "crosscutting Twitter streams" in coordinating protests at the 2009 Copenhagen climate summit (Segerberg and Bennett, 2011, p. 197).

Importantly, for outsider groups, new media offer "the potential for independent information distribution devoid of the mediating effect of news journalists and the established news media industries" (Lester and Hutchins, 2009, p. 579). Lester and Hutchins argue that this potential resides in the capacity of the web for "sustainable self-representation"; such self-representation "promises to avoid both the fickleness of changing news agendas, the vicissitudes of reporting and editorial practices, and the contending corporate interests of large-scale news conglomerates" (p. 591).

The promises of new media, nevertheless, may be somewhat premature in the strategic terrain in which environmental NGOs function. Recent studies suggest the relationship between such groups and traditional news media are more complex. In a study of Sea Shepherd Conservation Society's campaign against Japanese whaling, Lester (2011) found that, while the campaign's visibility involved "a complex flow of information and meanings across various 'old' and 'new' media form," it still remained reliant on traditional media to showcase its images and news feeds via distributed news and cable television (p. 124; see also, Lester and Hutchins, 2009, pp. 580 ff.).

Other scholars have reported very different findings as environmental NGOs increasingly deploy new media technologies. DeLuca et al. (2012), for example, reported in their study of the Occupy Wall Street movement that activists' uses of Twitter, Facebook, and YouTube created "new contexts for activism that do not exist in old media"; such media, they argued, "foster an ethic of individual and collective participation, thus creating a norm of perpetual participation" (p. 483).

Part of the challenge in identifying the strategic significance of such digital media, Segerberg and Bennett (2011) propose, is to determine *"how such technologies infuse specific protest ecologies,"* that is, how such media interact, not only with events, but with each other in mediating actors' relationships and the movement's visibility (p. 197; emphasis in original). In their study of activists' uses of Twitter at the 2009 Copenhagen summit, they reported that Twitter profile feeds, hyperlinks, and community-generated hashtags connected "diverse users, uses and different temporal and spatial regions of the protest space" (pp. 197, 203).

Contingency, oppositional media, and hypermediacy

Most studies acknowledge, implicitly, that environmental NGOs' targeting of media is far from being determinative of a group's desired outcomes. Wallack et al. (1999) cautioned that the strategic challenge for pressure groups' uses of media is more than simply "raising awareness" (p. 13). Ultimately, their success is also contingent upon multiple, intervening events and constraints.

Contingency and oppositional media

Environmental communication scholars have identified several recurring challenges and obstacles that potentially constrain the effectiveness of a group's media efforts:

1. Environmental groups' messaging may lack salience or fail to adapt to audiences' interests or motives; for example, use of fear-inducing appeals may be ineffectual in prompting desired behaviors (O'Neill and Nicholson-Cole, 2009).
2. Some journalists resist being "stage-managed" by environmental NGOs' reliance on "stunts" or dramatic events to draw media attention (Hutchins and Lester, 2006).
3. News media may frame an environmental group's message or identity unfavorably (Ihlen and Nitz, 2008). Such unfavorable coverage occurs, particularly, when issues are framed within

dominant discursive norms that marginalize environmentalists' rhetorical choices (Takahashi and Meisner, 2012).

4. Media norms may elide environmental NGOs' cultivation efforts due to news cycles, sourcing practices, etc. Mainstream media, for example, is often "authority-oriented," i.e., media are less likely to cover environmental sources than industry or government sources (Sovacool, 2008).

5. Environmental groups' publicity may have "a negative influence on [the] groups' claims ... particularly if the main effect is to galvanize opposition" or counter-media (Hansen, 2010, p. 51). As a result, environmental groups may have difficulty in controlling the reception of their messages as they "operate in an increasingly crowded discursive landscape, as campaigners and counter-campaigners articulate ... frames that resonate differently across changing social and cultural contexts" (Dauvergne and Neville, 2011, p. 192).

The "public screen" and hypermedia

Environmental NGOs' uses of media are further impacted by dynamics introduced by new, digital technologies. Increasingly, scholars have begun to address the complex processes in which multiple platforms and pathways for circulation, exchange, and remediation affect user/receivers' understanding and participation.

Uses of new media is made challenging both by the remediations enabled by the interconnected, open structures of such media as well as the instability of meaning this openness creates. DeLuca and Peeples (2002) have urged communication scholars to take seriously the claims of media theorists that "new technologies introduce new forms of social organization and new modes of perception" (p. 131).

In their study of the WTO protests in Seattle in 1999, DeLuca and Peeples (2002) introduced the "public screen" to name activists' practices adapted to "a wired society" (p. 125). Importantly, they argued, "the public screen is a scene of hypermediacy" (p. 132), that is, "a heterogeneous space," in which representations "open on to other representations or other media" in which mediations themselves are multiplied (Bolter and Grusin, 1999, quoted in DeLuca and Peeples, p. 132). Such multiplications are seen in Segerberg and Bennett's (2011) analysis of the hyperlinks and retweets of Twitter streams at the Copenhagen protests. And Wolfe (2009), in his study of the video *The Meatrix*, attributed its ability to attract and engage 20 million viewers online to a "rhizomatic view of *The Meatrix* assemblage," that is, "a distributed system, an open and shifting constellation of intertextual, disseminating, and user producing relationships" (p. 329).

As a consequence, strategic studies of environmental NGOs' uses of media now face another challenge—the instability of meaning and related difficulties of social actors in managing the flow of messages in media campaigns. Under such conditions, Lester and Hutchins (2012) argue, "Particular political positions and practices are alternatively legitimized, delegitimized and/or bypassed in highly unpredictable ways in the process. This instability in the social production of meaning then feeds back into multi-directional digital information flows, and public opinion(s) that are difficult to manage and predict (Cottle, 2011, quoted in Lester and Hutchins, 2012, p. 859).

Future challenges

While the multidisciplinary literature of environmental groups' uses of media/communication is quite broad, there still remain noticeable gaps and new challenges that invite further study. Despite such research, for example, NGOs' strategic rationales per se remain, surprisingly, understudied. By strategic rationale, we mean activists' accounts of the role or "work" their use of media is expected to accomplish in a campaign's broader advocacy efforts. Why (and how), for example, do activists believe *this* appeal and *these* modes of dissemination to *these* audiences are likely to align ultimately with certain outcomes or effects?

Understanding such accounts could yield new insights into factors contributing to successful as well as failed media strategies, enhancing both the theory and the practice of environmental communication.

Complicating such investigations, we believe, are two thorny challenges for environmental communication scholars. First, as we've observed, environmental NGOs are turning to new media platforms as pivotal components in their advocacy. Yet, few studies have fully traced the strategic implications of such advocacy, particularly within complex, open networks in which multiple sources, diffuse audiences, user participation, and remediation characterize the communication milieu of environmental campaigns. In what sense, for example, are concepts such as "targeting" of audiences or strategic alignment comprehensible within such distributed systems?

Relatedly, we believe environmental communication scholars must begin to tackle a second challenge—the question of what constitutes "effects" within new communicative systems in which linear assumptions of strategy are increasingly outmoded. Studies of environmental NGOs' uses of media have traditionally assumed a relatively uncomplicated relationship among message, media, audience, and outcomes. Within such an understanding of strategy, determinations of effects assume an alignment with, and impacts upon, an identifiable audience's attitudes or behavior. Yet, within increasingly open, distributed systems, scholars now must assess the strategic reach of environmental NGOs' media/communicative uses, its alignment and impacts, on diffused or unspecified audiences.

We believe tackling such challenges is critical to advances in our understanding of environmental pressure groups and NGOs' communicative actions in an increasingly changing media environment. Such analyses would begin to identify more rigorously the strategic contributions of media/communication in the wider designs of environmental NGOs' campaigns.

Further reading

Cottle, S. and Lester, L., 2011. Protesting ecology and climate. In *Transnational Protests and the Media*. New York: Peter Lang.

DeLuca, K. M., 2005. *Image Politics: The New Rhetoric of Environmental Activism*. New York and London: Routledge.

Hansen, A., 2010. *Environment, Media and Communication*. New York and London: Routledge.

Lester, L., 2010. *Media and Environment*. Cambridge: Polity.

Lester, L. and Hutchins, B. (eds), 2013. *Environmental Conflict and the Media*. New York: Peter Lang.

References

Anderson, A., 1991. Source strategies and the communication of environmental affairs. *Media, Culture and Society*, *13*(4), pp. 459–476.

Anderson, A., 1997. *Media, culture and the environment*. New Brunswick, NJ: Rutgers University Press.

Bolter, J. and Grusin, R., 1999. *Remediation*. Cambridge, MA: MIT.

Boudet, H. S., 2011. From NIMBY to NIABY: Regional mobilization against liquefied natural gas in the United States. *Environmental Politics*, *20*(6), pp. 786–806.

Brady, M. J. and Monani, S., 2012. Wind power! Marketing renewable energy on tribal lands and the struggle for just sustainability. *Local Environment*, *17*(2), pp. 147–166.

Brockington, D., 2008. Powerful environmentalisms: Conservation, celebrity and capitalism. *Media, Culture and Society*, *30*(4), pp. 551–568.

Buehler, D. O., 1998. Permanence and change in Theodore Roosevelt's conservation jeremiad. *Western Journal of Communication*, *62*(4), pp. 439–458.

Ceccarelli, L., 2011. Manufactured scientific controversy: Science, rhetoric, and public debate. *Rhetoric and Public Affairs*, *14*(2), pp. 195–228.

Chesters, G. and Welsh, I., 2004. Rebel colours: "Framing" in global social movements. *The Sociological Review*, *52*(3), pp. 314–335.

Cottle, S. 2011. Taking global crises in the news seriously: Notes from the dark side of globalization. *Global Media and Communication*, *7*(2), pp. 77–95.

Cox, R., 2010. Beyond frames: Recovering the strategic in climate communication. *Environmental Communication: A Journal of Nature and Culture*, 4(1), pp. 122–133.

Cox, R., 2013. *Environmental communication and the public sphere*. Los Angeles, CA: Sage.

Cracknell, J., 1993. Issue arenas, pressure groups and environmental agendas. In: A. Hansen, ed. *The mass media and environmental issues*. Leicester: Leicester University Press, pp. 3–21.

Dauvergne, P. and Neville, K.J., 2011. Mindbombs of right and wrong: Cycles of contention in the activist campaign to stop Canada's seal hunt. *Environmental Politics*, 20(2), pp. 192–209.

DeHanas, D. N., 2009. Broadcasting green: Grassroots environmentalism on Muslim women's radio. *The Sociological Review*, 57(s2), pp. 141–155.

Della Porta, D. and Rucht, D., 2002. The dynamics of environmental campaigns. *Mobilization: An International Quarterly*, 7(1), pp. 1–14.

DeLuca, K. M., 1999a. *Image politics: The new rhetoric of environmental activism*. London: Routledge.

DeLuca, K. M., 1999b. Unruly arguments: The body rhetoric of Earth First! Act-Up, and Queer Nation. *Argumentation and Advocacy*, 36(1), pp. 9–21.

DeLuca, K. M. and Peeples, J., 2002. From public sphere to public screen: Democracy, activism, and the "violence" of Seattle. *Critical Studies in Media Communication*, 19(2), pp. 125–151.

DeLuca, K. M., Lawson, S., and Sun, Y., 2012. Occupy Wall Street on the public screens of social media: The many framings of the birth of a protest movement. *Communication, Culture and Critique*, 5, pp. 483–509.

de Onís, K. M., 2012. "Looking both ways": Metaphor and the rhetorical alignment of intersectional climate justice and reproductive justice concerns. *Environmental Communication: A Journal of Nature and Culture*, 6(3), pp. 308–327.

Doyle, J., 2007. Picturing the clima(c)tic: Greenpeace and the representational politics of climate change communication. *Science as Culture*, 16(2), pp. 129–150,

Doyle, J., 2011. Acclimatizing nuclear? Climate change, nuclear power and the reframing of risk in the UK news media. *International Communication Gazette*, 73(1–2), pp. 107–125.

Endres, D., 2009. Science and public participation: An analysis of public scientific argument in the Yucca Mountain controversy. *Environmental Communication: A Journal of nature and Culture*, 3(1), pp. 49–75.

Entman, R. M., 1993. Framing: Toward clarification of a fractured paradigm. *Journal of Communication*, 43(4), pp. 51–58.

Eyerman, R. and Jamison, A., 1989. Environmental knowledge adds an organizational weapon: The case of Greenpeace. *Social Science Information*, 28(1), pp. 99–119.

Foust, C. R. and Murphy, W., 2009. Revealing and reframing apocalyptic tragedy in global warming discourse. *Environmental Communication: A Journal of Nature and Culture*, 3(2), pp. 151–167.

Fox, R. L. and Frye, J. J., 2010. Tensions of praxis: A new taxonomy for social movements. *Environmental Communication: A Journal of Culture and Nature*, 4(4), pp. 422–440.

Frederick, H. H., 1992. Computer communications in cross-border coalition-building: North American NGO networking against NAFTA. *International Communication Gazette*, 50(2–3), pp. 217–241.

Gandy, Jr., O. H., 1982. *Beyond agenda setting: Information subsidies and public policy*. Norwood, NJ: Ablex.

Greenberg, D., 1985. Staging media events to achieve legitimacy: A case study of Britain's Friends of the Earth. *Political Communication and Persuasion*, 2(4), pp. 347–62.

Haluza-DeLay, R. and Fernhout, H., 2011. Sustainability and social inclusion? Examining the frames of Canadian English-speaking environmental movement organisations. *Local Environment*, 16(7), pp. 727–745.

Hansen, A., 2006. Tampering with nature: "nature" and the "natural" in media coverage of genetics and biotechnology. *Media, Culture and Society*, 28(6), pp. 811–834.

Hansen, A., 2010. *Environment, media and communication*. London and New York: Routledge.

Hayes, G., 2006. Vulnerability and disobedience: New repertoires in French environmental protests. *Environmental Politics*, 15(5), pp. 821–838.

Hopke, J. E., 2012. Water gives life: Framing an environmental justice movement in the mainstream and alternative Salvadoran press. *Environmental Communication: A Journal of Nature and Culture*, 6(3), pp. 365–382.

Hutchins, B. and Lester, L., 2006. Environmental protest and tap-dancing with the media in the information age. *Media, Culture and Society*, 28(3), pp. 433–451.

Ihlen, O. and Nitz, M., 2008. Framing contests in environmental disputes: Paying attention to media and cultural master frames. *International Journal of Strategic Communication*, 2(1), pp. 1-18.

Kinsella, W. J., 2008. Forum: Narratives, rhetorical genres, and environmental conflict— Responses to Schwarze's "environmental melodrama." *Environmental Communication: A Journal of Nature and Culture*, 2(1), pp. 78–109.

Lange, J. L., 1990. Refusal to compromise: The case of Earth First! *Western Journal of Communication*, 54(4), 473–494.

Lester, E. A. and Hutchins, B. 2012. The power of the unseen: Environmental conflict, the media and invisibility. *Media, Culture and Society*, 34(7), pp. 847–863.

Lester, L., 2010. *Media and environment: Conflict, politics and the news.* Cambridge: Polity.

Lester, L., 2011. Species of the month: Anti-whaling, mediated visibility, and the news. *Environmental Communication: A Journal of Nature and Culture, 5*(1), pp. 124–139.

Lester, L. and Hutchins, B., 2009. Power games: Environmental protest, news media and the internet. *Media, Culture and Society, 31*(4), pp. 579–595.

Lin, T. C., 2012. Cross-platform framing and cross-cultural adaptation: Examining elephant conservation in Thailand. *Environmental Communication: A Journal of Nature and Culture, 6*(2), pp. 193–211.

Maibach, E. W., Nisbet, M., Baldwin, P., Akerlof, K., and Diao, G., 2010. Reframing climate change as a public health issue: An exploratory study of public reactions. *BMC Public Health,* 10: 299. Available at: www.ncbi.nlm. nih.gov/pubmed/20515503 [Accessed 30 August, 2013].

Moore, M. P., 2009. The Union of Concerned Scientists on the uncertainty of climate change: A study of synecdochic form. *Environmental Communication, 3*(2), pp. 191–205.

Nerlich, B. and Koteyko, N., 2009. Carbon reduction activism in the UK: Lexical creativity and lexical framing in the context of climate change. *Environmental Communication, 3*(2), pp. 206–223.

Nisbet, M. C. and Huge, M., 2006. Attention cycle and frames in the plant biotechnology debate: Managing power and participation through the press/policy connection. *The International Journal of Press/Politics, 11*(2), pp. 3–40.

O'Neill, S. and Nicholson-Cole, S., 2009. "Fear won't do it": Promoting positive engagement with climate change through visual and iconic representations. *Science Communication, 30*(3), pp. 355–379.

Opie, J. and Elliot, N., 1996. Tracking the elusive jeremiad: The rhetorical character of American environmental discourse. In: J. G. Cantrill and C. L. Oravec, eds, *The symbolic earth: Discourse and our creation of the environment.* Lexington, KY: University Press of Kentucky, pp. 9–37.

Pal, M. and Dutta, M. J., 2012. Organizing resistance on the internet: The case of the international campaign for justice in Bhopal. *Communication, Culture and Critique, 5*(2), pp. 230–251.

Patterson, R. and Lee, R., 1997. The environmental rhetoric of "balance": A case study of regulatory discourse and the colonization of the public. *Technical Communication Quarterly, 6*(1), pp. 25–40.

Peeples, J. A., 2011. Downwind: Articulation and appropriation of social movement discourse. *Southern Communication Journal, 76*(3), pp. 248–263.

Pezzullo, P. C., 2003a. Resisting "national breast cancer awareness month": The rhetoric of counterpublics and their cultural performances. *Quarterly Journal of Speech, 89*(4), pp. 345–365.

Pezzullo, P. C., 2003b. Touring "Cancer Alley," Louisiana: Performances of community and memory for environmental justice. *Text and Performance Quarterly, 23*(3), pp. 226–252.

Rosteck, T. and Frentz, T. S., 2009. Myth and multiple readings in environmental rhetoric: The case of *An Inconvenient Truth. Quarterly Journal of Speech, 95*(1), pp. 1–19.

Russill, C., 2008. Tipping point forewarnings in climate change communication: Some implications of an emerging trend. *Environmental Communication, 2*(2), pp. 133–153.

Ryan, C., 1991. *Prime time activism: Media strategies for grassroots organizing.* Boston, MA: South End Press.

Salvador, M. and Norton, T., 2011. The flood myth as archetype in the age of global climate change. *Environmental Communication: A Journal of Nature and Culture, 5*(1), pp. 45–61.

Schwarze, S., 2003. Juxtaposition in environmental health rhetoric: Exposing asbestos contamination in Libby, Montana. *Rhetoric and Public Affairs, 6*(2), pp. 313–335.

Schwarze, S., 2006. Environmental melodrama. *Quarterly Journal of Speech, 92*(3), pp. 239–261.

Segerberg, A. and Bennett, W. L., 2011. Social media and the organization of collective action: Using Twitter to explore the ecologies of two climate change protests. *The Communication Review, 14*(3), pp. 197–215.

Short, B., 1991. Earth First! and the rhetoric of moral confrontation. *Communication Studies, 42*(2), pp. 172–188.

Singer, Ross, 2010. Neoliberal style, the American re-generation, and ecological jeremiad in Thomas Friedman's "code green". *Environmental Communication: A Journal of Nature and Culture, 4*(2), pp. 135–151.

Sovacool, B. K., 2008. Spheres of argument concerning oil exploration in the Arctic National Wildlife Refuge: A crisis of environmental rhetoric? *Environmental Communication: A Journal of Nature and Culture, 2*(3), pp. 340–361.

Takahashi, B. and Meisner, M. (2012). Environmental discourses and discourse coalitions in the reconfiguration of Peru's environmental governance. *Environmental Communication, 6*(3), 346–364.

The people formerly known as the audience, 2011, 9 July. Special report. *The Economist, 400*(8741), pp. 9–12.

Väliverronen, E. and Hellsten, I., 2002. From "burning library" to "green medicine": The role of metaphors in communicating biodiversity. *Science Communication, 24*(2), pp. 229–245.

Van Gorp, B. and van der Goot, M. J., 2012. Sustainable food and agriculture: Stakeholder's frames. *Communication, Culture & Critique, 5,* pp. 127–148.

Wagner, T., 2008. Reframing ecotage as ecoterrorism: News and the discourse of fear. *Environmental Communication: A Journal of Nature and Culture, 2*(1), pp. 25–39.

Waisbord, W. and Peruzzotti, E., 2009. The environmental story that wasn't: Advocacy, journalism and the *asambleísmo* movement in Argentina. *Media, Culture and Society, 31*(5): 691–709.

Wall, D., 1999. Mobilising Earth First! In Britain. *Environmental Politics, 8*(1), pp. 81–100.

Wallack, L., Woodruff, K., Dorfman, L., and Diaz, I., 1999. *News for a change.* Thousand Oaks, CA: Sage Publications.

Wolfe, D., 2008. The ecological jeremiad, the American myth, and the vivid force of color in Dr. Seuss' "The Lorax". *Environmental Communication: A Journal of Nature and Culture, 2*(1), pp. 3–24.

Wolfe, D., 2009. The video rhizome: Taking technology seriously in *The Meatrix. Environmental Communication: A Journal of Nature and Culture, 3*(3), pp. 317–334.

Zoller, H. M., 2012. Communicating health: Political risk narratives in an environmental health campaign. *Journal of Applied Communication Research, 40*(1), pp. 20–43.

7

RESISTING MEANINGFUL ACTION ON CLIMATE CHANGE

Think tanks, 'merchants of doubt' and the 'corporate capture' of sustainable development

David Miller and William Dinan

The chapter examines various corporate and elite responses to climate change. In particular it notes the tensions within global elite networks between those who take a proactive response to climate issues and have aimed to provide leadership on climate by shaping (arguably dominating) the political and public discourse on this issue, and the defensive movement of climate change contrarianism and denial. These two tendencies have resulted in differing sorts of lobbying, public relations and corporate responses to climate issues.

It is important to understand that the role of think tanks and lobby groups is multidimensional. They aim to dominate the information environment in a number of distinct public and private arenas. Thus, it is important to examine the relative success of denial not simply in relation to media reporting or in relation to governmental decision making, but in relation also to a wider range of arenas. The economic, social and scientific networks within which decision makers are located is significant. However, it is clear that these networks span multiple levels of governance, so the intersection of the various think tanks and policy planning groups with the global, regional, national and sub-state levels is discussed. This holistic focus makes the empirical task of revealing the role of the think tanks and policy planning organisations more complex, but ultimately, makes understanding of the dynamics and drivers of the climate contrarian movement easier.

Introduction

There is a well-established scientific consensus that 'the warming of the Earth over the last half-century has been caused largely by human activity' (Royal Society 2010: 1). Yet meaningful political action on climate is painfully slow. Among the reasons for this are the activities of those corporations that stand to lose most from rational policy decisions. These corporations, from extractive and other industries with heavy environmental footprints, have in the main attempted to frustrate meaningful progress. As is well known, some corporations have attempted to foster doubt about the scientific consensus on climate (Michaels 2008; Oreskes and Conway 2010) – a strategy often referred to as climate change 'scepticism', 'denial' or 'contrarianism' (see O'Neill and Boykoff 2010).[1] One study concludes that contrarianism on climate, led by conservative think tanks 'is a tactic of an elite-driven counter-movement designed to combat environmentalism, and that the successful use of this tactic has contributed to the weakening of US commitment to environmental protection' (Jacques *et al.* 2008: 365). Less well known has been the strategy of a range of other corporations in the oil and associated industries, which has not denied the evidence that climate change is largely caused by human activity, but has sought to manage responses to protect their interests. We refer to this strategy as the attempted corporate capture of environmental policy.

The climate contrarian strategy is perhaps better known; however, it would be a mistake to focus only on the former. This is because, empirically, the oil industry has pursued both strategies and in terms of outcomes both have been effective in delaying or stopping meaningful climate action.

This chapter will examine how contending factions of corporate and policy elites have organised, constructed and communicated climate issues. The chapter will look specifically at the role of elite policy-planning groups, think tanks and other lobbying organisations that have played a significant role in communicating climate change and practically frustrating progress. Taking an approach that recognises the crucial role of ideas and communication in power relations, this chapter grounds analysis in an understanding that ideas must be put into practice to be effective (powerful), and therefore addresses the role of key agents such as think tanks in mediating between social interests, the realms of ideas and concrete policy outcomes. Our analysis suggests the centrality of communication to how the environment is constructed and contested. We advance a distinctive approach that sees communication in a wider context than just in terms of the mass media and the internet. Communication is fundamentally linked to social interests and, therefore, the material world. By this we mean, first, that 'environmental communication' is an irreducible component of environmental politics. This is not just a question of the centrality of mass media or the internet to environmental politics, but of communicative processes 'outside' of the media and 'inside' social institutions (such as the state/policy networks, corporations and civil society) and fields (national and international politics, environmental policy, the legal system, journalism etc.).

Second, we mean that ideas about the environment and their communication spring from social interests, or are related to them. We do not mean that this occurs automatically in a simple reflection of economic interests, since the human intellect and processes of judgement, strategy and assessment necessarily mediate how interests are conceived and are negotiated or contested within social institutions. Thus we do not adhere to the 'treadmill of production' approach in environmental sociology, which tends to reduce interests to a purely economic level (just like opposing approaches such as neoclassical economics and the rational actor and public choice models associated with it). Nor, however, do we agree with ecological modernisation approaches that are overly optimistic about the possibility of market solutions to environmental crises (Simionis 1989; Mol 2001). We are more sympathetic to the model advanced by Pulver (2007) which sees contest within and between economic and other actors as the context in which conceptualisation of, and decisions about, interests and therefore communicative strategies are made. The economic, social and scientific networks within which corporate decision makers are located is significant. We would extend this by noting the constitutive importance of ideas and their communication to processes of contestation in relation both to networks and outcomes. In this context inter-elite communication between different corporate factions (disembedded elites in some literatures) is too often underplayed.

Understanding collective ideas

Our perspective insists that ideas emerge from social interests and their communication is part of the process by which people 'become conscious of conflict and fight it out', as Marx put it (Miller 2002). But in the case of environmental communication more generally (as with most other areas of political struggle) it is necessary also to understand how ideas spread vertically and horizontally in society, and temporally and geographically. It is useful to consider theories that focus specifically on how ideas become popular and turn into collective phenomena. It is important to understand this process as one that can happen at many different levels in society, in relative divorce or conformity with other levels. In particular, because of the strong role of science in policy argument about climate change we need a concept that understands and explains how scientific theories emerge, are tested and either falsified or supported. In this respect, concepts such as the 'invisible college' (de Solla Price 1963, 1986) and 'epistemic communities' (Haas

1992) are useful in understanding how elite scientific ideas cohere. They are less able to explain how such ideas may spread more widely (i.e. in public debate or on policy) or conversely how ideas from elsewhere may influence science. The Polish microbiologist Ludwig Fleck (1979) argued that the development of scientific concepts is associated with the ideas and relative power of competing professional or ideological groups. A 'thought collective' is, says Fleck, 'even more stable and consistent than the so-called individual, who always consists of contradictory drives' (p. 44). German sociologist Karl Mannheim (1927) shared the sense of ideas having what he called an 'objective mental structure' that transcends the individual. 'In most of our intellectual responses', he wrote, 'we are not creative but repeat certain statements the content and form of which we have taken over from our cultural surroundings' (1927: 132).

But Mannheim's conception differs from that of Fleck, whose thought collectives are seen as hermetically sealed – not allowing for agreed information between contending perspectives – as if evidence not only might not make a difference, but could not. Fleck 'seems to preclude (productive) disagreement' within a thought collective (Plewhe 2009: 35). Mannheim, by contrast, notes that 'if thought developed simply through a process of habit-making, the same pattern would be perpetuated for ever, and changes and new habits would necessarily be rare' (1927: 133). Changes in thought, Mannheim suggested, are 'produced' by 'social causes' (p. 137), they are 'socially determined' (1927: 142). The 'sudden breakdown of a style of thought ... will generally be found to correspond to the sudden breakdown of the groups which carried it' (p. 135).

We draw attention, therefore, to the social interests that undergird ideas and their communication. These are condensed and crystallised in organisations such as think tanks, policy planning groups and lobby firms, all of which require financial and logistical support to enable their ideas to flourish in practice. This suggests the need to examine how ideas are produced and made effective in addition to engaging with the ideas themselves. In that sense we offer a materialist perspective on communicative power (Miller 2002).

The chapter reviews the existing literature on the mediation of climate issues, and argues that understanding the dynamics of climate change communication not only necessitates a critical examination of the sources the media rely on in their reporting, but also requires an analysis of the communication of climate outside of mass media. We note tensions within global elite networks between those that have aimed to provide leadership on climate by shaping and arguably dominating policy and public discourse and the defensive movement of climate change contrarianism. These two tendencies have resulted in differing sorts of lobbying, public relations and elite planning organisations and also some 'churn' in corporate responses to climate issues.

It is important to understand that the role of the think tanks and lobby groups is multidimensional. They aim to dominate the information environment in a number of distinct public and private arenas. Thus, it is important to examine the relative success of climate denial not simply in relation to media reporting or governmental decision making, but in relation also to a wider range of arenas including: the production of scientific knowledge; mainstream media reporting; elite policy planning; and the level of government and executive decision making. However, it is clear that climate policy and deliberation spans multiple levels of governance, so the intersection of the various think tanks and policy planning groups with the global, regional, national and sub-state levels will be discussed, considering how each arena is interpenetrated by actors operating at multiple levels of governance.

The mediation and communication of climate science

According to Boykoff and Yulsman, 'research spanning the past three decades has consistently found that the general public gains understanding of science (and more specifically climate change) largely through mass media accounts' (2013: 2). They correctly place public reliance on media reporting in the context of the political economy of the mass media, pointing to disinvestment in news-gathering and a decline in specialist correspondents. This, it is argued, has a negative impact on the ability of the media to scrutinise science and evaluate scientific controversy, thus making news media more reliant on 'information subsidies' from PR and official sources, or 'churnalism' (Davies 2008; Lewis *et al.* 2008; Miller and Dinan 2008; Dinan and Miller 2009). It also means that there is less capacity to analyse strategic communication campaigns targeting the media. Thus, understanding climate communication requires a wider frame of reference than simply analysing media reporting. However, we will begin this analysis by establishing how findings in relation to media coverage of climate issues are consistent with our more holistic approach, which sees the media as one (albeit important) social arena for climate communication.

The literature on media coverage of climate change provides broad agreement on a number of issues. For example, that media coverage of climate has increased since the beginning of the century, and the global patterns of media coverage are similar in that they tend to map onto key events such as intergovernmental conferences, IPCC assessment reports and controversies such as 'Climategate' (see Figure 18.1 in Chapter 18 by Boykoff, McNatt and Goodman).

There appears to be strong support for agenda setting effects in relation to climate issues, with public concern strongly correlated with media attention (Brulle *et al.*, 2012). The journalistic norms of balance and conflict have created opportunities for climate contrarian voices to acquire a prominence in the media that is at odds with the marginalisation of their ideas in expert arenas (Oreskes 2004; Boykoff and Boykoff 2007). However, it is difficult to conclude that journalistic norms alone could account for the prominence and efficacy of climate contrarianism. We need in addition to examine how such opinion is organised and disseminated, which is consistent with research on public opinion on this topic that finds that 'science-based information is limited in shaping public concern about the climate change threat. Other, more directly political communications appear to be more important' (Brulle *et al.*, 2012: 185). This connects with another finding in the literature, which suggests an increased role for strategic communication (by think tanks, policy planning groups and non-governmental sources) in the direct publication of news, commentary and analysis on the internet (Boykoff and Yulsman 2013). Such material often has more detailed analysis of the networks and connections of organised climate contrarianism than in the mainstream media. However, once we acknowledge that non-mainstream media are part of this communications complex we also must notice the communicative infrastructure and propaganda capacity (lobbying and public relations) marshalled by corporate (and other) interests in this debate.

Once we move beyond the analytical privilege given to the mainstream media in much communication scholarship, we can refocus on the communicative activities and strategies of social interests. In the case of climate, corporate interests have adopted two main diverging strategies with significant consequences for their communicative activities.

The merchants of doubt and the corporate capture of the climate debate

It was not until the late 1980s that transnational business responded significantly to the threat of climate change. There were different factions and interests within what has been called the corporate 'sustainable development historical bloc' (Sklair 2000) – that set of key corporations that take leadership, planning and influence on sustainable development policy as a mission for themselves, and on behalf of the wider business class. The extractive and automotive industries have interests in climate change policy, given their potential impact on business-as-usual practices and strategies. So do the insurance and reinsurance sectors, though their interests are obviously different. As Pulver puts it, 'competition between firms over conceptions of profitable firm action in the face of an environmental challenge, such as climate change, is a site through which the possibilities and limits of greening capitalism are constituted' (Pulver 2007: 50).

Though there were many initiatives, two are of note. The first was the creation of the contrarian Global Climate Coalition (GCC) in 1989 (see Figure 7.1). The GCC, though part of the attempt by

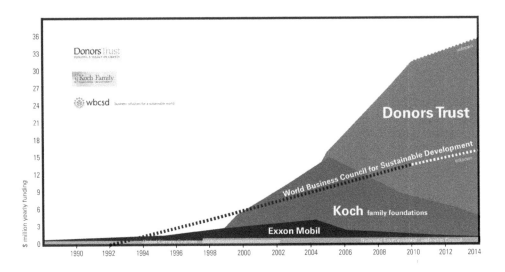

Figure 7.1 Corporate capture vs climate contrarians: timeline of funding

Credit: J. Boehnert, Ecolabs, 2014.

Original source: Climate contrarian funding data from Goldenberg (2013) and Greenpeace (2013). On the WBCSD, Najam (1999) estimates that minus in-kind and other forms of support, membership fees alone amounted to US$3.78 million in 1998. According to data in Corporate Europe Observatory (2010), BASF (€47,539) and E.On (70,000 Swiss francs) paid on average US$70,000 for membership of the WBCSD. By assuming that all members pay the same and multiplying by the number of members (198 in 2010 according to the WBCSD), we estimate tentatively that membership income may have risen to almost US$14 million by 2010.

business to exert environmental leadership, was a short-lived venture. GCC lobbying and PR strategy in the early to mid 1990s was undertaken in the full knowledge that the science and forecasting underpinning climate policy was sound. The GCC's own internal scientific assessment had concluded the threat was real (Revkin 2009) and that 'contrarian theories ... do not offer convincing arguments against the conventional model of greenhouse gas emission-induced climate change' (Bernstein, cited in Powell, 2012: 96).

Nevertheless, the GCC helped to stymie progress on climate issues in the early 1990s. But it was not long before cracks began to appear in the coalition. A key moment came when BP, Shell, Ford and DuPont withdrew in 1997, just as the IPCC was issuing its second report warning of increased concerns over the role of humanity in causing climate change.

The GCC disbanded in 2002, but its demise did not mark the end of contrarianism. Corporates (such as ExxonMobil and Koch Industries) continued to pursue them, but not always too publicly.

The cause of the split is argued by some to mirror the fundamental economic interests of the firms concerned. However, Pulver shows that economically the interests of ExxonMobil, Shell and BP and the balance of their investments (in extraction versus refining for example) were similar. The difference was that European headquartered corporations (BP, Shell) came to a different calculation of what might work politically than did the US headquartered Exxon. As Pulver notes:

> ExxonMobil executives were confident that regulation was unlikely and that opposition to regulation was a viable political strategy. In contrast, for BP and Shell managers, regulation was considered a

foregone conclusion, and the strategy choice centered on the extent to which the companies would participate in shaping the regulation.

(Pulver 2007: 63)

This analysis strongly supports our argument of the importance of ideas and communication in the assessment and identification of corporate interests and the planning of strategies. We can note that Exxon Mobil was never a member of the World Business Council on Sustainable Development (WBCSD), while Shell and BP were involved from the early days.

Shell and BP (and others) did not vacate the policy field when they left the GCC. They repositioned themselves as responsible and enlightened corporate citizens, joining the Business Environmental Leadership Council (BELC) in 1998. Shortly afterwards the elite global policy planning group, the World Economic Forum, identified climate change as the 'most important issue facing business *and the issue where business could most effectively play a leadership role*' (Levy 2005: 78, emphasis added).

More far reaching was the WBCSD. This emerged as a response to the UN initiated Brundtland Report *Our Common Future* (1987). Among a variety of responses the Business Council for Sustainable Development (BCSD) was created in 1990 (Timberlake 2006). It represented corporate interests at the Rio Summit in 1992, securing important industry friendly outcomes. The World Industry Council for the Environment was created by the International Chamber of Commerce in 1993 and then merged with the BCSD in 1995 to form the WBCSD (Najam 1999).

We suggest that the specific outcomes at Rio pale beside the most significant victory which was the corporate capture of the term 'Sustainable Development', altering how it was understood and used in elite debate and practice (Sklair 2000). The environmental movement had posed a challenge: in essence, that the emerging global ecological crisis was caused by global capitalism and that any solution had to confront the capitalist system. In response, leaders of globalising corporations fashioned the idea of sustainable development with the accent not on sustaining the planet and the human species – 'conservation' – but on sustaining development, which came to mean specifically sustaining capitalism with an environmental tinge. As Sklair (2000: 85) describes it: 'From this powerful conceptual base big business successfully recruited much of the global environmental movement in the 1990s to the cause of sustainable global consumerist capitalism.'

It is important to understand this capture of 'discourse' and the realm of ideas is not divorced from practice. The new definition of 'sustainable development' was henceforth the operating assumption of international policy and action. This illustrates the argument we made earlier that ideas and practice are intimately related.

It is difficult to tell how much the corporate capture strategy has cost. There is very little information in the public domain on the budget and spending of the WBCSD (Najam 1999: 76). Capturing sustainable development for the corporate interest requires planning, and active agents who implement strategy. The WBCSD played exactly that leadership and organisational role. Najam estimates that, minus in-kind and other forms of support, membership fees alone amounted to US\$3.78 million in 1998. Our calculations suggest this had risen to almost US\$14 million by 2010,[2] almost half of the known total spent on contrarianism by the opposing corporations (Goldenberg 2013).

Manufacture of doubt

It is important to understand that the role of think tanks, policy planning and lobby groups is multi-dimensional. They aim to dominate the information environment in a number of distinct public and private arenas and to capture policy. Thus, it is important not simply to examine the relative success of contrarianism in relation to media reporting (for example) but in relation also to a wider range of arenas such as the production of scientific knowledge; civil society and the legal system (Miller 1998; Miller and Harkins 2010). In the case of climate change contrarianism, rather than attempt to capture policy, the aim has been to manufacture doubt in order to dissipate pressure for progress and delay meaningful policy decisions (McCright and Dunlap 2010). As one study concludes: 'scepticism is a tactic of an elite-driven counter-movement designed to combat environmentalism … [T]he successful use of this tactic has contributed to the weakening of US commitment to environmental protection' (Jacques *et al.* 2008: 365).

It is important to recognise that climate contrarians are not a collection of disgruntled or alienated individuals who have come together to support each other and engaged in debate about climate science. Instead what we see is a 'movement' of myriad organisations and groups that has been bankrolled by corporations with direct material interests in frustrating climate action, together with a range of conservative foundations funded by individuals, connected to those corporations.

The strategy of fostering doubt is of course familiar from other science related public policy issues, most obviously the debate over the health effects of smoking tobacco (Oreskes and Conway 2010). David Michaels (2008: xii) argues:

> Product defense consultants … have increasingly skewed the scientific literature, manufactured and magnified scientific uncertainty, and influenced policy decisions to the advantage of polluters and the manufacturers of dangerous products. To keep the public confused about the hazards posed by global warming, second-hand smoke, asbestos, lead, plastics, and many other toxic materials, industry executives have hired unscrupulous scientists and lobbyists to dispute scientific evidence about health risks.

Since the split in the corporate community over climate change, around the turn of the century, the contrarian strategy has been developed. It has, however, grown significantly more intense since around 2006 as can be seen in the sheer volume of output from conservative think tanks, which are the overwhelming producers of contrarian books (see Figure 7.2).

It is clear that significant sums of money have been ploughed into the contrarian movement. It is difficult to tell how much, because the funding relations are not transparent. However, the increase in book publication does mirror the increase in known spending.

One of the most significant early funders of climate sceptic think tanks was Exxon Mobil, which published the names of organisations it supported and the amounts it gave them over the years on its website. It was on the receiving end of a barrage of negative publicity and as a result in 2008 stated 'we will discontinue contributions to several public policy research groups whose position on climate change could divert attention from' discussion on how to 'secure the energy required for economic growth in an environmentally responsible manner' (Adam 2008). Exxon did cut some funding streams as a result, though not all (Adam 2009). However, recent data suggests that the decline in Exxon funding has been made up many times over by the oil executives the Koch Brothers and by the hitherto little known Donors Trust, a secretive organisation that seems to exist to attempt to disguise the sources of funding going into climate contrarian causes (and other conservative preoccupations).

Since its creation in 1999, Donors Trust (and the affiliated Donors Capital Fund) has given nearly $400 million to support climate contrarianism. The donors use the Trust as a 'pass-through', according to Marcus Owens, the former director of the IRS Exempt Organizations Division, now in private legal practice. 'It obscures the source of the money', he notes. 'It becomes a grant from Donors Trust, not a grant from the Koch brothers' (Abowd 2013). According to the Centre for Public Integrity, 'donors can open an account and protect their identity from the public and even the recipient of their grants' (ibid.). All these funding connections feed through into a very large-scale effort to foster doubt on the science of climate.

There are a myriad think tanks and other organisations all of which appear to be separate from each other but which are singing from the same hymn sheet. The web of contrarianism is most developed in

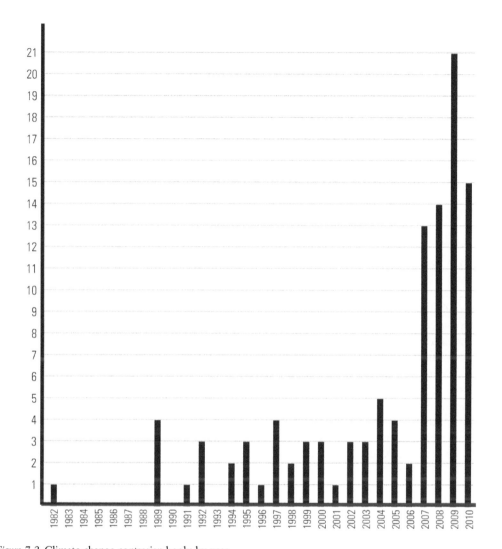

Figure 7.2 Climate change contrarian books by year

Credit: J. Boehnert, Ecolabs, 2014.

Source: Dunlap and Jacques (2013).

the US where large and well-known think tanks such as the American Enterprise Institute, the Heritage Foundation and the CATO Institute work alongside a whole host of lesser known bodies including the Heartland Institute and the Committee for a Constructive Tomorrow. The reach of climate contrarianism is, however, worldwide, with think tanks receiving funding in Australia and all across the EU. The organisations involved try to present themselves as basing their arguments on science. For instance, the UK based Global Warming Policy Foundation claimed that it had found 900+ peer reviewed papers supporting scepticism on climate change and refuting 'concern relating to a negative environmental or socio-economic effect' of climate change 'usually exaggerated as catastrophic' (Global Warming Policy Foundation 2011). However, analysis by the blog *Carbon Brief* (2011) showed that 'nine of the ten most prolific authors cited have links to organisations funded by ExxonMobil, and the tenth has co-authored several papers with Exxon-linked contributors'.[3]

Effectiveness and outcomes

To be successful the strategy of climate contrarianism does not need to convince scientists, policy makers or even a majority of the public. It needs only to foster the conditions under which meaningful action on climate are seen as too difficult or too politically costly. In other words, the strategy is largely elite focused, rather than mainly aiming to influence public opinion. Nevertheless, it does involve relentless advocacy that seeks to influence the news media, public opinion, the scientific debate and most obviously the decision making process. It is notoriously difficult to pin down specific policy effects, but the case of climate contrarianism is unusually clear because of the clarity of the scientific consensus. This is emphasised by the fact that the climate contrarian movement is almost entirely the product of funding from corporations and conservative foundations. We can see this in the finding that some 92 per cent of climate contrarian books surveyed between 1982 and 2010 were published by or through conservative think tanks (Dunlap and Jacques 2013).

As a result when we turn to measures of media coverage or public opinion we can be reasonably sure that climate sceptic views in the US and UK (where the movement is the most active) are in part the product of contrarian communications. Thus Painter and Ashe (2012) found in their examination of coverage in five countries that the USA and UK are 'particularly notable for the presence of sceptics who question the need for strong climate change policy proposals', representing 'more than 80% of [sceptic] voices' in the sample (see Figure 7.3).

We can reasonably conclude that contrarian campaigns in the UK and US have had some effect on popular opinion. It is important to note that this is by no means a majority and polls show that climate scientists are the most credible sources for a significant majority of the population in the US and the UK (as they are in other countries). It needs to be additionally emphasised that there is no clear relationship between public opinion and national, far less international decision making (Miller 1998, 1999).

Turning to policy questions, we can see that the general drift of international policy making is undergirded by the scientific consensus. Whether and to what extent the slow pace of progress is attributable at least in part to contrarian campaigns requires careful analysis as there are a variety of other factors including inertia, geopolitical interests and corporate decision making. However, we can note that some scholars claim that 'the overall activities of the conservative think tanks appear to have played a central role in generating congressional opposition to the Kyoto protocol' (Dunlap and McCright, 2010: 247).

But, considering the impact of contrarian strategies on climate is only to consider one of the two main corporate/conservative strategies we identify. What of the other major approach adopted by corporate actors, namely the corporate capture of environmental policy? Sklair (2000) charts how environmental activism by leading TNCs, dating back to the early 1970s, intensified throughout the 1980s and 1990s, resulting in an important ideological and practical victory wherein the radical 'limits to growth' thesis was first reformulated as sustainable development (1987), then partnered with sustainable consumption, fusing into the common sense, and highly business friendly, notion of sustainability. This discourse is now thoroughly emptied of its original charge (that there are limits to growth and capitalist led development, that growth trumps all other policy, moral and ecological considerations, etc.). Sustainability is now understood as continued growth, but with some optional environmental extras. Establishing this understanding in policy circles is the outcome of enormous communicative effort by corporations and their peak business associations, targeted at key decision making and policy planning fora (such as the World Summit on Sustainable Development and COP conferences), and transmitted via policy planning networks and think tanks.

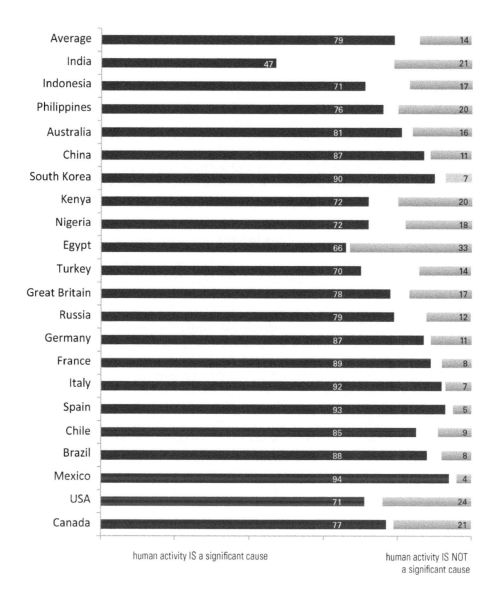

Figure 7.3 Global opinion on human causation of climate change

Credit: J. Boehnert, Ecolabs, 2014.

Source: World Public Opinion (2007).

The corporate response to climate change is shaped by a number of interrelated factors:

> [S]trategy was decided based on socially generated assessments of the state of climate science, the likelihood of greenhouse gas regulation, and the level of public interest in the climate issue. Moreover, these assessments reflect the embedded-ness of oil company executives in company-specific scientific

networks and national policy fields and not a global outlook commensurate with the companies' operational reach.

(Pulver 2007: 64)

What the major oil companies hold in common is the pursuit of the most profitable policy on climate – their different strategies reflect their differentiated assessments of how policy, legislation and stakeholder sentiment are likely to move on this issue in the medium term. Levy's (2005: 74–75) research on the oil and automotive industries supports this reading, suggesting that the post 2000 corporate accommodation of climate policy represents the ongoing assembly of an historical bloc involving key corporations, government agencies, NGOs and other intellectuals and experts to establish the norms and policies of a new (and clearly neoliberal) climate regime. The emerging worldview is one where climate mitigation is understood in terms of ecological modernisation, allowing for 'win–win' scenarios for those businesses best able to adapt.

Conclusions

Our examination of the communication of corporate climate strategy has focused on the communicative strategies adopted by the oil industry in particular over the question of climate change and renewable energy. We have not discussed how the activities of the industry have been modulated by other factors including pressures from government and international decision making fora, or from civil society and popular opinion. We do not dismiss these as irrelevant but our brief to examine corporate communication activities meant we focused on the two major strategies adopted by the oil industry following the split in the industry in the 1990s. What we see is the determined attempt by one faction (represented by ExxonMobil and the Koch Brothers in particular) to deny the science on climate, presumably making the calculation that this has a chance of political success in the US where they are headquartered. By contrast, the other industry faction notably associated with Shell and BP (and many others) has adopted a strategy of some investment in renewables and an acceptance that climate change is happening. They have devoted their attention to inflecting environmentalism so that it does not threaten profit making. This is evident in the corporate capture of the practical meaning of the term sustainable development and the widespread adoption of market based 'solutions' to climate by governments. The heightened role for corporate social responsibility among these companies is not an aberration but strongly related to their strategic attempt to avoid regulatory impacts on their business model.

Our key conclusion is that it is important to examine communication throughout environmental economics, politics and culture and not just in relation to the mass media or the field of journalism. For us, communication power is about both process and outcomes that are not independent of each other but intimately related precisely by circuits of communicative power (Philo *et al.* 2014).

Notes

1   In this paper, following O'Neill and Boykoff (2010), the term 'contrarian' will be used since it more adequately and specifically refers to those 'who critically and vocally attack climate science' as opposed to those who are misinformed, unconvinced or properly 'sceptical' about matters of public debate.
2   Based on averaging disclosures of their membership payments by two member companies for the year 2010 and multiplying by the number of members in that year (198) (Corporate Europe Observatory 2010).
3   Note that this analysis has been 'rebutted' by the compilers of the list. They conclude that 'The scientists unjustly attacked in the Carbon Brief article are not "linked to" [funded by] ExxonMobil. The Carbon Brief and any other website perpetuating this smear should issue a retraction' (Are Skeptical Scientists Funded by ExxonMobil?, *Popular Technology*, 10 May, 2011, www.populartechnology.net/2011/05/are-skeptical-scientists-funded-by.html). It can be noted that in the cases cited in this article, there is no reason to doubt that both Idso and Michaels have been funded by ExxonMobil since they have both admitted it and the evidence for this was included via links in the original Carbon Brief report.

Further reading

Jacques, P. J., Dunlap, R. E. and Freeman, M. (2008) 'The organisation of denial: Conservative think tanks and environmental scepticism', *Environmental Politics*, 17(3): 349–385.

This article examines of the role of think tanks in promoting climate scepticism in the United States, highlighting the role of conservative think tank networks in fostering doubt and denial of climate science.

Miller, D. and Dinan, W. (2008) *A Century of Spin*, London: Pluto Press.

This book offers an original analysis of the growth of corporate public relations and examines the role of policy planning elites and corporate propagandists in promoting and advancing neoliberal ideas.

Pulver, S. (2007) Making sense of corporate environmentalism', *Organization and Environment*, 20(1): 44–83.

This article compares the different strategies adopted by key transnational corporations in response to climate issues, and argues persuasively for a nuanced reading of corporate policy that considers the professional, political and regulatory cultures that corporate decisions makers operate in.

Sklair, L. (2001) *The Transnational Capitalist Class*. Oxford: Blackwell.

This book develops a conceptual model for understanding corporate led globalisation and offers a penetrating analysis of the corporate capture of sustainability discourse.

References

Abowd, P. (2013) 'Donors use charity to push free-market policies in states', Centre for Public Integrity, 14 February. Available at: www.publicintegrity.org/2013/02/14/12181/donors-use-charity-push-free-market-policies-states [Accessed 6 September 2013].

Adam, D. (2008) 'Exxon to cut funding to climate change contrarian groups', *Guardian*, 28 May. Available at: www.theguardian.com/environment/2008/may/28/climatechange.fossilfuels [Accessed 5 September 2013].

Adam, D. (2009) 'ExxonMobil continuing to fund climate sceptic groups, records show', *Guardian*, 1 July. Available at: www.guardian.co.uk/environment/2009/jul/01/exxon-mobil-climate-change-sceptics-funding [Accessed 5 September 2013].

Boykoff, M. and Boykoff, J. (2007) 'Climate change and journalistic norms: A case-study of U.S. mass-media coverage', *Geoforum*, 38:1190–1204.

Boykoff, M. and Yulsman, T. (2013) 'Political economy, media, and climate change: Sinews of modern life', WIREs Climate Change 2013, John Wiley & Sons. Available at: http://sciencepolicy.colorado.edu/admin/publication_files/2013.19.pdf [Accessed 4 September 2013].

Brulle, R. J., Carmichael, J. and Jenkins, J.C. (2012) 'Shifting public opinion on climate change: An empirical assessment of factors influencing concern over climate change in the U.S., 2002–2010'. *Climatic Change*, 114(2): 169–188.

Carbon Brief (2011) 'Analysing the "900 papers supporting climate scepticism": 9 out of top 10 authors linked to ExxonMobil', 15 April, 14: 24. Available at: www.carbonbrief.org/blog/2011/04/900-papers-supporting-climate-scepticism-exxon-links [Accessed 15 January 2015].

Corporate Europe Observatory (2010) Concealing their sources – who funds Europe's climate change deniers? December. Available at: http://corporateeurope.org/sites/default/files/sites/default/files/files/article/funding_climate_deniers.pdf [Accessed 15 January 2015].

Davies, N. (2008) *Flat Earth News*, London: Chatto & Windus.

de Solla Price, D. J. (1963) *Little Science, Big Science*. New York: Columbia University Press.

de Solla Price, D. J. (1986) *Little Science, Big Science – and Beyond*. New York: Columbia University Press.

Dinan, W. and Miller, D. (2009) 'Journalism, public relations and spin'. In Karin Wahl-Jorgensen and Thomas Hanitzsch (eds), *Handbook of Journalism Studies*. New York: Routledge, pp. 250–264.

Dunlap, R. E. and Jacques, P. J. (2013) 'Climate Change Denial Books and. Conservative Think Tanks: Exploring the Connection,' *American Behavioral Scientist*, 57(6): 699–731.

Dunlap, R. E. and McCright, A. M. (2010) 'Climate change contrarian: Sources, actors and strategies'. In Constance Lever-Tracy (ed.), *Routledge Handbook of Climate Change and Society*. Abingdon: Routledge, pp. 240–259.

Fleck, L. (1979) *The Genesis and Development of a Fact*. Chicago, IL: University of Chicago Press (reprint: original 1935).

Global Warming Policy Foundation (2011) '900+ Peer-Reviewed Papers Supporting Skepticism Of "Man-Made"' Global Warming Alarm' Thursday, 14 April, Retrieved from the Internet Archive of 4 January 2012. http://web.archive.org/web/20120104041631/http://www.thegwpf.org/science-news/2816-900-peer-reviewed-papers-supporting-skepticism-of-qman-madeq-global-warming-agw-alarm.html [Accessed 5 September 2013].

Goldenberg, S. (2013) 'Secret funding helped build vast network of climate denial thinktanks'. *Guardian*, 14 February. Available at: www.theguardian.com/environment/2013/feb/14/funding-climate-change-denial-thinktanks-network?INTCMP=SRCH [Accessed 5 September 2013].

Greenpeace (2013) 'Donors Trust: The shadow operation that has laundered $146 million in climate-denial funding', *Greenpeace Briefing*. Available at: www.greenpeace.org/usa/Global/usa/planet3/PDFs/DonorsTrust.pdf [Accessed 15 January 2015].

Haas, P. M. (1992) 'Banning chlorofluorocarbons: Epistemic community efforts to protect stratospheric ozone'. *International Organization*, 46: 187–224.

Jacques, P. J., Dunlap, R. E. and Freeman, M. (2008) 'The organisation of denial: Conservative think tanks and environmental scepticism'. *Environmental Politics*, 17(3): 349–385.

Lewis, J., Williams, A and Franklin, B. (2008) 'Compromised Fourth Estate? UK news journalism, public relations and news sources'. *Journalism Studies*, 9(1): 1–20.

Levy, D. L. (2005) 'Business and the evolution of the climate regime: The dynamics of corporate strategies'. In D. L. Levy and P. J. Newell (eds), *The Business of Global Environmental Governance*, Cambridge, MA: MIT Press, pp. 73–104.

McCright, A. M. and Dunlap, R. E. (2010) 'Anti-reflexivity: The American Conservative movement's success in undermining climate science and policy'. *Theory, Culture and Society*, 27(2–3): 100–133.

Mannheim, K. (1927) 'Conservative thought', in K. Wolff (ed.) (1971) *From Karl Mannheim*. Oxford: Oxford University Press, pp. 132–222.

Michaels, D. (2008) *Doubt is Their Product: How Industry's Assault on Science Threatens Your Health*. Oxford: Oxford University Press.

Miller, D. (1998) 'Mediating science: Promotional strategies, media coverage, public belief and decision making'. In E. Scanlon, E. Whitelegg and S. Yates (eds), *Communicating Science: Contexts and Channels*. London: Routledge, pp. 206–226.

Miller, D. (1999) 'Risk, science and policy: Definitional struggles, information management, the media and BSE'. *Social Science and Medicine*, 49(9): 1239–1255.

Miller, D. (2002) 'Media power and class power: Overplaying ideology'. In L. Panitch and C. Leys (eds), *Socialist Register 2002: A World of Contradictions*. 38. London: Merlin Press.

Miller, D. and Dinan, W. (2008) *A Century of Spin*. London: Pluto Press.

Miller, D. and Harkins, C. (2010) 'Corporate strategy and corporate capture: Food and alcohol industry lobbying and public health'. *Critical Social Policy*, 30(4): 564–589.

Mol, A. P. I. (2001) *Globalisation and Environmental Reform: The Ecological Modernization of the Global Economy*. Boston, MA: MIT Press.

Najam, A. (1999) 'World Business Council for Sustainable Development: The greening of business or a greenwash?'. In Helge Ole Bergesen, Georg Parmann and Øystein B. Thommessen (eds), *Yearbook of International Co-operation on Environment and Development 1999/2000*. London: Earthscan, pp. 65–75.

O'Neill, S. J. and Boykoff, M. (2010) 'Climate denier, skeptic, or contrarian?' *Proceedings of the National Academy of Science, USA. September 28: 107(39)*. Available at: www.ncbi.nlm.nih.gov/pmc/articles/PMC2947866/ [Accessed 5 September 2013].

Oreskes, N. (2004) 'The scientific consensus on climate change'. *Science*, 306(5702): 1686.

Oreskes, N. and Conway, E. M. (2010) *Merchants of Doubt: How a Handful of Scientists Obscured the Truth on Issues from Tobacco Smoke to Global Warming*. London: Bloomsbury Press.

Painter, J. and Ashe, T. (2012) 'Cross-national comparison of the presence of climate scepticism in the print media in six countries, 2007–10', *Environmental Research Letters*, 7(4). Available at: http://iopscience.iop.org/1748-9326/7/4/044005/article [Accessed 15 January 2015].

Philo, G., Miller, D. and Happer, C. (2014) 'Circuits of communication and structures of power: The sociology of the mass media'. In G. Philo, D. Miller and C. Happer (eds), *Contemporary Sociology*. Cambridge: Polity Press.

Plewhe, D. (2009) 'Introduction'. In P. Mirowski and D. Plewhe (eds), *The Road from Mont Pelerin: The Making of the Neoliberal Thought Collective*. Cambridge, MA: Harvard University Press.

Powell, J. L. (2012) *The Inquisition of Climate Science*. New York: Columbia University Press.

Pulver, S. (2007) Making sense of corporate environmentalism'. *Organization and Environment*, 20(1): 44–83.

Revkin, A. C. (2009) 'Industry ignored its scientists on climate'. *New York Times*. 23 April. Available at: www.nytimes.com/2009/04/24/science/earth/24deny.html?_r=0 [Accessed 15 January 2015].

Royal Society (2010) *Climate Change: A Summary of the Science*. September. London. Available at: http://royalsociety. org/uploadedFiles/Royal_Society_Content/policy/publications/2010/4294972962.pdf [Accessed 5 September 2013].

Simionis, U. E. (1989) 'Ecological modernisation of industrial society: Three strategic elements'. *International Social Science Journal*, 121: 347–361.

Sklair, L. (2000) 'The transnational capitalist class and the discourse of globalization'. *Cambridge Review of International Affairs*, 14(1): 67–85.

Sklair, L. (2001) *The Transnational Capitalist Class*. Oxford: Blackwell.

Timberlake, L. (2006) *Catalyzing Change: A Short History of the WBCSD*, Geneva: WBCSD.

World Public Opinion (2007) 'Developed and developing countries agree: Action needed on global warming'. *World Public Opinion.org*, September 24. Available at: www.worldpublicopinion.org/pipa/articles/btenvironmentra/412. php?lb=bte [Accessed 5 September 2013].

8

TRANSNATIONAL PROTESTS, PUBLICS AND MEDIA PARTICIPATION (IN AN ENVIRONMENTAL AGE)

Libby Lester and Simon Cottle

In Turkey, protests that began over trees and the redevelopment of an Istanbul park escalate: tear gas, water cannons, molotov cocktails, armoured police. Officials blame 'terrorist elements' for the violence. The Prime Minister says protesters are 'undemocratic'; Twitter is a 'menace'. Staff at television station NTV resign over the station's lack of coverage of the protests, and management apologizes to viewers (*Guardian* 2013a; BBC News 2013). Thousands are injured across Turkey, and deaths are reported. International news outlets begin their stories on the protests with 'On social media …'. The magazine website, *Vice*, tackles what it describes as a 'hard' issue by live streaming reports from its Istanbul correspondent wearing an adapted version of Google Glass. The correspondent, however, encounters a quiet night. The *Guardian* reports this journalistic experiment on its UK pages and website: 'Protest Power', hands-free reporting, immediate uploading (when the network is reliable) (*Guardian* 2013b). The story links to news sites across the world, including the new *Guardian Australia*, financed by the creator of Australia's most popular last-minute travel website.

In Chile, water cannons are also in use as 'Patagonia without Dams' marchers walk towards the presidential palace. The movement against the multi-billion dollar HidroAysen dam complex proposed for the wild southern-most regions of the Americas spreads from Santiago to other cities in Chile (*New York Times* 2011). Advertisements showing lights failing during a surgical operation or saturation media images of Japan's post-tsunami nuclear disaster have failed to convince these protesters that the additional 18,000-plus gigawatts a year of hydro-electric power is necessary or desirable for their country. In Spain, a small group of protesters armed with a megaphone and smart phones provide modest support.

In China, the government backs down on its approval of a copper plant in Sichuan province after 'thousands of protesters' 'reportedly' take to the streets of Shifang. China is one of the world's biggest polluters. Its citizens are the biggest users of the web. For one day, 'Shifang' is the most searched term on microblogging site Weibo. Three days of street protests: teargas, stun grenades, stormed buildings, smashed vehicles. Chinese authorities and protest organisers disagree on crowd size (BBC News 2012); it might be possible to quantify protest-related activity on Weibo, but it is still impossible to accurately count the number of people who physically take to the streets. Later, the government backs down again following mass protests over a coal-fired power station near Hong Kong, and local officials announce a petrochemical plant in Dalian will close after 12,000 'reportedly' protest.

In Russia, the 'Arctic 30' – as Greenpeace has dubbed the crew of its vessel *Arctic Sunrise* boarded by Russian authorities while protesting against oil drilling – are imprisoned. The international crew, including a photographer and 'videographer', faces a range of charges, including piracy. News media circulate claims and counter-claims about Greenpeace activities, including descriptions of the drugs found on board – morphine for medical supplies, says Greenpeace (Darby 2013). Three weeks after the crew's arrest, protest events are held in 48 countries in a 'Global Day of Solidarity'. YouTube, Google Hangout, Facebook, Twitter are among the deployed media. Greenpeace International's website provides blow-by-blow accounts and audiovisual material describing events – 'OCT 03 2013 14.15 Crew members Colin Russell from Australia and Andrey Allakhverdov from Russia are officially charged with piracy'. This is reused and recirculated by journalists. More than 10,000 people like a photograph posted on Facebook of a giant banner displayed at the football 'Don't Foul the Arctic – #Free the Arctic 30'. More than 400,000 like 'Save the Arctic'. Ingrid comments: 'F... the Russians'. Angelique responds: 'This is not going to help'.

These are a few of the many faces of mediated environmental protest in the second decade of the twenty-first century. There are recognizable practices and patterns: the labels of deviancy and damage, of illegal and uncivil activity, of conflict even over crowd size. There are powerful symbols that carry across media, evoking engagement, emotion and response. There are elite interests, hegemonic discourses, and political actors and lay voices that are able to break through to promote change. But there are also new political configurations and media practices. These, we argue in this chapter, concern the politics of connectivity and representation and how they are playing out across the still emerging transnational networks and publics. The following first outlines our understanding of the relationship between protest, transnational publics and changing media practices and technologies, before illuminating some of these complex shifts in a case study of environmental protests concerning three decades of conflict over native forest logging in Tasmania, Australia's southern island state. This case exemplifies not only the complex and shifting deployments of communications power across this period but also how new communication interplays can now scale-up protests from the local and national to the international and transnational levels (see also Lester, Chapter 19, this volume).

Protest, representation and connectivity

Across recent decades, as we have heard, protests and demonstrations have increasingly sought to highlight issues of global scope and transnational concern (Cottle and Lester 2011a, 2011b). Global justice and inequalities of trade, war and peace, human rights and humanitarian catastrophes as well as the environment and climate change have all, for example, been the focus of major protests in recent years. The crises and conflicts spawned by today's globalizing late modernity, it seems, can also summon into being new social movements, coalitions of opposition and voices of dissent worldwide. These can be globalizing in their communicative intent or forms of political action, lending substance to claims about emergent 'global civil society' and 'global citizenship'.

Mass protests, coordinated and conducted simultaneously on different continents, in different countries and across major cities have proved capable of mobilizing hundreds of thousands of people, sometimes millions of people, around the globe. Such protests – for example, 'days of action' on climate change – are designed to influence elite decision making and challenge the deliberation of policies behind closed doors, policies that if implemented will impact populations internationally or even globally. And some protests, though involving a handful of protesters only, can also now resonate nationally, internationally and transnationally – whether, for example, local environmentalists camping in trees in ancient Tasmanian forests threatened by the logging industry, or activists aboard the Sea Shepherd attempting to disrupt the slaughter of whales in the 'global commons' of the Southern Ocean.

Protests today can both engage and instantiate global forces of change – even when they are enacted locally or directed against national institutions and governments, as we shall pursue further below (see also della Porta and Tarrow 2005; Tarrow 2006). Through creative protest repertoires, campaigners seek to bring recognition to issues of global concern, secure legitimacy for their cause and mobilize identities and voices of support worldwide. The 'transnational' in the contemporary politics of protest, however, is not only characterized by political reach, motivating ethics or geographical scale – all of which invariably extend beyond sovereign national borders and parochial frames of national understanding. Crucially the 'transnational' in transnational protests and demonstrations, we argue, fundamentally inheres within *how* they become communicated and mediated around the globe.

It is in and through the flows and formations of worldwide media and communication networks that transnational demonstrations today principally become *transacted* around the planet. Though physically enacted in particular locales, cities, countries or indeed simultaneously on different continents, it is by means of contemporary communication networks and media systems that they effectively become coordinated, staged for wider audiences and disseminated around the world. And it is here too that they often discharge their affect and effects on supporters, wider publics and different decision makers; whether by redefining the terms of public discourse, bolstering solidarities and mobilizing identities of opposition, or shifting cultural horizons and seeking to influence political elites and government policies.

Through the internet and new social media, protests and demonstrations today can be coordinated and communicated internationally and it is by these same means that some also become exclusively conducted – whether for example, through online petitions, the mobilization of consumer boycotts or digital *hacktivism*. New digital means of recording, storing and disseminating images of protest have also eased the practical, if not ideological, cross-over of scenes of dissent into wider communication flows and mainstream media. And traditional print and broadcasting media continue to perform a critical role in defining, framing and dramatizing protests and demonstrations and, thereby, helping to publicly legitimize or de-legitimize them for mass audiences and readerships. It is in and through this fast-evolving complex of interpenetrating communication networks and media systems, then, that protests and demonstrations today principally become transacted around the world.

The transnational, as ethico-political imaginary (of what should be) and as collective political action (the struggle to bring this about), becomes *instantiated* within and through the communicative enactments of protest and demonstration – if only momentarily or imperfectly. Protests and resistance increasingly serve to focus political and public opinion on crises and issues of international and global concern. Their impacts, courtesy of global media and communication networks, can also register internationally and transnationally. Whether the 'scale shift' is upwards from the local/national to the transnational and global, or downwards from the transnational to the national/local (Tarrow and McAdam 2005), it is in and through communication networks and the pervasive and overlapping media ecology that oppositional currents and movements for change are principally conveyed. And it is here too that protests effectively register more widely. Their political efficacy need not always, however, be measured in terms of exerting measurable, direct and radical effects on political elites or power structures. Such moments of decisive impact, though not without historical precedents, are in fact historically rare. Transnational protests, like protests more generally, can be deeply implicated in processes of change nonetheless: building and mobilizing support, redrawing cultural and political horizons, influencing corporate and industrial behaviour, or even helping to create a pre-figurative politics based on participatory practices and ethical norms of equality and justice.

Lest we should succumb to either blanket pessimism about the power of protests and demonstrations or naïve celebration of the same, we need to situate them in relation to wider historical forces of change. We need to contextualize their mobilizing force within civil society(ies) and attend to the permeability of surrounding political structures and institutions to the voices of dissent. It would be simplistic therefore to approach, much less seek to explain or theorize, transnational protests as a simple reflex of available media technologies. Contemporary communication and delivery systems can certainly facilitate and shape the communicative forms and enactments of protests, but this should not be interpreted as a straightforward causality, much less media and communications determinism. Composed of overlapping media formations, both old and new, and accelerating, horizontal and vertical communications flows incorporating virtual, interactional and user-generated capabilities, today's complex media and communications ecology offers unprecedented opportunities for the wider enactment and diffusion of political protest – from the local to the global. But how these opportunities become realized, negotiated, challenged and contested in practice and with what repercussions on protest movements and coalitions, on their goals and tactics, supporters, publics and relevant elites, clearly demands detailed empirical exploration and careful theorization as the discussion below will demonstrate.

The complex and evolving interplay between protests and media and communications cannot be reduced to one-size fits all. Coordinated, networked actions spearheaded by new transnational movements and loosely affiliated campaign groups, such as the global justice movement, for example, can stage sieges and spectacular 'image events' (DeLuca 1999; Opel and Pompper 2003; de Jong, Shaw and Stammers 2005). These are often deliberately performed in front of the world's media or uploaded to them shortly thereafter on the basis of a media-savvy understanding of journalist news values and news organizations' predilections toward the dramatic, spectacular and revelatory. But mass actions can now also be recorded and communicated directly by protesters through new social media and alternative news sites, as well as by members of the general public who 'bear witness' via mobile phones, digital cameras or pocket videocams. When uploaded and circulated via the internet, these new forms of participation can serve to alter the balance of communicative power (McNair 2006; Allan and Thorsen 2009; Gowing 2009).

Though protests and their informing politics continue to be 'symbolically annihilated' in some national media, given their perceived assault on institutions of governance and the legitimacy of authorities, they nonetheless often 'leak out' into the wider circuits of global communications when recorded clandestinely. First-hand accounts and incriminating images of repressive state actions, corporate violence or individual acts of brutality can now all be captured and communicated almost instantaneously via the internet. These, moreover, can be circulated to émigré and diasporic communities, publics and political leaders around the globe and, importantly, can also re-enter the communications environment of the country from whence they secretly came, documenting the latest events on the ground and contributing to processes of protest diffusion and the mass mobilization of civil societies. In such ways local and global communications can interpenetrate, conditioning each other and intensifying political pressures for change. They facilitate the new 'civilian surge' (Gowing 2009) and the wider 'transformation of visibility' (Thompson 1995) of late modern, mediated, societies, and by these means they can seemingly empower the formerly communications disenfranchised.

The power relations that constitute the foundation of all societies, argues Manuel Castells, 'are increasingly shaped and decided in the communications field' (Castells 2007: 239, 2009). The conflicts and crises that have become characteristic of our global age can all become the focal point of protests. These often seek to move beyond and above 'the national' and register transnationally. But how exactly they become transacted and reverberate from the local to the global demands detailed and theoretically informed examination. Our next section helps to illuminate the changing nature of protest communications and the challenges and opportunities involved in today's more complex communication environment.

Environmental protests and transnational publics

Our case centres on Australia's longest running environmental dispute – the three-decade-long Tasmanian forests conflict – and, specifically, events that saw major Japanese companies stop buying Tasmanian wood products in 2012 following pressure from environmental activists.[1] This pressure was multifaceted, involving direct action and other protests, cyber-action campaigns, face-to-face meetings, and the production and circulation of data and images that stood as evidence of contested logging practices. Our case also locates environmental politics, protests and publics within the broader contexts of shifting relations between global and local risks and public concerns, the emergence of transnational and global governance and decision making, the breaking down and reconstruction of nationally bounded civil societies, and the pressures on landscapes and communities brought about by expanding resource procurement and global trade.

The period from early 2010 is notable for including the biggest downturn in demand in the history of the Tasmanian forestry industry. While initially the slowdown was blamed solely on the global financial crisis, it eventually became clear that international discomfiture over the procurement of woodchips sourced from native forests was a contributing cause. Tasmania's largest company and land owner, and the world's largest exporter of hardwood native woodchips, Gunns Ltd, belatedly replaced its hardline chief executive and board members, and announced that it was withdrawing from all native forest logging and woodchip exports in order to win a 'social license' for its $2 billion-plus pulp mill proposal (Stedman 2010). Nevertheless, the company failed to find a financial backer and suffered a steep decline in share value until it was eventually placed in receivership in September 2012. Meanwhile, industry and environmental groups began historic roundtable 'peace talks' with the stated aims of putting the forestry industry on a sustainable footing and ending community and political conflict over the forests (Lester and Hutchins 2012).

As Gunns retreated from centre stage, another company emerged to take a lead role in the Tasmanian forestry industry and accompanying environmental conflict. Like Gunns, Ta Ann Tasmania enjoyed strong support from the federal and Tasmanian Government and Opposition, including AUD10.4 million in establishment grants for eucalypt veneer mills and a 20-year guaranteed resource supply (Forestry Tasmania 2012). Via a relatively complex supply chain, Ta Ann Tasmania – an offshoot of Malaysian company Ta Ann Holdings, one of six major forest companies in Sarawak – supplied wood from Tasmanian regrowth and plantation eucalypt forests as veneer to Japanese manufacturers and retailers of flooring. This market comprised approximately two-thirds of Ta Ann Tasmania's business, which it claimed contributed a total of AUD45 million annually to the Tasmanian economy (Ta Ann Tasmania 2012).

By 2011, Ta Ann Tasmania had the full attention of environmental protest groups. The first issue at the core of these protests was that Ta Ann's wood supply was being sourced from forests earmarked for protection under the terms of agreement being negotiated as part of 'peace talks' between industry and major environmental groups (Tasmanian Government 2011). The second related to certification and definitions surrounding forests and wood supply. Ta Ann Tasmania's preferred certification (PEFC) is considered 'certification-lite' by environmental groups, compared to the Forests Stewardship Certification (Gale and Haward 2011). Industry-accepted definitions of 'regrowth' are also contested: defined as including a 'majority of trees less than 110 years old', they 'may contain scattered individuals or stands of ecologically mature trees' (Forestry Tasmania n.d.). In Tasmania, this can mean the logging of trees almost 80 metres in height and hundreds of years old.

Protest activity throughout this period was designed for news media and web delivery via, in particular, YouTube, and included spectacular raids on loading equipment at commercial wharves and on logging machinery in the forests (see, for example, ABC News 2012a). The period was also notable for the production of a report 'Behind the Veneer: Forest Destruction and Ta Ann Tasmania's Lies', by the Huon Valley Environment Centre, a direct action protest group outside the formal peace talks process. The report, based in part on information obtained via Right To Information laws, made a number of specific claims, including that Ta Ann was 'misleading' its customers by describing its 'eco-plywood' products as environmentally friendly when they were not sourced from plantations as claimed. As such, Ta Ann Tasmania was a 'major driver of forest destruction in Tasmania' (Huon Valley Environment Centre 2011). The report was released in early October 2011 to a small group of supporters in a hired conference room in central Hobart. Interestingly, no Tasmanian media covered the event (interview 17 April 2012).

The report, however, was greeted with some interest among environmental groups in Japan – in particular, the Japan Tropical Forest Action Network (JATAN) – which had been monitoring the situation in Tasmania with increasing focus since Ta Ann had expanded its operations. These groups organized for Tasmanian activists to meet with contacts in Japanese companies buying and selling Ta Ann Tasmania products, including Panasonic, Daiwa House, Sekisui House and Edai. According to the activists, the company representatives asked only 'simple questions', such as: 'Is it really not plantations?' (interview 17 April 2012). The visit prompted limited news coverage from the major Japanese news agency, Kyodo News, published under a heading translated as 'Eco-certification is dubious' (*Shizuoka News* 2011) and later in the influential CSR publication *Nikkei Ecology* (2012).

Meanwhile, physical direct action protests and market campaigns, including cyber protests, ramped up. The focus of this activity was Miranda Gibson, who on 14 December 2011 began a tree-sit 60 metres above ground in a forest earmarked for logging under the security-of-supply clause of the interim peace talks agreement (see http://observertree.org; Lester and Hutchins 2013). The choice of location for the protest was also determined in part by the construction of a telecommunications tower nearby, allowing Gibson access to the internet and mobile communications. Gibson, who eventually spent 14 months in the tree, contributed her many spare hours to online campaigning. This included a cyber-action campaign that targeted the Japanese companies selling Ta Ann products. The website generated letters in Japanese to undisclosed 'influential' contacts within the companies (interview 23 May 2012). By May 2012, there had been 2000 responses, largely from Australian supporters (interviews 23 May 2012, 12 June 2012). Gibson's protest action was also reported by major news media outlets, including *The Guardian* and CNN.

In February 2012, Ta Ann Tasmania issued a media release announcing that it had been forced to shed a shift of up to 40 jobs across its two mills as sales of its products to Japan had fallen by 50 per cent (Ta Ann Tasmania 2012). It blamed 'persistent market attacks on our customers by environmental groups'. Industry lobby groups went further:

> The head of Tasmania's Forest Industries Association, Terry Edwards, says the jobs were slashed because of the international campaigns. 'We've now seen Ta Ann, who was attracted down here to Tasmania by the Tasmanian and Commonwealth Governments, who were asked to invest here, and they too have been subject to quite a dishonourable campaign, in Japan by extremist environmental groups. They lost their jobs as a direct result of that campaign,' he said.
>
> (ABC News 2012b)

In the following weeks, forest workers and their supporters held a series of angry protests outside activist offices, and accusations of 'eco-terrorism', 'blackmail' and 'sabotage' increased in regularity (see, for example, FIAT 2012; O'Shea 2012; ABC Tasmania 2012). Edwards also withdrew his group from the 'peace' negotiations until 'environmentalists stop protesting' (FIAT 2012; ABC PM 2012).

While Tasmanian environmentalists were facing the impact of their actions in Japan, they were also expressing some confusion about how events had played out (interviews 17 April 2012, 23 May 2012). A lack of information emanating from Ta Ann's Japanese customers was a chief cause. When a UK flooring company had pulled out of an agreement to use Ta Ann Tasmania's products for an Olympic practice basketball court in 2011, its managing director made widely reported comments about its decision to 'buy products in a responsible manner' and in acknowledgement of 'the controversy' (ABC AM 2011; see also *The Independent* 2011). In contrast, there had been only silence from Japan. Environmentalists could only guess which companies had pulled out and why. Japanese corporate culture is notable for its closeness and privacy (Freeman 2000: 12–13; Kingston 2004); corporate silence is accepted more so than in a Western context where commercial-in-confidence often needs to be incited. This is magnified in situations of environmental conflict, not least because Japan with a relatively high population and small land size is reliant on maintaining stable and long-term resource procurement arrangements across a range of countries and regions, and companies will therefore seek to maintain trust in their relationships with foreign governments and investors (interviews 12 June 2012, 13 June 2012, 14 June 2012).

The relatively low response rate to the cyber-action campaign caused further confusion. Australian campaigners considered that, at 2000, the number of form letters emailed to the Japanese companies was too low to place significant pressure on their targets, yet activists had received feedback through their contacts in Japan that they had 'hit the right note' (interview 23 May 2012):

> We still don't know this but it seems from sort of anecdotal evidence that cyber actions aren't … as big in Japan as they are here, so that may have something to do with it, in terms of just an incredibly different cultural landscape and business landscape.
>
> (Interview 23 May 2012)

Japanese environmental NGOs have very small memberships: Friends of the Earth, for example, has 500 members in Japan; JATAN has 200; and Greenpeace claims 5000 Japanese supporters (interviews 12 June 2012, 13 June 2012a, 13 June 2012b). Many environmental NGOs are dependent on foundation and corporate grants for their continued operations, and rely on developing direct relationships with corporations to lobby for environmental change (interviews 12 June 2012, 13 June 2012a). Apart from an occasional banner targeted at a government office or corporate headquarters, very few physical protests are held. Even the post-Fukushima protests that attracted numbers in the tens of thousands could be considered poorly attended, given the magnitude of social disquiet caused by government and corporate handling of the disaster (interview 14 June 2012; Kato 2013). Within this context, the cyber-action campaign rallied a large number of supporters. The letters were also highly targeted, directed in Japanese at individual decision makers within the companies. Likewise, while there was a relatively low amount of mainstream media interest, local Japanese activists saw the *Nikkei Ecology* story as a major achievement. Japanese media and journalists are often accused of disinterest in pricking government or corporate power (Rausch 2012: 63-65). So when non-elite sources do break through, the counter information that is presented is both noticeable and noticed (interview 14 June 2012).

Back in Australia, the terms for acceptable protest and its transnational carriage were debated in early April when the national public broadcaster, the ABC, made Hobart the location of its live political panel show, *Q&A*. Among the guests were industry lobbyist Terry Edwards, three Tasmanian members of federal parliament, a 'change agent' and Brian Ritchie, the bass player in the 1980s US rock band the Violent Femmes, who had settled in Tasmania and become a prominent member of its arts community (ABC Q&A 2012). Edwards again claimed that the Ta Ann workers – 49 at final count – had lost their jobs as a direct result of a 'quite dishonourable campaign in Japan by extremist environmental groups' that had cost 'two contracts to date'. Independent MP Andrew Wilkie responded:

> Terry, I agree with you that we need to get our industry back on a sustainable footing but you can't blame a handful of activists overseas for the fact that the industry is on its knees at the moment. There are many reasons why the industry…

Minister Collins interrupted to warn of the consequences of allowing the debate to spread beyond Tasmania:

> That's true but you can say to people … that are going overseas that they're actually destroying jobs in Tasmania and the Government has said that we shouldn't have people going overseas trashing the brand of Tasmania because Tasmanians have got a lot to be proud of on a whole range of issues and I don't think it helps the Tasmanian economy to have anybody overseas saying things about Tasmania like that.

The content and style of local debate was also the subject of criticism. 'Change agent', Natasha Cica, speaking directly to the audience, said:

What really worries me about the toxic nature of forestry debate in Tasmania at the moment and it's not the first time it's been toxic is the way that we Tasmanians are turning against each other. I am worried for our society – for our civil society – and for the way we are speaking to and not listening to each other … We need to get better at having a grown-up conversation about this.

Musician Brian Ritchie's intervention was notable for its acceptance of broader parameters for debate:

Terry, I heard you on the radio last week and you said you would not negotiate with the environmental groups unless they stopped campaigning against you … It seems to me in a democracy, protest, free speech, these things are treasured … And I find it problematic for you to refuse to engage in any conversation.

(ABC Q&A 2012)

Here, then, a group of prominent citizens speaking on an elite media forum negotiated the bounds of acceptable debate and protest within the context of the state 'brand' and collective 'good'. In this case, the brand and collective are clearly connected to the economic interests of a geographically and politically defined 'local' – the island state of Tasmania – within the context of transnational representation.

Following the Japanese companies' decisions to stop buying Ta Ann Tasmania products, Tasmania's Deputy Premier and Leader of the Opposition embarked on an eight-day $57,000 trade mission to Japan, China and Southeast Asia (ABC Northern Tasmania 2012; Arndt 2012), in a direct example of formal 'public diplomacy' that attempted to reassure Tasmanian customers and restore damaged markets. In November 2012, Ta Ann Tasmania's executive director admitted that this attempt to 'reassure the markets' was 'clearly not sufficient to change the opinion of the companies involved' (ABC Tasmania 2012).

Conclusion

Around the world, increased resource extraction and procurement are producing flow-on effects in terms of protest activity and conflict over environmental futures. Deploying new forms of media and communication technologies and practices, translocal and transnational communities of concern are emerging to demand a voice, as well as legitimacy and influence, in the negotiations and decisions related to resource and landscape use. Consumer and media power coalesces transnationally to produce market and industrial change, which in turn effectively leads to local political impacts. A lack of resources and poor intercultural knowledge, including cross-language skills and understanding of media and political logics and systems, still hamper political attempts to harness this power, but environmental activists as we have seen are using shared concerns and issues to build networks that operate increasingly and effectively across previous boundaries.

In our case study, Japanese companies on one hand acted as part of a transnational community formed through shared environmental concerns and responsibilities. These companies at the end of complex supply and consumer chains became both, as Nancy Fraser has predicted (2007), targets of the affected (with the affected conceived through both local and global risks), and key decision makers affecting Tasmania's economic and environmental future. On the other hand, government and forest industry representatives in Australia refused to recognize the legitimacy of these emerging transnational publics and thus misunderstood the likely basis of Japanese corporate decision making.

A traditional and expensive bipartisan public diplomacy mission failed – in part, we would suggest, because its focus was on the corporate decision makers and their governments rather than recognizing and acknowledging as legitimate this transnational community of concern. As Melissen (2005) suggests in relation to the 'new public diplomacy':

The explosive growth of non-state actors in the past decade, the growing influence of transnational protest movements and the meteoric rise of the new media have restricted official diplomacy's freedom of manoeuvre. Non-official players have turned out to be extremely agile and capable of mobilizing support at a speed that is daunting for rather more unwieldy foreign policy bureaucracies.

(Melissen 2005: 24)

This failure to recognize the transnational community of protest and concern was evidenced by Australian government and industry condemnation of debate that flowed beyond national boundaries, and the threat to withdraw from 'peace' negotiations if the debate was not actively contained and controlled. Terms such as 'terrorism' and 'sabotage' are still deployed to undermine the transnational media and political flows, just as notions of 'civility' and 'toxicity' can act to delegitimize local debate. Paradoxically, however, these efforts make the transnational an even more powerful ambition for local activists (Lester and Cottle 2011). And the outcomes of such efforts are likely to become even less predictable in the multi-directional and layered communication flows that characterize the new political spaces of environmental protest.

Our study outlined above serves to underline how both 'old' and 'new' media are now at work within contemporary environmental conflicts and how these involve simultaneously a politics of representation and politics of connectivity, with both necessarily involved in the scaling-up and scaling-down of local–global concerns. Media systems and communication networks have become inextricably infused in and are indispensable to protests, publics and media participation.

Note

1 Our data is drawn from a larger Australian Research Council-funded project (DP1095173), 'Changing Landscapes: Online Media and Politics in an Age of Environmental Conflict', conducted 2010–2013 with research partner Brett Hutchins. The research included twelve interviews with environmental activists and CSR representatives in Australia and Japan, direct observation, and the tracing and analysis of a variety of media, political and activist-generated texts as they have flowed translocally, regionally and transnationally. For a more detailed account see Lester (2014), and for an overview of the broader project see Lester and Hutchins (2012). An earlier version of the section, 'Protest, representation and connectivity', appeared in Cottle and Lester (2011b), and the case study draws on the journal article, Lester (2014). Thank you to Peter Lang and *Media International Australia*.

Further reading

Castells, M. (2009). *Communication Power*. Oxford: Oxford University Press.
Castells highlights how social movements, the 'long march of environmentalism' and networked political action are reprogramming cultural codes and political values across the globe.

Cottle, S. and Lester, L. (eds) (2011) *Transnational Protests and the Media*. New York: Peter Lang Publishing.
The authors analyse the relationship between new communications technologies, new forms of protest and the possible emergence of a 'global civil society'.

Kato, K. (2013) 'As Fukushima unfolds: Media meltdown and public empowerment'. In L. Lester and B. Hutchins (eds) *Environmental Conflict and the Media* (pp. 201–214). New York: Peter Lang.
Examining mediated responses to the Fukushima disaster in Japan, Kato draws attention to the differences in cultural and media logics that can hinder transnational media research.

References

ABC AM (2011) 'London Olympic builders boycott Tassie timber', 22 December, accessed 30 August 2012 from www.abc.net.au/am/content/2011/s3396101.htm

ABC News (2012a) 'Anti-logging protesters charged', 12 January, accessed 29 August 2012 from www.abc.net.au/news/2012-01-12/protesters-charged/3769864

ABC News (2012b) 'More jobs go at Ta Ann', 3 April 2012, accessed 30 August from www.abc.net.au/news/2012-04-03/jobs-go-at-ta-ann/3928794

ABC Northern Tasmania (2012) 'Selling Tasmania to South East Asia', 20 February, accessed 16 November 2012 from www.abc.net.au/local/audio/2012/02/20/3435007.htm

ABC PM (2012) 'Tas Timber Talks on Knife Edge', 10 May, accessed 30 August 2012 from www.abc.net.au/pm/content/2012/ s3500188.htm.

ABC Q&A (2012) 'Live From Hobart', 2 April, accessed 30 August 2012 from www.abc.net.au/tv/qanda/txt/s3464006.htm

ABC Tasmania (2012) 'Mornings', 16 November, accessed 16 November from http://blogs.abc.net.au/tasmania/2012/11/mornings-on-demand-131112.html?site=hobart&program=hobart_mornings

Allan, S. and Thorsen, E. (eds) (2009). *Citizen Journalism: Global Perspectives*. New York: Peter Lang.

Arndt, D. (2012) 'Trade mission could be a trip of mixed messages', *The Examiner*, 20 February. Accessed 16 November 2012 from www.examiner.com.au/story/430835/trade-mission-could-be-a-trip-of-mixed-messages/

BBC News (2012) 'China "scraps" Shifang plant after violent protests', accessed 28 November 2013 from www.bbc.co.uk/news/world-asia-china-18700884

BBC News (2013) 'Turkey protests resume in Istanbul after apology', accessed 28 November 2013 from www.bbc.co.uk/news/world-europe-22776946

Castells, M. (2007) 'Communication, power and counter-power in the network society', *International Journal of Communication*, 1, 238–266.

Castells, M. (2009) *Communication Power*. Oxford: Oxford University Press.

Cottle, S. and Lester, L. (eds) (2011a) *Transnational Protests and the Media*. New York: Peter Lang Publishing.

Cottle, S. and Lester, L. (2011b) 'Transnational protests and the media: An introduction'. In S. Cottle and L. Lester (eds) *Transnational Protests and the Media* (pp. 3–16). New York: Peter Lang Publishing.

Darby, A. (2013) 'Russia stirs anger with Greenpeace drugs claim', Sydney Morning Herald, accessed 28 November 2013 from www.smh.com.au/federal-politics/political-news/russia-stirs-anger-with-greenpeace-drugs-claim-20131010-2vbfw.html

de Jong, W., Shaw, M. and Stammers, N. (eds) (2005) *Global Activism, Global Media*. London: Pluto.

della Porta, D. and Tarrow, S. (eds) (2005) *Transnational Protest and Global Activism*. Oxford: Rowman & Littlefield.

DeLuca, K. (1999). *Image Politics: The New Rhetoric of Environmental Activism*. New York: The Guildford Press.

FIAT (2012) 'Media Release: Ta Ann job losses – Government must act', Forest Industries Association of Tasmania, 13 February, accessed 30 August from www.taanntas.com.au/userfiles/Documents/FIAT.pdf

Forestry Tasmania (n.d.) 'Old growth, regrowth, high conservation value – what do they all mean?' accessed 29 August 2012 from www.forestrytas.com.au/international-desk/old-growth-regrowth-high-conservation-value-what-do-they-all-mean

Forestry Tasmania (2012) 'Forest management in Tasmania: The truth', accessed 29 August 2012 from www.forestrytas.com.au/uploads/File/pdf/pdf2012/fm_the_truth_2012_web.pdf

Fraser, N. (2007) 'Transnationalizing the public sphere: On the legitimacy and efficacy of public opinion in a post-Westphalian world', *Theory, Culture and Society*, 24(4), 7–30.

Freeman, L. A. (2000) *Information Cartels and Japan's Mass Media*. Princeton, NJ: Princeton University Press.

Gale, F. and Haward, M. (2011) *Global Commodity Governance: State Responses to Sustainable Forest and Fisheries Certification*. Basingstoke: Palgrave Macmillan.

Gowing, N. (2009) '*Skyful of Lies' and Black Swans: The New Tyranny of Shifting Information Power in Crises*. Oxford: Reuters Institute for the Study of Journalism, University of Oxford.

Guardian (2013a) 'Social media and opposition to blame for protests, says Turkish PM', accessed 28 November 2013 from www.theguardian.com/world/2013/jun/02/turkish-protesters-control-istanbul-square

Guardian (2013b) 'How Vice's Tim Pool used Google Glass to cover Istanbul protests', accessed 28 November 2013 from www.theguardian.com/technology/2013/jul/30/google-glass-istanbul-protests-vice

Huon Valley Environment Centre (2011) 'Behind the veneer: Forest destruction and Ta Ann Tasmania's lies', accessed 30 August 2012 from www.theguardian.com/world/2013/jun/02/turkish-protesters-control-istanbul-squarehttp://mps.tas.greens.org.au/2012/02/behind-the-veneer-report-by-huon-valley-environment-centre/

Independent (2011) 'Olympic athletes to train on timber from "endangered" forests', 8 November, accessed 30 August 2012 from www.independent.co.uk/environment/green-living/olympic-athletes-to-train-on-timber-from-endangered-forests-6258751.html

Kato, K. (2013) 'As Fukushima unfolds: Media meltdown and public empowerment'. In L. Lester and B. Hutchins (eds) *Environmental Conflict and the Media* (pp. 201–214). New York: Peter Lang.

Kingston, J. (2004) *Japan's Quiet Transformation: Social Change and Civil Society in the Twenty-First Century*. New York: RoutledgeCurzon.

Lester, L. (2014) 'Transnational publics and environmental conflict in the Asian century', *Media International Australia, 150*: 67–78.

Lester, L. and S. Cottle (2011) 'Transnational protests and the media: Toward global civil society'. In S. Cottle and L. Lester (eds) *Transnational Protests and the Media* (pp. 287–292). New York: Peter Lang Publishing.

Lester, L. and Hutchins, B. (2012) 'The power of the unseen: Environmental conflict, the media and invisibility', *Media, Culture and Society, 34*(7), 832–846.

Lester L. and Hutchins, B. (eds) (2013) *Environmental Conflict and the Media*. New York: Peter Lang.

McNair, B. (2006) *Cultural Chaos: Journalism, News and Power in a Globalised World*. London: Routledge.

Melissen, J. (ed.) (2005) *The New Public Diplomacy: Soft Power in International Relations*. Basingstoke: Palgrave Macmillan.

New York Times (2011) 'Plan for hydroelectric dam in Patagonia outrages Chileans', accessed 28 November 2013 from www.nytimes.com/2011/06/17/world/americas/17chile.html?pagewanted=all&_r=0

Nikkei Ecology (2012) 'Australian certified wood products accused of environment destruction: eNGOs develop boycott campaign against Japanese customer companies – SMKC, Eidai, Panasonic, Daiwa and Sekisui', May: 45–47.

Opel, A. and Pompper, D. (eds) (2003) *Representing Resistance: Media, Civil Disobedience and the Global Justice Movement*. Westport, CT: Praeger.

O'Shea, D. (2012) 'Logged out', SBS Dateline, accessed 29 August 2012 from www.sbs.com.au/dateline/story/related/aid/617/id/601522/n/The-Last-Frontier.

Rausch, A. S. (2012). *Japan's Local Newspapers*. London: Routledge.

Shizuoka News (2011) 'Eco-certification is dubious', *Shizuoka News*, 21 November 2011.

Stedman, M. (2010) 'Gunns in dramatic shake-up: Gay gets pulp mill role', *The Mercury*, 24 April: 7.

Ta Ann Tasmania Pty Ltd (2012) 'Media release: Green market campaign costs jobs', accessed 29 August 2012 from www.taanntas.com.au/userfiles/Documents/20120213%20job%20losses%20due%20to%20green%20campaign.pdf

Tarrow, S. (2006). *The New Transnational Activism*. Cambridge: Cambridge University Press.

Tarrow, S. and McAdam, D. (2005). Scale shift in transnational contention. In D. della Porta and S. Tarrow (eds) *Transnational Protest & Global Activism* (pp. 121–149). Oxford: Rowman & Littlefield.

Tasmanian Government (2011) 'Tasmanian Forests Intergovernmental Agreement', accessed 29 August 2012 from www.environment.gov.au/land/forests/pubs/tasmanian-forests-intergovernmental-agreement.pdf

Thompson, J. B. (1995). *Media and Modernity*. Cambridge: Polity.

9

PUBLIC PARTICIPATION IN ENVIRONMENTAL POLICY DECISION MAKING

Insights from twenty years of collaborative learning fieldwork

Gregg B. Walker, Steven E. Daniels, and Jens Emborg

Introduction

At the June 2014 United Nations climate change negotiations in Bonn, Germany, Manuel Pulgar-Vidal, Environment Minister of Peru, talked about the importance of public participation "to retain the confidence of civil society" and to demonstrate transparency. Peru's Environment Minister, who would later serve as the "Conference of the Parties" (COP 20) President in Lima in December 2014, stated that "people around the world want a solution, a climate change agreement. People want the opportunity to try to change the narrative" (Walker, field notes, 2014).

Peru's commitment to include public participation activities mirrors public participation policies that many nations and multinational organizations have adopted. Public participation has become increasingly important internationally through sustainable development and environmental management projects that address governance concerns. For example, as part of its governance strategy, the World Bank Institute promotes "multi-stakeholder collaborative action," recognizing that "successful development requires more than technical solutions. It requires getting individuals, groups, and organizations to work together to achieve a complex set of objectives" (World Bank Institute, 2014).

Just as Roger Fisher and colleagues have written in their best-selling negotiation text, *Getting to Yes*, that "without communication there is no negotiation," (Fisher et al., 2011, p. 32), a similar claim applies to public participation. Communication activity is an essential feature of public participation and environment and natural resource management decision situations (Cox, 2013). Communication as part of public participation strategies comes in numerous forms (as this *Handbook* illustrates); such as media (e.g., television, radio); technology (e.g., cell phones, social media); and face-to-face (e.g., public hearings, workshops, dialogues, appeals) (Burgess and Burgess, 1997; Senecah, 2004; Phillips et al., 2012).

This chapter considers communication and public participation as part of this latter area—face-to-face involvement. The discussion draws on over two decades of the authors' field experience, primarily collaborative learning training and field projects in the United States and Scandinavia.[1] The chapter begins by summarizing the significance of the United States' National Environmental Policy Act as an archetype for public participation policies throughout the world. Following that discussion, the essay presents the context in which collaborative stakeholder engagement has evolved, particularly in the United States.

With the context set, the authors draw on their field experiences and years as "pracademics" to offer a number of significant insights about communication, participation, and stakeholder engagement in the environmental and natural resource policy arenas.

Public participation and the United States National Environmental Policy Act

In United States environmental history 1969 and 1970 were eventful years. 1970 is perhaps best known for the first "Earth Day," when people celebrated the Earth and held activities to draw attention to threats to the environment (Earth Day Network, n.d.). Yet another watershed event occurred months earlier in 1969; passage of the National Environmental Policy Act, or NEPA. Passed with strong bipartisan Congressional support, NEPA was landmark legislation that established a comprehensive process for making environmental decisions.

NEPA scholar Lynton Caldwell observes that "Congress faced a need not only to respond to the values underlying the growing concern over a deteriorating environment but as much as possible to recognize and harmonize the diversity of values and concepts present in American society that related to the environment," noting that "not everyone saw environment in the same light or valued it in the same way" (Caldwell, 1999, p. 2). Caldwell surmises that "the reconciliation of differences regarding the place of the environment in public policy thus became – and remains – a problem that is political, juridical, administrative, and at its base, ethical" (Caldwell, 1999, p. 2).

Public concerns about the environment provided an impetus for NEPA. The architecture of the Act and the various regulations and executive orders that operationalized it responded to public values and interests in a groundbreaking way; it integrated public participation into the environmental planning process. Through its website, the Natural Resources Defense Council explains:

> NEPA is democratic at its core. In many cases, NEPA gives citizens their only opportunity to voice concerns about a project's impact on their community. When the government undertakes a major project such as constructing a dam, highway, or power plant, it must ensure that the project's impacts—environmental and otherwise—are considered and disclosed to the public ... It also gives members of the public a voice in project design by letting them suggest alternatives, which promotes collaboration in planning and buy-in on final decisions ... EISs [Environmental Impact Statements] are first released in draft form, allowing the public and other agencies and levels of government to comment on decisions they care about, provide outside scientific opinion, and ask for improvements.
>
> (Natural Resources Defense Council, 2014)

NEPA elevated the importance of public participation in environmental planning and decision making, particularly through Council of Environmental Quality or CEQ (created as a part of NEPA) policies. In the 1950s and 1960s countries were beginning to address environmental issues, but NEPA had no equivalent in 1969 nor was there a model or precedent for this action (Caldwell, 1999). NEPA has since become a model both internationally and nationally, with nations and U.S. states establishing environmental planning policies patterned after it.

While countries in all regions of the world aspire to make public participation a meaningful component of environmental planning (even with NEPA as a model), it is often more powerful on paper than in practice. Some nations, particularly those with weak democratic institutions, often fall short of providing citizens/stakeholders with a meaningful voice (Kakonge 2006).

In the United States, where detailed public participation policies have been operating for decades, the record is mixed. Although public participation strategies in environmental policy situations may ideally seek to improve decision making and strengthen legitimacy, they often fall short of achieving these goals

(Dietz and Stern, 2008). Public participation is one factor in a complex planning and decision process, with citizen expectations, decision space, and regulatory and technical demands seemingly at odds, a part of what Daniels and Walker (2001) have discussed as the "paradox of public involvement." As we discuss in this chapter what we have learned from the field, we can reflect on the paradox and the ways in which communication-related factors and tasks generate meaningful public participation.

The initial context

As the International Association of Public Participation (IAP2) illustrates with its popular "Public Participation Spectrum," public participation work may be presentational (inform and educate, through an activity such as an open house), feedback or comment-oriented (consultative, through an activity such as a public hearing), or collaborative, through community workshops, study groups, project teams, and the like. Although there are a variety of "spectrums" (e.g., "ladders of citizen participation," Arnstein, 1969; Connor, 1988; "levels of co-management," Berkes, 1994; "spectrum of collaboratives," Margerum, 2011), the IAP2 Spectrum is the most cited and adapted. The Spectrum can be viewed on the IAP2 website (www.iap2.org).

Collaboration, while often heralded by agencies and stakeholders, represents just one approach to public participation. It is, though, often the preferred approach. Collaborative approaches can take a variety of forms, such as habitat conservation planning, community-based consensus work, and stewardship contracting. Regardless of the approach, collaborative public participation generally displays a number of key elements. In her seminal work, *Collaborating*, Barbara Gray identifies five:

> Collaboration involves a process of joint decision making among key stakeholders of a problem domain about the future of that domain. Five features are critical to the process: (1) the stakeholders are interdependent, (2) solutions emerge by dealing constructively with differences, (3) joint ownership of decisions is involved, (4) stakeholders assume collective responsibility for the future direction of the domain, and (5) collaboration is an emergent property.
>
> (Gray, 1989, p. 11)

To Gray's set we add an essential sixth: participatory communication. Participatory communication is active and inclusive, fostered through activities that encourage parties to share ideas, learn, and influence decisions (Ramirez and Quarry, 2004; Walker, 2007). Participatory communication embraces a "social construction" view, that parties as both engaged audiences (listeners, receivers) and message creators will construct meaning from communicative acts (Daniels and Walker, 2001). Shared understanding through communication interaction and meaning creation is the core of participatory (and collaborative) communication (Walker, 2007).

The IAP2 developed its Spectrum in the early 1990s, as agencies, stakeholders, and the conflict management/public participation professional communities were thinking critically about the public's role in policy decision making. A number of consulting organizations emerged to address public policy conflict and decision situations, including how public participation might be incorporated and conducted in environmental matters.

During this same period in the United States, the "spotted owl war" was intense in the Pacific Northwest, illustrated both in actions and rhetoric (Moore, 1993; Yaffee, 1994; Lange, 1993). Public forest lands that had been harvested heavily in the post-World War Two period were now tied up in lawsuits often based on the Endangered Species Act. Legal actions limited timber production, and the spotted owl and other endangered species became symbols of the "jobs versus the environment" discourse and debate.

As public lands conflicts intensified, we (Walker and Daniels) began a partnership to develop and apply an approach for working through environmental and natural resource policy conflict and decision situations that would draw ideas from systems, learning, and conflict management areas. We imagined a "collaborative learning" approach that would incorporate a variety of stages and tasks in a collaborative process, including assessment, training, facilitation, and evaluation (Daniels and Walker, 2001; Walker et al., 2006, 2007). Since that time we have conducted dozens of comprehensive place-based collaborative work projects and training programs in the United States and Scandinavia, and introduced collaboration concepts and techniques to organizations such as the Nature Conservancy and programs such as the National Collaboration Cadre of the United States Forest Service.

Two decades of collaborative work: insights gained and lessons learned

As we have applied and presented our collaboration approach in the field and the classroom, our ideas about collaboration have evolved. We share key lessons here, recognizing that they have emerged from a community of people who have invested significantly in natural resource and environmental management collaborative efforts.

Insight one: collaboration should be appropriate

When considering the public's role in an environmental or natural resource management situation, collaboration is not everything and everything is not collaboration. Although collaborative approaches have received considerable attention in recent years (Wondelleck and Yaffee, 2000; Margerum, 2011; Dukes et al., 2011; Emborg et al. 2012; Cox, 2013), some situations provide restricted decision space, a short time frame, or legislative or judicial direction that limit options.

Commentaries on community-based collaborations (e.g., Buckles, 1999; Wondelleck and Yaffee, 2000; Brick et al., 2001; Brunner et al. 2002; Weber, 2003; Koontz et al., 2004; Sabatier et al., 2005; Dukes et al. 2011; Margerum, 2011) present examples that illustrate varying degrees of success. As collaborative efforts, formats, and venues have increased, a question emerges center stage, one that the cited authors and other natural resources decision making analyses have not necessarily addressed in depth. The question seems fundamental to conflict resolution and decision making: *When should collaboration occur?* And specific to discussion, when should public participation activities be conventional (e.g., a public hearing) and when should they be more interactive and collaborative (e.g., a community workshop)? This general question encompasses two specific points.

1. Are there conflict and decision situations in which a collaborative effort seems particularly useful?
2. Are there situations where collaboration does not seem warranted?

Two trajectories: tech-reg and appropriate collaboration

As we have combined our fieldwork, teaching, and reading of relevant literature, we have determined that environmental and natural resource decision making is bracketed by two trajectories: "technical-regulatory" and "appropriate collaboration." The trajectories are illustrated in Figure 9.1. Examining the situation in terms of these trajectories may be useful when addressing questions of when (and when not) to collaborate.

Along with colleagues, we outlined two fundamental approaches to decision making, conflict resolution and public participation in the natural resource and environmental policy arenas, a few years ago (Walker, 2004; Daniels and Cheng, 2004). These two approaches were initially called "techno-reg" and "discourse-based" (Daniels and Cheng, 2004) or "traditional" and "innovative" (Walker, 2004).

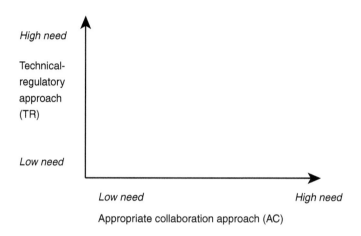

Figure 9.1 Tech-reg and appropriate collaboration

Source: authors.

Tech-reg

The "tech-reg" trajectory accommodates the view that the management of environmental concerns, natural resources, landscapes, and ecosystems has been defined to a great extent by two driving forces: the value placed on *technical* solutions to problems and the perceived need for *regulations* to implement and enforce those solutions. Combined, these two forces have dictated a decision making approach that has privileged agencies over communities and technical expertise over citizen input and traditional (local, indigenous) knowledge.

"Tech" represents the assumption that there are technically correct or preferable approaches to environment and natural resource issues. This view regards natural resource management situations essentially as scientific/engineering problems with technical solutions. "Tech" also presumes that when people do not agree with a proposal, "they just don't understand the science" (Daniels and Cheng, 2004, p. 128). The "tech" way of thinking "also tends to underestimate the likelihood that differences in preferences can arise from deeply-seated value differences that exist largely independent of technical issues that therefore defy technical solutions" (Daniels and Cheng, 2004, p. 128).

"Reg" refers to the dominant means through which various "Tech" solutions have become standard practice. "A long series of court cases, legislative solutions, and regulations have dominated the evolution of natural resource management," Daniels and Cheng explain. This generates a "Reg" presumption that regulations can be crafted consistent with the "best available science," a dominant "tech-reg" goal. The regulations can then be implemented in a standardized manner across the entire defined jurisdiction. "Tech-Reg" is the integration "of two different rationalities—science and law—and was the unifying cultural mindset that drove natural resource management throughout the 20th Century" (Daniels and Cheng, 2004, pp. 128–129).

Appropriate collaboration

In contrast to tech-reg approaches, "appropriate collaboration" (AC) refers to frameworks and methods that emphasize authentic collaboration. AC approaches emphasize access, dialogue, deliberation, mutual learning, and meaningful decision-space. Appropriate collaboration features constructive communication interaction, characterized by inquiry and advocacy (Senge, 2006) as part of both dialogue and deliberation

(Daniels and Walker, 2001). Daniels and Cheng have explained "discourse-based" work with ideas relevant to "appropriate collaboration":

> The unifying characteristic of discourse-based approaches is an emphasis on multi-party communication among the stakeholders in a decision. These processes are undertaken to promote creativity, to resolve misunderstandings of fact, to surface value differences, and to seek mutually acceptable outcomes … They are sometimes facilitated or mediated, and sometimes not. They often strive for specific implementable agreements, and sometimes not. But in all cases, the emphasis is on the discourse—the thoughtful process of deliberating on complex and often controversial issues. Listening and speaking is done as much to learn as to convince.
>
> (2004, pp. 127–128)

Appropriate collaboration approaches are decision-oriented. They link deliberation to decisions, reflecting Yankelovich's (1991) conception of stakeholders "working through" divergent goals and values, judiciously evaluating information, and coming to public judgment as a community of citizen learners and decision makers. Citing various scholars (Press, 1994; Gunderson, 1995; Moote et al., 1997), Daniels and Cheng note that "deliberative democratic theory has been applied to examinations of various natural resource management contexts" (2004, p. 131).

Table 9.1 compares tech-reg and appropriate collaboration approaches according to a variety of variables. These variables have been drawn from the planning legislation literature (e.g., NEPA; the National Forest Management Act or NFMA) and public policy decision making, conflict resolution, and citizen participation scholarship (e.g., Gray, 1989; Dukes, 1996; Wondelleck and Yaffee, 2000; Dukes and Firehock, 2001; Daniels and Walker, 2001; Weber, 2003; Koontz et al. 2004; McKinney and Harmon, 2004; Gastil and Levine, 2005; Dukes et al., 2011; Margerum, 2011).

As Table 9.1 indicates, the nature of communication is fundamentally different in these two approaches. In terms of communication philosophy, tech-reg approaches often feature an information processing or transmission model of communication (Daniels and Walker, 2001; Griffin et al., 2015). Conventional (and historical) public involvement techniques, typical of tech-reg, "usually provide highly controlled, one-way flows of information, guard decision-making power tightly, and constrain interaction between interested groups and decision makers" (Wondelleck and Yaffee, 2000, p. 104). Conventional methods for problem solving and public participation are often part of a strategy of command and control (Weber, 1998), in which political decision making and technical information are guarded, and centralized power and hierarchy are maintained. Weber calls this strategy a conflict game that shuns pluralism and collaboration. In contrast, appropriate collaboration decision making and public participation activities are likely cooperative, accessible, and inclusive, emphasizing learning and informal, open interaction. AC frameworks recognize that communication involves meaning creation and the goal of shared understanding (Daniels and Walker, 2001).

Conventional public participation activities, such as issue "scoping" meetings, public hearings, letter writing "comment" periods, open houses, and the like stem from and maintain decision authority/maker control. These techniques provide people and organizations with opportunities to communicate their concerns to a decision authority. The techniques seek to "inform and educate" and "invite feedback" while offering no guarantee of meaningful citizen input. In these settings, citizens do not know whether and how their ideas will be used. Whether or not their comments influence the decision may depend on the benevolence of the decision authority (Daniels and Walker, 2001).

A consultative strategy and conventional public participation techniques have limitations: the uncertainty over how citizen comments are used, the limited impact that comments have on the outcome, the quasi-arbitration authority of the deciding official (the agency as arbitrator), the formality of the communication environment, and the correspondent perceptions of a zero-sum game (Daniels and Walker,

Table 9.1 Comparing tech-reg and appropriate collaboration public participation and decision making approaches

Element	Technical-regulatory	Appropriate collaboration
Substance goal	Implement a decision based on sound science	Implement a decision based on sound science, local knowledge, and community interests and values
Procedural goal	Information gathering and feedback	Fair, inclusive process; respectful interaction; mutual gains outcome
Decision space	Low, limited, vague	Significant and clear
Decision authority	Rigid	Flexible
Collaborative potential	Low	Moderate to high
Prospects for consensus	Not likely or sought	Possible, emergent
Power	Centralized	Shared
Valued knowledge	Technical	Integrated; technical and traditional
Communication philosophy	Command and control	Dialogue and deliberation; inquiry and advocacy
Communication activity	Inform and educate; gather feedback, consultation	Interaction, mutual learning, idea development and refinement
Access	Structured and controlled by the decision authority	Multifaceted, open, and inclusive; possibly designed by parties
Inclusiveness	Not important beyond key parties; preference given to stakeholders who are "reasonable"	Valued; efforts made to include marginalized voices, diverse communities, skeptics, and critics
Negotiation	None likely without appeals or litigation	Fostered; mutual gains interaction
Primary public participation methods	Public hearings, comment letters, open houses, web sites	Workshops, roundtables, forums, dialogues, field trips, summits, community events
Transparency	Consistent with minimum legal requirements	A high priority; exceeding legal requirements
Measure of success	Quantitive; number of participant contacts	Qualitative; quality of participants' interaction and contributions

Source: adapted from Walker (2004).

2001; Senecah, 2004). Citizens may conclude that decisions were made before public comment was taken, leading the public to believe that the planning approach was one of "decide, announce, defend" (Hendry, 2004). In her case study of a Bureau of Land Management NEPA process about gravel mining, Hendry notes that the BLM, despite over 500 community members providing opposing comments at public scooping meetings and in letters, decided to issue a mitigated FONSI (finding of no significant impact). In this New Mexico case, BLM officials consulted but offered little evidence that public input made any difference (Hendry, 2004).

In contrast, public participation activities within an appropriate collaboration strategy feature open constructive communication. Wondelleck and Yaffee observe that the most successful collaborative efforts fostered two-way, interactive flows of information, and decision making occurred through an open, interactive process rather than behind closed agency doors. Such efforts actively involved people throughout a planning or problem-solving process so that they learned together, understood constraints, and developed creative ideas, trust, and relationships. Direct face-to-face interaction between stakeholders and decision making authorities was critical (2000, p. 105).

Table 9.1 may imply that tech-reg and appropriate collaboration approaches are oppositional. This is not the case. AC methods may include a technical emphasis and limited decision space on some aspects of a given situation, while simultaneously offering more decision flexibility and reliance on traditional knowledge in other parts of the situation.

Figure 9.2 illustrates that the prominence of tech-reg or appropriate collaboration varies depending on the nature of the decision situation. Figure 9.2 locates water quality (and tasks such as determining total maximum daily load or TMDL) compliance toward the tech-reg trajectory. Development of the national park tends more toward appropriate collaboration while a wind turbine farm project might fall somewhere in the middle.

Insight two: assessment of the situation is essential

In any given environmental conflict or decision situation, determining the mix of tech-reg and appropriate collaboration (or whether there is any collaboration at all) begins with assessment. Assessment is fundamentally a research activity; those conducting an assessment collect relevant data and analyze that information. The analysis guides the development of the planning, decision making, and public participation strategies.

Gathering assessment information

Conventional data gathering activities are typically part of an assessment. One can learn about an environmental or natural resource management situation through surveys, interviews, focus groups, and artifact

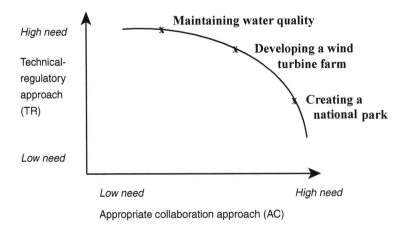

Figure 9.2 Tech-reg and appropriate collaboration integration

Source: authors.

analysis. Stakeholders can be surveyed via mail, telephone, or internet. Local organizations and agencies can identify potential participants in focus groups or interviews. Artifact analysis can include community, organization, and agency websites; blogs, newspapers, and documents.

Assessment work can be innovative, seeking out persons of influence, both supporters and critics. Interviews, for example, may start with influential or prominent community members. When working on a community-based project, we have often talked first with local school administrators, clergy, elected officials (e.g., mayor, county commissioners), health care professionals, newspaper publishers, staff of environmental groups, and industry personnel. These local citizens are knowledgeable about community issues (past and present) and the spirit or morale of the area. They may be an excellent source for referrals; recommending people with whom we should talk who may not be on an agency's mailing list.

Insight three: determine collaborative potential

Conducting an assessment is essential to understanding the situation. An important aspect of that situation is its collaborative potential. If parties believe that a decision situation needs collaboration, assessing collaborative potential determines (1) whether collaboration is feasible, and (2) what areas need to be addressed (e.g., trust) to increase collaborative potential and the likelihood of a best, collaboratively produced outcome.

Any party, whether the decision authority, a key stakeholder, or a citizens' group, that seeks to implement an innovative public participation strategy will likely perceive some collaborative potential (CP). Collaborative potential can be defined as the opportunity for parties to work together assertively in order to make meaningful progress in the management of a controversial, complex, and conflict-laden policy situation. This perception is based on three factors. First, the party determines that the nature of the situation exhibits a high or compelling need for collaboration. Second, the party believes that there is a possibility for meaningful, respectful communication interaction between the disputants. Third, the party surmises that a mutual gain or integrative outcome is possible, that is, that the fundamental structure of the conflict or decision situation offers the potential for both or all sides to achieve more of their objectives than would be likely in some other venue (Lewicki et al., 2014).

A critical factor: decision space

Central to assessing the situation and its collaborative potential is decision space. What in the conflict or decision situation is open to negotiation and influence? What aspects of the situation are "on the table" for discussion and what matters are not? For example, one of our earliest field projects involved a national recreation area managed by the U.S. Forest Service. As we designed and facilitated community workshops about this recreation area, we asked Forest Service officials to clarify what was "within" and "outside" the decision space. Recreation area curfews were within the decision space, while threatened and endangered species were not (Daniels and Walker, 1996).

Influence sharing, mutual learning, and participatory access and inclusiveness are indicators of "decision space." The greater the decision space, the greater the potential for meaningful public participation. Decision space is an important element that differentiates limited or traditional participation such as tech-reg approaches from more innovative and interactive participation, i.e., appropriate collaboration methods.

When assessing a controversial and complex environmental situation, issues of decision space and decision authority should be addressed. Assessment needs to reveal who has jurisdiction in the public policy decision situation; who has legal imperative to make or block a policy decision in that situation. Jurisdiction is related to decision authority—the individual or organization that has the legal or organizational duty to manage or regulate the situation (Walker, 2004).

Sharing decision space involves sharing a form of power and potential influence. While the deciding agency retains its authority by law to make the decision (e.g., under NEPA a forest supervisor signs a record of decision), citizens can participate actively in the construction of that decision. Significant and clear decision space is critical to a meaningful and innovative public participation process. Traditional public participation processes and tech-reg decision making do not necessarily include any shared decision space. Any agency can consult with the public (e.g., invite comments in writing or at a hearing) without any assurance of how those comments might be part of the decision process. A traditional public participation may embody a decision space façade (Walker, 2004).

Insight four: apply an assessment framework

Collaborative potential and decision space are determined through assessment. Natural resource and environmental conflict and decision situations can be appraised via any of a number of frameworks (Emborg et al., 2012) such as the "conflict map" (Wehr, 1979); the conflict dynamics continuum (Carpenter and Kennedy, 1988); "conflict management assessment" or CMA (Warner and Jones, 1998). Highly regarded public policy dispute resolution firms (e.g., CDR Associates, the Center for Collaborative Policy, the Consensus Building Institute, the Keystone Center, the Meridian Institute, Resolve, and Triangle Associates) offer assessment services and have developed varied approaches for doing so. There is no single or simple "formula" to this assessment process. Rather, the analyst has to assess the situation as comprehensively as possible given available resources to do so, such as time, access to people for interviews, review of documents, and so on.

These frameworks draw attention to many important factors, but may not emphasize situation elements that we have found to be central to decisions about appropriate collaboration and tech-reg approaches. Consequently, we have applied two frameworks for assessing conflict and decision situations and their collaborative potential. We began using the first, "The Progress Triangle," in the mid 1990s. More recently we have worked with the new "Unifying Negotiation Framework." The Progress Triangle, displayed in Figure 9.3, has been taught to thousands of students and professionals and applied to numerous projects (Daniels and Walker, 2001; Walker and Daniels, 2005).

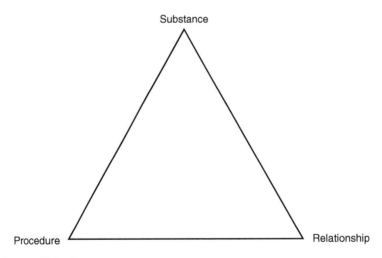

Figure 9.3 The Progress Triangle

Source: Daniels and Walker (2001, p. 36).

The Progress Triangle

Given the complexity, controversy, and uncertainty of many environmental and natural resource conflict situations, they are rarely if ever "resolved." Rather, they are managed. Management of complex and controversial conflict situations can be evaluated in terms of progress; improving the situation and making resilient decisions. As part of improving the situation, progress can include such ideas as reaching consensus, developing mutual gains, learning, resolving a dispute, achieving agreement, and laying a foundation for future negotiations. Progress is a way of thinking about a conflict situation that recognizes that conflicts are inevitable and ongoing, and that the competent management of those conflicts comes from continual improvements in three areas: substance, procedure, and relationship.

Items of *substance* are the "tangible" or concrete aspects of a conflict, such as the issues about which the disputants negotiate. They can be thought of as the agenda items of a decision process. The *procedure* dimension of the triangle includes those elements that pertain to the process or procedural aspects of how conflicts are managed and decisions made. It also includes the rules, both regulative and generative, that parties adhere to in working through the conflict situation. While environmental and natural resource policy conflicts are overtly about substantive matters, progress on them often hinges on the quality of the *relationships* that exist among the stakeholders or conflict parties. The relationship dimension includes the parties in the conflict and their history with one another. It also includes the "intangibles" of any conflict situation, such as trust, respect, legitimacy, identity, and face (Daniels and Walker, 2001).

Assessing an environmental conflict or decision situation via a triangle of three interrelated dimensions—substance, procedure, and relationship—offers three entry points into that situation. The assessment can begin with any dimension; one's understanding of context may guide that decision. All dimensions are important and interconnected; an assessment should consider all three. The procedural dimension may relate most directly to choices about public participation, but assessments of substance and relationship factors guide public participation decisions as well. Assessment via the three dimensions, taken together, reveals collaborative potential.

The unifying negotiation framework

The Progress Triangle (and similar frameworks) can be applied initially to provide a preliminary assessment of an environmental conflict or decision situation. It can offer a "snapshot" of collaborative potential, by identifying decision space (procedure), the degree of trust among parties (relationship), and available technical and traditional sources of information (substance). Some situations, though, benefit from an analysis that is richer and more comprehensive than a framework such as the Progress Triangle. To that end, the "unifying negotiation framework" has been developed recently.

The Unifying Negotiation Framework (UNF) "is a cognitive structure that aids in managing the intellectual complexity" of policy conflict, negotiation, and decision situations (Daniels et al., 2012). Examining a negotiation situation by focusing on its discourse (in all forms) provides unique analytical insights into power, influence, and prospects for common ground. As Figure 9.4 portrays, the UNF "can be visualized as a matrix consisting of three scales (the rows macro, meso, and micro) and six seminal factors in a complex policy situation (the columns of culture, institutions, agency, incentives, cognition, and actor orientation and experience)." Although in Figure 9.4 the UNF may appear static and descriptive, the UNF is designed to account for the systems complexity of policy situations.

A full articulation of the structure and uses of the UNF is beyond the scope of this chapter. The UNF is richly grounded in a social constructivist/discursive turn that public administration theory has taken in recent years (as epitomized by Fischer, 2009). It recognizes that policy decisions result from highly communicative processes by which various claims, rights, and facts are negotiated.

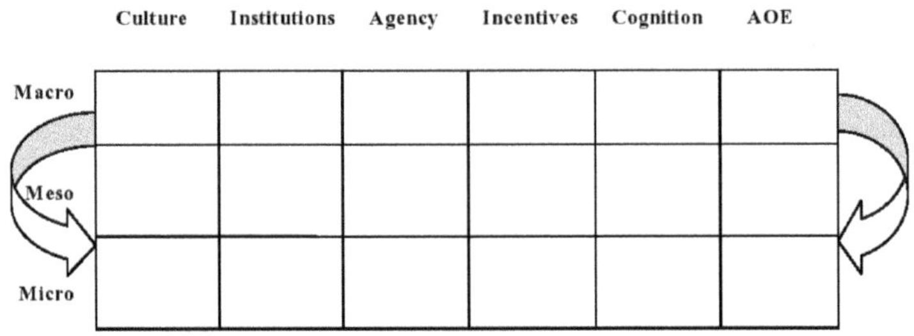

Figure 9.4 The Unifying Negotiation Framework

Source: Daniels et al. (2012).

The Unifying Negotiation Framework for policy discourse had been designed to feature and integrate seminal variables in complex and controversial policy conflict and decision situations. Its utility as an assessment framework stems from the scales and column variables, drawing attention to critical aspects of a situation. Furthermore, the UNF can illuminate concepts that cut across the major areas of a situation. Possible cross-cutting concepts include (1) decision space (integrating institutions, agency, and incentives), (2) collaborative leadership (integrating institutions, agency, and actor orientation), (3) scientific/ technical and traditional knowledge and information (integrating culture, institutions, and cognition), and (4) stakeholder investment (integrating culture, agency, cognition, incentives, and actor factors). Additionally, the UNF should assist practice beyond the assessment phase of a collaborative project. It can (1) identify sideboards and constraints, (2) highlight opportunities for intervention and change (e.g., capacity building, improvements, actions), and (3) guide organizational and procedural design.

Insight five: employ a systems view

As we developed our ideas about collaboration and designed a method for collaborative work, we recognized that we needed concepts and techniques to deal with complexity, controversy, and uncertainty. We turned to the conflict management, negotiation, and mediation literature for ideas and tools germane to controversy. To address uncertainty we drew on work in the fields of adult and experiential learning.

But what about complexity? We had observed public hearings at which citizens/stakeholders generally talked about a single issue (e.g., trails for off-highway vehicles, opposition to clear cuts, the need for more wilderness, development of oil and gas). Rarely did a citizen's comments recognize multiple issues and the complex nature of the situation. Consequently, we sought ways to engage complexity. We turned to the systems literature, particularly "soft systems" and "human activity systems" (Checkland and Scholes, 1990; Wilson and Morren, 1990).

The emphasis our collaborative work has placed on systems thinking has set it apart from other methods and frameworks for multi-party collaboration in environmental conflict and decision situations. While there are a variety of approaches and tools for stakeholder collaboration that rely on the conflict management field, our approach does so within a systems view. Systems thinking activities are essential and foundational. Our community-based collaboration workshops on natural resource management issues, for example, feature systems thinking activities early in the workshop process. Working through controversy, communicating competently, and developing a shared view of the situation begin with thinking systemically and comprehensively (Daniels and Walker, 2012).

In a recent essay, we have highlighted "lessons from the trenches" regarding systems thinking (Daniels and Walker, 2012). These lessons are:

1. Systems thinking promotes holistic learning and understanding.
2. Systems thinking transforms a single issue focus into a multi-issue view.
3. Systems thinking is accessible and pluralistic.
4. Systems thinking corresponds well to the natural resource management paradigms that have emerged in the past twenty-five years.
5. Systems thinking clearly illustrates that complex situations cannot be fully managed/controlled.
6. Systems thinking can encourage agencies to think beyond their default formulation of the situation.
7. Citizens can and do think systemically; they can understand complexity and handle systems thinking tasks. (Adapted from Daniels and Walker, 2012.)

Public participation can incorporate systems thinking; building a shared system understanding can help citizens achieve their goals. Systems thinking work offers a way for citizens to expand the discussion beyond a merely agency-centric formulation, such as a tech-reg approach. We have witnessed a situation in which a federal agency purported to be conducting a "landscape analysis," but the agency staff had omitted a critical drainage—owned by a state agency—that was the focal point of much of the public use and interest in the larger watershed. Since that land was not part of the agency's ownership map, it was also not part of the staff members' cognitive map. System thinking activities brought citizens'/stakeholders' broader interests into sharp focus and compelled the federal agency to expand its view, and arguably resulted in a more comprehensive and insightful analysis (Daniels and Walker, 2012).

Insight six: incorporate civic science

Our community-based collaboration fieldwork has incorporated a "science" approach that gives voice to both technical knowledge (e.g., an agency wildlife biologist) and traditional knowledge (e.g., a village elder, a local horseman). Our work strives to enact civic science. In *Compass and Gyroscope*, Kai Lee explains his view of civic science. "Managing large ecosystems should rely not merely on science," Lee writes, "but on civic science; it should be irreducibly public in the way responsibilities are exercised, intrinsically technical, and open to learning from errors and profiting from successes" (1993, p. 161). Civic science should integrate the idealism of science with the pragmatism of politics.

This integration draws on both technical knowledge and traditional knowledge (sometimes referred to as indigenous or local). Appropriate collaboration methods value ideas from a variety of sources: physical/biological science, political/social science, the local community, and indigenous cultures. Within a civic science orientation, natural resource decision making honors traditional knowledge (including traditional ecological knowledge or TEK) just as it seeks scientific and technical knowledge; voices from nonscientific communities are heard alongside those of the scientists (Walker and Daniels, 2001, 2004; Folke, 2004).

In a concise and accessible discussion of knowledge and environmental decision making, Adler and Birkhoff emphasize the importance of accommodating both technical and traditional knowledge, or what they refer to as knowledge from "away" and "here." They note:

> When people disagree over environmental issues, as they inevitably will, we forget this most fundamental principle: different people and different groups think in different ways. Some people value knowledge that is experientially or intuitively derived. Others treasure expert knowledge or knowledge revealed from spiritual sources. Some people prize stores and aphorisms passed down from [elders]. Others give priority to data and scientific principles.
>
> (2002, p. 4)

Agencies, communities, and organizations can engage environmental policy and natural resource management issues through civic science forums that emphasize constructive communication interaction: dialogue, deliberation, and learning. As Lee notes, civic science combines "a political strategy of bounded conflict with ecological learning based upon experimentation" (1993, p. 185).

The civic science forum is one public participation option of our place-based collaborative projects (Walker and Daniels, 2004) and is consistent with other appropriate collaboration approaches, such as mediated modeling (van den Belt, 2004), co-management (Borrini-Feyerabend et al., 2000), consensus-building (e.g., the Resolve model, www.resolve.org; for a general model see McKinney and Harmon, 2004), public dialogues (Littlejohn and Domenici, 2000), constructive confrontation (Burgess and Burgess, 1996), search conferencing (Diemer and Alvarez, 1995), participatory decision making (Sirmon et al., 1995), collaborative planning (Margerum, 2011; Selin and Chavez, 1995) and collaborative network structures (Innes and Booher, 2004, 2010). Each of these approaches respects technical and traditional knowledge and strives for decisions based on the inclusive politics of community and the best available knowledge and information.

Public participation can include "participatory science" or "citizen science" as well. In addition to contributing local knowledge, citizens may participate in relevant science-related work. The National Wildlife Foundation, for example, defines citizen science as members of the public volunteering their time to assist scientists in their research. They can support professional researchers in a lot of ways—by submitting data, sharing experiences, or spreading valuable information (National Wildlife Foundation, 2014). In its 2009 report to the Puget Sound Partnership, Washington Sea Grant notes that "the term citizen science has emerged to describe projects and programs involving the public in making observations, and collecting and recording data" (Litle et al., 2009, p. 2). Similarly, Dickinson and Bonney assert that citizen science is "public participation in organized research efforts ... hundreds of thousands of individuals around the world are 'citizen scientists,' people who have chosen to use their free time to engage in the scientific process" (2012, p. 1). To the extent that "citizen science" work is collaborative, it should welcome local, indigenous knowledge, qualitative interpretation, and action research agendas.

Insight seven: communication and participation activities should be priorities; both as "new school" and "old school"

Investing in communication work, and understanding the nature and importance of communication, are essential to productive and appropriate public participation. We have emphasized constructive communication activity in our field work; in assessment, training, and facilitated place-based events.

Clarifying communication

In our training programs, for example, we have introduced the "3M Model" of communication. We explain that communication in environmental and natural resource management situations features a triad of messages, methods, and meanings. The natural resource professionals who attend our trainings think about communication primarily as messages, but often do not consider methods and meanings, particularly as they relate to different audiences. Some public affairs staff members of federal agencies with whom we have worked, for example, have relied on standard mailing lists of the organization's website to communicate with stakeholders. The messages they post are often jargon-laden and technical. While messages get sent, the methods may not be relevant to a lot of citizens and the meanings stakeholders create may differ significantly from what the agency staff intended.

New school communication and participation

During our twenty-plus years of place-based collaborative field work, communication technology has changed substantially. We now recommend and employ social media (e.g., Facebook, Twitter), email, websites, and blogs to both communicate with stakeholders and invite their participation. For example, National Forests now have access to "collaborative mapping" (also called special places mapping), a new tool for public participation that accommodates both place-based (e.g., local) and interest-based (e.g., distant) stakeholders. As the Willamette National Forest website states: "Using this interactive map, you can make comments tied to a specific place. You can also reply to other people's comments—a great way to share and discuss ideas" (Willamette National Forest, 2014).

Old school communication and participation

While newer technologies may increase access, communication and participation plans should include traditional methods as well. Recent forest planning focus groups in northern New Mexico offered remarks about communication and public participation, many of which referred to traditional, non-techological activities. They urged the U.S. Forest Service staff to provide opportunities for written responses, especially for those who do not want to speak in front of large groups; to communicate regularly and notify folks about upcoming meetings through email and flyers at public places such as post offices; to be clear on the purpose of meetings and how comments are used; and to educate people about the forest plan so they can participate appropriately.

Consistency, transparency, and follow-through

Public participation approaches, whether aligned principally with appropriate collaboration or tech-reg, influence expectations. Regardless of how they participate, stakeholders expect that their comments will be read and accounted for in the planning and decision making process. In assessment sessions we have conducted (individual and group conversations), citizens have talked about the importance of consistent and ongoing communication and transparency in the public participation, planning, and decision making. If citizens believe that public participation activities are pro forma, that their ideas do not really matter, or they see no evidence that their contributions were considered, they may either become adversaries or believe that "collaboration" and "participation" activities are neither.

Investing in communication includes follow-up and follow-through. Decision authorities (e.g., government agencies) need to communicate consistently during planning and project work that includes significant public participation. Citizens ask for updates and progress reports, even when planning efforts are delayed by unforeseen events (such as a government shutdown or catastrophic weather). Furthermore, stakeholders appreciate seeing decisions carried out and are willing to contribute to implementation and monitoring work (Fernandez-Gimenez and Ballard, 2011).

Conclusion

Public participation is not a new phenomenon in the environmental and natural resource management arenas; it traces back to the National Environmental Policy Act (and even earlier). Early participation efforts relied on comment periods, public hearings, and open houses; conventional methods that were, at best, consultative, and at their worst, adversarial and intractable. In the United States, multi-party collaborative public participation efforts emerged significantly in the 1990s and offered agencies, organizations, and citizens an alternative to the "command and control" approaches that encouraged "us vs. them" perceptions and actions.

Public participation as part of environmental management and decision making has been embraced through the world on a wide range of environmental issues such as climate change, forest management, water policy, and mining practices. How public participation has been operationalized and utilized, though, varies significantly. Even nations with strong democratic traditions often struggle to incorporate stakeholder contributions into environmental and natural resource management decisions; enacting appropriate collaboration methods and collaborative governance poses challenges greater than more conventional, tech-reg approaches (Daniels and Cheng, 2004; McDermitt et al., 2011; Walker and Senecah, 2011).

Although we present key insights here, there are certainly more than these seven. We have learned that language matters, history and culture run deep, competent leadership is essential (Walker and Daniels, 2012), agencies and stakeholders need to clarify their commitments, and more. Throughout our two decades of fieldwork, we have learned to listen and adapt. Many parties prefer appropriate collaboration to tech-reg; they want to be involved meaningfully in the natural resource and environmental management decisions that affect their lives. Doing public participation well is both important and challenging; no doubt we will gain more insights as our public participation work continues to evolve.

Note

1   We have been involved in the field of public participation and environmental policy for over two decades. We (Daniels and Walker initially, with Emborg joining as a colleague a decade ago) work as both practitioners and academics, developing ideas through research and theory building and taking those ideas into the field. Our teaching, research, and practice all inform one another. What we learn in the field we share with our students; what we learn with our students we take into communities.

Further reading

Margerum, R. D. (2011). *Beyond consensus: Improving collaborative planning and management.* Cambridge, MA: MIT Press.

Drawing on extensive case studies of collaborations in the United States and Australia, Margerum examines the full range of collaborative enterprises in natural resource management, urban planning, and environmental policy and explains the pros and cons of collaborative approaches.

Daniels, S. E. and Walker, G. B. (2001). *Working through environmental conflict: The Collaborative Learning approach.* Westport, CT: Praeger.

Daniels and Walker present collaborative learning, a method designed for complex and controversial policy situations. As they illustrate, collaborative learning is a hybrid of systems thinking, is grounded explicitly in experiential, team—or organizational—and adult learning theories, and is a theory-based framework through which parties can make progress in the management of controversial environmental policy situations.

Dietz, T. and Stern, P. C., eds. (2008). National Research Council. *Public participation in environmental assessment and decision making.* Washington, DC: The National Academies Press.

This book presents the view that, when done correctly, public participation improves the quality of U.S. federal agencies' decisions about the environment. Well-managed public involvement can increase stakeholders' perceptions of the legitimacy of decisions and their implementation. This book recommends that agencies recognize public participation as valuable to their objectives, not just as a formality required by the law.

Depoe, S. P., Delicath, J. W., and Aelpi Elsenbeer, M-F., eds. (2004). *Communication and public participation in environmental decision making* (pp. 113–136). Albany, NY: State University of New York Press.

Contributors to this volume explore the communication practices of various stakeholders—interested citizens, grassroots and public interest groups, industry representatives, scientists and technical experts, government agencies, federal regulators—engaged in a variety of environmental decision making contexts in the U.S. and elsewhere.

References

Adler, P. S. and Birkhoff, J. E. (2002). Building trust: When knowledge from "here" meets knowledge from "away." Portland, OR: The National Policy Consensus Center.

Arnstein, S. R. (1969). A ladder of citizen participation. *Journal of the American Institute of Planners, 35*, 216+.

Berkes, C. (1994). Co-management: Bridging the two solitudes. *Northern Perspectives, 22*, 2–3. Retrieved from http://home.cc.umanitoba.ca/~berkes/berkes_1994.pdf (accessed October 15, 2013).

Borrini-Feyerabend, G., Taghi M. F., Nguinguiri, J. C., and Ndangang, V. A. (2000). *Comanagement of natural resources: Organizing, negotiating and learning-by-doing.* Heidelberg Germany: Kasparek Verlag – GTZ and IUCN.

Brick, P., Snow, D., and Van de Wetering, S. eds. (2001). *Across the great divide: Explorations in collaborative conservation and the American west.* Washington, D.C.: Island Press.

Brunner, R. D., Colburn, C. H., Cromley, C. M., Klein, R. A., and Olson, E. A. (2002). *Finding common ground: governance and natural resources in the American west.* New Haven, CT: Yale University Press.

Buckles, D., ed. (1999). *Cultivating peace: Conflict and collaboration in natural resource management.* Ottawa/Washington, D.C.: International Development Research Centre/The World Bank.

Burgess, H. and Burgess, G. (1996). Constructive confrontation: A transformative approach to intractable conflicts. *Mediation Quarterly, 13*, 305–322.

Burgess, H. and Burgess, G.M. (1997). *Encyclopedia of conflict resolution.* Santa Barbara, CA: ABC-CLIO.

Caldwell, Lynton K. (1999). *The National Environmental Policy Act: An agenda for the future.* Bloomington, IN: Indiana University Press.

Carpenter, S. and Kennedy, W. J. D. (1988). *Managing public disputes.* San Francisco, CA: Jossey-Bass.

Checkland, P. and Scholes J. (1990). *Soft systems methodology in action.* New York: John Wiley & Sons.

Connor, D. D. (1988). A new ladder of citizen participation. *National Civic Review, 77*(3), 248–257.

Cox, J. R. (2013). *Environmental communication and the public sphere,* 3rd ed. Thousand Oaks, CA: Sage.

Daniels, S. E. and Cheng, A. S. (2004). Collaborative resource management: Discourse-based approaches and the evolution of TechnoReg. In M. L. Manfredo, J. J. Vaske, B. L. Bruyere, D. R. Field, and P. J. Brown (eds), *Society and natural resources: A summary of knowledge* (pp. 127–136). Jefferson, MO: Modern Litho.

Daniels, S. E. and Walker, G. B. (1996). Collaborative learning: Improving public deliberation in ecosystem-based management. *Environmental Impact Assessment Review, 16*(6), 71–102.

Daniels, S. E. and Walker, G. B. (2001). *Working through environmental conflict: The Collaborative Learning approach.* Westport, CT: Praeger.

Daniels, S. E. and Walker, G. B. (2012). Lessons from the trenches: Twenty years of applying systems thinking to environmental conflict. *Systems Research and Behavioral Science, 29*, 104–115.

Daniels, S. E., Walker, G. B., and Emborg, J. (2012). The unifying negotiation framework: A model of policy discourse. *Conflict Resolution Quarterly, 30*(1), 1–14.

Depoe, S. P., Delicath, J. W., and Aelpi Elsenbeer, M-F., eds. (2004). *Communication and public participation in environmental decision making* (pp. 113–136). Albany, NY: State University of New York Press.

Dickinson, J. L. and Bonney, R. eds. (2012). *Citizen science: Public participation in environmental research.* Ithaca, NY: Comstock Publishing Associates.

Diemer, J. A. and Alvarez, R. C. (1995). Sustainable community–sustainable forestry: A participatory model. *Journal of Forestry, 93*(11): 10–14.

Dietz, T. and Stern, P. C., eds. (2008). National Research Council. *Public participation in environmental assessment and decision making.* Washington, D.C.: The National Academies Press.

Dukes, E. F. (1996). *Resolving public conflict: Transforming community and governance.* New York: St. Martin's Press.

Dukes, E. F., and Firehock, K. (2001). *Collaboration: A guide for environmental advocates.* Institute for Environmental Negotiation, The Wilderness Society, National Audubon Society.

Dukes, E. F. Firehock, K. E., and Birkhoff, J. E., eds. (2011). *Community-based collaboration: Bridging socio-ecological research and practice.* Charlottesville, VA: University of Virginia Press.

Earth Day Network. (n.d.). *Earth Day: The History of a Movement.* Retrieved from: www.earthday.org/earth-day-history-movement. Retrieved on September 15, 2013.

Emborg, J., Walker, G., and Daniels, S. (2012). Forest landscape restoration decision-making and conflict management: Applying discourse-Based approaches. In J. Stanturf (ed.), *Forest landscape restoration: Integrating natural and social sciences*. Dordrecht, The Netherlands: Springer.

Fernandez-Gimenez, M. E., and Ballard, H. L. (2011). How CBCs learn: Ecological monitoring and adaptive management. In E. F. Dukes, K. E. Firehock, and J. E. Birkhoff (eds), *Community-based collaboration: Bridging socio-ecological research and practice*. Charlottesville, VA: University of Virginia Press.

Fischer, F. (2009). *Democracy & expertise: Reorienting policy inquiry*. New York: Oxford University Press.

Fisher, R., Ury, W., and Patton, B. (2011). *Getting to yes: Negotiating agreement without giving in*, rev. ed. New York: Penguin Books.

Folke, C. (2004). Traditional knowledge in social–ecological systems. Editorial and introduction to a special issue on traditional knowledge. *Ecology and Society*, 9(3), 7. Retrieved from: www.ecologyandsociety.org/vol9/iss3/art7/ (accessed October 15, 2013).

Gastil, J. and Levine. P., eds. (2005). *The deliberative democracy handbook: Strategies for effective civic engagement in the twenty-first century*. San Francisco, CA: Jossey-Bass.

Gray, B. (1989). *Collaborating*. San Francisco, CA: Jossey-Bass.

Griffin, E., Ledbetter, A., and Sparks, G. (2015). *Communication theory: A first look*, 9th ed. New York: McGraw-Hill.

Gunderson, A. G. (1995). *The environmental promise of democratic deliberation*. Madison, WI. University of Wisconsin Press.

Hendry, J. (2004). Decide, announce, defend: Turning the NEPA process into an advocacy tool rather than a decision-making tool. In S. P. Depoe, J. W. Delicath, and M-F. Aelpi Elsenbeer (eds), *Communication and public participation in environmental decision making*. Albany, NY: State University of New York Press.

Innes, J. E. and Booher, D. E. (2004). Reframing public participation: Strategies for the 21st century. *Planning Theory & Practice*, 5(4), 419–436.

Innes, J. E. and Booher, D. E. (2010). *Planning with complexity: An introduction to collaborative rationality for public policy*. New York: Routledge.

Kakonge, J. O. (2006). *Environmental planning in Sub-Saharan Africa: Environmental impact assessment at the crossroads*. Working Paper Number 9. New Haven, CT: Yale School of Forestry and Environmental Studies.

Koontz, T. M., Steelman, T. A., Carmin, J., Korfmacher, K. S., Moseley, C., and Thomas, C. W. (2004). *Collaborative environmental management: What roles for government?* Washington, D.C: Resources for the Future Press.

Lange, J. I. (1993). The logic of competing information campaigns: Conflict over old growth and the spotted owl. *Communication Monographs*, 60(3), 239–257.

Lee, K. (1993). *Compass and gyroscope: Integrating science and politics for the environment*. Washington, D.C.: Island Press.

Lewicki, R. J., Saunders, D. M., and Barry, B. (2014). *Negotiation*, 7th ed. Boston, MA: McGraw-Hill Irwin.

Litle, K., Wainstein, M., Dalton, P., and Meehan, D. (2009). *Harnessing citizen science to protect and restore Puget Sound*. Washington Sea Grant and Washington State University Extension.

Littlejohn, S. W. and Domenici, K. (2000). *Engaging communication in conflict: Systemic practice*. Thousand Oaks, CA: Sage.

McDermitt, M. H., Moote, M. A., and Danks, C. (2011). Effective collaboration: Overcoming external obstacles. In E. F. Dukes, K. E. Firehock, and J. E. Birkhoff (eds), *Community-based collaboration: Bridging socio-ecological research and practice*. Charlottesville, VA: University of Virginia Press.

McKinney, M. and Harmon, W. (2004). *The western confluence: A guide to governing natural resources*. Washington D.C.: Island Press.

Margerum, R. D. (2011). *Beyond consensus: Improving collaborative planning and management*. Cambridge, MA: MIT Press.

Moore, M. P. (1993). Constructing irreconcilable conflict: The function of synecdoche in the spotted owl controversy. *Communication Monographs*, 60(3), 258–274.

Moote, M. A., McClaran, M. P., and Chickering, D. K. (1997). Theory in practice: Applying participatory democracy theory to public land planning. *Environmental Management*, 21(6), 877–889.

National Wildlife Foundation. (2014). Citizen Science. Retrieved February 10, 2014 from: www.nwf.org/Wildlife/Wildlife-Conservation/Citizen-Science.aspx (accessed October 15, 2013).

Natural Resources Defense Council. (2014). *Environmental Issues: U.S. Law & Policy. Never Eliminate Public Advice!* Retrieved January 10, 2014 from www.nrdc.org/legislation/nepa-success-stories.asp.

Phillips, L., Carvalho, A., and Doyle, J., eds. (2012). *Citizen voices: Performing public participation in science and environment communication*. Chicago, IL: University of Chicago Press.

Press, D. (1994). *Democratic dilemmas in the age of ecology: Trees and toxics in the American West*. Durham, NC: Duke University Press.

Ramirez, R. and Quarry, W. (2004). *Communication for development: A medium for innovation in natural resources management.* Ottawa and Rome: International Development Research Centre and the Food and Agriculture Organization of the United Nations.

Sabatier, P. A., Focht, W., Lubell, M., Trachtenburg, Z., Vedlitz, A., and Matlock, M. (2005). *Swimming upstream: Collaborative approaches to watershed management.* Cambridge, MA: MIT Press.

Selin, S. and Chavez, D. (1995). Developing a collaborative model for environmental planning and management. *Environmental Management, 19*, 189–195.

Senecah, S. L. (2004). The trinity of voice: The role of practical theory in planning and evaluating the effectiveness of environmental participatory processes. In S. P. Depoe, J. W. Delicath, and M-F. Aelpi Elsenbeer (eds), *Communication and public participation in environmental decision making* (pp. 13–33). Albany, NY: State University of New York Press.

Senge, P. (2006). *The fifth discipline: The art and practice of the learning organization,* rev. ed. New York: Doubleday.

Sirmon, J., Shands, W. E., and Liggett, C. (1995). Communities of interests and open decisionmaking. *Journal of Forestry, 91*(7), 17–21.

van den Belt, M. (2004). *Mediated modeling: A systems dynamics approach to environmental consensus building.* Washington, D.C.: Island Press.

Walker, G. B. (2004). The Roadless Areas Initiative as national policy: Is public participation an oxymoron? In S. P. Depoe, J. W. Delicath, and M-F. Aelpi Elsenbeer (eds), *Communication and public participation in environmental decision making* (pp. 113–136). Albany, NY: State University of New York Press.

Walker, G. B. (2007). Public participation as participatory communication in environmental policy decision-making: From concepts to structured conversations. *Environmental Communication: A Journal of Nature and Culture, 1*, 64–72.

Walker, G. B. (2014, June). Field notes. Bonn Climate Change Conference. United Nations Framework Convention on Climate Change. Bonn, Germany.

Walker, G. B. and Daniels, S. E. (2001). Natural resource policy and the paradox of public involvement: Bringing scientists and citizens together. *Journal of Sustainable Forestry, 13*(1/2), 253–269.

Walker, G. B. and Daniels, S. E. (2004). Dialogue and deliberation in environmental conflict: Enacting civic science. *Environmental Communication Yearbook,* Vol. 1. Mahwah, NJ: Lawrence Erlbaum Associates.

Walker, G. B. and Daniels, S. E. (2005). Assessing the promise and potential for collaboration: The Progress Triangle framework. In G. B. Walker and W. J. Kinsella (eds), *Finding our Way(s) in Environmental Communication: Proceedings of the seventh biennial conference on communication and the environment* (pp. 188–201). Corvallis, OR: Oregon State University Department of Speech Communication.

Walker, G. B. and Daniels, S. E. (2012). The nature and role of agency leadership: Building and sustaining collaboration in natural resource management and environmental policy decision-making. In D. Rigling Gallagher (ed.), *Environmental leadership: A reference handbook* (pp. 147–158). Thousand Oaks, CA: Sage.

Walker, G. B. and Senecah, S. L. (2011). Collaborative governance: Integrating institutions, communities, and people. In E. Dukes, K. E. Firehock, and J. E. Birkhoff (eds), *Community-based collaboration: Bridging socio-ecological research and practice.* Charlottesville, VA: University of Virginia Press.

Walker, G. B., Senecah, S. E., and Daniels, S. E. (2006). From the forest to the river: Citizen views of stakeholder engagement. *Human Ecology Review, 13*, 193–202.

Walker, G. B., Senecah, S. L., and Daniels, S. E. (2007) Reflections from the road: new and improved concepts, tools, and lessons for community-based collaboration. In L. S. Volkening et al. (eds) *Proceedings of the 8th biennial conference on communication and the environment* (pp. 213–234). Athens, GA: University of Georgia Department of Speech Communication.

Warner, M. and Jones, P. (1998). Assessing the need to manage conflict in community-based natural resource projects. *ODI Natural Resource Perspectives,* 35 (July), 1–11.

Weber, E. P. (1998). *Pluralism by the rules: Conflict and cooperation in environmental regulation.* Washington D.C.: Georgetown University Press.

Weber, E. P. (2003). *Bringing society back in: Grassroots ecosystem management, accountability, and sustainable communities.* Cambridge, MA: MIT Press.

Wehr, P. (1979). *Conflict regulation.* Boulder, CO: Westview.

Willamette National Forest (2014). Roads Investment Plan—Special Places Mapping. Retrieved March 10, 2014 from: www.fs.usda.gov/detail/willamette/landmanagement/resourcemanagement/?cid=stelprd3791685 (accessed October 15, 2013).

Wilson, K. and Morren, G. E. B. (1990). *Systems approaches to improvements in agriculture and resource management.* New York: MacMillan.

Wondelleck, J. M., and Yaffee, S. L. (2000). *Making collaboration work: Lessons from innovation in natural resource management*. Washington D.C.: Island Press.

World Bank Institute (2014). *Multi-stakeholder collaborative action*. Collaborative governance brochure insert. Retrieved from: http://wbi.worldbank.org/wbi/Data/wbi/wbicms/files/drupal-acquia/wbi/multistakeholder_insert_fy12.pdf (accessed October 15, 2013).

Yaffee, S. L., (1994). *The wisdom of the spotted owl: Policy lessons for a new century*. Washington, D.C.: Island Press.

Yankelovich, D. (1991). *Coming to public judgment: Making democracy work in a complex world*. Syracuse, NY: Syracuse University Press.

10

TO ACT IN CONCERT

Environmental communication from a social movement lens

Charlotte Ryan and Kimberly Freeman Brown

In spring 2013, ExxonMobil's Pegasus pipeline ruptured in Mayflower, Arkansas spilling 5,000+ barrels of crude oil. A network of regional and national environmental organizations, journalists, legal and technical advocates swung into action supporting residents' efforts to document exposure and damage. Environmental chemist and Louisiana Environmental Action Network staff member, Wilma Subru explained, "If the community doesn't speak up, it's like it doesn't exist."

(Koon 2013)

Environmental disasters occur regularly in more vulnerable global south communities as well as in poor, immigrant, and/or communities of color in the U.S. (Bullard, Johnson, and Torres 2005). Cumulatively, these inequalities constitute environmental apartheid (Shiva 1999) and climate apartheid (Tutu 2007). Since patterns of inequalities tend to replicate across institutional arenas, environmental communicators find environmental/climate apartheid routinely matched with "communication apartheid" (Hirschkop 1996: 93), the constrained communication opportunities experienced by less powerful nations, ethnicities, classes, and constituencies.

Communication is central to the democratic right "not just to act but to act in concert" (Arendt 1969: 44). And yet, environmental reformers have insufficiently addressed how communication apartheid sidelines large constituencies directly affected by climate change and other environmental crises, reducing their access to redress and weakening their engagement with the environmental movement as a whole. In this chapter, we consider how this might be remedied.

Like other structural inequalities (Powell 2007), communication inequalities permeate organizational practices that privilege short-term progress. For instance, two widespread communication models, social marketing and media advocacy, do not address the cumulative communication barriers faced by marginalized communities. Focused on individual behavioral change or a time-sensitive legislative campaign, neither model prioritizes building relationships that could challenge disparities in communication access, infrastructure, skills, and resources.

To reduce these deficits, communicators working with marginalized constituencies embrace, but also expand social marketing and media advocacy models by integrating insights from social movements. The resulting model works to address Subru's worry that, "*If the community doesn't speak up, it's like it doesn't exist.*"

We first review social marketing and media advocacy, two communication models widely used by environmental campaigns, and provide an illustrative case of each. We argue that social marketing's individual behavioral focus obscures power inequalities faced by marginalized communities. In contrast, the media advocacy model recognizes institutionalized inequalities, but still minimizes marginalized communities' critical roles. Noting these models' commonalities, divergences, and shared limitations, we turn to social movement communication, flagging this model's attention to expanding opportunities for marginalized constituencies to speak and act in concert. We illustrate the approach with a case study of Green For All's Stormwater community of practice. In the discussion, we summarize how each model conceptualizes the role of communication in civil society, drawing conclusions for environmental communicators.

Social movements and communication models

Recently, movement communicators and scholars (Frey and Carragee 2007, 2012; Napoli and Aslama 2011) have focused scholarly attention on the efforts of marginalized constituencies to reclaim communication as a democratic right. To date, social movement communicators grounded in communities of color have advanced the most consistent and developed communication model centered on justice as a value guiding both daily interactions and the selection of strategic priorities (www.centerformediajustice.org). The resulting communication approach interchangeably labeled, "social movement" or "justice" communication (Cutting and Themba-Nixon 2006; Cyril 2005) highlights the critical role of communities directly experiencing and challenging inequalities.

Linking global south communication theories (Barndt 1989; Riaño and Alcalá 1994; Rodriguez, Kidd, and Stein 2010) with theories emerging in the global north (Cyril 2005; Cutting and Themba-Nixon 2006; Powell 2007), social movement communicators argue that marginalized constituencies face communication barriers insufficiently addressed by social marketing and media advocacy approaches (www.centerformediajustice.org). We describe social marketing and media advocacy approaches below, and then contrast them with social movement communication.

Social marketing

Social marketing ranks among the most widespread communication models used in environmental campaigns. Social marketing pioneered the application of marketing concepts to social issues—so that intangible concepts such as brotherhood can be marketed as easily as soap (Wiebe 1951). Like traditional marketing, social marketing stresses a consumer orientation and voluntary exchange in a free market. Walsh, Rudd, Moeykens, and Moloney (1993) describe social marketing's rigorous attention to audience, market research, strategic planning, and evaluation. Social marketing proponents stress its flexibility and added value in a mass mediated society. Social marketers engage in systematic analysis of their desired audiences and translate their messages into formats that will easily reach and resonate with those audiences. Smokey the Bear (www.smokeythebear.com), the longest-running PSA campaign in U.S. history, exemplifies a sustained social marketing campaign aimed at encouraging individuals and peer groups to change fire safety practices. None of the practices require institutional or policy change (McNamara, Kurth, and Hansen 1981; Rice 2001). Boasting a 90 percent recognition rate among U.S. adults, the seventy-year campaign credits itself with reducing human-caused fires by 90 percent (Ad Council 2011).

Critiques of the social marketing framework

Social marketing's critics argue that its individual behavior focus diverts attention from systemic inequalities and the need to address these via institutional changes (Dorfman, Wallack, and Woodruff 2005; Wallack and Dorfman 1996). Critics further argue that social marketing reduces audiences to passive message consumers. A review of 100 public interest campaigns demonstrated that most failed to present or engage targeted audiences as active citizens who could shape civil society (Weiss and Tschirhart 1994).

Social marketing proponents reject such criticisms as extreme. They acknowledge that social marketing has been practiced in an individualized manner, but argue that the model need not function this way. Rather, they stress the model's market research tools and grounding in social scientific methods such as focus groups, surveys, polls, and rigorous evaluation (Walsh et al. 1993). Thus, social marketing's limitations can be resolved by applying its tools to institutional reforms rather than individual behavior changes (Siegel and Doner 2004).

Media advocacy

In contrast to social marketing, media advocacy campaigns call for policy changes that improved conditions for a whole class of individuals (Wallack and Dorfman 1996). At times, these reforms also target structural inequalities. The three-decade (1965–1997) media advocacy campaign that forced the tobacco industry to adopt cautionary labeling, restrict advertising, and fund anti-smoking campaigns (Pertschuk 2001) stands as a model.

The tobacco campaign also illustrates the limits of media advocacy as a mobilization model. The word *advocate* comes from the Latin, advocare—to speak for another. Despite the collective nature of media advocacy campaigns, the policy emphasis often encourages experts to retain control to maximize programmatic success. Media advocacy campaigns, for example, may urge citizens to sign petitions or vote on a referendum. There is nothing inherent in the media advocacy model that attends to base building or to including the base in setting the agenda, planning strategic actions, or reflecting on their efficacy. Policy reform is institutional change, but the citizen's role remains circumscribed.

We agree with social marketers who conceptualize media advocacy as an improved second-generation social marketing model that applies social marketing tools to the policy arena. In so doing, media advocates shift from an exclusive focus on individual behavioral change to urge individual citizens and existing civil society organizations to press for a specified change in public policy. But media advocacy falls short of providing the supports needed if marginalized constituencies are to communicate as empowered civil actors.

Shared assumptions of social marketing and media advocacy

Despite the differences noted above, social marketing and media advocacy share more commonalities than is usually acknowledged. Both models recognize that mass media form critical convening systems essential for public discourse. Both are mission-driven, setting clear strategic goals, and both involve expert use of social science tools such as market research, polling data, and focus groups to analyze targeted audiences, plan, execute, and evaluate strategic interventions. And both generally encourage individuals to act.

Shared limitations and omissions of social marketing and media advocacy

While media advocacy extends social marketing into the policy arena, the models share several problematic assumptions, limitations, and silences. First, both de-emphasize the "culture of silence" (Freire 1970: 16) by underestimating the institutional barriers that prevent marginalized groups from communicating their points of view in political and media venues. In failing to recognize these barriers, neither model seeks to dismantle them or to proactively build the communication capacity of those directly affected. Second, neither model incorporates an empowerment lens (Barndt 1989) that identifies the ways in which marginalized constituencies are essential actors in social change strategies. Third, both models define their fields of action to exclude long-term capacity building. While both individual behavior change and policy change are invaluable components of any strategic campaign, both social marketing and media advocacy tend to treat these as the end goals. As a result, movements can pursue a series of campaigns without addressing the continued marginalization of the very groups organizers hoped to engage. In short, without building an inclusive base, it is not possible to create, much less advance an inclusive message and strategy.

StepItUp2007 (www.StepItUp2007.org) illustrates both strengths and limitations of media advocacy models. Launched in January 2007 by environmentalist Bill McKibben and six Middlebury College seniors, StepItUp2007 used a web-centric organizing strategy to mobilize thousands over twelve weeks, and played a critical role in pushing 2008 Democratic Presidential candidates to support a call for 80 percent reduction in carbon emissions by 2050 (McKibben 2009). StepItUp2007's open-source structure also encouraged local groups to adapt the campaign to their needs while advancing the shared call to cut carbon emissions. As such, StepItUp2007 illustrated how the media advocacy model could expand social marketing's narrower focus on behavioral change. Finally, StepItUp2007 suggested how a small group with limited resources could leverage Internet technologies to link local, face-to-face engagements into a national campaign.

StepItUp2007 also illustrates the limits shared by media advocacy social marketing models. Given twelve weeks, limited resources and labor power, StepItUp2007's campaign primarily targeted constituencies easily reached by the Internet, that is, communities with pre-existing communication capacity. While the campaign's open structure could accommodate diverse approaches, the campaign's central messages, images, spokespersons, programmatic thrust, venues, etc. did not reflect the specific interests of the environmental justice (EJ) wing of the environmental movement. The resulting campaign revealed little about climate change's impacts on politically, socially, economically, and culturally under-represented communities:

> When we interviewed the national SIU organizers … they claimed they had thought about the EJ implications of climate change when planning the campaign. Despite this awareness, however, our analysis reveals that there was little presence from the EJ movement or marginalized peoples in the SIU campaign. First, we observed a lack of EJ groups at SIU actions … Second, our analysis of speeches and messages of the events indicates limited inclusion of EJ principles. Instead, we observed that the SIU event messages focused more on what behaviors people could perform to slow global warming (i.e., buying light bulbs and pressuring Congress) than on the consequences of global warming that disproportionately affect marginalized people and on the injustices of climate change.
>
> (Endres, Clark, Garrison, and Peterson 2009: 184)

Participants were largely white and middle-to-upper-class (Endres et al. 2009: 186), reproducing the racial and class fragmentation that has characterized the U.S. environmental movement (Sandler and Pezzullo 2007). The individual consumerist approaches commonly used by social marketing and media advocacy models may resonate with these young, white environmentalists but prove less effective with marginalized constituencies. These constituencies may feel recruited to support predetermined demands that lack redress of their priorities and interests. Rather than participating as critical actors shaping strategies, leaders may feel reduced to poster children (Endres et al. 2009: 195–196). Moreover, campaign efficacy may decline if organizers fail to include, mobilize, and leverage a critical constituency and framing.

A final concern rests with the fact that neither the behavioral focus of social marketing, nor the policy focus of media advocacy tackles deep communication disparities between middle-class white communities and communities marginalized by race, class, language, and/or geography.

Following these approaches, StepItUp2007 mobilized primarily via the Internet without addressing a widely recognized digital divide.

> Understanding the digital divide can help to explain why heavily relying on the Internet for movement participation (i.e., online sign-ups to host events, dissemination of information about the

campaign through the SIU website, blogs, and *Grist*, etc.) can result in a lack of participation of marginalized people who may suffer the most from climate change.

(Endres et al. 2009: 187)

Without downplaying the magnitude of StepItUp2007's accomplishments, Endres et al. conclude that the campaign's reliance on digital platforms suggest that StepItUp2007's organizers did not grasp the depth or significance of the digital divide (2009: 187). Nor did the programmatic emphasis on consumers attend to how poor communities and communities of color might conceptualize priorities vis-à-vis climate justice (2009: 193).

The accomplishments of StepItUp2007 remain impressive. A StepItUp2007's three-month campaign had successfully mobilized the constituencies easily reached with its existing communication capacity and it embedded calls for individual behavioral changes in broader calls to reduce carbon emissions. Using a media advocacy model, however, the campaign had underestimated the barriers to environmental activism faced by marginalized constituencies.

Social movement communication model

Social movement communicators recognize that environmental conflicts are inherently power-laden and contentious. In planning environmental campaigns, therefore, social movement strategists map inequalities as they plan organizing, policy, and communication strategies. They assess constituencies' relative strengths and identify supports needed for these constituencies to collaborate in designing, executing, and reflecting on strategies (Barndt 1989). Often defined as learning communities, social movement mobilizations include among their tasks establishing communication skills, communication infrastructure, and the capacity to reflect on experience (Cryan 2011; Hinson and Healey 2011). Social movement communicators also weigh existing assets—for instance, whether groups have previously established a shared vision, strategic and communication plan, and reflective practices. These preconditions are routinely flagged by independent media organizations (Downing 2001; Howley 2005; Independent Media Center 2004). What social movement or justice communicators add is willingness to augment independent media with mainstream media.[1]

We focus here on organizations that embed their organizing and communication practices in theories that critique structural inequality (Bullard et al. 2005; Powell 2007; Salazar 2009). Environmental communicators in general, and environmental justice communicators, in particular, do not operate in a power vacuum. Institutional hurdles (Endres et al. 2009) limit their access to the specialized knowledge, labor, skills, and communication infrastructure. Without these, marginalized groups have limited chances to gain standing as credible actors in environmental decision making, and to mobilize supporters and allies (Ryan and Alexander 2006).

Social movement theory acknowledges that those confronting cumulative, multi-arena, inequalities are disadvantaged when navigating institutions including mass media. At best, marginalized individuals and their grassroots organizations extend their standing and resources within existing structural arrangements, but rarely can they change those arrangements. For sustained, structural changes to occur, individual and group efforts must accrue so that expanding coalitions form networks capable of working toward structural change over time (Morris 1986).

Social movement communication is grounded in independent media (Downing 2001) but absorbs tools and tactics from social marketing and media advocacy. Operating in the tradition pioneered by justice communicators in communities of color (Cutting and Themba-Nixon 2006; Cyril 2005), the national green economy advocacy organization Green For All (GFA), and environmental justice groups (Bullard, Warren, and Johnson 2005) embrace social marketing and media advocacy models but extend them. Specifically, these groups work to augment and strengthen marginalized constituencies' ability to operate as "strategic meditative actors" (Cox 2010). This entails sustained attention to power inequalities and organizational development: learning communities and networks come together as communities of practice to share strategies, messages, skills, and lessons. To illustrate this model, we describe the communication work of GFA, fully recognizing that all three models represent simplified ideal types.

The Sierra Club, *Beyond Coal Campaign* (www.sierraclub.org/coal) and GFA's Green Jobs campaign (www.greenforall.org/resources/green-collar-job-overview) represent strategically focused advocacy campaigns that triangulated grassroots mobilizing, media advocacy, and institutional reforms that permit engagement on local, state, and/or national levels. The Sierra Club with 1.3 million members and staffed state chapters has a standing infrastructure capable of executing a higher level of strategic campaign compared to newer or more provisional social movement organizations. In contrast, GFA, as a relatively new collective actor, had to devote more attention to building its own capacity to help its constituents to act strategically and reflexively.

GFA's Green Jobs campaign provides a case in point. The rapid citizen embrace of the 2008 call for Green Jobs was not taken lightly by the fossil fuel industry, which launched a massive public relations campaign to "destroy the myth of green jobs" (http://blog.heritage.org/2009/03/18/the-green-job-myth-exposed/). Other tactics included Congressional lobbying, and outreach to community and faith-based groups in poor communities to suggest that environmental action would worsen joblessness.

GFA prioritizes organizing among communities disproportionately affected by environmental degradation. They are especially invested in poor communities and communities of color with special attention to youth of color and other constituencies historically under-represented in the environmental and climate change movements. Concretely, GFA's green jobs campaigns mapped and challenged marginalization of those communities in political, economic, and media arenas. In terms of political platform, GFA reworked its program to attend to the disproportionate impact of climate change on poor communities and communities of color (Agyeman, Bullard, and Evans 2003). In general, strategic campaigns demonstrate that a green economy will benefit marginalized communities in multiple ways—to create jobs, improve the community, and protect the environment. For instance, many groups call for and offer resources for energy efficiency. GFA, however, specifically focused on energy efficiency in multi-unit dwellings common in many urban, low-income communities. Their construction is cost-effective, energy efficient, and provides long needed housing.

Mass media strategies constitute a second area of intensified movement building. GFA and environmental justice groups are still working to gain the kind of media standing already attained by the 1.3 million-member Sierra Club and several other mainstream environmental organizations. While every organization working for environmental change struggles to maintain standing with news media, poor people and people of color face added barriers to news access. Because they are GFA's base, the organization shares their marginalization.

To counter this trend, GFA integrates media work throughout organizing so that organizers develop the expertise of a communicator, including focusing communication work to advance strategy. For instance, to raise the issue of Green Jobs in targeted audiences, GFA's outreach is constituency driven. This involved several conceptual and practical shifts with the goal of making environmental issues "cool," that is, popular with GFA's prioritized constituencies. Communication tactics for accomplishing this include:

- youth-led branding via hip hop—touring with Black Eyed Peas, Drake, and Whiz Khalifa, and appearances on Black Entertainment Television's hit show, 106 & Park;
- short YouTube videos rather than formal papers;
- tailoring media (community-controlled, local, national) to audience and event;
- embedding activities and messages in events, stories, etc. suiting priority constituencies;
- supporting current charismatic leaders while recruiting and training the next generation.

These tactics reflect cutting-edge market research and media technologies, but their selection was driven by strategy, in GFA's case a strategy that acknowledges both power inequalities and under-recognized assets possessed by prioritized constituencies.

Finally, GFA dedicates resources to building leadership networks that knit together and strengthen "relatively aware constituencies" (Carvalho and Peterson 2009) already sympathetic to the issues at hand.

- The College Ambassador program develops relationships among environmentally concerned students at fifteen Historically Black Colleges and Universities (HBCUs), linking them horizontally with each other and vertically to national environmental groups and political leaders and cultural leaders.
- The Green For All Fellowship program builds comparable horizontal and vertical networks drawing in environmental activists in local communities.
- Local business allies—primarily small businesses—enrich and benefit from similar networks.
- Communities of practice around sub-issues or campaigns facilitate collaboration across these groups.

GFA's Stormwater Infrastructure Working Group described below exemplifies how communities of practice strengthen leadership networks.

Stormwater Infrastructure Working Group

In 2011, in collaboration with American Rivers, Economic Policy Institute, and Pacific Institute, GFA published a report highlighting the strategic benefits of a sweeping renovation of the wastewater systems for U.S. communities, particularly poor communities and communities of color with high unemployment:

> We find that an investment of $188.4 billion spread equally over the next five years would generate $265.6 billion in economic activity and create close to 1.9 million jobs. We argue that maximizing the use of green infrastructure—infrastructure that mimics natural solutions—is essential to meet the stormwater management needs of our communities while also providing a number of additional co-benefits.

The report proposed state-by-state job projects as well as suggestions for 'high road standards' that would provide a living wage and benefits.

GFA tapped its national networks to form the Stormwater Infrastructure Working Group composed of public facilities professionals, technical experts, non-profits, and environmental activists. The emerging collaboration was intended to benefit communities, local economies, as well as the environment. The Stormwater Infrastructure Working Group illustrates how GFA as a bridging institution (Goodson 2005) links national, regional, and local policy makers and administrators directly with marginalized constituencies and their grassroots organizations. Participants in the Stormwater Infrastructure Working Group, for example, may have the power, the resources, and the desire to advance the goals that GFA and grassroots communities had targeted. Without the intervention of GFA and the opening of dialog with GFA's networks, however, Stormwater professionals, most often white, had generally lacked relationships, venues, and persuasive narratives that would influence leaders in marginalized communities to adopt these goals. GFA represented those perspectives and established linkages with local groups working on green job development and community environmental issues.

The net effect has been to build a sustained national network that can share lessons about environmentally sustainable approaches to waste water. Public officials in one region meet their counterparts in other public works departments and develop working relationships with actors in other sectors. They receive help with messaging so that they can convey the importance of the work. Participants in other sectors also build peer and cross sector networks. Cumulatively, these build the field by establishing common language, practices, infrastructure, lessons, and by gathering fresh, local stories for media and lobbying.

The community of practice provided an organizing mechanism through which working group members could connect with each other and build the field. Two-day gatherings of groups of twenty-five allowed working group members to share lessons via a series of conversations and structured dialogs. Follow-up national working calls allow participants to get concrete help and share best practices. GFA continues to play a catalytic role by providing technical resources, functioning as the "glue" that keeps working group participants in touch, creating communication materials that translate the technical complexity of greener stormwater management into everyday language.

Discussion

Several signature elements of social movement/justice communication stand out in the Stormwater community of practice. First, the community of practice recognizes the strategic importance of communities directly affected by environmental equalities while recruiting practitioners who exemplify good institutional actors who can plan strategic directions: in selecting storm water and job creation as joined foci, for instance, Green For All and its allies emphasize the need for jobs that do not require college degrees, which are in particularly high demand in low-income communities and communities of color.

Second, in forming the Stormwater community of practice, GFA demonstrated its bridging role. GFA links policy makers with community leaders, but as importantly, brings grassroots perspectives on environmental issues into ongoing dialogs with policy makers at the national level. This shifts the terms of the debate while heightening awareness among mainstream environmental groups and environmental reporters of the often under-recognized importance of Congressional Black Caucus members as strong climate equity advocates. GFA structured the Stormwater community of practice to build both horizontal and vertical networks.

Third, the gatherings, whether face-to-face or phone conference, are not uni-directional webinars but rather, are relational. Participants have recurring opportunities to share lessons and reflect together on the next strategic steps. This encourages their ongoing engagement in a field of action that provides mutual benefit to the economy, the community, and the environment.

We do not suggest here that communication was seamless, but simply that beyond mobilization, GFA's communicators factor into their work the need to strengthen participating groups' capacity to act and communicate in concert. Also, in contrast to the mainstream environmental movement, environmental justice organizers and communicators prioritize issue campaigns that address the structural dislocations experienced by marginalized communities. Green Jobs, for instance, more directly addresses the needs of poor communities and communities of color than Green Consumerism (Endres et al. 2009: 193; Salazar 2009).

GFA's justice communication supports field-building by sharing resources, knowledge, connections, and skills with partnering groups. In the Stormwater campaign, engaged organizations build relationships and share lessons in structured dialogs as part of a community of practice. GFA plays an essential bridging role between two sets of long-term strategic relationships—bottom-up grassroots groups and top-down influential policy makers. Policy makers need the stories generated by local activists, community fellows, communities of practice, and student ambassadors. GFA local fellows, ambassadors, and communities of practice benefit from exposure to national debates and to each other. This exposure expands their policy visions, strengthens their communication capacity and relations with comparable groups nation-wide.

As a bridging institution, GFA ensures that grassroots experiences, expertise, and points of view are brought into the national policy arena. For instance, in June 2013, President Obama speaking at Georgetown University elaborated on his administration's climate change policy (Obama 2013a). While noting the particular climate vulnerability of developing nations, the June unveiling of the President's Climate Action paid little attention to the particular vulnerability of poor communities and communities of color in the United States. Over the next six months, GFA and other national advocates such as the NAACP, the Joint Center for Political and Economic Studies, and many others collaboratively worked to include more explicit language highlighting the unique vulnerabilities of marginalized communities and tribal nations. The administration's Executive Order, "Preparing the United States for the Impacts of Climate Change" (Obama 2013b) demonstrated growth in that regard:

> The impacts of climate change—including an increase in prolonged periods of excessively high temperatures, more heavy downpours, an increase in wildfires, more severe droughts, permafrost thawing, ocean acidification, and sea-level rise—are already affecting communities, natural resources, ecosystems, economies, and public health across the Nation. These impacts are often most significant for communities that already face economic or health-related challenges, and for species and habitats that are already facing other pressures. Managing these risks requires deliberate preparation, close cooperation, and coordinated planning by the Federal Government, as well as by stakeholders, to facilitate Federal, State, local, tribal, private-sector, and nonprofit-sector efforts to improve climate preparedness and resilience; help safeguard our economy, infrastructure, environment, and natural resources; and provide for the continuity of executive department and agency (agency) operations, services, and programs.
>
> (Obama 2013b)

While less visible than media campaigns, these dialogs created an opening for under-represented constituencies' needs, perspectives, and experiences to be included at the agenda-setting stage (Lentz 2013).

Justice communication's added value

To challenge communication apartheid, justice-driven communication models map power inequalities, then create inclusive strategies that add communication capacity building to each campaign element. In other words, while advancing a specific campaign, justice communicators work to establish sustainable communication practices and infrastructure, position their organizations' as routine, credible news sources, and influence communication policies and technologies (Lentz 2010; www.MAG-NET.org; www.centerformediajustice.org). This allows marginalized constituencies to become fuller participants in social movements (Cutting and Themba-Nixon 2006; Cyril 2005); GFA's presence, for instance, extends the broader environmental movement's strategic goals, and builds the capacity of disenfranchised constituencies to communicate as empowered citizens.

To accomplish this, GFA and other justice communicators adopt elements of social marketing and media advocacy, but expand attention to power inequalities that undermine the participation of global south, poor communities, and communities of color. This entails building the capacity of directly affected constituencies to participate more fully in all phases of campaign planning, mobilizing, and evaluation. We summarize the key features of each model and the added value of justice communication in Table 10.1.

Conclusion

Victories by StepItUp2007, Sierra Club, Green For All, and other environmental groups are achieved with great difficulty. Movement theorists have always cautioned, however, that successful organizing is not easy or assured (Gamson 1975). Power, including communication power, varies by constituency (Couldry 2003), with political and discourse opportunities more plentiful for some. Even theorists of power often understate this inequality of communication power (Lukes 2005; Gaventa and Cornwall 2008). Yet, communication inequalities remain one of the fundamental reasons that opportunities for civil engagement remain more accessible to some constituencies than others.

As communities of practice (Wenger 2000) or learning communities (Senge, Kleiner, Roberts, Ross, and Smith 1994), movement organizations accrue knowledge in the course of designing, executing,

Table 10.1 Comparing communication mobilization models

Comparison	Social marketing approach	Media advocacy approach	Movement building approach
Underlying theoretical perspective	Individuals act in self interest	Pluralist model of democracy Interest groups compete on own behalf	Consistent democracy; disenfranchised join across inequalities to challenge communication apartheid, systemic inequalities
Mode of change	Individuals changing behavior	Individuals form interest groups to pressure elites for policy changes	Collective actors build relationships that prefigure more equitable relations
Target of change	Individuals	Elite policy makers	Create sustainable counterpractices that shift power toward the marginalized
Agent responsible for change	Individuals or peer groups	Experts/professionals including non-profits and NGOs "advocare" = to speak for another	Collective actors that include those directly affected by power inequalities

Source: authors.

and critiquing strategies (Brulle 2010; Jamison 2010). But this is far from automatic. Sustainable vision, knowledge-production, and relationships can grow from a campaign focused around a single event, but only if organizational capacity building is built into organizing. Movement building depends on collective actors' ability to share understandings across campaigns and generations in the interest of advancing a common goal.

Marginalized communities live with relentless economic, social, and political pressures that undermine capacity building and transfer of learning from one group and generation to others. To address these pressures, justice communication offers environmental and climate justice movements three valuable contributions—systematic attention to power inequalities and capacities among constituencies, intentional development of marginalized communities as strategic collective actors, and organizational forms that encourage reflection, knowledge sharing, and retention.

Acknowledgments

We thank Joseph Christiani and Jeremy Hays for extensive feedback.

Note

1   We do not address development and participatory communication in this chapter. For critical assessments, see Bah (2008); Ryan and Jeffreys (2012); Steeves (2001); and Waters (2000).

Further reading

Cox, J. R. (2010). Beyond frames: Recovering the strategic in climate communication. *Environmental Communication: A Journal of Nature and Culture, 4*(1), 122–133.

Cox demonstrates how two environmental organizations focused communication strategies to build the public will to address climate change.

Dunlap, R. E. and McCright, A. M. (2011). Organized climate change denial. In Dyzek, J., Norgaard, R. and Schlosberg, D. (eds), *The Oxford Handbook of Climate Change and Society* (pp. 144–160). Oxford: Oxford University Press.

The authors map how conservative movement networks worked to stall the Kyoto Protocol and subsequent climate change initiatives.

Pastor, M., Bullard, R., Boyce, J. K., Fothergill, A., Morello-Frosch, R., and Wright, B. (2006). Environment, disaster, and race after Katrina. *Race, Poverty and the Environment*, summer, 21–26.

The authors analyze how institutionalized race and class inequalities translate into systemic environmental injustices and provide accessible examples.

Sandler, R. and Pezzullo, P. (eds) (2007). *Environmental Justice and Environmentalism: The Social Justice Challenge to the Environmental Movement*. Cambridge, MA: MIT Press.

In addition to well-chosen chapters, the editors define environmental justice, provide Sierra Club guidelines for environmental justice organizing, and *Principles of Working Together* developed by the 2002 People of Color Leadership Summit.

References

Ad Council (2011). *Wildfire Prevention Case Study*, www.adcouncil.org/Impact/Case-Studies-Best-Practices/Wildfire-Prevention (accessed January 2, 2014).

Agyeman, J., Bullard, R. D., and Evans, B. (eds) (2003). *Just Sustainabilities: Development in an Unequal World*. Cambridge, MA: MIT Press.

Arendt, H. (1969). *On Violence*. New York: Harcourt Brace & World.

Bah, U. (2008). Daniel Lerner, Cold War propaganda and US development communication research: An historical critique. *Journal of Third World Studies, 25*(1), 183–198.

Barndt, D. (1989). *Naming the Moment: Political Analysis for Action. A Manual for Community Groups*. Ontario: The Moment Project. Jesuit Centre for Faith and Social Justice. Available at: www.popednews.org/downloads/naming the moment.pdf (accessed November 12, 2013).

Brulle, R. J. (2010). From environmental campaigns to advancing the public dialog: Environmental communication for civic engagement. *Environmental Communication, 4*(1), 82–98.

Bullard R., Johnson, G., and Torres, A. (2005). Addressing global poverty, pollution, and human rights. In Bullard, R., Warren, R., and Johnson, G. (eds), *The Quest for Environmental Justice: Human Rights and the Politics of Pollution* (pp. 279–297). San Francisco, CA: Sierra Club Books.

Bullard, R., Warren, R., and Johnson, G. (2005). *The Quest for Environmental Justice*. San Francisco, CA: Sierra Club Books.

Carvalho, A., and Peterson, T. R. (2009). Discursive constructions of climate change: practices of encoding and decoding. *Environmental Communication, 3*(2), 131–133.

Couldry, N. (2003). Beyond the Hall of Mirrors? Some theoretical reflections on the global contestation of media power. In Couldry, N., and Curran, J. (eds), *Contesting Media Power: Alternative Media in a Networked World* (pp. 39–54). Lanham MD: Rowan & Littlefield.

Cox, J. R. (2010). Beyond frames: Recovering the strategic in climate communication. *Environmental Communication, 4*(1), 122–133.

Cryan, P. 2011. *Strategic Practice for Social Transformation*. Available at: www.strategicpractice.org (accessed January 9, 2011).

Cutting, H. and Themba-Nixon, M. (2006). *Talking the Walk: A Communication Guide for Racial Justice*. San Francisco, CA: AKD Press.

Cyril, M. A. (2005). Media and marginalization. In McChesney, R., Newman, R., and Scott, B. (eds), *The Future of Media: Resistance and Reform in the 21st Century* (pp. 97–104). New York: Seven Stories Press.

Dorfman, L., Wallack, L., and Woodruff, K. (2005). More than a message: Framing public health advocacy to change corporate practices. *Health Education and Behavior, 32*(3), 320–336.

Downing, J. (2001). *Radical Media: Rebellious Communication and Social Movements*. Thousand Oaks, CA: Sage.

Dunlap, R. E., and McCright, A. M. (2011). Organized climate change denial. In Dyzek, J., Norgaard, R., and Schlosberg, D. (eds), *The Oxford Handbook of Climate Change and Society* (pp. 144–160). Oxford: Oxford University Press.

Endres, D., Clarke T., Garrison, A., and Peterson, T.R. (2009). Toward just climate-change coalitions: Challenges and possibilities in the Step It Up 2007 Campaign. In Endres, D., Sprain, L. and Peterson (eds), *Social Movement to Address Climate Change: Local Steps for Global Action* (pp. 179–209). Amherst, NY: Cambria Press.

Frey, L. R. and Carragee, K. M. (eds) (2007). *Communication Activism* (2 Vols). Cresskill, NJ: Hampton Press.

Frey, L. R. and Carragee, K. M. (eds) (2012). *Communication Activism*: Vol. 3. *Struggling for Social Justice amidst Difference*. Cresskill, NJ: Hampton Press.

Freire, P. (1970). Cultural freedom in Latin America. In Alba, V. and Colonnese, L. (eds), *Human Rights and the Liberation of Man in the Americas*. Notre Dame, IN: University of Notre Dame Press.

Gamson, W. A. (1975). *Strategy of Social Protest*. Homewood, IL: Dorsey Press.

Gaventa, J. and Cornwall, A. (2008). Power and knowledge. *The Sage Handbook of Action Research: Participative Inquiry and Practice* (pp. 172–189). Los Angeles, CA: Sage Publications.

Goodson, A. (2005) Building bridges, building leaders: Theory, action, and lived experience. In Croteau, D., Hoynes, W. and Ryan, C. (eds), *Rhyming Hope and History: Activists, Academics, and Social Movement Scholarship*. Minneapolis, MA: University of Minnesota Press.

Green For All. Available at www.greenforall.org (accessed December 15, 2013).

Hinson, S. and Healey, R. (2011). *Strategic Practices*. Available at: www.strategicpractices.org (accessed January 9, 2011).

Hirschkop, K. (1996). Democracy and the new technologies. *Monthly Review*, July–August, 86–98.

Howley, K. (2005). *Community Media: People, Places, and Communication Technologies*. Cambridge: Cambridge University Press.

Independent Media Center (2004). *The IMC: A New Model*. Available at: www.hedonistpress.com/pdfs/v1.1.pdf (accessed December 10, 2013).

Jamison, A. (2010). Climate change knowledge and social movement theory. *Wiley Interdisciplinary Reviews: Climate Change, 1*(6), 811–823.

Koon, D. (2013). A Q&A with chemist and MacArthur "Genius Grant" recipient Wilma Subra, on the Mayflower oil spill. *Arkansaw Reporter*, July 17, 2013. Available at: www.arktimes.com/arkansas/a-qanda-with-chemist-and-macarthur-genius-grant-recipient-wilma-subra-on-the-mayflower-oil-spill/Content?oid=2978391 (accessed February 14, 2014).

Lentz, R. (2010). Media infrastructure policy and media activism. (October 2010). In Downing, J. D. (ed.), *Sage Encyclopedia of Social Movement Media* (pp. 323–236). Thousand Oaks, CA: Sage.

Lentz, R. (2013). Excavating historicity in the US network neutrality debate: An interpretive perspective on policy change. *Communication, Culture and Critique, 6*(4), 568–597.

Lukes, S. (2005). *Power: A Radical View*, 2nd ed. Hampshire and New York: Palgrave Macmillan.

McKibben, Bill. (2009). Foreword. In Endres, D., Sprain, L. and Peterson, T.R. (eds), *Social Movement to Address Climate Change: Local Steps for Global Action*. Amherst, NY: Cambria Press.

McNamara, E., Kurth T., and Hansen, D. (1981). Communication efforts to prevent wildfires. In Rice, R. and Paisley, W. (ed.), *Public Communication Campaigns* (pp. 143–160). Beverly Hills, CA: Sage.

Morris, A. D. (1986). *Origins of the Civil Rights Movements*. New York: Simon & Schuster.

Napoli, P. M. and Aslama, M. (eds) (2011). *Communications Research in Action: Collaborations for a Democratic Public Sphere*. New York, NY: Fordham University Press.

Obama, B. (2013a). Remarks by the President on Climate Change. June 25, 2013, Georgetown University. Available at: www.whitehouse.gov/the-press-office/2013/06/25/remarks-president-climate-change (accessed January 18, 1914).

Obama, B. (2013b). Executive Order—Preparing the United States for the Impacts of Climate Change. November 1, 2013. Available at: www.whitehouse.gov/the-press-office/2013/11/01/executive-order-preparing-united-states-impacts-climate-change (accessed January 18, 2014).

Pastor, M., Bullard, R., Boyce, J. K., Fothergill, A., Morello-Frosch, R., and Wright, B. (2006). Environment, disaster, and race after Katrina. *Race, Poverty and the Environment*, summer, 21–26.

Pertschuk, M. (2001). *Smoke in Their Eyes: Lessons in Movement Leadership from the Tobacco Wars*. Nashville, TN: Vanderbilt University Press.

Powell, J. (2007). *Structural Racism: Building upon the Insights of John Calmore*. 86 N.C. L. Rev. 791 (2007). Available at: http://scholarship.law.berkeley.edu/facpubs/1637 (accessed December 15, 2013).

Riaño, P. and Alcalá, P. R. (1994). *Women in Grassroots Communication: Furthering Social Change*. Thousand Oaks, CA: Sage Publications.

Rice, R. E. (2001). Smokey Bear. *Public Communication Campaigns*, *3*, 276–279.

Rodriguez, C., Kidd, D., and Stein, L. (eds) (2010). *Making Our Media. Global Initiatives toward a Democratic Public Sphere. Volume One: Creating New Communication Spaces*. Cresskill, NJ: Hampton Press.

Ryan, C. and Alexander, S. (2006). Reframing the presentation of environmental law and policy. *Boston College Environmental Affairs Law Review*, *33*, 563.

Ryan, C. and Jeffreys, K. (2012) Challenging domestic violence: Trickle-up theorizing about participation and power. In Frey, L. and Carragee, K. (eds), *Communication Activism Volume Three: Struggling For Social Justice Amidst Difference*. Cresskill, NJ: Hampton Press.

Salazar, D. J. (2009). Saving nature and seeking justice: Environmental activists in the Pacific Northwest. *Organization and Environment*, *22*(2), 230–254.

Sandler, R., and Pezzullo, P. (eds) (2007). *Environmental Justice and Environmentalism: The Social Justice Challenge to the Environmental Movement*. Cambridge, MA: MIT Press.

Senge, P., Kleiner, A., Roberts, C., Ross R. B., and Smith, B. J. (1994). *The Fifth Discipline*. New York: Currency Doubleday.

Shiva, V. (1999). Ecological balance in an age of globalization. In Low, N. (ed.), *Global Ethics and Environmentalism* (pp. 47–69). London: Verso Books.

Siegel, M. and Doner, L. (2004). *Marketing Public Health: Strategies to Promote Social Change*. Boston, MA: Jones & Bartlett Publishing.

Steeves, H. L. (2001). Liberation, feminism, and development communication. *Communication Theory*, *11*, 397–414.

Tutu, D. (2007). We do not need climate change apartheid in adaptation. In United Nations Development Programme (ed.), *Human Development Report 2007/2008* (pp. 166–186). New York: UNDP.

Wallack, L. and Dorfman, L. (1996). Media advocacy: A strategy for advancing policy and promoting health. *Health Education and Behavior*, *23*(3), 293–317.

Walsh, D. C., Rudd, R. E., Moeykens, B. A., and Moloney, T. W. (1993). Social marketing for public health. *Health Affairs*, *12*(2), 104–119.

Waters, J. (2000). Power and praxis in development communication discourse and method. In K. G. Wilkins (ed.), *Redeveloping Communication for Social Change: Theory, Practice, and Power* (pp. 89–101). Lanham, MD: Rowman & Littlefield.

Weiss, J. A. and Tschirhart, M. (1994). Public information campaigns as policy instruments. *Journal of Policy Analysis and Management*, *13*(1), 82–119.

Wenger, E. (2000). *Communities of Practice*. New York: Cambridge University Press.

Wiebe, G. D. (1951). Merchandising commodities and citizenship on television. *Public Opinion Quarterly*, *15*(4), 679–691.

11

THE CHANGING FACE OF ENVIRONMENTAL JOURNALISM IN THE UNITED STATES

Sharon M. Friedman

Exactly what constitutes environmental journalism has become more difficult to define in today's crowded multimedia and social media landscape. While environmental journalism is a relatively new reporting specialty, having started in the 1960s (Friedman, 1999, 2004; Wyss, 2008), within its relatively short span, it has undergone many changes. These changes continue today at a faster pace, driven by journalism's changing business model, media convergence and the rise of the Internet. This chapter discusses many of the changes that have affected environmental journalism over time in the United States. It focuses on the United States because, while some of these changes also have occurred in other countries, different media structures, ownership histories and styles have provided varying results in other nations. For example, while a decline in the traditional newspaper business model has been evident in the United States, Canada and the United Kingdom, it does not seem to be affecting large Asian countries such as China and India, where newspaper readership is expanding. And no journalism crisis was felt by science journalists—who often cover environmental issues—in Latin America, Asia and North and Southern Africa, in contrast to science journalists' crisis concerns in the United States, Europe and Canada (Bauer et al., 2013).

Before tracing changes in environmental journalism over a long timeline, this chapter reminds readers that the mass media are not the only source of environmental information for people. It then introduces some early predecessors of environmental coverage and tracks varying cycles of public interest and environmental coverage in the United States from the 1970s to the early years of the new decade. Moving closer to the present, it discusses the impacts that media convergence and downsizing have had on environmental coverage in newspapers and on television and focuses on actions to eliminate the environmental "beat," using the *New York Times* as an example.

The chapter then examines some of the positive and negative effects that the Internet has had on U.S. environmental coverage and presents examples of selected traditional media websites, online publications and news aggregators, and some of the changes they have faced. After briefly reviewing several prize-winning multimedia examples of investigative environmental reporting, the chapter looks to the future and additional changes that might be needed to make environmental journalism a more compelling topic for both traditional media and Internet readers and viewers in the United States.

Changing environmental information sources

For years, newspapers and magazines were the mainstay of environmental journalism in the United States, with less emphasis on television and radio. Now, online media publications, blogs, Twitter feeds, YouTube videos and Facebook pages about environmental issues—with global reaches—are produced not only by reporters but also by nonprofit organizations, government and industry groups, and private individuals. Some of the Internet news content does not follow traditional journalistic rules for balance and fairness; some would not have been called journalism 20 or so years ago. Yet, whether labeled journalism or not, digital and social media sources have had a tsunami-like impact in providing environmental information that cannot be discounted, or more importantly, should not be overlooked, particularly in light of major declines in traditional media coverage of environmental issues, particularly in the United States.

The media and the Internet, however, are not the only sources of environmental information. Even before the Internet, books such as *Sand County Almanac* by Aldo Leopold, *Silent Spring* by Rachel Carson and *The End of Nature* by Bill McKibben, and television documentaries and films such as *The Toxic Assault on our Children* by Bill Moyers and Al Gore's *An Inconvenient Truth* discussed important environmental issues. Novels including *Flight Behavior* by Barbara Kingsolver or *Nature's End* by Whitley Strieber and James Kunetka, and movies such as *Erin Brockovich* and *Avatar* also helped spread environmental ideas.

Many other aspects of popular culture have also played a major role in disseminating environmental information in the United States. The average middle-class family's "knowledge and understanding of the environment is constructed and maintained by a constant stream of language and images derived from popular culture," according to Meister and Japp (2002). Among the many instances they describe as part of an environmental information montage seen in one day by a typical family could be beautiful mountain pictures on calendars, Weather Channel reports about hurricanes or floods, a 30-second television spot on the morning news about climate change, an advertisement showing an SUV being driven up mountain trails and "perching atop the Grand Canyon," school projects involving recycling, and evening television programs that emit "a steady stream of advertising and programming, all depicting the environment as a backdrop for human activity" (Meister and Japp, 2002, p. 3). With the advent of the Internet, these "popculture" environmental exposures have only increased.

Still, one cannot discount the importance of mass media and Internet coverage of environmental issues. Communication researchers have found that the "mass media are an integral part of the process of development of political and social opinions and, as such, they serve as a forum for debate on matters of public interest" (Dunwoody and Peters, 1993, p. 297). Media coverage draws people's attention to salient issues, serving an agenda-setting function. This means that while the media may not be successful in telling people what to think, they are successful in telling readers what to think about (Cohen, 1963). And by calling attention to various issues, the media serve as a watchdog, focusing public attention on important problems. Environmental journalism in the United States has often played this watchdog role since its earliest days.

Cycles of environmental journalism coverage

Whether public attention has been focused on preserving land for national parks, protecting communities from chemical dumping, contending with droughts and wildfires or developing policies and programs to deal with a changing climate, media interest in these topics has been closely linked with this public interest – spurring and responding to it. However, both media and public attention to environmental issues is cyclical and often short term, drawn by specific events or controversial issues that demand attention.

While conservation of the nation's natural resources was an important topic from the late 1800s through World War II (Frome, 2004), *Silent Spring* helped shift the nation's environmental focus to the harmful effects of chemicals, particularly pesticides such as DDT. Many people believe that by presenting a view of nature where man-made pesticides threatened birds, insects, fish and humans, the book helped spur the environmental movement that subsequently developed (Grizwold, 2012).

Increased public interest and discussion of environmental pollution and chemical dumping, coinciding with environmental incidents such as the Santa Barbara oil spill and Ohio's Cuyahoga River fire in 1969, prompted an increasing number of articles in the nation's media. The first Earth Day and the creation of the Environmental Protection Agency in 1970, and the passage of major environmental laws several years later focused even more public attention on environmental concerns, which did not escape the nation's media (Wyss, 2008).

In 1969, *Time, Life, Look* and *National Geographic* magazines all increased their environmental coverage, and the *New York Times* created an environmental beat (Sachsman, 2003). Some larger newspapers soon added specialized environmental reporting, as did a few local newspapers such as the *Niagara Gazette*. The *Gazette*'s coverage of toxic dumping in the Love Canal in Niagara Falls, New York, helped make Love Canal a symbol of the major hazards of careless industrial chemical dumping (Wyss, 2008).

Although the quantity of environmental journalism coverage increased during the 1970s, journalists recognized problems in its quality. According to a study by William Witt of the University of Wisconsin in 1972–1973, environmental reporters felt that their beat was inadequately covered with not enough staff, time and space, and that there was too much concentration on environmental problems, crisis reporting and sensationalism. They also complained about their editors' lack of interest in following the complexities and development of most environmental issues (Friedman, 1991).

The frequent but often superficial coverage of chemical dumping and accidents in the 1970s was followed by a period in the early and mid-1980s when environmental coverage slumped in numbers and environmental threats became less visible such as the loss of biodiversity or slow-onset chemical contamination. Even though scientists could detect many more health and ecological impacts, reporting about them was more difficult in the traditional U.S. news format (Cox, 2006). Consequently, as in the 1970s, environmental reporting in the 1980s often focused on describing events but not the root causes or multifaceted aspects of these events. Superficial reporting dominated media coverage of some of the most famous environmental disasters of the time such as the Bhopal chemical accident in India where thousands were harmed by poisonous gas (Wilkins, 1987), the Exxon Valdez oil spill in Prince William Sound in Alaska, and the Chernobyl nuclear accident in the Soviet Union (Friedman, Gorney and Egolf, 1987). The evacuation of the town of Times Beach, Missouri, because of residents' exposure to dioxin, a highly toxic organic pollutant, was often covered without discussing the numerous scientific and political aspects of the issue (Carmody, 1995; Friedman, 1999).

Environmental beats began to disappear in the United States in the 1980s, just as environmental issues were becoming more complex involving health effects, scientific uncertainty, economic and regulatory issues. Without knowledgeable reporters to write them, many articles, often written by general assignment journalists, lacked depth and context about these complex issues and they confused readers and viewers about environmental health risks.

Beginning in 1988, when a severe drought and summer heat wave helped create a resurgence of public environmental interest, the number of reporters assigned to the environmental beat and the amount of pages and airtime devoted to environmental topics increased considerably. In 1988, *Time* once again made a commitment to increase its environmental coverage as did a number of major U.S. newspapers and magazines. This spate of greatly increased media attention lasted for about four years and included the 1992 Rio Earth Summit, which generated reams of copy. By 1993, however, public and media interest in environmental issues began to decline again as people's interest turned to other issues (Friedman, 2004).

Without headline-grabbing environmental disasters or massive Earth Day celebrations, the amount of space or airtime given to environmental topics continued to shrink throughout the middle 1990s and the 2000s. A 1996 survey found that 23 percent of newspaper and 44 percent of television environmental journalists were spending less time reporting about environmental issues than they had the year before (Detjen, Fico and Li, 1996). There was some uptick in environmental coverage with President George W. Bush's election, but the September 11, 2001, terrorist attack focused U.S. media coverage on terrorism and national security, reducing space and airtime for specialized news including the environment. In 2002 interviews, 12 senior environmental journalists agreed that the amount of space or airtime at some publications and most broadcast stations had decreased overall with fewer resources for enterprise or investigative stories (Friedman, 2004).

While air, water and ground pollution were the dominant topics in the early years of U.S. environmental journalism, as the field matured, other more complex and wide-ranging issues were added including endocrine disruption, land management, sustainability, overfishing, invasive species, farm practices, genetically modified organisms and various energy sources, among others. Climate change, with all of its complexity and controversy, had a mixed coverage pattern, and it is important to remember that "the ups and downs in environmental issues coverage ... rarely if ever, reflect just a single influential factor, but likely result from complex interaction of multiple factors" (Hansen, 2011, p. 15). The impact of consolidation and downsizing in the traditional media was one major factor that made pursuing many environmental topics in depth harder to do.

The impacts of U.S. media downsizing

A shrinking U.S. economy, loss of advertising revenues, and audiences migrating to the Internet on tablets and mobile devices produced a "perfect storm" that forced traditional mass media outlets to downsize and change the way they dealt with news. By 2004, a growing number of news outlets faced declining profits, leading to newsroom cutbacks in staff and in the time reporters had to gather and report the news (Pew Research Center's Project on Excellence in Journalism, 2004). Specialty beat reporters were particularly vulnerable. To attract more readers and viewers, the media began to emphasize more local reporting, features about living and style, and celebrity news.

Journalists faced many pressures trying to maintain quality. Perry Beeman, an environmental journalist then with the *Des Moines Register*, expressed the concerns of many of his colleagues:

> Is there going to be room for serious environmental journalism in an O.J., Robert Blake, Janet Jackson, Michael Jackson world in which television networks have resigned themselves to presenting as news not-so-hard-hitting interviews about a star's latest CD? Will chain journalism, in which top leaders are at locations far removed from some of a company's media outlets, leave room for the independent thinking, probing pieces we offer? Can we expect column inches and air-time slots to expose the dangers of air pollution or the effects of the latest endangered-species battle when a certain element thinks the way to get readers to buy the paper is to tell them how to buy a prom dress, cook a Thanksgiving turkey, or carve a Jack o'Lantern?
>
> (Beeman, 2005)

As the decade progressed, the number of U.S. media buyouts and layoffs grew, particularly among newspapers, and to a lesser extent in network television. According to the Pew Center's State of the Media Report for 2008, the role of news was shifting from being a product to becoming a service to help and empower readers. All of these developments worked against original reporting and led news people to question whether the core values of accuracy and verification could be maintained when fewer people were being asked to do more (Pew Research Center's Project on Excellence in Journalism, 2008).

Downsizing continued throughout the rest of the decade, and estimates of newsroom cutbacks in 2012 had the industry with fewer than 40,000 full-time professional employees, down 30 percent from its peak in 2000. In local television, already brief stories were shortened even more, and sports, weather and traffic now accounted on average for 40 percent of the content produced. Many of these changes resulted in a "news industry that is more undermanned and unprepared to uncover stories, dig deep into emerging ones or to question information put into its hands." The Pew Center report noted that the continuing erosion of news reporting resources provided growing opportunities for "those in politics, government agencies, companies and others to take their message directly to a public" (Pew Research Center's Project

147

on Excellence in Journalism, 2013, Overview). And this public importantly includes "a generation raised on the idea that information is free of financial burdens, and, like the Pony Express, will get through even if the so-called mainstream media is forced out of business" (Cowell, 2014).

Reflecting on the dire financial situation that led to many of these changes, environmental journalist and *New York Times* Dot Earth blogger Andrew Revkin (2013) offered an environmental analogy:

> These background financial pressures, building around the industry the same way that heat-trapping greenhouse gases are building in the atmosphere, are what will erode the ability of today's media to dissect and explain the causes and consequences of environmental change and the suite of possible responses.

Mainstreaming environmental coverage

Environmental journalism, like many other specialty beats, was hit hard by this downsizing, which transformed numbers of staff reporters into freelancer writers. Environmental reporters "became an endangered species of sorts" (Dykstra, 2013) as many long-standing environmental newspaper and television beats and newer blogs at major newspapers disappeared. Mainstreaming,[1] a perceived need to spread environmental coverage among various topic areas such as politics, business, science, health and technology, has been part of a long-running disagreement over whether it is more effective to concentrate environmental reporting within a specialized beat or spread out its coverage because it affects many different subjects. Media top brass used mainstreaming, which conveniently fit their cost-cutting needs, as one of the main reasons behind the loss of environmental beats and blogs.

In 2008, CNN cut its science, technology and environmental news staff and six executive producers. "We want to integrate environmental, science and technology reporting into the general editorial structure rather than have a stand-alone unit," a CNN spokesperson said. Also that year, NBC cut the entire staff of the "Forecast Earth" environmental program during staffing cuts at The Weather Channel (Brainard, 2008). In 2010, the *Wall Street Journal's* energy and environmental blog, "Environmental Capital," ended after a two-year run (Brainard, 2010c), and the *Christian Science Monitor* canceled its "Bright Green" blog after 22 months. Explaining the decision, the *Monitor's* editor said environmental coverage needed more mainstream coverage in recent years and "all reporters are expected to be well versed in environmental issues and pursue these as a matter of course" (Brainard, 2010b).

The *New York Times* also used mainstreaming as a reason to shut down its three-year-old environmental desk, or "pod," in January 2013, redistributing its two editors and seven reporters. These reporters, who had exclusively covered the environment, comprised the largest single environmental staff of any daily U.S. newspapers (Bagley, 2013). Two months later, the *Times* also ended its Green blog. With the end of the Green blog came the loss of "$40,000-a-year worth of freelance reporting … that extended the sweep and scope of environmental coverage" (Sullivan, 2013b). According to Media Matters, a progressive media watchdog group, environmental coverage by the *Times*' Green blog often proved to be more thorough than that of the print edition. For example, a Media Matters report found that the Green blog covered a report on the rapid decline of the Great Barrier Reef while the print edition, along with several other mainstream media outlets, ignored it. The same was true for a 2012 World Bank report warning of the calamitous effects of climate change (Fitzsimmons, 2013).

The *Times*' managing editor Dean Banquet explained that while these changes were done in part to cut costs, they were done more for coverage reasons. "I think our environmental coverage has suffered from the segregation—it needs to be more integrated into all of the different areas, like science, politics and foreign news," he said (Sullivan, 2013a). Although the *Times* still has an environmental section on its online masthead, it is now subsumed under the science section.

Supporting the *Times*' mainstreaming decision, senior editor Jonathan Thompson of the regional environmental publication *High Country News*, said: "to de-silo-ize/de-ghettoize the environmental coverage is a good thing, potentially." He said he was hoping that without a dedicated team and editor, the reporters at the *Times* will continue to prioritize environmental stories (Guerin, 2013). The *Times'* Dot Earth blogger Revkin (2013) agreed: "The *Times* excelled at environmental coverage before there was an environment pod, continued during that phase and, I predict, will do so going forward, within the financial constraints facing all journalism."

However, Margaret Sullivan (2013a), public editor of the *Times*, said she was not convinced that the *Times'* environmental coverage "will be as strong without the team and the blog. Something real has been lost on a topic of huge and growing importance."

Even before these changes occurred, the *Times* ran fewer environmental stories in 2011 and 2012 than in its peak year in 2010, according to a survey of environment-related stories aggregated by Environmental Health Sciences (EHS). The publisher of EHS noted, however, that the total number of stories in the *Times* still remained higher than those of other U.S. daily newspapers (Anonymous, 2013).

Trying to summarize the effects of the downsizing and mainstreaming media changes in early 2013, Curtis Brainard, who analyzed science, environmental and health coverage for the *Columbia Journalism Review*, estimated that only between 12 and 15 full-time environmental reporters remained at the nation's five top newspapers: two at the *Wall Street Journal*, three at *USA Today*, three at the *Washington Post* and three at the *Los Angeles Times*, along with probably five *New York Times* reporters who would continue to write about the environment (2013). Shortly after his assessment appeared, one of the main environmental reporters for the *Washington Post* was reassigned to the White House beat and replaced by a sports editor with no environmental or science background (Raeburn, 2013).

In November 2013, the *Times'* public editor Sullivan returned to the issue of the *Times'* removal of the environmental "pod" and the Green blog, observing that the quantity of climate change coverage had decreased as had the amount of deep, enterprising coverage of climate change. "In the six months since these changes, there were only three front-page stories in which climate change was the main focus compared with the year before," she said, explaining that what was missing was the number and variety of fresh angles from the previous year (2013b).

She pointed out, however, some good news for environmental journalism. The *Times* had moved to strengthen its environmental reporting by hiring a Washington-based reporter to cover the U.S. Environmental Protection Agency, while two other reporters had begun covering environmental issues. Also, a science desk editor had been appointed to coordinate environmental coverage in addition to other duties (Sullivan, 2013b).

Changing television network coverage

Losses of environmental coverage were not as severe with U.S. television network news as they were with the major newspapers. One reason was that television network news had not lost as much of its audience and advertising, and its reach was aided by its cable-network partners and live streaming to Internet and mobile devices. Still, as in the past, viewers did not find much environmental news on the weekday evening news broadcasts of the three major U.S. networks. Evening newscast minutes devoted to environmental issues from 2008 to 2012 were minimal with the exception of two major events that drove up coverage in 2010 and 2011.

The environmental event that generated the most network television coverage during that period was the BP oil spill in the Gulf of Mexico: 1,410 minutes of airtime in 2010 and 51 minutes in 2011 on the three evening weekly newscasts. The Japanese earthquake, tsunami and resulting Fukushima nuclear plant meltdown received 389 minutes in 2011 and 70 minutes in 2012 on the three evening weekly newscasts. However, much of the Japanese coverage was focused more on the state of the disaster and its impacts on people, the economy and the nation than on the environment. From 2008 through 2012, excluding these disasters and others related to weather events such as tornados and hurricanes, the most covered environmental subjects concerned a small variety of energy issues. Climate change concerns were allotted 153 minutes—about two and one-half hours—from 2008 to 2012 on national network news (Tyndall, 2008–2012).

In 2013, coverage of climate change received more airtime in nationally broadcast news than in the previous few years, according to a study by Media Matters. Climate change coverage increased on the Sunday morning talk shows and on nightly news broadcasts to a total of 27 minutes and 102 minutes respectively for the year, but was nowhere near its peak year of 2009. The study described overall broadcast coverage of climate change in 2013 as "tepid" despite the occurrence of a major report, Presidential speech and scientific milestone on climate change during the year (Santhanam, 2014).

Internet coverage of environmental journalism

While many of the financial impacts resulting from the rise of the Internet and digital and social media led to downsizing and mainstreaming in the world of traditional journalism, many positive effects also resulted from these new developments, particularly in enlarging environmental journalism in range, audience and participants.

In the years before the media downsizing, U.S. environmental journalists were enthusiastic about the possibilities offered by the Internet. One environmental reporter said the Internet drastically changed the way journalists did their jobs, mostly for the better, because it was now "much easier to find a broad range of voices for stories and to get background material quickly and efficiently" (Friedman, 2004, p. 183).

Pointing out many important online capabilities for reporters, an environmental journalist explained: "We can load whole databases on our websites. Let people look up their town and find out how pollution there ranks with the rest of the state or country." He added that reporters can augment their stories with audio, video, Internet links, blogs, expanded photos, polls, questions and answers, music and quizzes. They can also use a digital camera to give the sights and sounds of the kayakers, the birds, the cleanup crews and wildlife (Beeman, 2006).

Numerous resources and databases are now available to environmental journalists of all types—staffers, freelancers, bloggers—and to people interested in environmental issues, including the multitude of websites provided by Google and Yahoo. Hundreds, if not thousands, of blogs about environmental issues are online for readers. Many are on Tumblr, written by staff and freelance environmental journalists, scientists, environmental activists, students, industry groups, political organizations and citizens. Including podcasts, Facebook, Twitter and YouTube posts and reader comments, there is no dearth of environmental information on the Internet. What follows are brief descriptions of several of the many components of environmental journalism on the Internet and some of the changes that have recently affected them.

Traditional news organization websites and multimedia reporting

As newspapers, magazines, radio and television stations, and cable networks developed websites, they had to change the way they processed news. Their websites now store thousands of images, stories and documents, allowing news consumers to access far more information than could be presented in newsprint or on the air and for much longer periods. Traditional reporting methods have also changed in many ways. News organizations have had to adapt to much faster news cycles online, particularly when news is breaking for major events. During such events, people looking for the most up-to-date information about a natural or environmental disaster can overwhelm websites with millions of hits (Wyss, 2008). Reporters and editors now need to tweet on Twitter and maintain Facebook pages, as well as write stories and blogs. They also need to have multimedia skills.

Tackling multimedia reporting is daunting for environmental and other journalists because it challenges them to figure out how to report a story using video, audio and print all at once. One reporter called it being a "one man band." Veteran print environmental journalist Mark Schliefstein, who now writes for his news company's website, described his typical assignment:

> When I cover a meeting, I'll set up my laptop, using my cell phone as the hot spot, and sometimes I'll be tweeting, basically using that as my note-taking process … and also taking general notes, while at the same time trying to get up and take a picture of the speaker, or I may actually do some video.
>
> (Archibald, 2014, p. 21)

Despite its daunting aspects, multimedia reporting is important, particularly for investigative projects and storytelling in the long form when reporters are often teamed with computer specialists, photographers, videographers, illustrators and others to tell the story both verbally and visually. While agreeing that multimedia reporting is an asset that draws audiences, many in the media and Internet world are waiting to see whether this trend can make the news pay for itself again and help enlarge and enhance online news coverage (Archibald, 2014).

Online publications

In addition to traditional print and broadcast media websites, many solely online publications use different formats to cover the environment. One concern about such websites is that as the Web evolves and creates new narrative vehicles, older websites have to strain to embrace innovation to lure readers, subscribers or advertisers (Cowell, 2014). The number of hits a website gets is critical, particularly for attracting advertising and indicating audience interest in a topic.

According to a study that used data from the Pew Research Center's Project on Excellence in Journalism, the leading online publication in providing environmental headlines among 12 nationally focused news organizations was *The Huffington Post*, which has a daily Green section and posts information from 42 environmental blogs. From January 2011 through May 2012, HuffPost devoted 3 percent of its headlines to the environment. While this is not a large number per se, it was 12 times the number of environmental headlines from Google News during the same time period (Project for Improved Environmental Coverage, 2013b).

At least 17 environmental magazines publish online in the United States, although some also have print editions. These include *E/The Environment Magazine*, *Earth First*, *National Geographic*, *Mother Earth News*, *World Watch* and *Yes!* (Environmental Magazines Online; Environmental Magazines/Journals). One veteran online magazine that stands out is *Grist*, which is known for its biting and frequently humorous takes on many environmental topics. *Grist* has been covering environmental news and providing environmental commentary since 1999—which, as it says on its website, "was way before most people cared about such things" (About *Grist*, n.d.).

News aggregators

Two of the most widely used general aggregators, Google News and Yahoo News, do not concentrate on environmental information as shown by their low number of environmental headlines. From January 2011 to May 2012, environmental headlines in Yahoo News comprised only 0.42 percent of its headlines while environmental headlines in Google News comprised 0.25 percent of its headlines (Project for Improved Environmental Coverage, 2013b).

This paucity of environmental news in general aggregators, however, is made up for by a number of online organizations that supply environmental news by aggregating articles they and others write. Environmental News Service is one of the oldest sources of such online news, publishing original content and articles from major wire services plus stories from numerous partners, contributors, freelance and citizen journalists, and research associates from around the world. It distributes a daily e-newsletter to 36,000 people.

Greenwire and ClimateWire are leading wire-service information sources with subscriber-only daily coverage of environmental and energy policy, politics and markets. Greenwire publishes more than 25 daily stories, some of which are republished in the online version of *Scientific American*.

Environmental Health Sciences, a nonprofit organization, publishes two widely used websites. One is Environmental Health News, which reports its own news stories and offers daily access to articles on environmental health topics published in the world press. Its free syndication services are used by more than 300 other websites. The other EHS website, the Daily Climate, writes and syndicates articles to other media outlets and provides a daily aggregation highlighting worldwide news on climate change.

From this brief survey, it should be clear that the amount of environmental information and number of viewpoints that news consumers can find on the Internet is significant. However, environmental journalism on the Internet is not without its problems, particularly if the information does not come from reliable sources. Problems of misleading information, source credibility and balance frequently arise. Questions about who is an information gatekeeper and what information is valid are particularly important in complex environmental issues and remain problematic for environmental journalism on the Internet.

Prize-winning investigative environmental series

While concerns about the validity of some environmental information on the Internet exist, environmental investigative reporting using Internet resources has been a strong point, earning a number of journalism prizes in recent years. New investigative journalism models have evolved, including partnerships between traditional media, online organizations and nonprofit foundations. Journalism societies and universities also have helped support these efforts and citizens have even donated funds through Kickstarter and other crowdsourcing mechanisms.

Internet resources have been used heavily in many hard-hitting investigative series on complex environmental topics to help cope with the need to tell more in-depth stories. A number of recent U.S. environmental series by both mainstream and online publications have won awards. These prize-winning series include "Toxic Waters," a series in the *New York Times* about the worsening pollution of American waters, and "The Smokestack Effect: Toxic Air and America's Schools," a major investigative series in *USA Today* about some of the most polluted schools in a number of states (Brainard, 2010a). "Fracking: Gas Drilling's Environmental Threat" is a continuing series by ProPublica, an online nonprofit investigative journalism organization, that has spotlighted fracking practices and regulations in the United States and other nations (ProPublica, n.d.).

In 2013, the biggest prize was won by the smallest online organization when "The Dibit Disaster: Inside the Biggest Oil Spill You've Never Heard of" was awarded the Pulitzer Prize for national reporting. The series was produced by InsideClimate News, a small nonprofit online news organization supported by grants from charitable foundations and reader donations. The series involved a seven-month investigation into a major spill of Canadian tar sands oil into the Kalamazoo River in Michigan in 2010, and an examination of U.S. pipeline safety (ICN Staff, 2013). This was only the third Pulitzer won by a solely online publication; the other two went to the larger *Huffington Post* and ProPublica.

One of the distinguishing features about these series, beyond their extensive investigations and excellent verbal storytelling, was the use of multimedia and visual elements to accompany the reporting. These series had photos, videos featuring individuals affected by the issues, maps, graphs and explanations of the methodologies used for the research. Another distinguishing feature was an invitation to readers to interact with the series and use databases and other graphics to glean information about their potential exposures and the impacts these exposures might have on their health and lives. Both interactive databases and visual storytelling are now a critically important part of environmental journalism on the Internet, helping to illustrate and explain important and complex information.

The future of environmental journalism

At a time when societies worldwide already face resource limitations, increasing human populations, shrinking food supplies, chemical threats to health, and climate change with its multitude of environmental and health impacts, more and better environmental coverage could help people seek solutions to meet the challenges ahead.

Unfortunately, from the earliest days of environmental journalism, the emphasis has not been on solutions, but rather on content that often scares readers, according to a veteran journalist, Mark Dowie. Climate change coverage, in particular, has been quite negative. A study by the Reuters Institute for the Study of Journalism found that eight in ten stories related to climate change focused on disaster and implicit risk. Dowie said, if he had to do it over again when he was publisher and editor of the magazine *Mother Jones*, "I would make sure that 20 percent of the stories we ran would be positive" (O'Connor, 2013).

According to a 2012 poll by the Opinion Research Corporation, 20 percent of 1,000 U.S. respondents wanted more media coverage of solutions to environmental problems. Twenty-one percent said that coverage should make the relationship between the environment and other issues more clear. Twelve percent thought coverage could be improved by making environmental stories more appealing to a larger cross-section of society, while 10 percent said environmental news needed to be more visible by being included in the top headlines. Overall, 79 percent of those surveyed believed news coverage of the environment should be improved (Project for Improved Environmental Coverage, 2013a).

These suggested improvements point to the need for additional changes in how environmental journalism is pursued in the United States and perhaps elsewhere. Dot Earth blogger Andrew Revkin had another suggestion: "blogs may be a better tool than conventional coverage for exploring the complex underlying issues shaping 21st-century environmental risks." He explained:

> I started the blog at *The New York Times* to explore these questions that don't fit into the normal news cycle. How you head toward 9 billion people with the fewest regrets is not a news story. It's an ongoing question. And anyone who says there's a single answer is not being truthful. So the only form of journalism that really captures that well is an ongoing conversation, which is a blog.
>
> (Revkin, 2014)

Along with blogs, use of social media to report breaking disaster stories has already had an impact on reporting strategies (Abano, 2013). And more changes are sure to be on the horizon in this age of rapidly developing technologies that can be easily adapted for environmental journalism such as small low-cost drones equipped with miniature cameras and other imaging devices (Allen and Shastry, 2014).

A strong base exists in the United States to help support such changes in environmental journalism, including the Society of Environmental Journalists, which provides networking and assistance for its approximately 1,500 journalist and academic members in the United States, Canada, Mexico and 27 other countries. The SEJ launched a Fund for Environmental Journalism in 2010 to support environmental reporting and entrepreneurial journalism projects. Since its inception, small grants totaling more than $78,000 have been awarded to 46 staff and freelance journalists to "cover the costs of travel, lab testing, graphics development, website costs, and other budget items without which journalists might have been unable to produce and distribute specific timely stories about important environmental issues" (Society of Environmental Journalists, a).

A number of U.S. universities and nonprofit organizations also offer fellowships and training for environmental journalists to provide for more in-depth reporting on various topics. And there are undergraduate and graduate university programs to educate new environmental journalists, some of which are led by SEJ academic members (Society of Environmental Journalists, b). Training programs also are being offered to empower and enable journalists from developing countries to cover the environment and global issues more effectively. One example is the Earth Journalism Network (EJN), an organization that provides training in environmental journalism through partner networks in 70 countries. From 2006 to 2012, the EJN trained more than 2,200 journalists from dozens of developing countries in a wide variety of environmental issues (About EJN, n.d.).

Environmental journalists in the United States and elsewhere will face many challenges in the years ahead. Responsible reporting that employs a variety of media methods and adapts to change is a way forward to helping improve public understanding of complex environmental issues.

Note

1   Mainstreaming in this context is not the same as it is used in Cultivation Theory, where it predicts an interaction between television viewing and direct experience, and direct experience is a moderating factor in the cultivation effect (Schrum and Bischak, 2001).

Further reading

Robert Cox, 2013. *Environmental Communication and the Public Sphere*, 3rd ed. Los Angeles, CA: Sage.
Bob Wyss, 2008. *Covering the Environment: How Journalists Work the Green Beat.* New York: Routledge.
Both books cover a wide range of environmental communication issues from different approaches. Wyss' book concentrates on environmental journalism and what reporters need to know to do it effectively. Cox's book has a wider purview and not only includes valuable information about the media but also about other aspects of environmental communication including risk communication, environmental justice, and citizen voices and public forums.

Mark Neuzil, 2008. *The Environment and the Press: From Adventure Writing to Advocacy*, Evanston, IL: Northwestern University Press.
This book presents a history of environmental journalism, looking at the field's roots in nature writing and an overview of the development of the field.

Julia B. Corbett, 2006. *Communicating Nature: How We Create and Understand Environmental Messages*, Washington, DC: Island Press.
Taking a nature-writing approach to environmental communication, this book deals with how the news media, advertising, public relations and social change, among other factors, help form public attitudes about nature and the environment.

Sharon M. Friedman, Sharon Dunwoody and Carol L. Rogers, eds, 1999, *Communicating Uncertainty: Media Coverage of New and Controversial Science*, Mahwah, NJ: Lawrence Erlbaum Publishing.
This edited book explores the impact of scientific uncertainty and controversy on a number of science and environmental issues and journalism processes.

References

Abano, I. (2013). Social media and its role in reshaping the future of environmental journalism, Philippine Network of Environmental Journalists, Inc. [Online]. Available at: http://pnej.org/?p=842 (accessed January 20, 2014).
About EJN (n.d.) Earth Journalism Network. [Online]. Available at: http://earthjournalism.net/about (accessed January 24, 2014).
About *Grist* (n.d.) Grist. [Online] Available at: http://grist.org/about/ (accessed 17 January 2014).
Allen, B. and Shastry, S. (2014). The promise of flight: Drones and environmental journalism. *SEJournal*, Winter, pp. 6–7, 18–19.

Anonymous. (2013). Database suggests downward trend in *Times* coverage. Yale Forum on Climate Change and the Media. [Online]. Available at: www.yaleclimatemediaforum.org/2013/04/database-suggests-downward-trend-in-times-coverage/ (accessed April 4, 2013).

Archibald, R. (2014). Photojournalism upheaval heralds multimedia's rise. *SEJournal*, Winter, pp. 20–21.

Bagley, K. (2013). About a dozen environment reporters left at top 5 U.S. papers. *InsideClimate News*. [Online]. Available at: http://insideclimatenews.org/news/20130114/new-york-times-dismantles-environmental-desk-climate-change-global-warming-journalism-newspapers-hurricane-sandy (accessed June 13, 2013).

Bauer, M.W., Howard, S.A., Ramos, Y.J.R., Massarani, L. and Amorim, L. (2013). Global science journalism report. *Science and Development Network* [Online]. Available at: www.scidev.net/global/evaluation/learning-series/global-science-journalism-report.html (assessed January 18, 2014).

Beeman, P. (2005). Journalism's struggle offers big challenges for SEJ. *SEJournal*, Winter, 25. [Online]. Available at: www.sej.org/publications/journalismmedia/journalisms-struggle-offers-big-challenges-for-sej (accessed June 16, 2013).

Beeman, P. (2006). This is our time amid media turmoil. *SEJournal*, Spring. [Online]. Available at: www.sej.org/publications/sejournal/this-is-our-time-opportunity-amid-media-turmoil (accessed June 16, 2013).

Brainard, C. (2008). CNN cuts entire science, tech team. *Columbia Journalism Review*. [Online]. Available at: www.cjr.org/the_observatory/cnn_cuts_entire_science_tech_t.php?page=all (accessed February 5, 2013).

Brainard, C. (2010a). USA Today wins Oakes award. *Columbia Journalism Review*. [Online]. Available at: www.cjr.org/the_observatory/usa_today_wins_oakes_award.php (accessed July 26, 2013).

Brainard, C. (2010b). *Monitor*-ing the environment. *Columbia Journalism Review*. [Online]. Available at: www.cjr.org/the_observatory/monitoring_the_environment.php?page=all (accessed February 5, 2013).

Brainard, C. (2010c). WSJ cancels energy/environment blog. *Columbia Journalism Review*. [Online]. Available at: www.cjr.org/the_observatory/wsj_cancels_energyenvironment.php?page=all (accessed February 5, 2013).

Brainard, C. (2013). Faded green. *Columbia Journalism Review*. [Online]. Available at: www.cjr.org/the_observatory/how_many_environmental_reporte.php (accessed February 5, 2013).

Carmody, K. (1995). It's a jungle out there: Environmental journalism in an age of backlash, *Columbia Journalism Review*, May/June, pp. 40–45.

Cohen, B. (1963). *The Press and Foreign Policy*. New York: Harcourt.

Cowell, A. (2014). Embracing the power of the Internet, *New York Times*. [Online]. Available at: http://nyti.ms/1aMkX7Q (accessed January 24, 2014).

Cox, R. (2006). *Environmental Communication and the Public Sphere*. Los Angeles, CA: Sage.

Detjen, J., Fico, F. and Li, X. (1996). Covering environmental news is becoming more difficult, new survey by Michigan State University researchers find. News release. 6 pp.

Dunwoody, S. and Peters, H.P. (1993). The mass media and risk perception. In B. Ruck, ed. *Risk as a Construct*. Munich: Knesebeck, pp. 293-317.

Dykstra, P. (2013) What we're seeing now: Media digs in, drills down on climate coverage, *The Daily Climate*. [Online]. Available at: wwwp.dailyclimate.org/tdc-newsroom/2013/09/trends-climate-media-coverage. (accessed January 17, 2014).

Environmental magazines/journals, Online Magazines. [Online]. Available at: www.onlinenewspapers.com/magazines/magazines-environment.htm. (accessed 24 January 2014).

Environmental magazines online, World Newspapers. [Online]. Available at: www.world-newspapers.com/environment.html (accessed January 24, 2014).

Fitzsimmons, J. (2013). How closing New York *Times'* green blog will hurt environmental coverage, Media Matters for America Blog. [Online]. Available at: http://mediamatters.org/blog/2013/03/04/how-closing-new-york-times-green-blog-will-hurt/192901 (accessed January 3, 2014).

Friedman, S.M. (1991). Two decades of the environmental beat. In C.L. LaMay and E.E. Dennis, eds *Media and the Environment*. Washington, DC: Island Press, pp. 17–28.

Friedman, S.M. (1999). The never-ending story of dioxin. In S. M. Friedman, S. Dunwoody and C.L. Rogers, eds *Communicating Uncertainty: Media Coverage of New and Controversial Science*. Mahway, NJ: Lawrence Erlbaum Associates, pp. 113–136.

Friedman, S.M. (2004). And the beat goes on: The third decade of environmental journalism. In S.L. Senecah, ed. *The Environmental Communication Yearbook*, vol. 1. Mahwah, NJ: Lawrence Erlbaum Associates, pp.175–187.

Friedman, S.M., Gorney, C.M. and Egolf, B.P. (1987). Reporting on radiation: A content analysis of Chernobyl coverage. *Journal of Communication*, Summer, pp. 58–78.

Frome, M. (2004). The roots of e-journalism or, life before Rachel Carson. *SEJournal*, Spring, p. 16.

Grizwold, E. (2012). How "Silent Spring" ignited the environmental movement. *New York Times Sunday Magazine*. [Online]. Available at: www.nytimes.com/2012/09/23/magazine/how-silent-spring-ignited-the-environmental-movement.html?ref= (accessed June 6, 2013).

Guerin, E. (2013). Reorganization or regression? *High Country News*. [Online]. Available at: www.hcn.org/blogs/goat/reorganization-or-regression (accessed January 25, 2013).

Hansen, A. (2011). Communication, media and environment: Towards reconnecting research on the production, content and social implications of environmental communication. *International Communication Gazette*. [Online]. Available at: http://gaz.sagepub.com/content/73/1-2/7 (accessed February 6, 2013).

ICN Staff (2013). InsideClimate News team wins Pulitzer Prize for national reporting. [Online]. Available at: http://insideclimatenews.org/news/20130415/insideclimate-news-team-wins-pulitzer-prize-national-reporting (accessed August 2, 2013).

Meister, M. and Japp, P.M. (eds) (2002) *Enviropop*, Westport, CT: Praeger.

O'Connor, M.C. (2013). Whither, environmental journalism? A panel of reporters weigh next steps, Yale Forum on Climate Change & the Media. [Online]. Available at: https://www.google.com/search?q=Whither%2C+environmental+journalism%3F+A+panel+of+reporters&ie=utf-8&oe=utf-8&aq=t&rls=org.mozilla:en-US:official&client=firefox-a&channel=fflb (accessed January 17, 2014).

Pew Research Center's Project on Excellence in Journalism. (2004). The state of the news media 2004: Eight major trends. [Online]. Available at: http://stateofthemedia.org/2004/overview/eight-major-trends/ (accessed April 4, 2013).

Pew Research Center's Project on Excellence in Journalism. (2008). The state of the news media 2008: Major trends. [Online]. Available at: http://stateofthemedia.org/2008/overview/major-trends/ (accessed April 4, 2013).

Pew Research Center's Project on Excellence in Journalism. (2013). The state of the news media 2013: Overview. [Online]. Available at: http://stateofthemedia.org/2013/overview-5/ (accessed March 21, 2013).

Project for Improved Environmental Coverage. (2013a). Poll Summary – Citizen Attitudes re: Environmental Coverage (2013) p. 7. [Online]. Available at: http://greeningthemedia.org/wp-content/uploads/Environmental-Coverage-in-the-Mainstream-News.pdf (accessed January 25, 2014).

Project for Improved Environmental Coverage. (2013b). Environmental Coverage in Mainstream News: We need More. [Online]. Available at: http://greeningthemedia.org/wp-content/uploads/Environmental-Coverage-in-the-Mainstream-News.pdf (accessed August 1, 2013).

ProPublica (n.d.) Focusing public attention—and staying with a story relentlessly. [Online]. Available at: www.propublica.org/about/focusing-public-attention-and-staying-with-a-story-relentlessly (accessed January 4, 2014).

Raeburn, P. (2013). *Washington Post* moves a sports editor/fitness columnist to the environment beat. Knight Science Journalism Tracker. [Online]. Available at: http://ksj.mit.edu/tracker/2013/03/washington-post-moves-sports-editorfitne#sthash.GgM726oH.dpuf (accessed March 27, 2013).

Revkin, A. (2013). The changing newsroom environment, *New York Times*. [Online]. Available at: http://dotearth.blogs.nytimes.com/2013/01/11/the-changing-newsroom-environment/?_r=0 (accessed March 6, 2013).

Revkin, A. (2014). Journalists on the environment beat look ahead, *New York Times*. [Online]. Available at: http://dotearth.blogs.nytimes.com/2014/01/27/journalists-on-the-environment-beat-look-ahead/?_php=true&_type=blogs&module=BlogPost-ReadMore&version=Blog%20Main&action=Click&contentCollection=Opinion&pgtype=Blogs&region=Body&_r=0 (accessed January 28, 2014).

Sachsman, D.B. (2003). The mass media and environmental risk communication: Then and now. In B.M. West, M.J. Lewis, M.R. Greenberg, D.B. Sachsman, and R.M. Rogers, eds *The Reporter's Environmental Handbook*, 3rd edn. New Brunswick, NJ: Rutgers University Press, p. 82.

Santhanam, L. (2014). Study: How broadcast news covered climate change in the last five years, Media Matters. [Online]. Available at: http://mediamatters.org/research/2014/01/16/study-how-broadcast-news-covered-climate-change/197612 (accessed January 24, 2014).

Schrum, L.J. and Bischak, V.D. (2001). Mainstreaming, resonance, and impersonal impact. *Human Communication Research*, April, 27, pp. 187–215.

Society of Environmental Journalists (a.) Fund for environmental journalism. [Online]. Available at: www.sej.org/initiatives/fund%20for%20environmental%20journalism/overview (accessed January 25, 2014).

Society of Environmental Journalists (b). Teaching tools. [Online]. Available at: www.sej.org/library/teaching-tools/overview (accessed January 25, 2014)

Sullivan, M. (2013a). For *Times* environmental reporting, intentions may be good but the signs are not, *New York Times*. [Online]. Available at: http://publiceditor.blogs.nytimes.com/2013/03/05/for-times-environmental-reporting-intentions-may-be-good-but-the-signs-are-not/ (accessed March 6, 2013).

Sullivan, M. (2013b). After changes, how green is the Times? *New York Times*. [Online]. Available at: www.nytimes.com/2013/11/24/public-editor/after-changes-how-green-is-the-times.html?_r=0 (accessed January 11, 2014).

Tyndall, A. (2008–2012) *Tyndall Report year in review*. [Online]. Available at: http://tyndallreport.com/yearinreview2012/ (accessed June 13, 2013).

Wilkins, L. (1987). *Shared Vulnerability: The Media and American Perceptions of the Bhopal Disaster*. New York: Greenwood Press.

Wyss, B. (2008). *Covering the Environment*, New York: Routledge.

12

ENVIRONMENTAL REPORTERS

David B. Sachsman and JoAnn Myer Valenti

The period from 1990 until the early twenty-first century can be regarded as a golden age of environmental journalism in the United States. However, the first decade of the twenty-first century also marked the financial decline of the American newspaper industry and a weakening trend in television, both of which were made much worse by the stock market crash and recession of 2008. By the end of the decade, many experienced environmental reporters had lost their jobs, and the future of American environmental journalism was in question.

Who are America's environmental reporters? Where do they work and what difficulties do they face? How has the environmental beat developed in the past half century and what is the future of the men and women who cover environmental news? This chapter reports the findings of a baseline study of environmental reporters in the United States at the beginning of the twenty-first century and then turns to three of America's finest environmental reporters for their personal descriptions of the changes facing environmental reporters because of the decline of the American newspaper business in the Internet age.

Environmental news coverage exploded in the US in the late 1960s and would become the norm in the 1970s. Television coverage of such environmental accidents as the Santa Barbara oil spill, with oil-soaked birds dying in the arms of weeping rescue workers, and the Cuyahoga River in Ohio actually on fire, caused the American people and the press to focus on environmental issues. By the end of the 1960s, conflicting public relations forces—representing government and environmental groups as well as business interests—were vying for the attention of American journalists, and some science reporters and others were adding the environment to their beat.

"Environmental journalism changed little by little throughout the 50s and 60s," said Peter Dykstra, the former head of the CNN environmental unit who currently serves as publisher of the online daily news aggregators Environment Health News (www.ehn.org) and Daily Climate (www.dailyclimate. org).[1] Science writing, with its focus on interpreting and explaining science, offered few policy or economic components. By 1970 and the first Earth Day, Rachel Carson's passionate writing about science had impacted the public's demand for more information and political response, he said. Environmental journalists' reporting included the missing economics and politics. Dykstra pointed to the back half of the 1980s: a disaster in Bhopal, nuclear crisis in Chernobyl, medical waste on New Jersey beaches, an ozone hole, and serious cleanup needs at the Hanford nuclear site in the state of Washington. By 1990 the environment had earned its own beat.

The number of environmental reporters grew steadily along with the economic success of the newspaper and television industries. In 1990, the Society of Environmental Journalists was formed in response to the growing number of environmental journalists. By the beginning of the twenty-first century, 534 daily newspapers (36.5 per cent) and 86 television stations (10 per cent) employed specialized environmental reporters (Sachsman, Simon, and Valenti 2010: 53–56). In addition, there were many more freelancers, magazine reporters, book authors, and Internet writers covering the environmental beat. By 2012, many newspapers were in severe economic distress, and many of the experienced environmental journalists were no longer working for newspapers. The Society of Environmental Journalists in 2004 boasted a membership of 968 active journalists, including 430 newspaper reporters. In 2012, SEJ active membership was 924, with the number of newspaper reporters down to 225, compared to 403 freelancers (Society of Environmental Journalists 2012).

The environmental reporters of the twenty-first century

The best baseline information on environmental journalists was collected from 2000 to 2004 in a series of regional studies that finally amounted to a census of those environmental reporters then working for daily newspapers and television stations. The resulting book, *Environment Reporters in the 21st Century*, analyzed the interviews that had been conducted with 652 of the 686 environmental journalists identified, including 577 of 603 newspaper writers (95.7 per cent) and 75 of 83 television reporters (90.4 per cent). Researchers David B. Sachsman, James Simon, and JoAnn Myer Valenti found that 78.7 per cent of newspapers with circulations greater than 60,000 employed one or more environmental reporters, compared to slightly more than one out of three newspapers overall and only one out of every ten television stations (Sachsman et al. 2010: 43, 53–6). Bigger newspapers may really be better newspapers in terms of newspaper coverage of the environment (Bogart 2004: 40–53; Gladney 1990: 58–72; Logan and Sutter 2004: 100–12; Meyer and Kim 2003).

Sachsman, Simon, and Valenti identified environmental journalists as those who said they covered the environment on a regular basis as part of their reporting duties. They found that only 29.0 per cent of these writers carried a title containing the word "environment." Nearly half were simply titled reporters, general assignment reporters, or staff writers, while the others were specialized editors, outdoor writers, specialized reporters, science writers, and health reporters. More than half spent less than a third of their time covering environmental issues, while only 26 per cent spent more than two-thirds of their time on the environmental beat. Overall, reporters spent an average of 43 per cent of their time on environmental stories. There were some regional differences; reporters in the two most western regions, the pacific west and the mountain west, spent more than 50 per cent of their time covering the environment (Sachsman et al. 2010: 42, 57–59).

Sachsman, Simon, and Valenti compared environmental journalists with U.S. journalists in general, using data collected by an Indiana University research team headed by David H. Weaver and published as *The American Journalist in the 21st Century: U.S. News People at the Dawn of a New Millennium* (Weaver et al. 2007). Overall, the environmental researchers concluded that environmental reporters working at daily newspapers and television stations shared many individual and work-related characteristics with each other and with U.S. journalists in general. The environmental reporters were *journalists* first, perhaps due in part to their similar backgrounds and to the basic professional training received by most journalists. While the most popular major among environmental journalists was journalism/communication, the environmental reporters differed slightly but significantly from U.S. journalists since many of the environmental reporters had minored, majored, or received advanced degrees in scientific fields (Sachsman et al. 2010: 70, 63).

Environmental journalists in the beginning of the twenty-first century were generally satisfied with their jobs, as were U.S. journalists in general. But three out of four environmental reporters believed they needed additional training and more than a third felt that environmental journalists generally were not well enough educated to cover news about environmental issues. Neither group saw cutbacks or layoffs as a serious threat at that time (Sachsman et al. 2010: 74–75, 63, 80).

Modern environmental journalism in the 1960s and early 1970s had been seen as an offshoot of the science beat, but by the time the Society of Environmental Journalists was created in 1990, the journalists who attended the annual convention offered a different specialized focus than science writers who attended the annual meeting of the American Association for the Advancement of Science. Their interests went beyond the particulars of science. Sachsman, Simon, and Valenti asked environmental reporters how often they used various story angles in addition to the "environment" in their coverage. Nine out of ten said they used government, human interest, business, pollution, and nature in their stories, and eight out of ten also included science, politics, and health. Finally, seven out of ten included the concept of risk assessment in their coverage (Sachsman et al. 2010: 95–96).

Scientists and industry leaders believe the concept of scientific degree of risk is central to environmental news reporting. For scientists, risk analysis, management, and education allow for the appropriate measurement and solutions of environmental problems, as well as for informing the public how to protect themselves. For industry leaders, the concept of scientific degree of risk is central to the idea that environmental issues are fundamentally health risk issues that should be addressed when there is a provable human health risk (Sachsman 1999: 88–95).

The basic news standards of journalism include human interest, proximity, timeliness, prominence, and consequence, which includes risk (MacDougall 1977: 56). In addition, a Rutgers University study found that television news about the environment also focused on visual impact and what the researchers called "geography," the cost and convenience of covering an environmental story (Greenberg, Sachsman, Sandman, and Salomone 1989: 267–76). By the beginning of the twenty-first century, environmental reporters generally understood and paid attention to the concept of risk, but because they framed environmental stories in terms of government actions, human interest, business, pollution, nature, science, politics, and health (as well as risk), they did not accept the industry argument that environmental issues became problems only when they posed a provable human health risk. The environmental beat had become as much about the politics behind the issues as about the underlying science.

Environmental reporters and their sources

The public relations efforts of environmental news sources have always had a significant influence on news coverage. In the first half of the twentieth century, environmental press releases often came from business interests, explaining their solutions to issues of public concern. In the 1960s, environmental anti-pollution activists joined the fray, and the federal government weighed in with a new recognition of a need to respond to environmental problems. By the beginning of the 1970s, one study found that public relations press releases accounted for some 20 per cent of environmental coverage and that public relations sources contributed to no fewer than 25 per cent and as many as 50 per cent of environmental stories (Sachsman 1976: 54–60). By that time, conflicting public relations sources were engaged in a public relations war to capture the attention of the mass media and set the environmental agenda. Public relations sources have continued to be very influential. A 2004 study of health reporting showed the ongoing influence of public relations in specialized news reporting (Tanner 2004: 350–363).

In the 1970s, government officials dominated the environmental discussion and played a major role in setting the environmental agenda, as they do today. Federal officials make environmental announcements on a daily basis; state government agencies are important sources for environmental reporters; and almost every local government meeting has at least one environmental issue on its agenda. The environment is now such a basic government story that every government reporter must be prepared to deal with it (Brown et al. 1986: 45–54; Gans 1979; Greenberg, Sandman, Sachsman, and Salomone 1989: 16–20, 40–4; Lacy and Coulson 2000: 13–25; Lovell 1993; Sachsman 1973: 54–60; Sigal 1973; Taylor, Lee, and David 2000: 175–192).

The news sources most commonly used by the interviewed environmental reporters were departments of environmental quality, local environmental groups, citizens active on the environment, state departments of natural resources, academics, state departments of health, local manufacturers, developers and other business leaders, state legislative offices, the US EPA, and local health departments. The next seven sources on the list were also government officials. The journalists said the most common barriers in environmental reporting were time constraints, financial or travel constraints, and the size of the news hole (Sachsman et al. 2010: 97–98, 85–86).

Traditionally, public relations influence meant press releases and press conferences aimed at newspaper and television reporters. Public relations sources also produced formal reports and sometimes even their own magazines. But in the Internet age, a whole new dimension has been added. Today, every news source has its own website, often containing the equivalent of thousands of pages of information, easily accessible by search engines. Thus, while news sources have always tried to bypass the traditional gatekeeping function of the news media and take their messages directly to the public, in the era of the Internet, they are able to do so, with Google currently among their most important gatekeepers.

Objectivity vs. advocacy among environmental reporters

When Sachsman, Simon, and Valenti asked environmental journalists a number of questions about their attitudes, they split on the question of whether environmental problems are generally better stories than environmental successes. More than half felt that environmental journalists generally concentrate far too much on problems and pollution. Nearly four out of five disagreed with the argument that environmental journalists generally have blown environmental risks out of proportion. And a like number disagreed that environmental journalism generally centers too much on personalities and not enough on actual findings (Sachsman et al. 2010: 105–108).

All but four of the reporters who answered these questions agreed that environmental journalists need to be as objective as reporters in general and need to be fair to sources such as environmental activist groups and corporations. Nevertheless, 36.7 per cent said that environmental journalists sometimes should be advocates for the environment and 32.9 per cent felt that environmental journalists should work with community leaders to help solve environmental problems (the definition of "civic journalism"). Discounting the four reporters who disagreed with the concept of objectivity, the environmental journalists could be characterized as belonging to one out of four groups. Nearly half (48.4 per cent) were objectivity purists, who believed that environmental reporters should never be advocates or civic journalists, while 18.9 per cent agreed that environmental reporters should be objective, should sometimes be advocates, and should work with community leaders to solve problems. The remaining journalists felt that reporters should sometimes be advocates, but never civic journalists (17.9 per cent), or believed that they should be civic journalists, but never advocates (14.1 per cent) (Sachsman et al. 2010: 122–126).

Some environmental reporters thus are ethically torn as journalists between their basic belief in objective reporting (and being fair to all sides) and the idea that everyone should support a healthy, clean environment, just as every health reporter should support good health. At the beginning of the twenty-first century, 38 per cent of environment reporters felt that some of their colleagues tended to be too "green," meaning slanted in favor of environmentalism. Overall, these reporters recognized the importance of the business community, generally including business angles and industry sources in their coverage (Sachsman et al. 2010: 120, 105).

Factor analysis suggests that environmental reporters have a nuanced, multilayered view of their profession. Some of the reporters, whom Sachsman, Simon, and Valenti called "objective/fair reporters," responded in a common, uniform way to questions about objectivity and fairness. A second grouping, the "workplace critics," agreed with several traditional complaints about the way environmental reporters perform their duties, including how stories are written, the focus of stories, and the education level of their peers. Finally, the group characterized as the "advocates/civic journalists" seemed willing to go beyond the concept of traditional objectivity and appeared to frame their roles differently from other environmental journalists (Sachsman et al. 2010: 127).

Three top environmental journalists reflect on a profession in transition

An interview with Peter Dykstra

Today, fewer environmental journalists are affiliated with large media. Even though all three US television networks (ABC, CBS, and NBC) finally have a full-time environment reporter, the dramatic overall change in media and journalists with the experience and knowledge to cover the beat can best be measured by SEJ's membership; the largest group of active members is now freelance writers. Some laid off or "retired" environmental reporters have moved to non-profit organizations and government agencies. "We call ourselves [at Environment Health News] the fossil collection," Peter Dykstra said in 2012 of the nonprofit's staff and content contributors; among them, Marla Cone, formerly of the *Los Angeles Times*; Rae Tyson, formerly environment editor at *USA TODAY*; Jane Kay from the *San Francisco Chronicle*; Doug Struck formerly at the *Washington Post*; and Doug Fischer, ex-*Oakland Tribune* reporter. The change reflected in media over the last twenty-some years is also reflected in political leadership, he said. "Twenty-four years ago the elder Bush [President George H.W. Bush] said he'd be 'the environmental president.'" Somewhere along the line, concern morphed into contempt for environment issues, he said. These days, you won't find a two-hour special aired as in the past on ABC for Earth Day. "The post 9-11 singular focus [on security and war] and the country's economic collapse led to contempt for government, journalists, educators … and an organized push back [for perceived liberal bias] by Tea Party types," he said. "What's left is horrible for an informed democracy." While Dykstra believes we are now entering a new cusp of environmental awareness, he feels "media are in a sink hole."

In spite of continuing, new disasters—climate change related weather crises, the Arctic meltdown, Japan's tsunami and nuclear meltdown—real news analysis has been shut down. As Dykstra sees it, Fox News, talk radio, and numerous conservative websites paint extremes and nothing but relentless attacks, building a message pleasing to those who favor resource extraction. It's all about "'gotcha' quotes and oops moments," he said, none of which indicate serious environmental coverage. To pander to a seemingly dumbed-down, altered public sentiment and interest—and to maintain advertising dollars—major news outlets are closing costly headquarters, cutting home delivery, laying off staff, and "balancing"—in the worst form of journalistic practice—even editorial cartoons in an effort not to offend.

"Personality driven trends" have replaced news, Dykstra lamented. "The conversion from information to entertainment is complete." Media have gone to a menu of reality shows—cheap to produce and a favorite of advertisers. None of this bodes well for an informed citizenry, he argues, unless you're linked in to selective online sites—government sites such as USGS or a growing span of ex-journalists who are now environment bloggers. Unfortunately, the widened diversity of sources carries the risk of leading users only to what makes them feel comfortable, he warned. "We don't know in what direction information will go," Dykstra said. Have we been dumbed down? Are we willing to accept an "enfotainment" media devoid of environment news? "Inertia is a serious problem for a democracy," Dykstra concluded.

An interview with Marla Cone

When she voluntarily left the *Los Angeles Times* in 2008 after watching an exodus of colleagues she most respected—some laid off, some taking offered buyouts—Marla Cone couldn't imagine leaving journalism.[2] Turns out, she was only moving on in the changing world of reporting news. She had covered environment issues locally, nationally, and internationally for the *Times*, with lots of front-page stories. But new management was less interested in environment stories even though the audience was still there, especially for information on how environmental problems impacted health, she said in 2012. Stories on the *Times* website got weaker; some excellent, hard-hitting stories were cut in half. The new management "tore apart the *Times*," she said. Space and resources to cover stories were greatly reduced, with the staff cut in half. "I was really sad leaving newspapers," she said, but she knew she wasn't leaving journalism. The nonprofit online daily news service Environmental Health News had always been her "home page," so an offer to serve as EHN editor-in-chief was a natural move.

"There's so much gray area in environmental health reporting," Cone said, and that requires an experienced journalist. She now believes that much of the best environment reporting is online and comes from foundation-funded groups, pointing to several nonprofit sites including EHN and ProPublica. Huge operating overheads and costly investments are problems for newspapers but less of a problem for nonprofit journalism. She said the Internet Age caused financial and logistical problems for newspapers. "It will take them another ten years to figure out how to come up with a sound business model," she said, adding, "It's ridiculous to have seasoned journalists doing three-sentence online coverage." Where's the informed news, she asks. The depth is missing, she says. Some posted stories are no more than headlines, empty shells. Environment news really suffers. Many experienced journalists are gone. She worries that too few reporters are watching what's happening in the environment. EHN, like other evolving online news sites, has stepped up to provide serious, enterprise, long-form journalism, particularly about the problems faced by low-income communities. EHN's journalists, many refugees from the pool of laid-off print reporters, are filling in the holes in critical information. In an ironic example, Cone points to successful, well-read, in-depth stories by former *San Francisco Chronicle* reporter Jane Kay, published online at EHN, then picked up by the *Chronicle*. Kay, an award-winning, veteran environment writer, left the *Chronicle* after a lengthy career there.

"My value to environmental journalism now," Cone said, "is to cover all aspects of a story, cover it for scientists, consumers, everyone." Her expertise is crucial. "The *LA Times* still does great journalism," she said, "but they're a business." EHN has added social media, Facebook, tweeting, and its audience is growing. The future looks solid as long as funding holds up for the nonprofit operation, a contrast to trimmed down newsprint and the blogosphere. She is concerned that most bloggers have no journalism background; she labels them as fast and dirty. "Environmental journalism requires experience, nuance, high journalistic standards," she said. She worries that there is a lot of environment information available "out there" and a lot of it is bad. Most in the audience don't know how to check the credibility of sources. "It's mattering less and less that real journalists are at newspapers," Cone concluded. What matters is that reporting is coming from "real" journalists, whatever media they work for.

An interview with Mark Schleifstein

"This is the future," Mark Schleifstein said in August 2012 of the *Times-Picayune* management's decision to lay off some 200 employees and shift the focus from its print product to an expanded website.[3] The newspaper was scheduled to be printed only three times a week: Wednesday, Friday, and Sunday, the three best ad days. The Pulitzer Prize-winning reporter made clear early on he would stay with the paper even though his LinkedIn profile reads, "In the rapidly changing world of newspaper journalism, I'm interested in learning of other opportunities." He's a realist—even optimistic—about the dramatic changes in the way news is now delivered.

Unlike many environment reporters around the country who reported spending only one-third of their time covering the environment according to research covering the first decade of the century (Sachsman et al. 2010: 59), Schleifstein has always spent at least two-thirds of his time reporting on environment issues. Although his workday has already expanded in order to fill the daily web feed, sticking with one of the country's leading papers is "the best way for me to get information [about the environment, science, and engineering] out to the public," he said. After nearly three decades covering coastal woes, hurricanes, the 2010 BP Deep Water Horizon oil spill in the Gulf of Mexico, and more, a major restructuring of news delivery systems doesn't faze him.

The merger of two separate vehicles—the reduced print product plus an expanded web presence— really doesn't change anything, said Schleifstein. "My beat will stay the same. I'm still the alleged environment reporter." The reality is if something comes up and the editors need a warm body, he may be called. "I'm actually covering all environment, science, and engineering issues." And he says, with some amount of enthusiasm, there will be others covering environment stories, for example, on the city hall team or government desk or when or wherever a science/environment/engineering issue arises.

As newspapers elsewhere evolve toward similar restructuring, Schleifstein sees individual beats also converging. "We've been converged without really understanding what that means," he said. For some time, "I've been the outlier saying environment belongs all over the paper."

The *Times-Picayune* daily circulation dropped from a pre-Hurricane Katrina high of around 210,000 to 135,000, with a recent, steady reader return. The changed delivery via the Web now offers expanded stories including aggregated input, some of which is already coming from Schleifstein's desk/computer/Droid operation. Aggregation takes time, but he believes it offers potential opportunity, including more video and more photo ops. The print editions will be "curated" from online stories. The possibility for having depth in stories may be hampered, but it's too early to tell, he said. Quality and quantity issues for environment reporting will be in need of scrutiny in the long run.

The *Times-Picayune*'s search for ways to monetize what's on the website has marketers working to link to products, begging the questions: What is journalism? What is advertising? Although he's not sure how all of this will set with reporting on the environment, Schleifstein notes that the U.S. Army Corps of Engineers has already bought ads on the newspaper's web page with links to their spin on stories. BP—and the federal government—set up stories online, on Facebook, and on Twitter during the oil spill crisis. BP had its own reporters on site putting out stories related to the oil cleanup operation in the Gulf alongside journalists trying to do some real journalism. It's understood that BP's purchase of daily full-page ads likely kept the paper going. Regardless, Schleifstein is confident "people aren't as dumb as some think." People are learning what is good and what's bad, he said.

"Five or ten years from now we'll all be doing what the *Times-Picayune* is doing now," he said in 2012. He feels New Orleans is a good test market for how this will all work. Print may no longer be a sustainable product. The public is moving to electronic delivery systems.

Soon after announcing the *Times-Picayune*'s move to digital, owner Advance Publications announced that it would be making similar changes to its newspapers in Birmingham, Mobile, and Huntsville, laying off a total of 600 employees in New Orleans and the three Alabama cities (Bazilian 2012). In April 2013, Advance announced that it also would be cutting its print editions of the Cleveland *Plain Dealer* to three days a week. Significant layoffs were expected (Haughney 5 April 2013). But that same month, the *Times-Picayune* reported that it would produce a three-day-a-week tabloid newsstand edition called *TPStreet* on the weekdays when there was no *Times-Picayune* home-circulation newspaper (Haughney 1 May 2013). *New York Times* media writer David Carr commented:

> The much ballyhooed unmaking of daily newspapering seems to be unmaking itself, and there's a reason for that. Most newspapers have hung onto the ancient practice of embedding prose on a page and throwing it in people's yards because that's where the money and the customers are for the time being.
>
> (Carr 2013)

Newspapers' decline in the Internet age

The newspaper business has traditionally been a factory business, producing a throwaway paper product. Its overhead expenses have included enormous, expensive buildings, giant printing presses, fleets of trucks, and large numbers of production and circulation employees, in addition to journalists. The factory business model has been problematic in the U.S. for some years, but despite declines in readership, the newspaper business was generally kept afloat by a steady stream of advertising revenues. The stock market crash and recession of 2008 cut these revenues virtually in half (Edmonds, Guskin, Rosentiel, and Mitchell 2012; Sachsman and Sloat 2014).

Since 2008, there have been job reductions in many media, from ABC News to Time, Inc., but the cutbacks in the newspaper business have been staggering. The Gannett Company alone has made thousands of buyouts and cuts, and newsrooms across the nation have many empty desks (Sonderman 2012). In 2012, the American Society of News Editors reported that newspapers employed 40,600 news professionals, a loss of 28 per cent from the beginning of the twenty-first century. In the same year, the Pew Research Center's Project for Excellence in Journalism reported that these cutbacks were continuing to result in

> much less coverage of government in suburbs or remote cities, pulling back on state government coverage, the decimation of specialty beats like science and religion, fewer feature stories and elimination of many weekday feature sections, a smaller business report, typically not a freestanding section anymore.
>
> (Edmonds et al. 2012)

Pew noted that the Federal Communications Commission in 2011 had concluded, "In very real ways, the dramatic newspaper-industry cutbacks appear to have caused genuine harm to American citizens and local communities" (Edmonds et al. 2012).

The fate of environmental reporting in the U.S. is only a single aspect of the fate of journalism in general and of fundamental changes in the delivery of news and information. As long as American newspapers continue to employ environmental reporters, and as long as these reporters continue producing in-depth stories, the newspaper business will continue to be an important source of environmental news. Television has never employed many environmental reporters, but environmental journalists are employed by many specialized magazines, and the numbers of environmental freelancers are growing steadily.

Finally, there is the Internet, where people often search for information, rather than looking for journalistic coverage of news, and where every news source and every niche publisher has its own web page. Today, the Internet is cutting into newspaper, magazine, and television audiences. In the long run, whether as information or as news, more and more environmental writing will find its way to the Internet, and if readers choose news and if an innovative economic model brings back advertising revenue, American journalism may witness another golden age.

Some signs are good. On April 15, 2013, the Pulitzer Prize for U.S. national reporting was awarded for a series of reports on the regulation of oil pipelines. This was not the first Pulitzer to be awarded for a work of environmental journalism, but it may have been an indicator of the future because the winner was a tiny Internet startup, *InsideClimate News*, a five-year-old nonprofit with only seven employees (Bercovici 2013). As newspapers in the US cut their budgets and close their environmental bureaus,[4] more and more American environmental reporters are finding their way to the Internet, a bright hope for the future of environmental journalism.

Notes

1   Peter Dykstra served as media director (1981–1991) for the international conservation organization Greenpeace before moving to CNN in Atlanta for the next 17 years heading up science, technology, environment and weather coverage for the cable network's broadcasting and Internet platform. When CNN changed ownership and direction—and the economy faltered—all but CNN's weather staff disappeared. (Telephone Interview with JoAnn Myer Valenti, September 5, 2012)

2   Marla Cone reported for newspapers for 30 years, 18 at the *Los Angeles Times* as an environmental writer, before becoming Editor in Chief of Environmental Health News (www.ehn.org). (Telephone Interview with JoAnn Myer Valenti, August 23, 2012)

3   Mark Schleifstein has covered environment stories in the Gulf Coast area for nearly three decades, landing numerous awards including Pulitzer Prizes for Public Service and Breaking News Reporting in 2006 after his coverage during and after hurricane Katrina. His first Pulitzer Prize came in 1997 for a co-authored series on the world's doomed fisheries, "Ocean of Trouble." (Skype Interview with JoAnn Myer Valenti, August 23, 2012)

4   On January 11, 2013, *InsideClimate News* reported that the *New York Times* was closing its environment desk and reassigning its nine-person staff to other departments. "We have not lost any desire for environmental coverage. This is purely a structural matter," one *Times* editor said. (Bagley, K. 2013 "New York Times Dismantles Its Environmental Desk," *InsideClimate News*, January 11, 2013. Online. Available HTTP: http://insideclimatenews. org/news/20130111/new-york-times-dismantles-environmental-desk-journalism-fracking-climate-change-science-global-warming-economy (accessed April 24, 2013))

Further reading

Sachsman, D.B., Simon, J., and Valenti, J.M. (2010) *Environment Reporters in the 21st Century*, New Brunswick, NJ: Transaction.

Sandman, P.M., Sachsman, D.B., Greenberg, M.R., and Gochfeld, M. (1987) *Environmental Risk and the Press: An Exploratory Assessment*, New Brunswick, NJ: Transaction.

Valenti, J.M. (1995) *Developing Protocol for Ethical Communication in Environmental News Coverage*, Provo, UT: Brigham Young University, Dept. of Communications.

Weaver, D.H., Beam, R.A., Brownlee, B.J., Voakes, P.S., and Wilhoit, G.C. (2007) *American Journalists in the 21st Century: U.S. News People at the Dawn of a New Millennium*, Mahwah, NJ: Lawrence Erlbaum.

West, B.M., Lewis, M.J., Greenberg, M.R., Sachsman, D.B., and Rogers, R.M. (2003) *The Reporter's Environmental Handbook*, 3rd edn, New Brunswick, NJ: Rutgers University Press.

References

Bazilian, E. (2012) "600 Employees Will Be Laid Off at '*Times-Picayune*,' Alabama newspapers, Advance Publications switching formats to digital," *AdWeek*, June 13, 2012. Online. Available HTTP: www.adweek.com/news/press/600-employees-will-be-laid-times-picayune-alabama-newspapers-141102 (accessed December 11, 2012).

Bercovici, J. (2013) "The Tiny News Startup That Crashed the Pulitzer Prizes," *Forbes*, April 16, 2013. Online. Available HTTP: www.forbes.com/sites/jeffbercovici/2013/04/16/the-tiny-news-startup-that-crashed-the-pulitzer-prizes/ (accessed April 24, 2013).

Bogart, L. (2004) "Reflections on Content Quality in Newspapers," *Newspaper Research Journal*, 25: 40–53.

Brown, J.D., Bybee, C., Wearden, S., and Straughan, D. (1986) "Invisible Power: Newspaper News Sources and the Limits of Diversity," *Journalism Quarterly*, 63: 45–54.

Carr, D., (2013) "Newspaper Monopoly that Lost its Grip," *New York Times*, May 1, 2013, B1.

Cone, M. (2012) Interviewed by JoAnn Myer Valenti [telephone] August 23, 2012.

Dykstra, P. (2012) Interviewed by JoAnn Myer Valenti [telephone] December 6, 2012.

Edmonds, R., Guskin, E., Rosentiel, T., and Mitchell, A. (2012) "Newspapers: By the Numbers," *The State of the News Media 2012: An Annual Report on American Journalism*, Pew Research Center's Project for Excellence in Journalism. Online. Available HTTP: http://stateofthemedia.org/2012/newspapers-building-digital-revenues-proves-painfully-slow/newspapers-by-the-numbers/ (accessed December 12, 2012).

Gans, H. (1979) *Deciding What's News*, New York: Random House.

Gladney, G. (1990) "Newspaper Excellence: How Editors of Small & Large Papers Judge Quality," *Newspaper Research Journal*, 11: 58–72.

Greenberg, M.R., Sachsman, D.B., Sandman, P.M., and Salomone, K.L. (1989) "Risk, Drama, and Geography in Coverage of Environmental Risk by Network TV," *Journalism Quarterly*, 66.2: 267–276.

——, Sandman, P.M., Sachsman, D.B., and Salomone, K.L. (1989) "Network Television News Coverage of Environmental Risks," *Environment*, 31: 16–20, 40–44.

Haughney, C. (5 April 2013) "Cleveland Paper to Trim Delivery and Cut Staff," *New York Times*, B7.

—— (1 May 2013) "*Times-Picayune* Plans a New Print Tabloid," *New York Times*, B7.

Lacy, S. and Coulson, D.C. (2000) "Comparative Case Study: Newspaper Source Use on the Environmental Beat," *Newspaper Research Journal*, 21: 13–25.

Logan, B. and Sutter, D. (2004) "Newspaper Quality, Pulitzer Prizes, and Newspaper Circulation," *Atlanta Economic Journal*, 32: 100–112.

Lovell, R.P. (1993) *Reporting Public Affairs: Problems and Solutions*, 2nd edn, Prospect Heights, IL: Waveland.

MacDougall, C. (1977) *Interpretative Reporting*, 7th edn, New York: Macmillan.

Meyer, P. and Kim, K. (2003) "Quantifying Newspaper Quality: 'I Know It When I See It,'" paper presented at Newspaper Division, Association for Education in Journalism and Mass Communication, Kansas City, MO, July.

Sachsman, D.B. (1973) "Public Relations Influence on Environmental Coverage (in the San Francisco Bay Area)," unpublished PhD dissertation, Stanford University.

—— (1976) "Public Relations Influence on Coverage of Environment in San Francisco Area," *Journalism Quarterly*, 53.1: 54–60.

—— (1999) "Commentary: Should Reporters Use Risk as a Determinant of Environmental Coverage?" *Science Communication*, 21.1: 88–95.

——, Simon, J., and Valenti, J.M. (2010) *Environment Reporters in the 21st Century*, New Brunswick, NJ: Transaction.

—— and Sloat, W. (2014) *The Press and the Suburbs: The Daily Newspapers of New Jersey*, 2nd edn, New Brunswick, NJ: Transaction.

Sandman, P.M., Sachsman, D.B., Greenberg, M.R., and Gochfeld, M. (1987) *Environmental Risk and the Press: An Exploratory Assessment*, New Brunswick, NJ: Transaction.

Schleifstein, M. (2012). Interviewed by JoAnn Myer Valenti [Skype] August 23, 2012.

Sigal, L.V. (1973) *Reporters and Officials*, Lexington, MA: DC Health.

Society of Environmental Journalists (2012) "SEJ Membership from 1992 to 2012," Microsoft Excel file, Society of Environmental Journalists, December 6, 2012.

Sonderman, J. (2012). "600 newspaper layoffs in one day is, unfortunately, not a record," *Poynter*, 13 June 2012. Online. Available HTTP: www.poynter.org/latest-news/mediawire/177145/600-newspaper-layoffs-in-one-day-is-unfortunately-not-a-record/ (accessed December 11, 2012).

Tanner, A.H. (2004) "Agenda Building, Source Selection, and Health News at Local Television Stations: A Nationwide Survey of Local Television Health Reporters," *Science Communication*, 25: 350–363.

Taylor, C.E., Lee, J., and David, W.R. (2000) "Local Press Coverage of Environmental Conflict," *Journalism and Mass Communication Quarterly*, 77: 175–192.

Valenti, J.M. (1995) *Developing Protocol for Ethical Communication in Environmental News Coverage*, Provo, UT: Brigham Young University, Dept. of Communications.

Weaver, D.H., Beam, R.A., Brownlee, B.J., Voakes, P.S., and Wilhoit, G.C. (2007) *American Journalists in the 21st Century: U.S. News People at the Dawn of a New Millennium*, Mahwah, NJ: Lawrence Erlbaum.

West, B.M., Lewis, M.J., Greenberg, M.R., Sachsman, D.B., and Rogers, R.M. (2003) *The Reporter's Environmental Handbook*, 3rd edn, New Brunswick, NJ: Rutgers University Press.

13

THE CHANGING ECOLOGY OF NEWS AND NEWS ORGANIZATIONS

Implications for environmental news

Curtis Brainard

Since the dawn of the Internet era twenty years ago, the media industry has been undergoing a profound and disruptive metamorphosis as consumers have shifted steadily from printed newspapers and magazines—and eventually, television and radio broadcasts—to new, digital platforms. What began with bulletin boards, websites, and blogs quickly evolved into a rich ecosystem of podcasts, streams, and social media groups. The U.S. news outlets of the twentieth century largely failed to adapt to their novel surroundings and suffered greatly for it as mighty publishers began to lay off employees in droves and make severe cutbacks in reporting capacity.

As advertising shrank and subscriptions dwindled, journalists of every stripe suffered the consequences, but the reductions hit some corners of the newsroom especially hard, including the environment beat, which had never been a particularly high priority to begin with. Politics, business, sports, and arts dominate the newshole, and public opinion polls have consistently found that environmental issues rank low among other social priorities. There are a few exceptions. Natural disasters and weather have always had high news value, and energy sometimes draws high levels of interest when politicians are blaming one another for high gas prices, or an oil company dumps a million gallons of crude into the Gulf of Mexico. Generally, however, the environment is a third-class concern at most traditional media outlets, and over the last ten years, many of those covering it for them have lost their jobs or been reassigned to other beats.

While the crisis in those once hallowed halls of journalism continues, there is a bright side. So-called "special interest" or "niche" topics such as the environment have faired remarkably well online. The Internet is an exquisite matchmaking service between producers and consumers, and a variety of native outlets, better adapted to the digital age, have started to pick up some of the slack in coverage of the environment. Nonetheless, the new ecosystem is a jungle compared to the well-maintained gardens of the old world—a bewildering place where there is greater need to differentiate between "journalism" and "media" and where the principal of *caveat emptor* is more important than ever.

There are many important questions about the transformation of the news industry that remain largely unanswered. Are the newcomers reaching as wide an audience as their forbearers? What is the balance between professional journalism and amateur reporting? Between impartiality and advocacy? Between investigations and entertainment? And most importantly, perhaps, are readers able to discern high-quality sources of information and the many charlatans that populate cyberspace?

★★★

Two events in 2013 illustrated the prevailing trends in environmental journalism. On January 11, InsideClimate News, a four-year-old online startup dedicated to covering stories pertaining to climate change and energy broke the news that the *New York Times* was dismantling its so-called environment "pod." The decision prompted an outcry from critics who accused the *Times* of abandoning its responsibility to a vital beat.

The dissolution of the pod was "purely a structural matter," then managing editor Dean Baquet (who later became executive editor) said, pointing out that the paper had created the pod only four years earlier by pulling reporters who'd been assigned to covering the environment on the National, Foreign, Metro, Business, and Science desks into an autonomous unit. By breaking it up, Baquet intimated, the *New York Times* was merely going back to the embed approach, and he claimed that the paper was devoting "more [resources] than ever" to coverage of the environment (Bagley, 2013, para. 2). While everyone from the environment group was reassigned to new desks, however, not all of them remained focused on the environment, and two months later, the other shoe dropped when the *Times* announced that it was canceling its Green blog. A second outcry followed, and Baquet promised readers that he was simply following a strategy of integration whereby items that would have appeared on the blog would be threaded into other parts of the paper and website, where casual readers would be more likely to see them. The logic makes sense, but even the paper's well-respected public editor sided with critics. "Here's my take," she wrote after talking it over with Baquet. "I'm not convinced that the *Times*'s environmental coverage will be as strong without the blog. Something real has been lost on a topic of growing importance" (Sullivan, 2013, para. 21).

Elsewhere, fortunately, the industry was making gains. In April, InsideClimate News, the same startup that had broken the story about the dismantling of the *New York Times*' environment pod, won a Pulitzer Prize for National Reporting for a series that it had published the summer before, titled "The Dilbit Disaster: Inside the Biggest Oil Spill You've Never Heard Of." Its detailed investigation revealed the inept response of industry and government to the 2010 oil spill in Michigan's Kalamazoo River. It was the most costly onshore spill in U.S. history, but at the time it occurred the media's and the nation's attention was focused on an even larger calamity—the ongoing oil spill that followed the Deepwater Horizon explosion in the Gulf of Mexico. As InsideClimate News would later reveal, however, the pipeline rupture in Michigan had special significance because it dumped a million gallons of bitumen, a thick, dirty oil from Canada's tar sands region that has to be thinned with chemicals in order to flow through oil pipelines. It was the same type of oil slated to run through the controversial Keystone XL pipeline from Alberta to the Gulf Coast, and it proved particularly difficult to clean up. When the "dilbit" was released into the environment, the chemicals used to thin it vaporized, creating a potential local air pollution hazard. Unlike normal "sweet" crude, which floats, the tarry bitumen sank to the river bottom where it could be neither skimmed nor contained by booms. When the authorities showed up, InsideClimate News reported, they had no idea what they were dealing with because the U.S. doesn't require oil companies to disclose which type of crude is coursing through their pipes at any given moment.

"The Dilbit Disaster" was not only public service journalism at its best; it was new media at its best, and the Pulitzer victory brought InsideClimate News some well-deserved recognition. It wasn't the first web-native outlet to win the vaunted prize. There was ProPublica in 2010 and The Huffington Post in 2012, but those were large operations from the get-go, and InsideClimate News may have been "the leanest news start-up ever to be presented" with journalism's highest honor, as a *New York Times* article put it (Stelter, 2013, para. 4). Almost all of the ensuing news reports included the amusing observation that InsideClimate News didn't even have an office in which to hang the award. Spread out across the country, the outlet's small staff worked from home or rented a small office like the one that publisher David Sassoon occupied in Brooklyn, New York. Sassoon co-founded InsideClimate News with managing editor Stacy Feldman in 2007. At first they called it SolveClimate News and were doing what Sassoon calls "derivative journalism," aggregating and commenting on the work of other outlets in order to boost web traffic. A year later, they made a strategic decision to begin producing only original reporting, however, and in 2009 SolveClimate News began sharing content with Reuters, which Sassoon described as a "watershed moment" for the site, which validated the change of direction. Over the following two years, he continued to expand the staff, hiring veteran journalists as well as rookies, and in 2011, he changed the name from SolveClimate News to InsideClimate News in order to counter the perception that it was an environmental advocacy organization. The Pulitzer, however, was the real turning point for InsideClimate News. Sig Gissler, the administrator of the Prizes at Columbia University's Graduate School of Journalism, told the *New York Times* that the victory was an indication that "the way journalism as we've always known it and loved it is being reconfigured" (Sig Gessler, quoted in Stelter, 2013, para. 7).

★★★

The operational blueprint for InsideClimate News isn't exactly new. The Center for Investigative Reporting, which was also in the running for a Pulitzer in 2013, was founded in 1977. Like its younger cousins, InsideClimate News and ProPublica, it uses a non-profit business model and relies largely on grants from charitable foundations to make ends meet, posting most of its content free on the Web, but also partnering with for-profit news organizations to mount and publish more ambitious reporting projects. As those for-profit newsrooms have cut back, however, small and medium-sized investigative news operations, often focused on a particular region, have proliferated. In 2009, twenty-seven of these groups, new and old, gathered on the former Hudson Valley estate of oil tycoon John D. Rockefeller to discuss the future of investigation journalism and issued the Pocantico Declaration, which created the Investigative News Network, an organization dedicated to promoting the non-profit model of journalism in an effort to address the ongoing crisis in the field. By 2014, the organization had grown to 100 members, many of which were founded by former newspaper journalists who had lost their jobs.

The green beat features prominently in the Investigative News Network. Not long before it was created, the *Seattle Post-Intelligencer* had laid off most of its staff following a decision to cease printing the daily newspaper and publish strictly online. The P-I's longtime environment reporter, Robert McClure, rallied a group of his colleagues and launched the non-profit InvestigateWest, which joined the network in 2011 and continues the focus on environmental issues in the Pacific Northwest. Aspen Journalism, which came together in the wake of severe cutbacks at the city's two local dailies and covers the Roaring Fork River watershed in western Colorado, also makes the environment stories a priority—as do many of the network's locally oriented members. In 2012, three outlets with an even more explicit environmental focus joined its ranks. One was FERN, the Food & Environment Reporting Network, which began operations the year before and won a prestigious James Beard award in 2013 for a story about the plight of farmworkers titled, "As Common as Dirt." The other two outlets were Environmental Health News and its sister site, The Daily Climate.

Environmental Health News was already ten years old at the time. Founded in 2002, it started in much the same way that InsideClimate News did, doing mostly aggregation. In 2008, however, the site hired an award-winning environment reporter from the *Los Angeles Times*, Marla Cone, who had recently taken a buyout from the paper amidst downsizing. In a leaked email that Cone had sent to her fellow members of the Society for Environmental Journalists, she wrote that she had "lost hope" in the *Times* and she said elsewhere, "Editors misjudge how important [environmental] issues are to consumers" (Marla Cone, quoted in Walker, 2008, para. 5). In her first editor's note (www.environmentalhealthnews.org/ehs/editorial/from-the-editor-filling-the-void) to readers at Environmental Health News, Cone explained that the site would begin to produce original content. "In the vein of ProPublica, a nonprofit newsroom for investigative reporters, Environmental Health News will publish independent, foundation-funded, non-advocacy journalism," she wrote. "We believe that coverage of environmental issues must grow, not shrink, and, unlike most media, we don't have to rely on advertising to survive." Cone lived up to her promise. Environmental Health News's enterprise articles have appeared in the *San Francisco Chronicle*, the *Seattle Times*, and Reuters among others. In 2013, it won an honorable mention in the Oakes Awards for Distinguished Environmental Journalism for a series co-published with *Scientific American* called, "Pollution, Poverty, People of Color," which revealed the disproportionate toxic burden born by low-income minorities in the United States.

Lean, non-profit newsrooms like those mentioned above have accomplished a tremendous amount in terms of compensating for cutback in labor-intensive, public service journalism in traditional newsrooms. The modern media ecosystem is a fickle place, however, and the existence of most small start-ups is precarious. Indeed, in November 2014, Environmental Health News announced that due to fundraising problems, it was cutting staff and that beginning in 2015, the focus of both its website and The Daily Climate would "shift away from generating in-depth reporting and toward commentary and perspective on important daily happenings in the world of environmental health, energy and climate change" (The Daily Climate, 2014, para. 1). The announcement, posed online, added that Cone would no longer serve as editor-in-chief. "Enterprise journalism as practiced by the two news sites will be spun off into a separate entity," it read, "and we are launching an effort to secure separate funding for the world-class work that she has led and produced here" (para. 7).

Without more scholarly research, it is hard to say exactly how much, or in which ways, the media start-ups of the last decade have compensated for the decline in traditional newsrooms. What is more, there is an incredibly wide array of websites that also contribute to public communications about environmental issues, and the field is no longer the uncontested domain of media companies. Universities have bid for a larger share of people's attention, for example. In 2008, Yale University's School of Forestry and Environmental Studies launched the foundation-funded non-profit, Yale Environment 360, or e360 as it is known, under the direction of veteran journalist Roger Cohn, a former editor of *Mother Jones* and *Audubon* who had helped those magazines transition into the digital world. Likewise, in 2013, the University of Minnesota's Institute on the Environment relaunched a three-year-old non-profit environmental magazine, which had already won a number of awards, expanding from a thrice-yearly print publication to a sleek new website called *Ensia*. Both the Yale and University of Minnesota publications have attracted top journalistic talent, but unlike ProPublica or InsideClimate, they do not focus strictly on news and investigations. Instead, they feature a heavy dose of commentary and analysis from scientists, environmentalists, academics, policy makers, and business people. Newspapers and magazines have always shared their pages with such figures, of course, but there is now a greater variety of institutional goals among publishers. While e360 endeavors simply to offer "opinion, analysis, reporting and debate on global environmental issues," *Ensia*'s mission is more explicit: to "showcase solutions to Earth's biggest environmental challenges."

Advocacy-oriented journalism has become more common in the media world in general. One of the Web's oldest and best-loved sources of environmental news and commentary is *Grist*. At one point, *Grist*'s "about" page said that it "has been dishing out environmental news and commentary with a wry twist since 1999—which, to be frank, was way before most people cared about such things." A newer version of the page tells readers: "Our goal is to get people talking, thinking, and taking action." Financially speaking, *Grist* remains a humble non-profit supported by foundation grants, readers' support, and advertising, but other outlets have amounted to more significant commercial and consumer-driven enterprises. One of the largest is Treehugger, which describes itself as "the leading media outlet dedicated to driving sustainability mainstream. Partial to a modern aesthetic, we strive to be a one-stop shop for green news, solutions, and product information." Founded in 2004, Treehugger was sold three years later to Discovery Communications, the company that owns the Discovery Channel, for $10 million.

There are many other non-profit and for-profit media outlets out there that contribute in some measure to the coverage of environmental issues, but more numerous still—and more disruptive—are the countless independent blogs, small and large, that populate cyberspace. The most significant change in the media ecosystem engendered by the creation of the Internet is the absolute democratization of publishing. Blogs are the penny press newspapers of the twenty-first century, allowing every citizen to share his or her views on the world. The impacts of this empowerment on environmental communications have been profound, and nowhere is that more evident than in the realm of climate change.

Global warming is likely the biggest environmental story of the last decade, inspiring a cascade of related stories about energy, ecosystems, and sustainability, but its importance has also made it a political, economic, and social flashpoint. While blogs have allowed scientists and other legitimate experts, in fields from politics to economics, to communicate more easily and directly with the media and the public, a vast cacophony of other voices make the Internet a bewildering place where the quality of information can be hard to judge. RealClimate.org, established by a group of nine prominent American and European climate scientists in 2004, is one of the most trusted sources. It aims to better inform "the interested public and journalists" by providing "a quick response to developing stories and provide the context sometimes missing in mainstream commentary." At the other end of the spectrum are influential sites for "climate skeptics," such as Watts Up With That?, a blog run by meteorologist Anthony Watts, whom scientists have repeatedly criticized for misleading readers on subjects such as the reliability of the U.S. surface temperature record. Often, however, the disputes go far beyond points of science.

In 2009, an unknown party acquired a large cache of private emails between climate scientists from computer servers at the University of East Anglia in the United Kingdom and published them online. Cherry-picking quotes in order to make the scientists appear as though they were discussing data manipulation, bloggers such as Watts whipped up a pseudo-scandal that reverberated for years despite the fact that a series of nine investigations in the U.S. and the U.K. cleared the scientists of any wrongdoing. Indeed, the level of vitriolic propaganda online is so intense that Michael Mann, one of the most accomplished—and maligned—climatologists in the U.S. wrote a book in 2012 that described the pitched battles between scientists and skeptics, many of which take place online, as "the climate wars."

It's unclear how much exposure the general public has to such fights, but it's clear that as more and more people go online for their news information they'll need to be on guard against purveyors of spurious information and ever more prepared to evaluate the quality of sources on a case-by-case basis.

★★★

Environmental reporting has not disappeared at traditional news outlets, thankfully. A quick glance at recent winners of the prestigious journalism and communications awards from organizations such as the American Association for the Advancement of Science, the National Academies, Columbia University, and the Society of Environmental Journalists will show inspiring work at newspapers, magazines, and broadcast outlets small and large. Still, there is reason for concern.

The Society of Environmental Journalists, founded in 1989, reached its highest ever membership in 2009, with 1,523 members, but that number declined to 1,243 in 2013—roughly where it had been a decade earlier. The greatest losses were in the newspaper, television, and student categories, according to a membership survey. While 314 people said they worked primarily for newspapers in 2009, only 197 did so four years later, a drop of 37 percent. In contrast, the number of freelancers remained relatively high, at 375 in 2009 and 365 in 2013. And, despite the overall decline in membership in those years, there was an increase in the number of people who checked off author, faculty, online media, and film. The tally doesn't necessarily reflect the make-up of the entire industry, but it offers some sense of the changes taking place.

Shortly after InsideClimate News broke the story that the *New York Times* was dismantling its environment desk, it followed up with an article announcing that there were less than ten full-time environment reporters left at the top five U.S. newspapers, but a week later, it issued a correction saying that there were "about a dozen." Regardless of the exact number, it's a small group. In 2012, *The State of the News Media*, an annual report from the Pew Research Center's Project for Excellence in Journalism, found that coverage of the environment fell to 1 percent of the newshole in 2011, down from 2 percent the year before.

There aren't many scholarly, quantitative analyses of that coverage. Public-service oriented investigations, often revolving around environmental pollution and human exposure to toxins, tend to win the big awards, but impressionistic evidence suggests that for the last ten years climate change has been the biggest story on the green beat, and one of the few that researchers have tracked in any detail. The University of Colorado's Max Boykoff has tracked climate coverage in the *Wall Street Journal*, the *New York Times*, *USA Today*, the *Washington Post*, and the *Los Angeles Times* since 2000 (http://sciencepolicy.colorado.edu/media_coverage). And Drexel University's Robert Brulle has tracked climate coverage on NBC, CBS, and ABC since the 1980s (www.drexel.edu/culturecomm/ccdept/faculty/brulle.asp). In print and on television, the subject received very little attention until about 2005, when it quickly rose to prominence, reaching its zenith in 2007 following the release of *An Inconvenient Truth* and the Intergovernmental Panel on Climate Change's seminal Fourth Assessment Report. The level of coverage remained fairly high until the end of 2009 when efforts to pass climate legislation in the U.S. and to draft an international treaty for reducing greenhouse-gas emissions globally both failed. Coverage fell for the three straight years and eventually saw an uptick in late 2012 after Hurricane Sandy made climate change an eleventh-hour issue in the presidential race between Barack Obama and Mitt Romney. In fact, whether or not the coverage will remain high likely depends on the amount of attention given to it by politicians, according to one study (Upton, 2014), and on extreme weather events, according to others.

Other, less formal surveys reveal a few other key players in climate change coverage. The Daily Climate—a website that produces and tracks news about climate change—aggregates upwards of 20,000 articles about climate change each year. According to annual analyses of that database produced by The Daily Climate, the world's leading newswires, Reuters and The Associated Press, have consistently ranked within the top five since at least 2009 in terms of volume of coverage. In 2013, however, a former staffer from Reuters charged that the outlet's management had instructed staff that climate change was not to be a priority. Shortly thereafter, Media Matters, a media watchdog organization, charged that Reuters' climate coverage had fallen by nearly 50 percent since 2011. Whatever the case may be, two other outlets have consistently ranked high in The Daily Climate's annual survey. One is *The Guardian*, a British newspaper, which has invested heavily in coverage of environmental issues worldwide and increases its offerings many times over via the Guardian Environment Network, a group of more than 30 content partners from across the media spectrum. The other outlet is Energy & Environment Publishing (E&E), founded in 1998, which produces Greenwire, ClimateWire, and four other online news services. Aimed at professionals with an interest in environmental policy, law, and science, E&E uses an institutional subscription model, high-priced subscriptions to government offices, companies and corporations, universities, and non-governmental organizations, rather than low-priced subscriptions to individuals.

In the world of text, magazines are also still relevant. Newsweeklies such as *TIME* and *The Economist* regularly feature environmental coverage, including cover stories and special reports on various subjects. Literary and monthly magazines such as *The New Yorker*, *The Atlantic*, *Harper's*, and *Mother Jones* have made similar investments. Science magazines such as *Scientific American*, *Discover*, and *New Scientist* have always made the environment a major facet of their work. Most magazines that focus specifically on the environment are backed by non-profit societies. This includes *Audubon*, *National Geographic*, *Orion*, and *OnEarth* (a publication of the Natural Resources Defense Council, which canceled its print issue and went online only in 2014). Other, independent environmental magazines, such as *High Country News*, also tend to be non-profits.

Sustainability was even fashionable at consumer magazines for a while. Beginning in 2007, publishing annual Green Issues around Earth Day in April became all the rage. Led by *Vanity Fair*, consumer magazines such as *Esquire* and *Outside* got on the bandwagon in addition to news magazines such as *The New Republic* and the *New York Times Magazine*. By 2010, however, the bubble burst and the Green Issues vanished.

Television news networks have traditionally been the weakest leak in coverage of the environment, however. Among the three major networks, NBC News, where Anne Thompson has been Chief Environmental Affairs Correspondent since 2007, has perhaps displayed the greatest commitment to environmental coverage. CBS News had an Emmy-nominated and award-winning journalist covering the environment from 2006 until 2010, but when he left, the position stayed vacant for two years. Finally, CBS announced that it hired M. Sanjayan, lead scientist at The Nature Conservancy, as its science and environmental contributor, but Sanjayan's simultaneous work for an advocacy organization was a source of concern for media watchdogs. Just a few months after he was hired, CBS News aired a report on Western wildfires wherein Sanjayan interviewed a fire ecologist at The Nature Conservatory, without ever disclosing his own ties to the organization. The risk is that such conflicts of interest will only become more common as networks move from traditional staff reporters toward a greater reliance on experts from the worlds of government, business, academia, and NGOs.

Environment reporters at the local affiliates of the three major networks are exceedingly rare, unfortunately, since that is where many people still tend to get their news. Often, however, it is the weather forecaster at those stations who, by default, covers subjects such as climate change, and somewhat ironically, surveys have found that many television forecasters tend to be so-called "climate skeptics." There have been numerous attempts to improve the quality of environmental reporting at local and regional television stations. For example, Climate Central, launched in 2008, is a hybrid newsroom of sorts, comprising journalists and scientists alike. Like so many other start-ups, however, its mission has changed over time. At one point, it was to "provide local broadcast meteorologists with scientific expertise, data and graphics on local weather events and their connection with climate change." Later, the objective became to "communicate the science and effects of climate change to the public and decision-makers."

As far cable channels go, most environmental coverage tends to come in the form of "info-tainment" epitomized by The Discovery Channel and Animal Planet, both of which are owned by Discovery Communications. In 2008, CNN bet on a similar approach, canceling its entire science, technology, and environment news staff. "Now that the bulk of our environmental coverage is being offered through the Planet in Peril franchise, which is produced by the Anderson Cooper 360 program, there is no need for a separate unit," a CNN spokeswoman said at the time (Brainard, 2008, para. 2), but the channel soon lost interest in Planet in Peril as well. Of far greater concern, however, is the most widely viewed cable news channel, Fox News, which regularly spreads misinformation about environmental topics such as climate change.

The same problem exists on radio, where popular voices such as Rush Limbaugh and Glenn Beck make a habit of pooh-poohing environmental science. Elsewhere, however, there is some very good reporting coming over the airwaves. Public broadcasting, in particular, has done some strong work on the green beat. PBS and NPR regularly feature stories and special series about the environment, and although they are rare, there are some local efforts that focus specifically on the environment. These include award-winning programs such as KQED's QUEST program in northern California and The Allegheny Front on WESA in western Pennsylvania.

★★★

As the media ecosystem has grown more intimidating and confusing, a variety of new media criticism outlets have opened shop, generally acting as both watchdogs and guides. Many focus either partially or entirely on environmental issues, but like other start-ups, they have struggled to hang on. One of the most respected was the *MIT Knight Science Journalism Tracker*, which was launched in 2006 and provided daily commentary on major science, environment, and medical news. Affiliated with the science-writing program at the Massachusetts Institute of Technology, it was run by a group of veteran science reporters, but, in August 2014, it was announced that the *Tracker* would cease "operations in their current form" at the beginning of 2015. Likewise, in 2007, Bud Ward, a longtime environmental journalist and educator, founded the *Yale Forum on Climate Change & the Media*, which focused specifically on coverage of global warming before morphing into *Yale Climate Connections* in 2014. And in 2008, I founded The Observatory, *Columbia Journalism Review*'s first full-time department dedicated to the analysis of science and environment news and the challenges and obstacles facing reporters on those beats. In addition to these outlets, which take a fairly traditional, journalistic approach to criticism, there are a slew of advocacy-oriented criticism sites, such as *DeSmogBlog*, founded in 2006, whose stated mission is, "Clearing the PR pollution that clouds climate science."

These watchdogs and guides notwithstanding, it has become more important than ever for media consumers to abide by the principle of *caveat emptor*, evaluating the credibility of sources and information for themselves on a case-by-case basis. But this is not to say that there is some inherent flaw in the modern media setup. While a skeptical eye has become more necessary, the environmental media ecosystem is as rich and diverse as it has ever been. Most encouragingly, perhaps, the web-native, general-interest outlets that dominate online news—from titans such as *The Huffington Post* and BuzzFeed to up-and-comers such as Vice, Vox, and FiveThirtyEight—are investing in high-quality environmental reporting. It is now possible to find news, information, and commentary about nearly every facet of the environment known to humankind, and while some important issues still get overlooked—with disastrous consequences, in some cases—in many other areas, the Internet provides a bottomless well. There are significant uncertainties about whether or not consumers with no particular interest in environmental issues would naturally encounter the best new sources of information that digital media has to offer, however. Likewise, it's unclear whether or not consumers are, in fact, becoming more discerning as the media ecosystem grows more ideologically fragmented. More scholarly research is needed to answer these questions. Only then will we have a better sense of whether the media ecosystem is becoming more hospitable or more forbidding.

References

Bagley, K. (2013, January 11). *New York Times* dismantles its environmental desk. *New York Times*. Retrieved July 11, 2014, from http://insideclimatenews.org/news/20130111/new-york-times-dismantles-environmental-desk-journalism-fracking-climate-change-science-global-warming-economy

Brainard, C. (2008, December 4). CNN cuts entire science, tech team. The Observatory. *Columbia Journalism Review*. Retrieved July 15, 2014, from www.cjr.org/the_observatory/cnn_cuts_entire_science_tech_t.php?page=all

Mann, M. E. (2012). *The Hockey Stick and the Climate Wars: Dispatches from the Front Lines*. New York: Columbia University Press.

Stelter, B. (2013, April 16). A Pulitzer Prize, but without a newsroom to put it in. *New York Times*. Retrieved July 11, 2014, from www.nytimes.com/2013/04/17/business/media/insideclimate-news-hopes-to-build-on-pulitzer.html?pagewanted=all

Sullivan, M. (2013, March 5). For *Times* environmental reporting, intentions may be good but the signs are not. *New York Times*. Retrieved July 11, 2014, from http://publiceditor.blogs.nytimes.com/2013/03/05/for-times-environmental-reporting-intentions-may-be-good-but-the-signs-are-not/?_php=true&_type=blogs&_r=0

The Daily Climate. (2014, November 17). Changes at The Daily Climate. Retrieved January 30, 2015, from http://www.dailyclimate.org/tdc-newsroom/2014/11/daily-climate-changes

Upton, J. (2014, January 15). Major newspaper coverage of climate change plummeted last year. *Grist*. Retrieved July 11, 2014, from http://grist.org/news/major-newspaper-coverage-of-climate-change-plummeted-last-year/

Walker, B. (2008, September 15). Marla Cone to lead expanded news operation. *Grist*. Retrieved July 11, 2014, from http://grist.org/article/ace-reporter-joins-environmental-health-news/

14

NEWS ORGANISATION(S) AND THE PRODUCTION OF ENVIRONMENTAL NEWS

Alison Anderson

Introduction to news production

News production studies contribute a vital lens to our understanding of environmental communication as they help to reveal the complexities that underlie news work. Participant observation and in-depth interviews can provide fascinating insights into routine processes and constraints. The classic news ethnographies of the 1970s and 1980s laid important ground for the development of news production studies in the environmental arena (see for example, Gans, 1979; Gitlin, 1980 and Schlesinger, 1987; Tuchman, 1978). The term 'ethnography' is often loosely applied to refer to almost any form of qualitative research, but in its strict sense entails observation and immersion in the field to produce what Geertz (1973) called 'thick description'. These intensive studies, sometimes undertaken over a period of years, examined the professional ideologies of journalists, news values and dependence on news sources, newsroom conventions and their everyday routines. While providing rich insights into 'behind-the-scenes' processes involved in news-making they tended to over-theorize structural factors to the extent that news was often seen as simply the outcome of a set of routine professional, organizational and bureaucratic processes (Cottle, 2009). Moreover, they presented an essentially static, snapshot account of the organizational factors that shape the news, given that they tended to be ahistorical and limited in their ability to capture shifts over time (see Anderson, 1997; Schudson, 2000). As Cottle observes, these 'first wave' ethnographies of news production have become outdated by economic, technological, regulatory and cultural change. The sheer range of different media platforms and formats means that it is time to question the received wisdom of earlier studies, with their tendency to generalize about the organizational nature of news production. Today's global 24/7 news culture is now much more complex and multilayered. It is characterized by round-the-clock deadlines, increased competition and 'multi-skilled' journalists who are forced to work more flexibly than ever before by the increased casualization of the workforce, short-term contracts and digital convergence (see Cottle, 2003; Cox, 2013; Mitchelstein and Boczkowski, 2009; PEW, 2013). Environmental journalism in the US and some European countries has been especially affected over recent years, as budgets for science and investigative reporting have been substantially cut and workloads significantly increased. As Boykoff and Yulsman note:

> Focused on efficiency, media organizations have forced journalists to cover an increasing range of beats under tighter deadlines. Moreover, content producers in publishing organizations that have survived newsroom cuts and shortfalls have faced increased multiplatform demands (video, audio,

and text, along with blogs, Twitter, Facebook, Tumblr, Reddit, 4chan, and YouTube). This has posed significant challenges even to the most skilled and experienced reporters, including the likes of environmental journalist Andrew Revkin, whose *Dot Earth* blog at the *New York Times* is one of the best known outlets for information and commentary on global environmental issues, including climate change.

(2013: 362)

PEW State of the Media Report (2013) concludes that: 'This adds up to a news industry that is more undermanned and unprepared to uncover stories, dig deep into emerging ones or to question information put into its hands.' This increases the likelihood of errors, as may be seen in recent examples involving Fox News in its reporting of climate change (Cox, 2013). A further consequence of the intensely competitive news ecology is the tendency for many outlets to place increasing emphasis upon personalization and drama through simplifying and exaggerating conflict between claims-makers (Anderson *et al.*, 2005; Weingart *et al.*, 2000).

If we look beyond the US and many parts of Europe there is a rather mixed picture since science journalism, and newspapers generally, appear to be thriving in many developing parts of the world such as Africa (Boykoff and Yulsman, 2013). Nevertheless the rise of digital media is clearly impacting on the ability of reporters across the world to cover complex environmental issues. It is often lamented that journalism has been replaced by churnalism; that reporters have become mere passive conduits for press releases and copy churned out by news agencies and public relations companies (Davies, 2008 cited in Boykoff and Yulsman, 2013: 363; Lewis *et al*, 2008), though it is important to note that this is by no means a new phenomenon. In the mid 1970s Sachsman, for example, found that more than half of environmental news reports in the San Francisco Bay area originated or drew directly from press releases from sources or PR copy – and in many cases they were virtually word for word the same (see Sachsman, 1976). However, with the growth of digital media, the organizational constraints referred to above and the increasingly desk-bound nature of journalism this tendency appears to have been exacerbated (see Lewis *et al.*, 2008; Sachsman *et al.*, 2010). As the PEW State of the Media 2013 Report observed:

> Efforts by political and corporate entities to get their messages into news coverage are nothing new. What is different now—adding up the data and industry developments—is that news organizations are less equipped to question what is coming to them or to uncover the stories themselves, and interest groups are better equipped and have more technological tools than ever.

These economic and organizational factors combined mean that daily working routines are more pressured and there is little opportunity to network and verify information face-to-face, leading journalists to be more reliant upon pre-packaged material from news sources. As such, the job of the journalist becomes more about sifting through the multitude of information they receive each day (via international news agencies, press releases, emails, phone calls, video news releases, social media, electronic bulletin boards, etc.) and deciding what merits a news story, as opposed to actively searching for news.

The power of elite sources to set the news agenda

The rise in digital media and mobile phone communications has changed not only how people in the developed world access and interact with information, but who has access and who produces content. As Cottle (2010) remarks: "Whose voices and viewpoints structure and inform news discourse goes to the heart of democratic views of, and radical concerns about, the news media" (2010: 427). In the late 1970s Stuart Hall and colleagues (Hall *et al.*, 1978) published a classic study, *Policing the Crisis*, which examined the crisis over mugging in the UK and argued that official sources or 'primary definers' (such as government ministers and corporate officials) gain advantaged access to the media. This was seen as the outcome of the professional ideologies governing journalism and shared news values that granted greater legitimacy and credibility to ruling elites, reflecting their institutional status in society. The media were seen as 'secondary definers' through their role in reproducing the views of the powerful. For Hall *et al.*:

> These two aspects of news production – the practical pressures of constantly working around the clock and the professional demands of impartiality and objectivity – combine to produce a systematically structured over-accessing to the media of those in powerful and privileged institutional positions.
>
> (1978: 58)

In this way official sources have long been observed to predominantly have the upper hand; they frequently set the agenda for all subsequent framing of the issue, leaving less powerful sources in the position of having to respond rather than introduce their own frame (see Carlson, 2009). This reflects the wider inequalities of power with society, as Carlson puts it: "News reaffirms the unequal distribution of knowledge within society by promoting some sources as authoritative while ignoring other voices" (2009: 536). The contest to gain favourable media coverage is not a level playing field since official sources tend to have greater financial resources and stocks of cultural capital (Anderson, 1997). Indeed, numerous studies have shown how news organizations rely most heavily upon government and corporate sources; a finding upheld in the general field of the sociology of journalism (e.g. Sigal, 1973; Ericson *et al.*, 1989), as well as studies focusing upon the environment more specifically (e.g. Einsiedel, 1988; Greenberg *et al.*, 1989; Trumbo, 1996). However, while still influential, the 'primary definers' model has been considerably qualified over time. In an influential piece in the 1990s Schlesinger argued that the model, while offering a number of useful insights, was overly static and media-centric (see Schlesinger, 1990). It underestimated the extent of competition that occurs between news sources and ignored the complexities of source–media relations, such as conflict and division among powerful news sources themselves, and how media access changes over time, reflecting broader transformations within society. Moreover, in focusing upon the processes through which official sources gain news access it directed attention away from looking at the question of how marginal sources attract coverage. In subsequent years a range of empirical studies added weight to such observations, including studies focusing on news production and environmental issues (Anderson 1991, 1993, 1997; Hansen 1993).

As Cottle observes, news sourcing has shown to be a much more complex process than previously thought, whereby: "complexity and contingency are found where once social dominance alone was assumed sufficient to guarantee successful news entry" (2000: 437). It should be noted that elite access is not automatically guaranteed. News entry for both official and non-dominant sources is dependent on numerous contingencies (internal and external to the media). Research shows that marginal groups can sometimes gain elevated news entry (e.g. Anderson, 1997; Hansen, 2000; Manning, 2001). Non-dominant sources may lack the status, finance and PR personnel advantages enjoyed by official sources, but they are often able to respond to media demands much more quickly because they are not held back by cumbersome bureaucratic procedures and political restrictions (Anderson, 1997). Or in the face of silence among 'official sources', journalists may more actively seek out the views of alternative sources (Anderson, 1997).

As traditional news has declined there have been a variety of new players, a cacophony of voices, all vying to make themselves heard, from 'A' list bloggers, to radio chat show hosts, to Twitter and YouTube users (PEW, 2014). Celebrities are increasingly being used by NGOs campaigning about environmental issues, such as climate change, as a means to catapult their actions into the headlines (see Anderson, 2011; Brockington, 2013). While citizens themselves have many more opportunities to contribute to news content, in the vast majority of instances this involves user-generated comment rather than news reporting. There has been a significant growth in the popularity of alternative, not-for-profit media outlets, including web-based magazines such as *Grist* and *Climate Central* (Spencer, 2010). Environmental NGOs make extensive use of online communication and transnational activist networks have proliferated in recent years (Hutchins and Lester, 2011; Schäfer, 2012). At the same time, however, climate change sceptics have shown themselves to be particularly skilled at taking advantage of the opportunities afforded by new media platforms (Cox, 2013). There is no clear-cut distinction between 'old' and 'new' media, the two often being intermeshed together, just as the difference between 'mainstream' and 'alternative media' is increasingly blurred (Cottle, 2013).

Yet a recent comparative study of general coverage non-specific to environmental issues in leading new websites in nine countries concluded:

> Online news emerges as being very similar to the news of other media in being heavily reliant on state representatives (such as government ministers, judges and public officials) and on experts. Together, these represent 70 per cent of the sources of online news compared with 64 per cent of press sources and 60 per cent of television news sources. On average, online news actually gives less of a hearing to the political opposition, civil society and individual citizen sources than either television or newspaper news in the nine countries.
>
> (Curran *et al.*, 2013: 886)

Similarly, McNutt's study of the Climate Change Virtual Network found that the Canadian Government controlled most of the flows, and there were few links to NGOs (McNutt, 2008).

A further layer of complexity is that gaining access to the media and achieving coverage is only half the battle. How news sources' claims are framed, and whether they are portrayed as credible and legitimate is critically important. As Ryan (1991: 53) argues, 'the real battle is over whose interpretation, whose framing of reality, gets the floor'. Frames are culturally specific and offer a particular window on the world but they tend to be taken for granted and accepted as self-evident rather than actively scrutinized. Framing involves selecting certain truth claims over others and, in the process, denying or silencing rival versions of reality. Frames are shaped by claims-makers including politicians, scientists and NGOs; they do not occur in a political vacuum (Olausson, 2009). A key aspect of framing, then, is selectivity which 'arises through the efforts of claims-makers to effect a particular definition of an issue or problem by establishing a frame that is likely to resonate with prevailing values or ways of understanding' (Allan *et al.*, 2010: 30).

Robert Entman aptly sums it up in this way:

> Framing essentially involves selection and salience. To frame is to *select some aspects of perceived reality and make them more salient in the communicating text, in such a way as to promote a particular problem definition, causal interpretation, moral evaluation and/or treatment recommendation* for the item described. Frames, then, *define* problems – determine what a causal agent is doing and costs and benefits, usually measured in terms of cultural values; *diagnose* causes – identify the forces creating the problem; *make moral judgments* – evaluate causal agents and their effects; and *suggest remedies* – offer and justify treatments for the problem and predict their likely effects.
>
> (1993, 55; emphasis in original)

Framing therefore involves contestation between various claims-makers who seek to impose their preferred definitions of 'common sense' reality through the use of particular tactical manoeuvres (including staged news releases) and rhetorical devices including imagery, language and metaphors (Allan *et al.*, 2010). Through processes of framing, some players are cast in a more credible light than others. It should also be noted that control over the media is as much about the power to suppress or silence issues (Anderson, 2009). Examining the degree of space devoted to environmental issues by different news outlets, and the amount of coverage given to different news sources, while useful, offers an inevitably partial account.

While the behind-the-scenes struggles among news sources competing for media attention has often been overlooked, we now know much more about environmental journalism and source strategies through interviews with journalists and sources, examination of press releases and policy documents, or through observational methods (see Anderson 1997; Hansen 1993; Hutchins and Lester, 2011; Lester and Hutchins, 2009). Most previous research has focused on analysing the strategies of environmental pressure groups rather than industry, politicians or scientists (see Anderson, 2009; Hansen, 2011). However, there is an emerging body of literature that examines how business and political parties are seeking to actively shape environmental news (e.g. Beder, 2002; Davis, 2007; Greenberg *et al.*, 2011; Schlichting, 2013).

In an influential article published in 2000, Simon Cottle called for a 'second wave' of news ethnographies, questioning the validity of earlier assumptions that the organizational nature of news production automatically results in ideological closure. Such calls are slowly beginning to be addressed, but there remains much work to be done. Most research examining online media has focused on content or reception rather than production, perhaps reflecting the practical difficulties in gaining access and the time and space to undertake such intensive fieldwork (Paterson and Domingo, 2008). Some scholars have begun to address the gap by studying news production in online outlets (e.g. Deuze, 2008; Domingo and Paterson, 2011; Weiss and Domingo, 2010) but there are few studies that specifically focus on environmental news. In the next section we turn to focus on key internal and external constraints that further shape media construction of environmental issues.

News values and issue attention cycles

Over the past forty years environmental communication research has demonstrated how a myriad of factors shape news media representation of environmental issues in addition to the role and power of news sources. Far from mirroring reality, the coverage of environmental affairs, as with news in general, is highly selective and reflects economic, political and cultural factors. News about the environment is the end product of a complex process of construction. Deciding which environmental issues are newsworthy and merit coverage is governed by journalists' and editors' taken-for-granted ideas about what constitutes 'news' and judgements about the relative appeal of competing news items. Shifting issue attention cycles over time may be explained by a combination of numerous different factors, both internal and external to the workings of the media (see Anderson, 1997; Hansen, 2010; Lester, 2010). Here I single out some of the most important criteria (note this list is not exhaustive) that have been shown to influence which environmental issues get covered and how they fare over time.

Internal factors

First, environmental news is highly event oriented and this often determines whether an issue attracts coverage. For example dramatic events such as major oil spills tend to attract intense media coverage, particularly when they involve elite nations (Anderson and Marhadour, 2007). The more rare or sudden the event, the more likely it is to gain novelty value and grab headline attention. News quickly becomes stale and the unexpected and new is valued.

Second, environmental issues that lend themselves to ready visualization are generally more likely to be picked up, and this is especially so for digital formats. In many cases the availability and quality of pictures becomes a central factor affecting broadcasters' judgements about the newsworthiness of a given environmental issue (Anderson, 1997; Smith, 2005).

Third, environmental issues tend to involve long, drawn-out processes and there is often a long period of scientific uncertainty, which sits uneasily with 24/7 news cycles (Schoenfeld *et al.*, 1979). Unless claims-makers are able to skilfully draw attention to such issues by packaging them in more attention-grabbing ways, they are likely to remain relatively invisible.

Fourth, the tendency for media to focus on conflict and controversy and to exaggerate points of divergence can influence story selection (see Anderson *et al.*, 2005; Lester and Hutchins, 2013). Stories that fit into a classic 'protest frame' may garner more attention since they include ingredients of drama, spectacle and disorder (see Lester, 2010).

Fifth, research suggests that editorial pressures can bear strongly upon journalists' coverage of environmental issues. For example, O'Neill, an Australian broadcaster seconded to the Reuters Institute, interviewed fourteen mainly UK based journalists and editors (broadcast and print) in May and June 2010 and found many of the reporters experienced considerable hostility from their editors in the wake of Climategate (see O'Neill, 2010). Similarly, Smith concluded:

> The negotiation between correspondents and editors is a critical point in the mediation of climate change knowledge. It often centres on the degree to which the proposed stories fit with dominant news frames. These negotiations take place in the context of immense time pressures and acute surveillance of the performance of individual editors ... The result is very likely to be stories that satisfy editorial standards much more satisfactorily than they communicate the social or scientific reality or significance of an issue as understood by specialists.
>
> (2005: 1477)

Sixth, different media formats affect the amount of space environmental issues receive and how they are framed. Different newspapers, for example, are governed by their own particular restrictions, professional cultures and distinctive ideological standpoints (Carvalho and Burgess, 2005). There are also important differences between different types of media outlets. We still know relatively little about the different factors that affect environmental coverage in local/regional media as opposed to national and international media (Anderson, 1997; Anderson, 2009; Hansen, 2010). However, studies suggest that there are significant differences between local/regional and national reporting of environmental issues. This includes the amount of space devoted to particular environmental issues, the types of news sources drawn upon and the framing of the issues (e.g. Brown *et al.*, 2011; Cottle, 2000; Crawley, 2007). For example, in the reporting of the Prestige oil disaster of 2002 the regional Spanish press focused on implications for the local economy rather than the effects on wildlife. By contrast, national newspapers in Spain, France and the UK framed the oil spill in terms of its ecological impacts and the political controversy over who was to blame (Anderson and Marhadour, 2007). Similarly, Cottle (2000) found significant differences between regional TV news reporting of environmental issues in the UK, which was more likely to air the voices of ordinary 'lay' people, compared with national coverage.

External factors

First, media ownership and the broader political economy clearly shape news content (Anderson, 2009; Boykoff and Yulsman, 2013; Carvalho, 2005). For example, UK based studies have shown a close relationship between the political agenda and the reporting of environmental affairs (e.g. Anderson, 1997; Carvalho, 2005). Olausson's study of climate change reporting in the Swedish national press found numerous similarities between media and international policy discourse on the issue of climate change (Olausson, 2009). Similarly McGaurr and Lester (2009) highlight how *The Australian*, Australia's only national newspaper, largely followed the Prime Minister, John Howard's lead, in its approach to viewing nuclear power as the solution to climate change.

Second, related to this are economic factors. Powerful business interests may exert pressure on the reporting of environmental affairs (see Beder, 2002; Boykoff and Yulsman, 2013). Editors' decisions may be influenced by the fear that running critical items may result in lost advertising revenue (Anderson, 2009). As discussed above, economic conditions can also impact on the capacity of journalists to undertake in-depth investigative reporting. Moreover, the global economic downturn itself may in part account for recent declining attention given to issues such as climate change in many countries, as the news media focus on more 'immediate' concerns.

Third, an increasing number of international comparative studies (most of which focus on climate change) suggest cultural factors are highly significant when explaining differences in news production and content (see Boykoff, 2007; Schmidt *et al.*, 2013). The framing of particular environmental issues may resonate more strongly than others with people in different social groupings and in different countries. That is to say, they may connect more closely with culturally deep-seated, historically rooted symbolic imagery. Several studies, for example, show how imagery associated with nuclear energy may link with common cultural narratives in different ways (see Anderson, 1997).

Concluding comments

News production research has certainly come a long way since the 1970s. The field is ripe for further development, given the increasingly complex media environment that we inhabit and the growing influence of strategic communication. Despite a number of calls to move beyond a media-centric approach, the field is still largely dominated by content-based studies. Further empirical work needs to step up in scale to adequately capture the dynamic and competitive processes deployed to frame environmental issues that cannot simply be captured by examining media content. This poses major practical challenges in the era of global digital media where flows reciprocally impact on one another in non-linear ways. As Cottle argues: 'The complex flows of news communications and dispersed productive activity requires international research collaborations as well as methodological ingenuity if we are to capture the online traces of journalist production activity before they evaporate into the virtual ether' (2007: 9).

We need to know much more about the impact of technological changes on news work and how this has affected journalists' reporting of environmental issues and their relationship with sources in different national contexts. There is also considerable scope to explore the extent to which claims-makers are successful, not just in gaining visibility through media coverage, but in terms of being portrayed in a credible light and achieving their goals (see Anderson, 2006; Hansen, 2011). Ethnographies can provide illuminating insights into the often hidden intricacies underlying the construction of environmental news and remain as important as ever.

Suggested further reading

Anderson, A. (2014) *Media, Environment and the Network Society*. Basingstoke: Palgrave Macmillan.
Cottle, S. (2007) 'Ethnography and news production: New(s) developments in the field', *Sociology Compass*, 1 (1): 1–16.
Cottle, S. (2009) 'New(s) times: Towards a "second wave" of news ethnography', in A. Hansen (ed.) *Mass Communication Research Methods*. Volume I, Chapter 19, 366–386. London: Sage.
Hansen, A. (2010) *Environment, Media and Communication*. London: Routledge, Chapter 4.
Lester, L. (2010) *Media and Environment: Conflict, Politics and the News*. Cambridge: Polity, Chapter 3 and 4.

References

Allan, S., Anderson, A. and Petersen, A. (2010) 'Framing risk: Nanotechnologies in the news', *Journal of Risk Research*, 13 (1): 29–44.
Anderson, A. (1991) 'Source strategies and the communication of environmental affairs', *Media, Culture and Society*, 13 (4): 459–476.
Anderson, A. (1993) 'Source–media relations: The production of the environmental agenda'. In A. Hansen (ed.) *The Mass Media and Environmental Issues*. Leicester: Leicester University Press, pp. 51–68.
Anderson, A. (1997) *Media, Culture and the Environment*. London: Routledge.

Anderson, A. (2006) 'Media and risk'. In S. Walklate and G. Mythen (eds) *Beyond the Risk Society*. Open University/ McGraw Hill, pp. 114–131.

Anderson, A. (2009) 'Media, politics and climate change: Towards a new research agenda', *Sociology Compass*, 3 (2): 166–182.

Anderson, A. (2011) 'Sources, media and modes of climate change communication: The role of celebrities', *Wiley Interdisciplinary Reviews: Climate Change*, 2 (4): 535–546.

Anderson, A. and Marhadour, A. (2007) 'Slick PR? The media politics of the Prestige oil spill', *Science Communication*, 29 (1): 96–115.

Anderson, A., Petersen, A., Allan, S. and Wilkinson, C. (2005) 'The framing of nanotechnologies in the British newspaper press', *Science Communication*, 27: 200–220.

Beder, S. (2002) *Global Spin: The Corporate Assault on Environmentalism*. Totnes: Green Books.

Boykoff, M.T. (2007) 'Flogging a dead norm? Newspaper coverage of anthropogenic climate change in the United States and United Kingdom from 2003 to 2006', *Area*, 39, 470–481.

Boykoff, M.T. and Yulsman, T. (2013) 'Political economy, media, and climate change: Sinews of modern life', *Wiley Interdisciplinary Reviews: Climate Change*, 4 (5): 359–371.

Brockington, D. (2013) 'Celebrity, environmentalism and conservation'. In L. Lester and B. Hutchins (eds) *Environmental Conflict and the Media*. Oxford: Peter Lang, pp. 139–152.

Brown, T., Budd, L. Bell, M. and Rendell, H. (2011) 'The local impact of global climate change: Reporting on landscape transformation and threatened identity in the English regional newspaper press', *Public Understanding of Science*, 20 (5): 658–673.

Carlson, M. (2009) 'Dueling, dancing, or dominating? Journalists and their sources', *Sociology Compass*, 3 (4): 526–542.

Carvalho, A. (2005) 'Representing the politics of the greenhouse effect', *Critical Discourse Studies*, 2: 1–2.

Carvalho, A. and Burgess, J. (2005) 'Cultural circuits of climate change in U.K. broadsheet newspapers, 1985–2003', *Risk Analysis*, 25 (6): 1457–1469.

Cottle, S. (2000) 'New(s) times: Towards a "second wave" of news ethnography', *Communications: The European Journal of Communication Research*, 25 (1): 19–41.

Cottle, S. (ed.) (2003) *Media, Organisation and Production*. London: Sage.

Cottle, S. (2007) 'Ethnography and news production: New(s) developments in the field', *Sociology Compass*, 1 (1): 1–16.

Cottle, S. (2009) 'New(s) times: Towards a "second wave" of news ethnography'. In A. Hansen (ed.) *Mass Communication Research Methods*, Volume I, Chapter 19. London: Sage, pp. 366–386.

Cottle, S. (2010) 'Rethinking news access', *Journalism Studies*, 1 (3): 427–448.

Cottle, S. (2013) 'Environmental conflict in a global media age: Beyond dualisms'. In L. Lester and B. Hutchins (eds) *Environmental Conflict and the Media*. Oxford: Peter Lang, pp. 19–33.

Cox, R. (2013) 'Climate change, media convergence and public uncertainty'. In L. Lester and B. Hutchins (eds) *Environmental Conflict and the Media*. Oxford: Peter Lang, pp. 231–243.

Crawley, C.E. (2007) 'Localized debates of agricultural biotechnology in community newspapers: A quantitative content analysis of media frames and sources.' *Science Communication*, 28 (3): 314–346.

Curran, J., Coen, S., Aalberg, T., Hayashi, K., Jones, P.K., Splendore, S., Papathanassopoulos, S., Rowe, D. and Tiffen, R. (2013) Internet revolution revisited: A comparative study of online news. Working paper.

Davies, N. (2008) *Flat Earth News: An Award-Winning Journalist Exposes Falsehood, Distortion and Propaganda in the Global Media*. London: Chatto and Windus.

Davis, A. (2007) *The Mediation of Power: A Critical Introduction*. London: Routledge.

Deuze, M. (2008) 'Toward a sociology of online news?' In C. A. Paterson and D. Domingo (eds) *Making Online News: The Ethnography of New Media Production*. New York: Peter Lang, pp. 199–209.

Domingo, D. and Paterson, C. (2011) *Making Online News Volume 2. Newsroom Ethnographies in the Second Decade of Internet Journalism. Digital Formations*. Oxford: Peter Lang.

Einsiedel, E.F. (1988) 'The Canadian Press and the Environment', paper presented to the XVIth Conference of the International Association for Mass Communications Research, Barcelona, July.

Entman, R.M. (1993) 'Framing: Toward clarification of a fractured paradigm', *Journal of Communication*, 43 (4): 51–58.

Ericson, R., Baranek, P. and Chan, J. (1989) *Negotiating Control: A Study of News Sources*. Toronto: University of Toronto Press.

Gans, H. (1979) *Deciding What's News*. London: Constable.

Geertz, C. (1973) 'Thick description: Toward an interpretive theory of culture'. In C. Geertz (ed.) *The Interpretation of Cultures: Selected Essays*. New York: Basic Books, pp. 3–30.

Gitlin, T. (1980) *The Whole World Is Watching: Mass Media in the Making and Unmaking of the New Left*. Berkeley, CA: University of California Press.

Greenberg, J., Knight, G. and Westersund, E. (2011) 'Spinning climate change: Corporate and NGO public relations strategies in Canada and the United States', *International Communication Gazette*, 73 (1/2): 65–82.

Greenberg, M.R., Sachsman, D.B., Sandberg, P.M. and Salome, K.L. (1989) 'Network evening television news coverage of environmental risk', *Risk Analysis*, 9 (1): 119–126.

Hall, S., Critcher, C., Jefferson, T., Clarke, J. and Roberts, B. (1978) *Policing the Crisis: Mugging, the State and Law and Order* (1st edn). Basingstoke: Palgrave Macmillan.

Hansen, A. (1993) 'Greenpeace and press coverage of environmental issues'. In A. Hansen (ed.) *The Mass Media and Environmental Issues*. Leicester: Leicester University Press, pp. 150–178.

Hansen, A. (2000) 'Claims-making and framing in British newspaper coverage of the Brent Spar controversy'. In S. Allan, B. Adam and C. Carter (eds) *Environmental Risks and the Media*. London: Routledge, pp. 55–72.

Hansen, A. (2010) *Environment, Media and Communication*. London: Routledge.

Hansen, A. (2011) 'Communication, media and environment: Towards reconnecting research on the production, content and social implications of environmental communication', *International Communication Gazette*, 73 (2): 7–25.

Hutchins, B. and Lester, L. (2011) 'Politics, power and online protest in an age of environmental conflict'. In S. Cottle and L. Lester (eds) *Transnational Protests and the Media*. Oxford: Peter Lang, pp. 159–171.

Lester, L. (2010) *Media and Environment: Conflict, Politics and the News*. Cambridge: Polity.

Lester, L. and Hutchins, B. (2009) 'Power games: Environmental protest, news media and the Internet', *Media, Culture and Society*, 31 (4): 579–595.

Lester, L. and Hutchins, B. (eds.) (2013) *Environmental Conflict and the Media*. New York: Peter Lang.

Lewis, J., Williams, A. and Franklin, B. (2008) 'A compromised fourth estate? UK news journalism, public relations and news sources', *Journalism Studies*, 9 (1): 1–20.

McGaurr, L. and Lester, L. (2009) 'Complementary problems, competing risks: Climate change, nuclear energy and the Australian'. In T. Boyce and J. Lewis (eds) *Climate Change and the Media*. Oxford: Peter Lang, pp. 174–185.

McNutt, K. (2008) 'Policy on the web: The climate change virtual policy network', *Canadian Political Science Review*, 2 (1): 1–15.

Manning, P. (2001) *News and News Sources*. London: Sage.

Mitchelstein, E. and Boczkowski, P.J. (2009) 'Between tradition and change: A review of recent research on online news production', *Journalism*, 10 (5): 562–586.

Olausson, U. (2009) 'Global warming – global responsibility? Media frames of collective action and scientific certainty', *Public Understanding of Science*, 18: 421–436.

O'Neill, M. (2010) *A Stormy Forecast: Identifying Trends in Climate Change Reporting*. Reuters Institute for the Study of Journalism: University of Oxford. Available online at: https://reutersinstitute.politics.ox.ac.uk/fileadmin/documents/Publications/fellows__papers/2009-2010/A_STORMY_FORECAST.pdf (accessed 14 August 2013).

Paterson, C. and Domingo, D. (eds) (2008) *Making Online News*. Oxford: Peter Lang.

PEW (2013) State of the News Media 2013 Report. Available online at: http://stateofthemedia.org (accessed 20 August 2013).

PEW (2014) State of the News Media 2014. Available online at: www.journalism.org/packages/state-of-the-news-media-2014/ (accessed 14 August 2013).

Ryan, C. (1991) *Prime Time Activism: Media Strategies for Grassroots Organising*. Boston, MA: South End Press.

Sachsman, D. (1976) 'Public relations influence on coverage of environment in San Francisco area', *Journalism Quarterly*, 53 (1): 54–60.

Sachsman, D., Simon, J. and Myer Valenti, J. (2010) *Environmental Reporters in the 21st Century*. Somerset, NJ: Transaction.

Schäfer, M. S. (2012) 'Online communication on climate change and climate politics: A literature review', *Wiley Interdisciplinary Reviews: Climate Change*, 3 (6): 527–543.

Schlesinger, P. (1987) *Putting 'Reality' Together*. London: Sage.

Schlesinger, P. (1990) 'Rethinking the sociology of journalism'. In M. Ferguson (ed.) *Public Communication*. London: Sage, pp. 61–83.

Schlichting, I. (2013) 'Strategic framing of climate change by industry actors: A meta-analysis', *Environmental Communication: A Journal of Nature and Culture*, 7 (4): 493–511.

Schmidt, A., Ivanova, A. and Schäfer, M.S. (2013) 'Media attention for climate change around the world: A comparative analysis of newspaper coverage in 27 countries', *Global Environmental Change*, 23 (5): 1233–1248.

Schoenfeld, A.C., Meier, R.F. and Griffin, R.J. (1979) 'Constructing a social problem—the press and the environment', *Social Problems*, 27: 38–61.

Schudson, M. (2000) 'The sociology of news production revisited (again)'. In J. Curran and M. Gurevitch (eds) *Mass Media and Society*. London: Edward Arnold, pp. 175–200.

Sigal, L.V. (1973) *Reporters and Officials: The Organization and Politics of Newsmaking*. Lexington, MA: D.C. Heath.

Smith, J. (2005) 'Dangerous news: Media decision making about climate change risk', *Risk Analysis*, 25 (6): 1471–1482.

Spencer, M. (2010) 'Environmental journalism in the greenhouse era,' *FAIR Extra*, 2 February. Available online at: http://fair.org/extra-online-articles/environmental-journalism-in-the-greenhouse-era/ (accessed 20 August 2013).

Trumbo, C. (1996) 'Constructing climate change: Claims and frames in US news coverage of an environmental issue', *Public Understanding of Science*, 5 (3): 269–283.

Tuchman, G. (1978) *Making News*. New York: Free Press

Weingart, P., Engels, A. and Pansesgrau, P. (2000) 'Risks of communication: Discourses on climate change in science, politics, and the mass media', *Public Understanding of Science*, 9: 261–283.

Weiss, A. and Domingo, D. (2010) 'Innovation processes in online newsrooms as actor-networks and communities of practice', *New Media and Society*, 12 (7): 1156–1171.

15

CITIZEN SCIENCE, CITIZEN JOURNALISM

New forms of environmental reporting

Stuart Allan and Jacqui Ewart

This chapter offers an evaluative assessment of the fluid, evolving interface between citizen scientists and citizen journalists where the news reporting of environmental issues is concerned. Citizen science, depending on how it is defined, dates back over centuries, and has often been described using a host of different terms – recent examples include 'do-it-yourself science', 'crowdsourced science' or 'democratised science', as well as 'mass scientific collaboration', 'participatory action research', 'volunteer monitoring' or even 'citizen cyberscience'. Just as the relationship between the 'amateur' and the 'professional' scientist can prove challenging at times, however, such has also proven to be the case between journalists and members of the public who feel compelled to adopt a journalistic role, either temporarily or on a more sustained basis (Allan, 2013; Thorsen and Allan, 2014).

Accordingly, we begin by exploring various definitions of citizen science, namely with a view to pinpointing how those definitions have been applied in practice, thereby illuminating corresponding tensions between professional and citizen scientists. We then turn our attention to how citizen journalists have helped to push environmental science onto the public agenda. In addition to scoping the academic literature in this area, we look at examples – including the Deepwater Horizon oil spill off the coast of the United States, and the Fukushima Daiichi nuclear plant meltdown in Japan – to consider how citizen journalists have covered environmental issues, hazards and crises. While many citizens engaged in this work find it personally fulfilling, often regarding it as a contribution to community service, others encounter significant difficulties – including when questions of risk and scientific uncertainty prove controversial. By its end, this chapter will have identified several issues worthy of attention for future efforts to enrich and deepen citizen environmental journalism.

Citizen science

Many scientists devote considerable time striving to create effective ways to engage ordinary members of the public (that is to say, 'laypersons' or 'non-scientists') in science, particularly where environmental issues have become contentious. Such efforts have acquired even greater impetus with the advent of digital technologies in recent years – ranging from the personal computer to the smartphone or tablet of mobile participatory cultures – leading some to herald a new age of 'citizen science' dawning on the horizon. While others are quick to challenge any assertion that this is a 'new' phenomenon, pointing out that although 'the internet has clearly increased opportunities for mass participation and "crowdsourcing" data, there is a long history of gathering scientific information from amateurs' (Kilfoyle and Birch, 2014), the growing prominence of nascent forms of collaboration is readily apparent.

Definitions of 'citizen science' vary, of course, depending on who is doing the defining in question. One of its early conceptual formulations pertinent to our discussion was introduced by Irwin (1995) in his book *Citizen Science: A Study of People, Expertise and Sustainable Development*. The use of the term in its title was intended to evoke 'a science which assists the needs and concerns of citizens', he explained, while at the same time implying 'a form of science developed and enacted by citizens themselves' (1995: xi). Pointing out that earlier examples of citizen-oriented science had met with limited success, such as the 'science for the people' movement, Irwin stressed the time was right for a re-evaluation. Of particular interest to him were the kinds of knowledges ordinary individuals develop and bring to bear 'in the face of the truth claims of science', as well as the active role they may choose to play in wider processes of knowledge dissemination. Using a series of case studies, he proceeded to discern the features of what he termed 'contextual knowledges', which are typically put together 'piecemeal' by citizens actively 'learning through doing', usually incorporating technical information in an ad hoc, selective manner. Such involvement may well produce tensions, of course – the relationship between the contextual knowledges of lay accounts and the more formalised ones of official science can be fraught, particularly where the latter seem 'impervious to renegotiation and revision on the basis of locally generated evidence' (1995: 128). Citizens struggling to have their knowledges recognised and substantiated by decision-making authorities (with the hope that they will be acted upon) may be all too aware of what Irwin termed 'the social gap' between these different forms of expertise, and thereby 'the inappropriateness of most enlightenment assumptions about the public understanding of science' (1995: 131).

Conceptions of citizen science placing an emphasis on the production of knowledges help to open up afresh familiar debates about expertise and how it is negotiated, even contested in particular circumstances. 'One of the dangers of an increasingly professional and specialized corps of "experts" is the mistaken belief that people who do not have academic credentials, research budgets, and fancy equipment lack the means to contribute to knowledge or discourse about environmental issues', Parris (1999) maintained in the wake of Irwin's intervention. Looking across a range of differing perspectives on what counts as citizen science in this regard, it soon becomes apparent that varied definitions tend to privilege certain recurrent rationales for public participation. More specifically, such efforts are perceived to revolve around one or more of the following imperatives:

- the enrichment of scientific understanding with accurate, cost-effective and often time-critical data-collection;
- fostering communities of practice where expertise in scientific inquiry is shared;
- enhancing civic pride and awareness with the promotion of values associated with 'science-literacy', citizenship and responsibility to the natural world;
- encouraging the development of numeracy as well as practical skills and techniques in observation, measurement or computational activities (including the use of online recording or smartphone apps);
- expanding productive links between communities and their local environments for sustainable development;
- raising the public profile of education or careers in environmental science;
- and, most important for some, representing an enjoyable hobby, consistent with an active lifestyle.

These imperatives are recurrently interwoven, typically signalling emphases rather than stark differences, but together they help to explain how priorities emerge under particular circumstances.

Members of the scientific community typically welcome the enthusiasm of dedicated 'amateurs', but some tend to be rather sceptical about whether the results being produced satisfy research-grade standards. 'Naysayers might chide, "The data are of poor quality, they cannot be trusted; they could be misleading or even dangerous; and they are certainly not admissible in court"', Schnoor (2007), editor of *Environmental Science and Technology*, points out. Certain scientific tasks are better than others for citizen science, he concedes, and findings always need to be interpreted with due attention to how they were achieved. Still, he adds, there 'is a sizable literature which attests that data collected by properly trained citizen volunteers are of as high a quality as those obtained by professionals with the same equipment' (2007: 5923; see also Bonney *et al.*, 2009; Caitlin-Groves, 2012). Further research suggests that projects must be made meaningful for prospective volunteers, which is to say enjoyable and rewarding. A 2012 report by the UK Environmental Observation Framework stressed the importance of enthusiasm for the goals being set, as well as feelings of control over the scientific process. 'Projects must be tailored to match the interests and skill-sets of participants', its authors state, 'and understanding the motivations and expectations of potential volunteers is crucial to developing successful projects (Roy *et al.*, 2012: 5; see also Bell, 2010). Equally important, other studies suggest, are efforts to ensure communities become active stakeholders in pursuing research of relevance to them, with local concerns and priorities oriented towards shaping policy-formation. Here it is worth noting that economically deprived communities are much more likely to be overlooked than prosperous ones in this regard, even though environmental problems may be that much more apparent (Rowland, 2012; see also Dickinson *et al.*, 2010).

The growing significance of online citizen science projects – striving to make the most of digital, web-based resources – underscores how the boundaries of professional science are being redrawn. Lending shape to the ethos of these 'new wave' projects is their commitment to moving beyond more traditional, deficit-model (top-down, zero-sum) conceptions of the 'public understanding of science' in order to emphasise meaningful engagement in co-operative ventures. Examples of projects attracting news headlines include:

- ClimatePrediction.net, which claims to be 'the world's largest climate modelling experiment', encourages volunteers – some 32,000 active participants in 147 countries – to download a climate model program to run in the background on their personal computer when it is otherwise idle. A distributed computing project, it draws on 'ensemble forecasting' to produce predictions of the Earth's climate up to the year 2300 while, at the same time, testing the accuracy of climate models. 'By looking at the uncertainties associated with each model', it explains, 'we can selectively use the best models to make predictions for the future, thereby giving policy makers an accurate picture to date of climate predictions'.

- Open Air Laboratories (OPAL) network is 'open to anyone with an interest in nature' inspired to 'explore, study, enjoy and protect their local environment'. Funded by a Big Lottery Fund grant in the UK, it aims to bring 'scientists, amateur-experts, local interest groups and the public closer together' to work on projects, including national surveys examining air, soil and water quality, as well as biodiversity and climate change. A 'Bugs Count' smartphone app, for example, may be used to submit photographs of one of six species of invertebrates being studied.

- The Atlas of Living Australia (Atlas) contains information on all the known species in the country, gathered from a wide range of 'data providers', including individuals and community groups. 'Harnessing the efforts of the thousands of people participating in citizen science will enhance the range and depth of data available in the Atlas', its website explains, encouraging volunteers to contribute their sightings of plants or animals they have seen in their chosen area using the FieldData software made available (aggregates data on flora and fauna can then be analysed geospatially). Funding support is provided by the Australian government.

- 'National Sampling of Small Plastic Debris' is a project involving nearly 1,000 children from 39 schools along the coastlines of Chile and Easter Island in documenting the distribution and abundance of small plastic debris found on beaches. Prior to the launch of the project, the data available were considered insufficient to offer an accurate evaluation of the problem. Data obtained by the students, the validity of which was confirmed when samples were recounted in the laboratory, showed the 'average abundance obtained was 27 small plastic pieces per m² for the continental coast of Chile, but the samples from Easter Island had extraordinarily higher abundances (>800 items per m²)'. This and related ecological evidence will be used to improve waste management (see Hidalgo-Ruz and Thiel, 2013).

- Fraxinus is a free Facebook genetic pattern puzzle game designed to help counter the Chalara fungus, responsible for an epidemic of disease in ash trees across Europe. 'Using genetic data taken from Chalara and ash tree samples, Fraxinus challenges you to manipulate patterns to match sample sequences', the instructions explain. 'The closer the match, the higher your score and the more useful the results will be to scientists.' This competition to claim patterns by recognising sequences in the genome for the ash tree (954 million nucleotides long) is an initiative supported by The Sainsbury Laboratory (TSL), the John Innes Centre and game developer Team Cooper in the UK.

To the extent these and related projects privilege collaboration over competition, they help to expand the scope of scientific institutions. Efforts to establish and sustain a symbiotic relationship between professional and citizen science will recognise that it is contingent upon a range of factors, such as flexibility in project leadership and decision making roles; shared involvement in the setting and refinement of research questions, priorities and protocols; and the necessity of securing participatory capacity for mutual learning, among other considerations. Greater openness, in principle, enhances transparency, thereby inviting public trust in knowledge production that is as beneficial as it is socially responsible.

In the next section, we turn to consider recent events in environmental journalism, paying particular attention to the ways in which citizen-initiated forms of reportage have sought to address certain shortcomings brought to light by their alternative ethos. Efforts to secure new strategies in this regard, we argue, have much to gain by fostering productive synergies between citizen journalism and citizen science, thereby making the most of their potential to reinvigorate environmental news coverage (see also Major and Attwood, 2004; Mazur and Lee, 1993; Sampei and Aoyagi-Usui, 2009). Citizen environmental journalism, as we shall see, can be differentiated from traditional news media reportage of environmental risks, threats and hazards at several levels, particularly where it is closely associated with activist-centred priorities. Research conducted by Lester and Hutchins (2012), for example, found that environmental activists were using technologies to highlight salient issues with a view to engaging with diverse publics. The interviews they conducted with journalists and environmental activists pinpointed the potential affordances to be secured by such strategies:

> Systematic and ongoing experimentation with self-representation via online communication promises to avoid both the fickleness of changing news agendas, the vicissitudes of reporting and editorial practices, and the contending corporate interests of large-scale news conglomerates. For instance, it is notable that few of the interviewees discuss in detail the possibilities for direct interaction with users (other than journalists) offered by websites, the potential of participatory media and online citizen journalism, and the use-value of community-driven wikis, blogs, vlogs and video-hosting sites.
>
> (Lester and Hutchins, 2012: 591)

Given the challenges news media organisations encounter when reporting environmental concerns, citizen environmental activists strive to make the most of opportunities to take up journalistic-style roles with the aim of communicating outside the boundaries characteristic of more mainstream news coverage.

Citizen environmental journalism

Citizens have long featured as sources in environmental news reports, typically by sharing their first-hand, eyewitness perspectives from the scene of a specific event (Greenberg *et al.*, 1989; Major and Attwood,

2004; Robinson, 2002). Recent years have seen journalist–source relationships dramatically recast, however, with citizens making the most of the internet and digital technologies to engage in their own forms of reportage. Such forms of 'accidental journalism' may include using a smartphone to capture an image or short video clip, craft a tweet, or update a blog or Facebook page in order to connect with distant friends or family – as well as with news organisations, which may be actively gathering such material to supplement their coverage. While these types of developments have begun to attract the attention of researchers, one aspect of citizen journalism that Mythen (2010: 45) proposes has been under-explored is 'the impact of citizen journalism on the reporting of risk'. There are several benefits to be gained by citizen journalists' reportage of risks and hazards, he maintains. These include the ability of citizen journalists to 'add to the plurality of discourses circulating about hazardous events, provide an alternate agenda from mainstream news and bring into question the political and cultural logics that underpin risk incidents' (p. 49; see also Bruns *et al.*, 2012; Lariscy *et al.*, 2009).

Prospective intersections between citizen journalism and citizen science, we would suggest, similarly warrant greater consideration from researchers than they have received to date. More specifically, we would suggest that such explorations might advantageously focus on the forging of co-operative partnerships between citizen scientists and citizen journalists. This section, in seeking to illustrate this potential, turns to examine two recent examples of environmental crises – the Deepwater Horizon oil spill off the coast of the United States and the Fukushima nuclear accident in Japan – that garnered sustained global news media attention. We consider how citizen scientists and citizen journalists contributed to the news coverage of these disasters, highlighting how their efforts helped to reveal the basis for new, collaborative approaches to environmental crisis reporting to emerge.

Deepwater Horizon oil spill

A gas leak on the British Petroleum (BP) oil drilling rig, Deepwater Horizon, situated in the Gulf of Mexico, led to an explosion on 20 April 2010. Eleven people died as a fireball was ignited and the rig sank, while oil continued to leak into the Gulf before it was finally capped almost three months later, on 15 July. Of particular interest for our purposes here was the crowdsourcing of information – generated, in part, by citizen scientists and citizen journalists alike – to help narrow information gaps, much to the benefit of traditional news media striving to comprehend the significance of what was transpiring (see also Aulov and Halem, 2012; McCormick, 2012; Starbird, 2011). As journalists scrambled to gather and interpret official assertions in the face of media blackouts, ordinary citizens – scientists and impromptu reporters alike – stepped in to provide valuable, near-instant information via websites and social media platforms, such as Twitter. News organisations were quick to draw on these accounts, especially where the eyewitness accounts of those directly affected by events were concerned (Veil *et al.*, 2011).

In the immediate aftermath of the Deepwater Horizon oil spill, ordinary individuals – such as residents and fishermen situated along the coastline – were providing 'real time monitoring of exposures to the spill and its observable impacts' (McCormick, 2012). In doing so, they performed recognisable scientific and journalistic roles in monitoring shoreline conditions, such as collecting samples, assessing impacts on fish stocks and giving first-hand accounts of the perceived damage caused by the spill. A variety of technological tools were pressed into service, including two apps – OilReporter and MoGO – developed for mobile telephones, both of which facilitated the logging of reports (Reiter, 2011). A key collection point for much of the data being gathered was the Louisiana Bucket Brigade's website The Oil Spill Crisis Response Map (see also Merchant *et al.*, 2011). Such crowdsourcing, McCormick (2012) observed, enables members of the public to drive data aggregation, thereby representing 'a new form of citizen science, where lay people engage in research design, data collection, and analysis. This crowdsourcing was meant to collect real time exposure data that was otherwise unmeasured' (2012: 30). Moreover, these kinds of activities, she adds, extended the capacity of data collection because they documented exposure on a much more intense basis than scientific studies would typically allow. During the Deepwater Horizon event, bloggers who had some scientific training also worked with advocacy groups to monitor the effects of the oil spill, posting news and informational updates on the website Deep Sea News (Thaler *et al.*, 2012).

While citizen scientists' efforts to gather and record data about the potential and actual hazards associated with Deepwater Horizon were significant, they were complemented by the endeavours of citizen journalists who provided important information about the disaster in the absence of 'on the ground' news media reportage. Partnerships established between local fishermen, citizen scientists and citizen journalists aimed to close vital gaps in the information chain, thereby helping to bring the extent of the oil spill to the attention of the broader public. The fishermen's vessels provided access to areas near the affected zone, enabling those on-board to see – and record – first-hand perspectives (Rowley, 2013). A case in point revolved around the use of digital cameras to create balloon maps of areas not yet contaminated by the oil spill (Griffith *et al.*, 2012). Balloon mapping in this instance involved taking multiple aerial photographs of the zones in question, and then combining the ensuing images to provide a visual survey of the fast-changing situation. The maps were then published on the free, open source Public Laboratory for Open Technology and Science website. Griffith and colleagues (2012) explained how this strategy worked:

> For less than $100 in parts, we used helium balloons and kites to send cameras to over a thousand feet, and stitched the resulting images into high-resolution maps using our free, open-source software. Over a hundred volunteers hit the beaches to take tens of thousands of photos, depicting slicks, oiled wetlands, and the birds, fish, and plants threatened by the disaster. Our efforts resulted in the largest repository of publicly archived oil spill data to date and it is for use without restriction.
>
> (Griffith *et al.*, 2012: 13773)

For Griffith and colleagues the value of initiatives such as the Public Laboratory lies in their portability and accuracy, as well as cost efficiency, while providing a picture of a crisis as it unfolds. In their words, it is a 'bottom up approach in data collection, management, and literacy' (2012: 13773). Moreover, these types of data were later collated by the Government Accountability Project, part of a 2013 report on the long-term health effects of the oil spill and the use of chemical agents in the clean-up.

Social media platforms played an important role in disseminating information, with citizens using them to report on the events associated with Deepwater Horizon. As Aulov and Halem (2012) pointed out:

> In the aftermath of the disaster, the public was very active in discussing its impact and implications across a range of SM outlets. Many people who witnessed firsthand the damage caused by the oil such as birds soaked in oil or tar balls washing up on the shore reported their accounts in different SM outlets. People posted photos and videos of oiled beaches, tweeted from their smartphones when they were prohibited from swimming because of oil pollution, and so on.
>
> (Aulov and Halem, 2012: 2812)

Although Aulov and Halem (2012: 2812) suggest that the 'primary purpose of such online activity is social interaction between friends', other researchers have identified that there are a variety of other reasons that motivate individuals to engage with social media during disasters (Allan, 2013; Pantti *et al.*, 2012). The social media site Twitter was a key point for information sharing after the oil spill. Starbird (2011) and her colleagues used an interactive mapping tool to track pertinent Twitter posts, not least those relaying information about the location of the oil. Although citizens were initially encouraged only to post Tweets about 'bad smells or fumes' due to concerns for their personal safety, the huge response led to a change in strategy. Starbird explained:

Later, after seeing many tweets with oil impact information and other efforts to allow for citizen reporting of oil through other channels, we added and promoted the ability for Twitterers to report oil impact on the shore, affected wildlife, response activities, volunteer opportunities, and area closures. Using TtT in concert with some behind-the-scenes, manual work to add location information to tweets, we collected and mapped over 800 oil impact reports between May 9 and August 9, 2010.

(Starbird, 2011: 111–112)

While citizen scientists and citizen journalists responded quickly in the aftermath of the oil spill, Hayworth *et al.* (2011) found that these groups continued their involvement well after the initial events. In the years after the disaster, citizen groups continued to collect data on the effects of the oil spill on beaches and marine life, although as Hayworth and colleagues point out the 'end result is a dataset comprised of public and private accumulations of various types of samples and associated results varying in representativeness and quality' (p. 3641). Still, bearing such caveats in mind, there is little doubt that the value of citizen contributions was considerable, not least in helping to bridge the gaps between official expertise – both scientific and journalistic – and that of lay publics committed to alternative forms of fact-finding.

Fukushima nuclear plant accident

The Fukushima Daiichi nuclear plant meltdown provides a second salutary example of how the citizen collection of scientific data can converge with journalistic engagement to report on a fast-breaking environmental crisis. Following a magnitude nine earthquake and tsunami on 11 March 2011, three of the six nuclear reactors melted down, forcing the rapid relocation of nearby residents. Kaigo's (2012) research found that the mainstream news media were unable to 'provide information about lifeline disruption or other necessary information for the vast majority of victims in disrupted areas' and thus were 'highly ineffective' (2012: 26) in the areas affected by the earthquake. Ng and Lean (2012) also criticised mainstream coverage of the crisis, namely for its focus on the damage to the nuclear power station, rather than the effects of the accident. In their view, citizens could have been used to complement both traditional science and journalism to a greater extent, not least with regard to social networking sites to help offset the shortcomings of official efforts to disseminate vital information under confusing, frequently contradictory circumstances. Dromey's (2012) research echoed this point, underscoring the value of new media technologies as 'invaluable assets in comparison to traditional news media and institutional scientific enquiry' (2012: 5).

In the absence of official data being made available in the immediate aftermath of the Fukushima accident, several websites sought to crowdsource information about the levels of radiation in nearby areas. Safecast co-founders Pieter Franken and Sean Bonner provided Geiger counters to volunteers for use on their vehicles as data gathering tools (Strickland, 2013). The website collated and posted the resulting datasets so that members of the public would have access to them in the form of data maps. Thanks to citizen journalists blogging about the citizen scientists' efforts in gathering this data, mainstream media quickly followed up with news stories that placed pressure on government officials. A Public Broadcasting Service (2011) report in the US, for example, framed the story as an attempt by citizen scientists to address the paucity of publicly available information about what was actually happening near the site of the disaster. Another website, called Pachube, similarly provided crowdsourced data about the fallout, collating radiation readings provided by local citizens (Dromey, 2012; Kera *et al.*, 2013; Mayer and Karam, 2012). These and related initiatives can be considered as citizen science, Dromey (2012) contends, even where the 'methods and the results obtained perhaps would not pass as "serious" science'. Rather, their 'focus was to use openly available tools to gather and communicate invaluable data to get as clear a picture as possible on the spread of radionuclides during the fallout' (2012: 32–33). New ways to tap into this type of impromptu citizen response, it follows, must be found.

Citizen science activities in response to the Fukushima accident were not isolated to Japan. Several people living in Canada, for example, gathered scientific data for The Woods Hole Oceanographic Institution. Senior scientist Ken Buesseler asked citizens to collect samples of seawater from sites along the west coat of North America and send them to his research centre so that the presence of radioactive materials could be analysed. The campaign involved two collection points, on Haida Gwaii and at Bamfield on the west coast of Vancouver Island (Suzuki, 2014). Data was made available on the website 'How Radioactive is our Ocean', where site users were able to link to a pertinent map showing levels of radioactivity. 'North Americans may have little cause for concern for now', Suzuki (2014) maintained in light of these findings, 'but without good scientific information to determine whether or not it is affecting our food and environment we can't know for sure. The Woods Hole initiative is a good start.'

Just as citizen scientists contributed to knowledge about radioactivity levels, the role of citizen reporters in extending scientific data also came to the fore. Kaigo (2012) identified how social networking sites such as Twitter became highly useful sources of real-time information for citizens in the aftermath of the earthquake. Even where this activity consisted of re-tweets – namely those of government authorities and local news media accounts – citizens were helping to extend the range and reach of official communicative strategies. Such efforts similarly served to mobilise citizen responses to the disaster, including bringing together volunteer groups to assist officials, including with respect to data-gathering. Still, citizen activity posed its own challenges for journalists reporting on the crisis. Friedman's (2011: 62) research, for example, noted the 'size of the Fukushima information explosion on the Internet, and the speed of transmission to readers and viewers worldwide', including how formidable these factors proved for 'gatekeepers' to manage. Despite the best intentions of many – albeit not all – of the citizens involved to be accurate, she observed, much of this reportage proved to be scientifically incorrect (see also Johansson *et al.*, 2012).

Before turning to our conclusions, it is worth pausing to note how these and related examples only begin to highlight the multiplicity of activities performed by citizens committed to blurring longstanding boundaries demarcating the realm of the professional from the amateur in environmental science as well as its corresponding forms of journalism. It is early days for efforts to maximise the potentials afforded by such partnerships, yet both the Deepwater Horizon and Fukushima disasters provide formative, if inchoate indications of how such partnerships may further consolidate to the betterment of crisis communication in the years ahead.

'Science by the people'

In this chapter we have highlighted certain aspects of citizen science with a view to discerning possible points of convergence with citizen journalism. Our aim has been to invite further explorations of their potential for rethinking how best to improve environmental journalism. The tensions between professional and citizen science will continue to spark lively, often diverging points of view, but as funding sources wax and wane depending on changing political, economic and cultural landscapes, few would dispute that the time is right to engage in dialogue. Despite the frictions that sometimes occur, there is much to be gained by mutually respectful deliberations over how best to encourage environmental journalism to reconsider its guiding tenets afresh. Chief among these concerns, as brought to light by the examples considered above, is its over-reliance on official voices to demarcate the parameters of newsworthiness. Precisely what counts as legitimate scientific evidence in this regard frequently proves to be much more important – and open to debate – than journalism's time-worn conventions tend to recognise, not least because the issues at stake do not fit easily within event-centred narratives revolving around unexpected novelty, drama, conflict or scandal.

Several obstacles loom large in the path of those committed to improving environmental journalism, as other contributors to this volume have similarly sought to make clear for further investigation.

At a time when most news organisations are coping with increasingly severe resource restrictions, with journalists working under intense time pressures to produce multiple versions of news stories across various platforms, commitments to specialist genres of investigative reporting become increasingly difficult to justify for cash-strapped news organisations. The demands placed upon environmental journalists have always been formidable, of course, yet they continue to intensify (Allan, 2011; Allan *et al.*, 2000; Anderson, 1997; Boyce and Lewis, 2009; Friedman, 2004, 2011; Hansen 2010, 2011; Lester, 2010; Reynolds, 2001). Still, while the internet and associated digital technologies are bringing to bear new demands and contingencies, we have sought to show that they also have the capacity to foster innovative alternatives. Conventional journalist-source dynamics are being recast, not just in the blogosphere where citizen scientists subject news stories to interrogation for accuracy's sake, but also where others partake in either scientific or journalistic activity of their own (and sometimes both), often motivated – as we have seen – by a personal desire to cultivate more open, responsive cultures for scientific discourses. The opportunities afforded news organisations to help facilitate such forms of public engagement with environmental issues are becoming increasingly apparent as citizen science grows in stature and influence, yet much work remains to be done to consolidate this potential in ways that encourage greater dialogue, transparency and trust.

Acknowledgements

Stuart Allan was a Gambrinus Fellow at the University of Dortmund, Germany, when first developing his research on citizen science. He is grateful for this support, as well as for discussions with Holger Wormer and Wiebke Rögener at the Institute for Journalism. Jacqui Ewart would like to acknowledge Bob Morrish of the Cooper's Creek Protection Group for his insights, as well as Griffith University's Key Centre for Ethics, Law, Justice and Governance for its on-going support.

Further reading

Dickinson, J.L., Zuckerberg, B. and Bonter, D.N. (2010) 'Citizen science as an ecological research tool: Challenges and benefits', *Annual Review of Ecology, Evolution, and Systematics*, 41: 149–172.
This article examines a number of the ways in which citizen science has 'increased the scale of ecological field studies with continent-wide, centralized monitoring efforts and, more rarely, tapping volunteers to conduct large, coordinated, field experiments'. It assesses both the relative strengths and limitations of crowdsourcing in ecological research, particularly with regard to habitat and climate change.

Friedman, S.M. (2011) 'Three Mile Island, Chernobyl, and Fukushima: An analysis of traditional and new media coverage of nuclear accidents and radiation', *Bulletin of the Atomic Scientists*, 67(5): 55–65.
This is a discussion of how traditional news media's coverage of environmental crises (e.g. *The New York Times*) compares to various new forms of online media reporting, including citizen contributions via blogs, Facebook, Twitter and YouTube. In adopting complementary approaches, it is argued, news organisations can secure improved ways to communicate complex technical information to enhance public understanding.

McCormick, S. (2012) 'After the cap: Risk assessment, citizen science and disaster recovery', *Ecology and Society*, 17(4), http://dx.doi.org/10.5751/ES-05263-170431
Taking the Deepwater Horizon oil spill in 2010 as its focus, this article examines how 'crowdsourcing is used as a new form of citizen science that provides real time assessments of health-related exposures'. Its findings suggest that crowdsourcing, when used effectively, facilitates online data gathering systems that allow 'a broader range of participation and the detection of a broader range of impacts'.

References

Allan, S. (2011) 'Introduction: Science journalism in a digital age', *Journalism*, 12(7): 771–777.

Allan, S. (2013) *Citizen Witnessing: Revisioning Journalism in Times of Crisis*, Cambridge: Polity Press.

Allan, S., Adam, B., and Carter, C. (eds) (2000) *Environmental Risks and the Media*, London and New York: Routledge.

Anderson, A. (1997) *Media, Culture and the Environment*, London: UCL Press.

Aulov, O. and Halem, M. (2012) 'Human sensor networks for improved modeling of natural disasters', *Proceedings of the IEEE*, 100(10): 2812–2823.

Bell, A. (2010) 'Citizen science still needs specialism', *The Guardian*, 20 June.

Bonney, R., Cooper, C.B., Dickinson, J., Kelling, S., Phillips, T., Rosenberg, K.V. and Shirk, J. (2009) 'Citizen science: A developing tool for expanding science knowledge and scientific literacy', *BioScience*, 59(11): 977–984.

Boyce, T. and Lewis, J. (eds) (2009) *Climate Change and the Media*, New York: Peter Lang.

Bruns, A., Burgess, J., Crawford, K. and Shaw, F. (2012) '#qldfloods and @QPSMedia: Crisis communication on Twitter in the 2011 South East Queensland floods', Published by the ARC Centre of Excellence for Creative Industries and Innovation, Queensland University of Technology.

Caitlin-Groves, C.L. (2012) 'The citizen science landscape: From volunteers to citizen sensors and beyond', *International Journal of Zoology*, Article ID 349630, doi:10.1155/2012/349630

Dickinson, J.L., Zuckerberg, B. and Bonter, D.N. (2010) 'Citizen science as an ecological research tool: Challenges and benefits', *Annual Review of Ecology, Evolution, and Systematics*, 41: 149–172.

Dromey, B.L. (2012) 'Understanding Fukushima: Designing for an embodied interaction with citizen science data', MA New Media Thesis, Aalto University.

Friedman, S.M. (2004) 'And the beat goes on: The third decade of environmental journalism', in S.L. Senecah (ed.) *The Environmental Communication Year Book Volume 1*, Mahwah, NJ: Lawrence Erlbaum Associates, pp. 175–187.

Friedman, S.M. (2011) 'Three Mile Island, Chernobyl, and Fukushima: An analysis of traditional and new media coverage of nuclear accidents and radiation', *Bulletin of the Atomic Scientists*, 67(5): 55–65.

Greenberg, M., Sachsman, D.B., Sandman, P.M. and Salomone, K.L. (1989) 'Network evening news coverage of environmental risk', *Risk Analysis*, 9(1): 119–126.

Griffith, A., Dosmagen, S. and Warren, J. (2012) 'Complete data lifecycles and citizen science integration via The Public Laboratory', *Geophysical Research Abstracts*, 14: 13773.

Hansen, A. (2010) *Environment, Media and Communication*, London: Routledge.

Hansen, A. (2011) 'Communication, media and environment: Towards reconnecting research on the production, content and social implications of environmental communication', *International Communication Gazette*, 73(1–2), 7–25.

Hayworth, J.S., Clement, T.P. and Valentine, J.F. (2011) Deepwater Horizon oil spill impacts on Alabama beaches', *Hydrology and Earth System Sciences*, 15: 36393649.

Hidalgo-Ruz, V. and Thiel, M. (2013) 'Distribution and abundance of small plastic debris on beaches in the SE Pacific (Chile): A study supported by a citizen science project', *Marine Environmental Research*, 87: 12–18.

Irwin, A. (1995) *Citizen Science: A Study of People, Expertise and Sustainable Development*, London: Routledge.

Johansson, F., Bryneilsson, J. and Quijano, M.N. (2012) 'Estimating citizen alertness in crises using social media monitoring and analysis', paper presented at the European Intelligence and Security Informatics Conference.

Kaigo, M. (2012) 'Social media usage during disasters and social capital: Twitter and the great East Japan earthquake', *Keio Communication Review*, 34: 19–35.

Kera, D., Rod, J. and Peterova, R. (2013) 'Post-apocalyptic citizenship and humanitarian hardware', in R. Hindmarsh (ed.) *Nuclear Disaster at Fukushima Daiichi: Social, Political and Environmental Issues*, London: Routledge, pp. 97–115.

Kilfoyle, M. and Birch, H. (2014) 'Placing citizens at the heart of citizen science', Political Science blog hosted by *The Guardian*, 6 January.

Lariscy, R.W., Avery, E.J., Sweetser, K.D. and Howes, P. (2009) 'An examination of the role of online social media in journalists' source mix', *Public Relations Review*, 35: 314–316.

Lester, L. (2010) *Media and Environment*, Cambridge: Polity Press.

Lester, L. and Hutchins, B. (2012) 'The power of the unseen: Environmental conflict, the media and invisibility', *Media, Culture and Society*, 34(7): 847–863.

McCormick, S. (2012) 'After the cap: Risk assessment, citizen science and disaster recovery', *Ecology and Society*, 17(4), online at: http://dx.doi.org/10.5751/ES-05263-170431

Major, A. and Attwood. L.E. (2004) 'Environmental risks in the news: issues, sources, problems, and values', *Public Understanding of Science*, 13: 295–308.

Mayer, S. and Karam, D.S. (2012) 'A computational space for the web of things', *Proceedings of the Third International Workshop on the Web of Things*, New York.

Mazur, A. and Lee, J.L. (1993) 'Sounding the alarm: Environmental issues in the US national news', *Social Studies of Science*, 23: 681–720.

Merchant, R.M., Elmer, S. and Lurie, N. (2011) 'Integrating social media into emergency-preparedness efforts', *The New England Journal of Medicine*, 28 July.

Mythen, G. (2010) 'Reframing risk? Citizen journalism and the transformation of news', *Journal of Risk Research*, 13(1): 45–48.

Ng, K.H. and Lean, M.L. (2012) 'The Fukushima nuclear crisis re-emphasizes the need for improved risk communication and better use of social media', *Health Physics Society*, 103(3): 307–310.

Pantti, M., Wahl-Jorggensen, K. and Cottle, S. (2012) *Disasters and the Media*, New York: Peter Lang.

Parris, T.M. (1999) 'Connecting with citizen science', *Environment*, 41(10): 3.

Public Broadcasting Service (2011) 'Safecast draws on the power of the crowd to map Japan's radiation', online at: www.pbs.org/newshour/bb/science-july-dec11-japanradiation_11-10/

Reiter, E. (2011) 'Citizen science and mobile applications', online at: www.elainereiter.com/itecportfolio/wpcontent/uploads/2013/04/Citizen-Science-and-Mobile-Applications_ereiter8.pdf

Reynolds, T. (2001) 'News headlines feed on fear of cancer risk, experts say', *Journal of the National Cancer Institute*, 93(1): 9–11.

Robinson, E.E. (2002) 'Community frame analysis in Love Canal: Understanding messages in a contaminated community', *Sociological Spectrum*, 22(2): 139–169.

Rowland, K. (2012) 'Citizen science goes "extreme"', *Nature*, 17 February.

Rowley, M.J. (2013) 'Open data and the next evolution in citizen science', online at: www.thetoolbox.org/articles/111-open-data-and-the-next-evolution-in-citizen-science#.UzDU9hy_7nV

Roy, H.E., Pocock, M.J.O., Preston, C.D., Roy, D.B., Savage, J., Tweddle, J.C. and Robinson, L.D. (2012) 'Understanding citizen science and environmental monitoring', Final Report, NERC Centre for Ecology & Hydrology and Natural History Museum on behalf of UK-EOF.

Sampei, Y. and Aoyagi-Usui, M. (2009) 'Mass-media coverage, its influence on public awareness of climate-change issues, and implications for Japan's national campaign to reduce greenhouse gas emissions', *Global Environmental Change*, 19: 203–212.

Schnoor, J.L. (2007) 'Citizen science', *Environmental Science & Technology*, 41(17): 5923.

Starbird, K. (2011) 'Digital volunteerism during disaster: Crowdsourcing information processing', Paper presented at the Conference on Human Factors in Computing, May 7–12, Vancouver, BC.

Strickland, E. (2013) 'Citizen scientists in the Fukushima fallout zone', online at: http://hereandnow.wbur.org/2013/08/23/citizen-scientists-fukushima

Suzuki, D. (2014) 'Citizen scientists can fill info gaps about Fukushima effects', online at: www.davidsuzuki.org/blogs/science-matters/2014/01/citizen-scientists-can-fill-info-gaps-about-fukushima-effects/

Thaler, D.A., Zelnio, K.A., Freitag, K., MacPherson R., Shiffman, D., Bik, H., Goldstein, M.C. and McClain, C. (2012) 'Digital environmentalism: Tools and strategies for the evolving online ecosystem', in D. Gallagher (ed.) *Environmental Leadership: A Reference Handbook*, London: Sage, pp. 364–373.

Thorsen, E. and Allan, S. (2014) *Citizen Journalism: Global Perspectives, Volume 2*, New York: Peter Lang.

Veil, S.R., Buehner, T. and Palenchar, M.J. (2011) 'A work in process literature review: Incorporating social media in risk and crisis communication', *Journal of Contingencies and Crisis Management*, 19(2): 110–122.

16

ENVIRONMENTAL NEWS JOURNALISM, PUBLIC RELATIONS AND NEWS SOURCES

Andy Williams

Introduction: environment journalists and their sources

Important to maintaining journalistic objectivity is the task of getting information – the raw materials of the news – from elsewhere. For this journalists usually turn to news sources: people with knowledge or expertise who can provide perspectives on a news event. But sources are not neutral purveyors of information; they have agendas, and try to construct and circulate their own (favourable) discourses about news events, as well as aiming to keep unfavourable stories out of the news. In public relations (PR), sources have developed an entire industry to tightly control the flow of information; PR operatives try to influence news agendas and coverage with pre-packaged materials such as press releases, news briefings, press conferences, persuasive personal communications, and sometimes manipulative and hidden media management tactics (Davis 2002). Journalistic practice deals with this by providing reporters with methods and routines that aim to minimise the dangers of sources dominating coverage. Notions of editorial independence are foregrounded (Franklin *et al.* 2010: 203), and norms of journalistic research advocate that a range of alternative news sources be consulted to provide a plurality of perspectives (Berkowitz 2009: 103). Despite this, much research into how sources access the mainstream news media has found that official, elite social actors tend to get more coverage than others, partly because of the resources they are able to devote to media management (Gans 1979; Tuchman 1978).

Research into environment journalism has often produced insights that overlap with broader studies of general news production. A Cardiff University study of news about science commissioned by the (then) UK Government's Office of Science and Innovation illustrates this well. By recording what they call the 'news hook' of each story, they were able to find from which sector of society newsworthy events emanate: 30 per cent of all stories dealt with university research, 18 per cent emanated from industry and 13 per cent from the UK Government. Only 7 per cent of news pieces originated from the efforts of NGOs or pressure groups (Boyce *et al.* 2007: 22). When such campaigning groups were quoted they were most often used as secondary 'reactive' sources, and were rarely allowed to set the agenda of articles (27).

Hansen writes that when covering the environment the 'mass media are notoriously authority oriented', and that studies of environment news:

have virtually without exception shown that the sources who get to be quoted ... and who get to define environmental issues are ... predominantly those of the public authorities, government representatives, industry and business, and independent scientists.

(2010: 56)

Dorothy Nelkin identifies a 'reverential attitude' among journalists dealing with scientists as news sources (Nelkin 1995: 98), and others have suggested that specialist reporters' relationships with sources are too cosy (Hargreaves *et al.* 2003). The principal objection to such close relationships is that they tend to mean journalists depend too much on powerful sources with efficient PR teams, something that reduces their capacity for independence, allows sources too much control over the news agenda and often over how specific stories are framed (Williams and Clifford 2010). In recent years journalistic scrutiny of environment news sources has become even more difficult because of cuts to journalism staffing levels, increasing workload demands in newsrooms and consequent falling editorial standards (Curran and Seaton 2010; McChesney and Nicholls 2010). As one recent commentator on science journalism puts it, the reporting process is now subject to 'intense pressures' (Allan 2009: 281). Such pressures, explored in more detail in the final section of this chapter, have not been so keenly felt in the PR offices of most key environment news sources.

Environment news sources and their PR

Journalism's contraction in the past two decades has been more than matched by an expansion in the field of PR (Cottle, 2003; Davies 2008; Miller and Dinan 2000). Not only have those in the energy, chemicals, agriculture, pesticide and biotechnology industries expanded their public relations efforts, but so have public and civil society players such as universities, research councils, specialist science publications, charities, NGOs and other activists (Dinan and Miller 2007; Göpfert 2008).

Scientists and PR

University scientists, their institutions, research funders and those who publish and disseminate their research are key sources of environment news and all of these actors have invested in increasing the volume and effectiveness of their communications activities (Anderson 2002: 331; Williams and Clifford 2010; Williams and Gajevic 2009; Williams *et al.* 2003). This trend reaches as far back as the late 1950s in the USA when the government initiated the lavishly funded 'Public Understanding of Science' programme in the wake of the Sputnik crisis (Lewenstein 1992: 60). In 1985 the UK followed with the instigation of the Committee on the Public Understanding of Science (COPUS), which marked the birth of a burgeoning public understanding of science industry (Gregory and Miller 1998: 7). By the late 1990s the UK government spent around £4.5 million annually on Public Understanding of Science initiatives, which included measures to improve public relations work of scientific institutions, media training for scientists and journalists, and prizes for successful science communicators (Göpfert 2008: 216). More recently, scholars who explore the 'medialisation' of science have shown how scientists, in order to legitimatise their work, build reputations and secure funding, have increasingly sought to communicate with mass publics by securing mainstream media coverage in recent years (Rödder, 2009: 453). They argue that growing media coverage of science has been accompanied by 'an increasing orientation of science towards the media' (ibid.). This has meant considerable further growth in the professional science communication sector with an emphasis on media relations (Schäfer, 2011: 402).

The tactics used by publicly funded environmental scientists and associated institutions when seeking to influence news coverage have remained fairly consistent over the last half century, even though the volume of science PR has increased significantly and the online channels for publishing and circulating information have become more efficient (for example since the inception of www.eurekalert.org, a PR newswire funded by the American Association for the Advancement of Science). Science communicators have adopted a set of (often defensive) tactics in order to control the flow of information about science to the news media. Most prominently, these include press briefings and press conferences, and a steady flow of press releases from universities and scientific journals. These 'information subsidies' (Gandy 1982) are highly valued by journalists working under difficult institutional and economic constraints because they package and translate news about highly complex science in an easily reproducible form (Nelkin 1995). But they also present challenges to journalistic autonomy, not least because of the well-established, and carefully policed, practice of placing embargoes on information (which determine when a press release can be used by reporters) (Kiernan 2006). Embargoed press releases are circulated to journalists and news organisations in advance of the proposed publication date, and this allows news workers to plan and thoroughly research their news pieces in good time. But they also afford much power over the nature and timing of coverage to sources of environment news. In choosing what research to write about in press releases editors, press officers and scientists highlight some research, while downplaying the importance of other projects (Kiernan 2003). Furthermore, journalists who break embargoes can be punished by their news sources, most often by temporarily or permanently blocking access to future press releases.

In recent years, communication of scientific research carried out in universities has moved beyond the supply of press releases, and has begun to engage more in the kind of media management previously the preserve of political and corporate PR. Talking of changes in science communication in the last two decades, former *Guardian* science editor Tim Radford explains that the 'conscious and manipulative media management that was [previously] a feature of city reporting and of political reporting has spread very quickly to science' (Williams and Clifford 2010: 54). The UK Science Media Centre (SMC) was set up in 2002 in the wake of a series of high profile, perceived public relations disasters for science (most prominently critical campaigning reporting of the environmental and health risks associated with genetically modified food). The SMC describes itself as an 'independent press office helping to ensure that the public have access to the best scientific evidence and expertise through the news media when science hits the headlines' (SMC 2013). It engages in a range of sophisticated and persuasive communication techniques such as: relationship management (managing relationships with specialist science, environment and health reporters, supplying them with information subsidies and putting them in contact with trained and confident 'media friendly scientists'); supplying press releases, briefing papers and organising press conferences; performing pre-emptive 'issues management' by preparing materials for release alongside potentially controversial scientific research and events; and by engaging in rapid reaction crisis management when needed (Williams and Gajevic 2013).

As well as raising concerns about eroding journalistic independence, scholars have also critiqued the effects of such communicative practices on science and its interactions with publics. Bauer and Gregory usefully theorise such developments when they describe them as part of a shift away from democratic, dialogic and public-centred models of science communication to an 'incorporated', one-way, business-influenced, persuasion-oriented model that they call 'public understanding of science incorporated', or 'PUS Inc.' (Bauer and Gregory 2007). Promotional science PR can contribute to hype, exaggeration and misinformation (Rödder and Schäfer 2010). Equally seriously, it has been argued that such persuasion-based, science-advocacy PR militates against more open, dialogic and democratic attempts by scientists to engage publics (Haran 2011). It may also endanger long-term future public trust in science. As Nisbet and Scheufele argue:

> [I]f the public simply feels like they are being marketed to, this perception is likely to only reinforce existing polarisation and perceptual gridlock. ... Anytime public engagement is defined, perceived, and implemented as a top-down persuasion campaign, then public trust is put at risk.
>
> (2009: 1776)

Environmental activist PR

Since the seminal studies on source dominance in news media referenced in the first section, further research in environmental communication has added nuance to this picture. Such work has tended to confirm the overall picture of elite source dominance, while adding insights gleaned from paying critical attention to the (often successful) attempts of politically marginalised groups to access the news (Manning 2001). The principal insight of such research is that sources do not gain access to the news media simply by dint of their power and wealth alone, but because of 'strategic' media relations efforts in competition with others (Schlesinger 1990: 77). For instance, early innovators in activist news management of environmental issues such as Greenpeace were able to bypass routine biases towards better-funded official PR sources because, among other factors, of their understanding of journalistic 'news values' such as conflict (Lowe and Morrison 1984) and the need for strong audio and visual content in their promotional materials (Anderson 2002: 9–10). Anderson identifies a growth in the number and influence of single-issue environmental pressure groups in the UK since the 1960s that have focused on matters such as nuclear power, genetically modified crops, road building and climate change. She outlines a number of factors in their relative media management successes including: the mobilisation of (tactical and financial) resources; targeting communications effectively; and paying particular attention to issue cycles and policy changes in order to mount interventions aimed at influencing key decisions (Anderson 2003: 120–121).

The tactics used by Greenpeace in the 1990s and 2000s offer a good example to explore the implications of some of these issues. The organisation, in many ways, operates more like a large corporation than an activist group. It has offices across the world that employ people with backgrounds in journalism and PR, as well as impressive resources for the production, editing and distribution of media content (Anderson 2003: 122). It has employed these resources most effectively in gaining media coverage of spectacular, visually arresting acts of protest designed to 'generate public outcry' and to 'force [issues] onto the public agenda' (123). Success at generating news coverage for attention-grabbing stunts and direct action was made more likely by the concomitant ability of such pressure groups to understand a range of different factors. For instance, they need to know what news organisations want and when (e.g. in terms of visually arresting publicity materials, offering sources willing to go on record and acting as spokespeople at the right time, etc.). It also helps if they are able to frame manufactured news events in relation to pre-existing policy and news cycles (Hansen 2010: 55–56).

Despite winning continued and often high-profile media coverage for the issues on which they campaign, better-resourced NGOs and grassroots activists alike have often had trouble gaining coverage for their own 'frames' or definitions of issues (Hansen 2010: 56). While they are often very good at getting contentious issues on the news agenda, numerous studies have shown that their influence does not routinely extend to commanding a 'prominent role' in continued debates (57). In addition to this, the success of normally marginal campaigning voices is often counterbalanced by a redoubling of communications efforts from richer and more powerful corporate or political players (Hansen 2011; Wallack *et al.* 1999). The limited efficacy of such spectacular episodic activist media management has been a factor in many campaigners' wish to eschew media-oriented activism entirely. Echoing social scientists' concerns that publicly funded science PR may damage the public legitimacy of science, some have argued that the need for media attention can have detrimental impacts on social movements themselves (Gitlin 1980). Indeed, a common theme in research about political and environmental protest suggests the more spectacular the protest, the more likely it is that protestors will be covered in a delegitimising way (Rosie and Gorringe 2009).

Corporate PR

Much corporate PR about issues relevant to environment news uses commonly applied and largely uncontroversial communications tactics. But worries over the persuasive methods used by communicators of *publicly funded* scientific research seem less significant when viewed in the context of the worst excesses of secret and manipulative media management by *private interests* seeking to influence news agendas around environmental issues such as climate change, genetically modified foods and environmental pollution. Studies have contributed much to our understanding of how climate sceptics linked to the fossil fuel and transport industries have concentrated their PR efforts on exploiting the journalistic norm of balancing sources in order to make it seem like the evidence for anthropogenic global warming is more uncertain than it actually is (Boykoff 2011). We have also learned much about the use of third party spokespeople (companies employing seemingly independent speakers in order to make their points more persuasively), astroturf organisations (fake 'grassroots' campaigning organisations that seem like they are bottom-up, democratic entities but which are actually confected and/or heavily funded by corporations to spread their own message) and other front groups in order to sew doubt and confusion about climate research inconvenient to industry (Beder 2002; McCright and Dunlap 2003; Rowell 2007). Similar work has been done to describe and analyse news media susceptibility to spin from the food biotechnology industry (Fenell 2009; Matthews 2007; Weaver and Motion 2002) and crucially the tobacco industry, where such practices were initiated and developed in the United States from the mid 1950s onwards (Cummings and Pollay 2002).

Most of these tactics involve (often secretively) putting industry's own message in the mouths of seemingly independent third parties in order to make them seem more credible and independent. As Sharon Beder explains:

> When a corporation wants to oppose ... regulations, or support [a] damaging development, it may do so openly, in its own name. But it is far more effective to have a group of citizens or experts – and preferably a coalition of such groups – which can publicly promote the outcomes desired by the corporation whilst claiming to represent the public interest. When such groups do not exist, the modern corporation can pay public relations firms to create them.
>
> (Beder 2002: 27)

The use of such 'front groups', then, lets corporations influence public debate (in the media, but also in policy circles) by proxy, and behind a carefully constructed veil of expert or grassroots concern. An indication of the bewildering scale of such webs of (often covertly-funded) industry spokespeople can be found at www.exxonsecrets.org, a US Greenpeace-created website that allows readers to trace ExxonMobil's donations to organised climate change sceptics, visualise links between the hundreds of groups and individuals who receive this cash, and drill down into each node in the network to view their remarkably similar positions on climate change. The messages of such media and policy players have been varied, but their goals are consistent: they exist to attack climate science as uncertain, doubtful or ideologically motivated 'junk science' (Michaels 2008; Oreskes and Conway 2010), and to oppose regulation on CO_2-producing industries in order to prolong profitable, but very damaging, industrial practices (Dunlap and McCright 2011).

There is a growing research base about the nature and extent of such corporate PR's influence on the news media in particular (e.g. Antilla 2005), but many studies have been broader in focus. Such valuable work has so far concentrated on mapping the connections between industry, conservative foundations, think tanks, contrarian scientists and front groups, and on examining the tactics used to manipulate public opinion and discredit climate science more generally. The media are a central forum for public debate, and more work needs to be done by journalism scholars to help us understand the scale and nature of secretive industry-backed PR on news. One small study of coverage of ExxonMobil-funded climate sceptic front groups on BBC News Online is suggestive of avenues for future research (Holmes 2009). Attempting to map the 'PR footprint' of industry-funded individuals or groups, Holmes found 88 articles in the BBC news archive that cite Exxon-funded individuals or groups as sources, only 20 of which disclose any possible conflict of interest (96). He also found 90 stories that contained web links to industry-funded organisations and front groups (such as the Heartland Institute and the Global Climate Coalition) in sidebars or at the bottom of articles, framed as resources for 'further reading'; only three of these disclosed information to readers about industry funding (96).

The emphasis on qualifying contextual information given to news audiences about sources is crucial. Such research, in identifying that industry-funded spokespeople are used as news sources at all, attests to the influence of the climate sceptic lobby. But more concerning is that (even BBC) journalists seem to rarely inform readers about the corporate backers of seemingly independent news sources, despite such information being freely available. Studies such as this can highlight corporate bias, but they also raise issues of journalistic accountability and transparency. The news media, according to theorists of the public sphere, are essential to the process of allowing publics to exercise formal and informal control over elites. They should distribute the information necessary for citizens to make informed choices and they should facilitate the formation of public opinion by providing an independent forum for debate (Curran 1991: 29). If the mainstream news media continue to quote corporate spokespeople as if they were independent commentators, their capacity for independence will be further reduced.

Environment journalism and PR

Maintaining journalistic independence and editorial standards in the face of such investment in tactical media management has been very difficult. High quality, independent and (when needed) critical report-ing is expensive: it costs in time, money and human resources, all of which are in increasingly short supply in newsrooms in the USA and much of Europe. Nowhere have economic factors affected journalism about science and the environment as much as in the USA, where 'large numbers of metropolitan daily newspapers have done away with their special science pages' (Kennedy 2010). In 1989 a total of 95 US newspapers had dedicated science sections (Brumfiel 2009). By 2012 this number had fallen to just 19 (Morrison 2013). In 2008 the cable news organisation CNN cut its entire science, technology and envi-ronment staff of seven news workers (Brainard 2008), and in 2013 the *New York Times* closed its specialist environment desk (Sheppard 2013). In the UK there was a significant expansion in the staffing of the UK national science, environment and health news beats in the 1990s (Williams and Clifford 2010: 21), but this growth tailed off after 2005, when a number of key news outlets started to make cuts (29).

In line with changes across the industry as a whole (Phillips 2010: 95-7), workloads for science special-ists have risen a lot and this has fuelled a number of problems (Williams and Clifford 2010: 36). Principal among them is that most journalists are simply pressured to produce far more news stories than their historic counterparts. Eighty-eight per cent of specialists surveyed said that their workloads had increased between 2005 and 2010 (37), and long-serving reporters bemoaned the fact that story counts had risen significantly since the 1990s (40–41). This change is partly down to pressure to produce more online and cross-platform news: as one reporter put it: 'the web is never full' (38). This leads to a newsroom envi-ronment where the same number, or fewer, journalists are asked to do far more with no extra resources. Basic, day-to-day tasks, such as finding original news, researching and fact-checking stories, are now under increasing pressure. Almost half of UK specialists claim that they now have 'less time' to check facts for accuracy, while almost a quarter say they don't have enough time to make what they regard as 'adequate' checks on their facts (49). This lack of time has exacerbated an already extant shift in the bal-ance of power between reporters and their sources.

An important element of the democratic value of any news is that it should be independent. Journalists and editors should decide what news to cover and in what way to present it to their audiences. The decline of journalism in general, and environment journalism in particular, is leading to elements of journalism practice being outsourced to powerful and efficient news sources with slick and well-resourced public relations teams. Long-serving journalists told Williams and Clifford that their job has been 'de-skilled', and has changed significantly for the worse over the last 20 years (12), so much so that actually finding original stories and fresh angles from which to report them has become less necessary. In the late 1980s Hansen and Dickinson found almost a quarter of stories covering science issues were triggered by sources contacting journalists rather than journalists contacting sources (1992). This trend persists today, with only 23 per cent of respondents reporting that 'most of their stories' originated with their own 'active journalistic investigation'; 46 per cent say they are more often than not the 'passive recipients' of news story ideas from sources. When it comes to relying directly on public relations in journalistic output, environment news has long been susceptible. In the mid 1970s Sachsman found that more than half of news pieces about the environment originated in, or drew on, public relations material (1976). More recently, Lewis *et al.* find that 60 per cent of general UK home news pieces 'rely wholly or mainly on pre-packaged information, and a further 20 per cent are reliant to varying degrees' (2008: 14–15). They argue their data portrays 'a picture of the journalistic processes of news gathering and news reporting in which any meaningful independent journalistic activity by the media is the exception rather than the rule' (18). Reporters often claim that PR's influence is mainly as an agenda-setter, providing initial ideas for stories and a starting point for later journalistic work. Nevertheless, the evidence suggests it also often facilitates cut and paste 'churnalism' (Davies 2008), which means that news stories are increasingly similar to institutional press releases, tellingly characterised by one specialist reporter as 'low-hanging fruit' (Williams and Clifford 2010: 42).

In aggregate, and considered alongside the research into the rise of environment PR discussed above, these findings suggest the prospects for high-quality, independent environment journalism in the mainstream news media are diminished. It seems that in some important respects much of the job of translating or conveying this news from the scientific community is being outsourced to a growing army of professional environment communicators, while journalists act more and more like stenographers to their sources. This has potentially serious consequences for the ability of science news to play the necessary role of holding such social actors to account. When changes in routine journalistic and public relations practice facilitate such a shift in power from journalists to their news sources, it is far less likely that reporters will be able to play a critical, democratic, watchdog role when needed.

Suggested further reading

Oreskes, N. and Conway, E. (2010) *Merchants of Doubt*, London: Bloomsbury.
Sourcewatch (www.sourcewatch.org) and Powerbase (www.powerbase.info): two online wiki spaces produced by academics and investigative journalists with transparently sourced information about a wide range of environment PR.
Franklin, B., Williams, A., and Lewis, J. (2010) 'Journalism, News Sources, and Public Relations', in S. Allan (ed.), *The Routledge Companion to News and Journalism Studies*, London: Routledge, pp. 202–211.
Nelkin, D. (1995) *Selling Science*, New York: Freeman.

References

Allan, S. (2009) 'Introduction: Science Journalism in a Digital Age', *Journalism*, 12: 771–777.
Anderson, A (2002) 'The Media Politics of Oil Spills', *Spill Science and Technology Bulletin*, 1–2: 7–15.
—— (2003) 'Environmental Activism and the News Media', in S. Cottle (ed.), *News, Public Relations, and Power*, London: Sage, pp. 117–132.

Antilla, L. (2005) 'Climate of Scepticism: US Newspaper Coverage of the Science of Climate Change', *Global Environmental Change*, 15: 338–352.

Bauer, M. and Gregory, J. (2007) 'From Journalism to Corporate Communication in Post-War Britain', in M. Bauer and M. Bucchi (eds), *Journalism, Science, and Society*, New York: Routledge, pp. 33–51.

Beder, S. (2002) *Global Spin: The Corporate Assault on Environmentalism*, White River Junction, VT: Chelsea Green.

Berkowitz, D. (2009) 'Reporters and Their Sources', in K. Wahl Jorgensen and T. Hanitzsch (eds), *The Handbook of Journalism Studies*, London: Routledge, pp. 102–115.

Boyce, T., Kitzinger, J. and Lewis, J. (2007) *Science Is Everyday News*, Project Report Funded by the UK Government's Office for Science and Innovation, Cardiff: Cardiff University.

Boykoff, M. (2011) *Who Speaks for the Climate?*, Cambridge: Cambridge University Press.

Brainard, C. (2008) 'CNN Cuts Entire Science, Tech Team', *The Observatory* (Columbia Journalism Review Blog), 4 December, available at: www.cjr.org/the_observatory/cnn_cuts_entire_science_tech_t.php (last accessed October 2009).

Brumfiel, J. (2009) 'Science Journalism: Supplanting the old media?', *Nature*, 458: 274–277.

Cottle, S. (2003) *News, Public Relations and Power*, London: Sage.

Cummings, K. and Pollay, R. (2002) 'Exposing Mr. Butts' Tricks of the Trade', *Tobacco Control*, 11: 1–14.

Curran, J. (1991) 'Rethinking the Media as a Public Sphere', in P. Dahlgren and C. Sparks (eds), *Communication and Citizenship: Journalism and the Public Sphere in the New Media Age*, London: Routledge, pp. 27–57.

Curran, J. and Seaton, J. (2010) *Power Without Responsibility*, London: Routledge.

Davies, N. (2008) *Flat Earth News*, London: Chatto & Windus.

Davis, A. (2002) *Public Relations Democracy*, Manchester: Manchester University Press.

Dinan, W. and Miller, D. (2007) *Thinker, Faker, Spinner, Spy: Corporate PR and the Assault on Democracy*, London: Pluto.

Dunlap, R. and McCright, A. (2011) 'Organized Climate Change Denial', in J. Dryzek, R. Norgaard and D. Schlossberg (eds), *The Oxford Handbook of Climate Change and Society*, Oxford: OUP, pp. 144–160.

Fennell, D. (2009) 'Marketing Science', *Journal of Communication Inquiry*, 33: 5–26.

Franklin, B., Williams, A. and Lewis, J. (2010) 'Journalism, News Sources, and Public Relations', in S. Allan (ed.), *The Routledge Companion to News and Journalism Studies*, London: Routledge, pp. 202–211.

Gandy Jr., O., (1982) *Beyond Agenda Setting*, Norwood, NJ: Ablex.

Gans, H. (1979) *Deciding What's News*, New York: Pantheon.

Gitlin, T. (1980) *The Whole World is Watching*, Berkeley, CA: Cambridge University Press.

Göpfert, W. (2008) 'The Strength of PR and the Weakness of Science Journalism', in M. Bauer and M. Bucchi (eds), *Journalism, Science and Society*, New York: Routledge, pp. 215–226.

Gregory, J. and Miller, S. (1998) *Science in Public*, New York: Plenum.

Hansen, A. (2010) *Environment, Media, and Communication*, London: Routledge.

—— (2011) 'Communication, Media and Environment', *International Communication Gazette*, 73: 7–25.

—— and Dickinson, R. (1992) 'Science Coverage in the British Mass Media: Media output and source input', *European Journal of Communication*, 17: 365–77.

Haran, J. (2011) 'Campaigns and Coalitions: Governance by Media', in S. Rödder, M. Franzen and P. Weingart (eds), *The Sciences' Media Connection: Communication to the Public and its Repercussions, Sociology of the Sciences Yearbook*, Dordrecht: Springer, pp. 241–256.

Hargreaves, I., Lewis, J. and Speers, T. (2003) *Towards a Better Map: Science, the Public and the Media*, report produced for the Economic and Social Research Council, available online at: http://cf.ac.uk/jomec/resources/Mapdocfinal_tcm6-5505.pdf (last accessed August 2013).

Holmes, T. (2009) 'Balancing Acts: PR "impartiality" and power in mass media coverage of climate change', in T. Boyce and J. Lewis (eds), *Climate Change and the Media*, New York: Peter Lang, pp. 92–102.

Kennedy, D. (2010) 'The Future of Science News', *Daedalus*, 139: 57–67.

Kiernan, V. (2003) 'Diffusion of News About Research', *Science Communication*, 25: 3–13.

—— (2006) *Embargoed Science*, Urbana, IL: University of Illinois Press.

Lewenstein, B. (1992) 'Public Understanding of Science in the USA after World War II', *Public Understanding of Science*, 1: 45–68.

Lewis, J., Williams, A. and Franklin, B. (2008) 'A Compromised Fourth Estate? UK News Journalism, Public Relations and News Sources', *Journalism Studies*, 9: 1–20.

Lowe, P. and Morrison, D. (1984) 'Bad News or Good News: Environmental politics and the mass media', *Sociological Review*, 32: 75–90.

McChesney, R. and Nichols, J. (2010) *The Death and Life of American Journalism*, New York: Nation.

McCright, A. and Dunlap, R. (2003) 'Defeating Kyoto: The conservative movement's impact on US climate change policy', *Social Problems,* 50: 348–373.

Manning, P. (2001) *News and News Sources: A Critical Introduction,* London: Sage.

Matthews, J. (2007) 'Biotech's Fake Persuaders', in W. Dinan and D. Miller (eds), *Thinker, Faker, Spinner, Spy,* London: Pluto, pp. 117–137.

Michaels, D. (2008) *Doubt is Their Product,* Oxford: Oxford University Press.

Miller, D. and Dinan, W. (2000) 'The Rise of the PR Industry in Britain, 1979–1998', *European Journal of Communication,* 15: 5–35.

Morrison, S. (2013) 'Hard Numbers: Weird Science', *Currents Magazine: Columbia Journalism Review,* 2 January, available at: www.cjr.org/currents/hard_numbers_jf2013.php (last accessed August 2013).

Nelkin, D. (1995) *Selling Science,* New York: Freeman.

Nisbet, M. and Scheufele, D. (2009) 'What's Next For Science Communication?', *American Journal of Botany,* 96: 1767–1778.

Oreskes, N. and Conway, E. (2010) *Merchants of Doubt,* London: Bloomsbury.

Phillips, A. (2010) 'Old Sources: New Bottles', in N. Fenton (ed.), *New Media, Old News: Journalism and democracy in the digital age,* London: Sage, pp. 87–101.

Rödder, S. (2009) 'Reassessing the Concept of Medialization of Science – A Story from the "Book of Life"', *Public Understanding of Science,* 18: 452–463.

—— and M. Schäfer (2010) 'Repercussion and resistance: An empirical study in the interrelation between science and mass media', *Communications,* 35: 249–267.

Rosie, M. and Gorringe, H. (2009) '"The Anarchist World Cup": Respectable protest and media panics', *Social Movement Studies,* 8: 35–53.

Rowell, A. (2007) 'Exxon's Foot Soldiers', in W. Dinan and D. Miller (eds), *Thinker, Faker, Spinner, Spy,* London: Pluto, pp. 94–116.

Sachsman, D. (1976) 'Public Relations Influence on Coverage of Environment in San Francisco Area', *Journalism Quarterly,* 53: 54–60.

Schäfer, M. (2011) 'Sources, Characteristics and Effects of Mass Media Communication on Science', *Sociology Compass,* 5: 399–412.

Schlesinger, P. (1990) 'Rethinking the Sociology of Journalism', in M. Ferguson (ed.), *Public Communication,* London: Sage, pp. 61–83.

Sheppard, K. (2013) 'The Heat is on as the *New York Times* closes its Environment Desk', *theGuardian.com,* 14 January, available at: www.theguardian.com/commentisfree/2013/jan/14/new-york-times-environment-climate-change (last accessed August 2013).

SMC (2013) Home Page of the Science Media Centre, available at: www.sciencemediacentre.org/ (last accessed August 2013).

Tuchman, G. (1978) *Making News,* New York: Free Press.

Wallack, L., Woodruff, K., Dorfman, L. and Diaz, I. (1999) *News for a Change: An advocate's guide to working with the media,* London: Sage.

Weaver, K. and Motion, J. (2002) 'Sabotage and Subterfuge', *Media, Culture and Society,* 24: 325–343.

Williams, A. and Clifford, S. (2010) *Mapping the Field: A Political Economic Account of Specialist Science News Journalism in the UK National Media,* Project Report Funded by the Science and the Media Advisory Group, UK Government Department for Business, Innovation, and Skills. Available at: http://cf.ac.uk/jomec/resources/Mapping_Science_Journalism_Final_Report_2003-11-09.pdf (last accessed August 2013), Cardiff: Cardiff University.

Williams, A. and Gajevic, S. (2009) *UK National Newspaper Coverage of Hybrid Embryos: Sources Strategies and Struggles.* Project Report Funded by the Academy of Medical Sciences, the Wellcome Trust, and the Science Media Centre, available at: www.cf.ac.uk/jomec/resources/hybridembryosreport.pdf (last accessed August 2013), Cardiff: Cardiff University.

Williams, A. and and Gajevic, S. (2013) 'Selling Science? Source struggles, public relations, and the UK press coverage of animal-human hybrid embryos', *Journalism Studies,* 14(4): 507–522.

Williams, C., Kitzinger, J., and Henderson, L. (2003) 'Envisaging the Embryo in Stem Cell Research: Rhetorical strategies and media reporting of the ethical debates', *Sociology of Health and Illness,* 25: 793–814.

## PART III

Covering the environment

News media, entertainment media and cultural representations of the environment

17

NEWS COVERAGE OF THE ENVIRONMENT

A longitudinal perspective

Anders Hansen

Introduction

Research on the media and the environment/environmental issues since the 1970s has overwhelmingly focused on the news media and their reporting on environmental controversies and problems. Most of this research on news in turn has tended to focus on news coverage of specific environmental issues, problems or disasters over a limited period of time, and in this respect has provided valuable evidence on the processes involved in the short-term public construction and representation of particular issues or problems. What these studies have been less adept at accounting for or explaining are the longer-term ups and downs or cyclical nature of news media attention to the environment and environmental issues. This chapter thus examines research that has studied the longer term trends in news media coverage of the environment generally or of particular environmental issues. The chapter reviews the evidence that these studies have provided on the factors and processes that drive and influence news media attention to the environment, and it explores how longitudinal studies of news coverage of environmental issues can contribute to our understanding not only of the politics and careers of environmental issues, but also of the privileged roles of media and communication processes in this context.

The potential and promise of longitudinal research

Like comparative communication research in general (see Esser and Hanitzsch, 2012),[1] longitudinal or diachronic studies in environmental communication research are exciting and interesting because they potentially – through tracking and analysis over extended periods of time – hold out the promise of elucidating classic concerns with the role of media and communication in the changing politics of the environment.

Longitudinal, diachronic studies of media and public communication about the environment thus have much to offer not only in terms of mapping the changes over time in society's attention to the environment, but also in terms of addressing and understanding the sociologically much more interesting questions about the dynamics driving such changes. In this chapter, I start with a brief account of some of the practical and methodological challenges of longitudinal research. I then look at what research has shown about longer term trends in environmental issues coverage, before proceeding to examine how longitudinal studies have helped in elucidating the dynamics of public communication and the politics of the environment.

Challenges of longitudinal research

Detailed studies of news media reporting of environmental issues have historically tended to focus on news reporting of single issues/controversies/disasters over a relatively limited period of time, i.e. weeks and months rather than years. The short-term focus has undoubtedly generally been well justified in terms of the framework, theoretical context and objectives of such research, and, indeed these are the types of studies that have provided us with detailed insights into the nature of media representation of environmental issues and into some of the complex dynamics of source–journalist interaction, news values, media organisational arrangements, etc.

Practical considerations to do with the cost of (media) research, available methods of analysis, access to media content etc. have undoubtedly also played a role in the popularity of short-term, synchronic studies over longitudinal, diachronic studies. Thus, for much of the early period of media research on environmental news coverage, the sampling and collection for analysis of news coverage of the environment was a laborious, time-consuming and labour-intensive process. The digital revolution since the 1980s–1990s has radically changed not only how news coverage is produced and stored, but importantly – for communication researchers – how it can be accessed, sampled and indeed analysed (e.g. with the help of computer-assisted analysis programmes). This, then, may in part explain why comprehensive longitudinal studies of news coverage of the environment have become much more prominent in the last couple of decades than they were in the early decades of environmental communication research. And this is a very welcome development, because a great deal is potentially to be gained from longitudinal research in terms of understanding the nature and role of news media reporting for the social and political construction of environmental issues.

While longitudinal studies have only in recent decades begun to gain prominence in environmental communication research, they have a rather longer pedigree in communication research generally. Indeed, some of the earliest arguments for large scale longitudinal research on news media coverage and on public opinion date back to the early parts of the twentieth century with prominent figures such as Max Weber, Harold Lasswell and Paul Lazarsfeld. In the 1970s key scholars such as Morris Janowitz (1976) and James Beniger (1978) advocated the idea of using longitudinal media monitoring/analysis as 'social indicators' or 'cultural indexes', i.e. as a way of 'taking the temperature' (Max Weber) of society and monitoring social and political change. The idea articulated by Beniger (1978) that media indicators can be used in similar ways to public opinion data to monitor changes in social, political and cultural values was also taken up by David Fan (1988) in his development of software for the computer-assisted analysis of media coverage. This approach, as well as reference to numerous studies conducted since the late 1980s by Fan, Bengston and other associates (see also Webb *et al.*, 2008, discussed below), has recently been re-articulated (Bengston *et al.*, 2009).

While it is no longer as difficult or challenging as it once was before the digital revolution to map the contours of media representation both cross-temporally and cross-nationally, there are significant theoretical and methodological challenges that remain both in terms of what to measure, and perhaps most particularly, in terms of interpreting the changes, trends and variations uncovered. Thus, while mapping the contours of media coverage over extended periods of time is now much less challenging than it once was, interpreting the drivers of such changes or indeed their social, cultural and political meaning remains a significant challenge. In other words, while changes in the extent and nature of media coverage of the environment can readily be mapped, what do they mean or signify? What are they evidence of? They might, for example – in the classic tradition of social indicators research – be seen as evidence of changes in the social/political values and concerns of society (media as reflection or engine of value change; content analysis as a route into 'reading' the 'cultural temperature' of society). Or they could – within a more narrowly defined media/communication research perspective – be seen as evidence of changing media practices (i.e. changing media formats; changing organisational arrangements; changing professional values and practices; changing economic and ideological pressures on media organisations). Or they might – drawing on linguistic/public discourse/cultural perspective – be seen as evidence of changes in public vocabulary and discourse, the way we as societies talk about 'the environment'.

The contours of news coverage of the environment

In very general terms, longitudinal studies of media coverage of environmental issues (e.g. Brookes *et al.*, 1976; Bowman and Fuchs, 1981; Strodthoff *et al.*, 1985; McGeachy, 1989; Einsiedel and Coughlan, 1993; Hansen, 1994; Ader, 1995; Brossard *et al.*, 2004; Boykoff, 2007 and 2008 – see also Boykoff *et al.*, Chapter 18 *Communicating in the Anthropocene* in this Handbook; Djerf-Pierre, 2012a, 2012b and 2013) have confirmed two important characteristics: a) that once introduced in the 1960s, the 'environment' has consolidated itself as a news-beat and category of news coverage, and b) that news coverage of the environment has gone up and down in regular cycles.

Broadly described, media interest in 'the environment' and in problems/issues characterised as 'environmental' began in the mid 1960s, increasing to an initial peak in the early 1970s, then declining through the 1970s and early 1980s, followed by another dramatic increase in the latter half of the 1980s, peaking around 1990, then receding again through the 1990s, only to experience a considerable new resurgence in the first decade of the 2000s. This broad-brush characterisation inevitably obscures the very significant variations from issue to issue as well as the much more frequent ups and downs that can be observed within each of these broad periods (Hansen, 2010).

The considerable and not trivial finding, however, demonstrated by early longitudinal analyses of media coverage of the environment is the finding that the new perspective or 'framework' of ecology and its associated more holistic view of the environment only emerged on the public arena in the 1960s (Hansen, 2010). Furthermore, they confirm that this perspective steadily consolidated in media coverage over the next couple of decades (Bowman and Fuchs, 1981), to the extent that Einsiedel and Coughlan were able to conclude in 1993 that 'What seems to differentiate the more recent period of the late 1970s through the early 1990s is that "the environment" has been vested with a more global character, encompassing attributes that include holism and interdependence, and the finiteness of resources' (p. 141).

What we might broadly call the environmental/ecological paradigm, once introduced in the 1960s, has remained firmly on the media and public agenda ever since. It could even be argued that the environmental paradigm – far from showing signs of increasing fragmentation – has become increasingly holistic through the rise of global concerns about climate change (Hansen, 2010).

In one of the most systematic and comprehensive longitudinal studies to date, Monica Djerf-Pierre (2012a, 2012b, 2013) – analysing Swedish television news coverage of the environment over the half century from 1961 to 2010 – confirms the indications from earlier studies, that while news coverage of the environment has gone up and down in cycles, the overall trend has been up and, despite numerous periods of low media attention to the environment, it has never vanished completely.

Evidence for the overall resilience – despite cycles of high and low prominence – has similarly been confirmed by a recent comprehensive meta-analysis of studies of climate change. In their meta-analysis of the findings of over 40 studies of climate change coverage across a broad variety of media and 27 countries in the period 1996–2010, Schmidt *et al.* (2013) show that while coverage of climate change goes up and down in cycles, media attention to climate change has increased very significantly in an overall upward trend across all countries. They are also able to demonstrate the consolidation of climate change as a topic for news coverage through showing that climate change coverage has been quite resilient in the face of increasing economic pressures on media organisations, resulting in an overall reduced news space (also referred to as the news hole, or in Hilgartner and Bosk's (1988) classic term, the 'carrying capacity' of news media).

The longitudinal mapping of environmental issues generally and, in the recent two decades, of climate change coverage in particular, then provide interesting and essential information about the extent and variation of media attention over time. They demonstrate for starters that media coverage has little to do with the real life seriousness or impact of these issues, and instead point to key questions about factors influencing media coverage and to questions about how media coverage interacts with other institutions, forums and events in society.

Broadly, longitudinal research on media coverage and representations of the environment can be considered under three general headings: social/cultural/value-indicators studies; issue-attention cycle studies; and agenda-building/setting studies. These are not necessarily entirely mutually exclusive, but they provide a way of examining the major emphases of these studies.

Social/cultural/value indicators studies

Drawing on the long tradition of (and fascination with) cultural indicators research in the tradition of Harold Lasswell and others, longitudinal research on media representations of the environment has been used to show how – over extended periods of time – societies' ways of viewing, defining and understanding the environment change. So too do the priorities and values that we attach to nature and the environment and to political questions about environmental protection, recreational versus resource-exploitation/economic uses of environments, energy policy etc.

Longitudinal studies of television documentaries (Wall, 1999), news magazines (Meisner and Takahashi, 2013), film (Mitman, 1999) and advertising (Howlett and Raglon, 1992; Rutherford, 2000; Kroma and Flora, 2003; Ahern *et al.*, 2012) have thus demonstrated how views and representations of the environment change over time, demonstrating particularly a macro shift from a romanticised, through a utilitarian/science-driven/resource-focused to a sustainability/nature-attuned view of our natural environment.

Glenda Wall (1999), in a detailed study of the Canadian documentary-series *The Nature of Things* over the extended period from 1960 to 1994, examines the changing discourses of science, nature and environment. While demonstrating the considerable diversity of available constructions of nature throughout the period, her analysis shows how the emphasis of discourses of nature in the television documentary changed over this period: broadly from a view of nature as a resource to be controlled and exploited, to a view of nature as fragile, under attack by technological growth, in need of protection and liable to 'respond with a vengeance to the abuses piled upon it' (p. 68). The study thus provides, due in part to its longitudinal perspective, a rare insight into changes in the moral/ethical discourses or perspectives that in turn provide context for the development of public and political approaches to the environment.

In an exemplary analysis of advertisements for pesticides in agricultural magazines, spanning the half century from the 1940s to the 1990s, Kroma and Flora (2003) show key changes from a 'science' discourse in the 1940s–1960s, through a 'control' of nature/the environment discourse in the 1970s–1980s, to a 'nature-attuned' (sustainability, protection, harmony) discourse in the 1990s. They conclude that 'changing images reflect how the agricultural industry strategically repositions itself to sustain market and corporate profit by co-opting dominant cultural themes at specific historical moments in media advertising' (Kroma and Flora, 2003: 21).

In a similarly exemplary visual analysis, Peeples (2013) examines the changing images associated with the toxicant Agent Orange (used extensively as a defoliant during the Vietnam War, with devastating long-term human and environmental consequences) over the extended period from 1979 to 2008. She demonstrates how the changing use of images is at once reflective of and plays into changes in wider "stories of national identity, of culture, of gender, of race, and, most significantly, of power" (p. 205).

Webb *et al.* (2008) – in a study spanning 1997–2004 – demonstrate changes in the values and orientations expressed in Australian news media discourse about the management of native forests, during a time of public debate and controversy about, broadly, resource-exploitation versus protection. They identify three major value orientations (commodity orientation, ecological orientation and moral/spiritual/aesthetic value orientation) and demonstrate a significant shift away from commodity value orientations towards values prioritising ecological protection and the 'moral/spiritual/aesthetic' value of forests. A change then, which is not dissimilar to some of the macro-changes identified in other longitudinal studies referred to above.

Increasingly, studies interested in tracking and mapping historical change through media analysis, have drawn on the concept of framing and closely associated concepts, such as Gamson and Modigliani's (1989) notion of 'cultural packages' or indeed the broader notion of environmental discourses (Dryzek, 1997; Hajer, 1995). Key to all of these is the analysis of how linguistic and discursive changes are indicative of wider social/political and cultural changes.

Bauer *et al.* (1999), in a comprehensive collection of studies of biotechnology in the European public sphere, show how significant lexical changes in how biotechnology and genetics are referenced are indicative of public and political attitudes, as well as – and perhaps more significantly – of deliberate attempts (by key stakeholders in the biotechnology debate) at changing/influencing such attitudes and perceptions. Condit *et al.* (2002) touch on similar points in a longitudinal analysis, 1919–1996, of changing meanings of the word 'mutation' in US mass magazine articles.

In a recent longitudinal study of British newspaper coverage of flooding and flood management over the 25-year period 1985–2010, Escobar and Demeritt (2014) show how changes in the amount of coverage bear little relationship to the frequency or severity of flooding, but instead are indicative of changes in stakeholder controversy and public concern about how to manage flood risk. Their analysis demonstrates a significant shift in how flooding is defined as a problem and associated changes in policies and strategies for remedying flood problems.

The longitudinal studies discussed above then provide valuable indications of how the longitudinal analysis of media coverage can provide insights into the changing values and definitions that we, as individuals and societies, associate with the environment. More importantly, however, many of them indicate how such changes are closely linked to the activities of claims-makers interested in promoting particular definitions or understandings over others.

Issue-attention cycles and longitudinal research

One of the most widely used frameworks for getting to grips with the longitudinally oscillating patterns of media attention to environmental issues, observed by many analyses of news reporting on the environment, has been and continues to be Anthony Downs's issue-attention-cycle framework. Downs (1972) – focusing on 'ecology'/environmental issues – argued that social problems have a typical career that follows a five-stage issue attention cycle, analysis of which can help in understanding the processes through which social problems suddenly emerge on the public stage, remain there for a time and 'then – though still largely unresolved – gradually [fade] from the centre of public attention' (Downs, 1972: 38).

The attraction of this framework to media and communication researchers is obvious, perhaps particularly when considering the oft-noted finding of longitudinal studies, that the ups and downs in media attention to environmental issues seem to bear little relationship to the severity of these issues, as measured by for example by scientific, economic or other indicators.

Trumbo (1996), McComas and Shanahan (1999), Mikami *et al.* (2002) and Brossard *et al.* (2004) have all usefully drawn on Downs's 'issue-attention' cycle for explaining the cyclical phases of media coverage of global warming/climate change, as have Nisbet and Huge (2006) in their comprehensive study of a different environmental issue, plant-biotechnology.

Trumbo's (1996) analysis of climate change coverage over the decade 1985–1995 provides a particularly clear early application of Downs's framework. Trumbo shows that the coverage 'fits Downs's five-stage model fairly well' (1996: 276), and for example confirms the prediction that while issues rise and fade in prominence, once introduced on the media agenda, attention – even when their prominence fades in the 'post-problem-stage' – remains above the low levels seen before (the 'pre-problem stage') the initial increase in attention. Trumbo also notes, however, the elasticity of the different stages in Downs's model, and thus concludes that it is possible that the variations in media attention to climate change observed in his analysis of the period 1985–1995 may in fact be – not the full five stages – but only the first three stages of the cycle.

While Downs's issue attention cycle was developed for the analysis and understanding of the overall career of social problems (i.e. not only the 'media' career of these problems), it provides a useful framework for analysing the factors and dynamics that can help account for the cyclical patterns of media coverage. More particularly, when combined with the sociology of news work on source activity, journalistic practices, media organisational arrangement, news values, etc., the issue-attention cycle framework can help in identifying the key factors that impinge on and propel news coverage and cause the peaks and troughs in such coverage.

The most comprehensive and systematic evidence to date on issue-attention cycles in media coverage of environmental issues comes from the studies by Monica Djerf-Pierre (2012a, 2012b, 2013) in her longitudinal analysis of Swedish television news coverage of the environment from 1961–2010, and by Mike Schäfer *et al.* (2014) in their analysis of climate change coverage in Australian, German and Indian print media from 1996 to 2010.

Schäfer *et al.* (2014) confirm that – country and media-system differences notwithstanding – the coverage of climate change follows remarkably similar trends, with clearly identifiable ups and downs, but also with an overall significant increase in media attention over the 15-year period examined. They are further able to demonstrate the relative contribution of selected factors/drivers on the prominence of climate change coverage in the newspapers. This analysis shows that, contrary to what has been suggested by many previous studies, 'weather and climate characteristics are not important drivers of issue attention'. Neither do they find evidence that scientific publications impact to any significant degree on the amount of media attention. By contrast, it is clear from their analysis that domestic political attention, NGO activity and international summit events are major drivers of media attention.

Djerf-Pierre's study of Swedish television news coverage of environmental issues over the half century from 1961–2010 is valuable for its comprehensive demonstration that despite multiple cycles of ups and downs, these become less frequent over time and the overall trend is up, and she is able to conclude that her analysis thus provides clear evidence of the 'long-term institutionalization of the environmental domain in television news journalism'. The longitudinal nature of the study and the differentiation between different environmental issues/themes in the coverage further enable Djerf-Pierre to advance Downs's issue-attention framework beyond the cycles of single issues, and to show how larger clusters of issues form 'meta-cycles' – with a rather larger time-interval than single issues – of peaks and troughs in media attention to environmental issues.

Equally importantly, the longitudinal nature of the study facilitates analysis of the dynamics of issue interaction as well as media-external impacts on the amount of coverage. Interestingly, and perhaps contrary to expectation, key 'real-life' events such as major disasters are not – with a few exceptions (Chernobyl and the Deepwater Horizon oil spill) – by themselves major contributors to the attention cycles at the expense of other issues. Instead, they are shown to have a focusing or galvanising effect (Djerf-Pierre refers to a 'crowding-in' phenomenon) on the overall coverage of environmental issues.

In more general terms, Djerf-Pierre's analysis provides evidence of the synergistic interaction of different categories of environmental issue, showing 'positive correlations … between the different categories of environmental issues on the news agenda. When one issue category receives a high level of attention, the attention to other environmental issues tends to increase as well' (Djerf-Pierre, 2012a: 299). In general terms then, enhanced media attention to one environmental issue brings with it in its wake enhanced attention to other environmental issues. This confirms findings from studies elsewhere, including Germany (Kepplinger and Habermeier, 1995) and Australia (McGaurr and Lester, 2009), and there are interesting parallels here – from the analysis of the impact of key events as well as from the analysis of issue-interaction – to the notions of 'threshold events' or 'tipping points' in media coverage. Mazur (1981), for example, in pioneering research showed how it took a major nuclear accident such as Three Mile Island to significantly sensitise and alert the news media to nuclear issues, but once alerted, media coverage of all things nuclear increased dramatically and continued to remain high for years following the accident.

By contrast, the study also confirms what has been suggested by many previous analyses (e.g. Hansen, 1993; Boykoff, 2009), that environmental issues generally lose out when competing for space against 'economic news and news on war and armed conflicts in times of crises' (Djerf-Pierre, 2012b: 499) in what Hilgartner and Bosk (1988) have aptly characterised as the limited 'carrying capacity' of the media.

Agenda-building/setting studies

Where the studies discussed in the previous section were focused predominantly on mapping the long-term changes in environmental values, definitions and themes as reflected in media coverage, other studies – those discussed in this section – have sought to examine, as part of their study design, how longitudinal changes in media coverage (the media agenda) interact with changes in public opinion or other key forums in society. The intriguing early observation that both media agendas and public opinion go through cycles of higher and lower degrees of concern about the environment was not lost on communication researchers. Even more intriguing perhaps, was the tantalising observation that such cycles appeared in very general terms to coincide, follow similar patterns (e.g. Funkhouser, 1973).

Christine Ader (1995), comparing the press and public agendas over the period 1970 to 1990, provides particularly strong evidence for the agenda-setting role of the media on the issue of environmental pollution. The study is noteworthy not only for the extended period of time analysed, but also for its careful control for the potential impact of 'real-world' conditions and events on the relationship between media and public agendas.

Sampei and Aoyagi-Usui (2009), echoing earlier findings from another Japanese study (Mikami *et al.*, 1995), in their agenda-setting study of newspaper coverage and public opinion in Japan from 1998–2007 found that news coverage of global warming correlated with public concern. They also found specific evidence that the influence of news coverage on public concern was relatively immediate, but short term.

Longitudinal studies within the agenda-setting framework have not confined themselves to examining the media and public opinion agendas, but have extended their investigations to longitudinal changes and interactions of a range of public 'agendas', including media agendas, political/policy agendas, 'real-world cues' and public opinion. In a longitudinal study of the agenda-setting process in relation to global warming, Trumbo (1995) found that the influence of media coverage on politicians (members of the US Congress) was considerably more pronounced than media influence on the public opinion agenda. In a particularly sophisticated design, Soroka (2002) similarly examined interactions of the agendas of Canadian newspapers, public opinion polls, formal political forums and legislative initiatives from 1985–1995. He found that the media were a powerful agenda-setter on environmental issues, but less so on other issues where the public could draw on personal experience or had access to more immediate and more trusted sources of information. Similar conclusions emerged from a longitudinal analysis in Belgium by Walgrave *et al.* (2008), who also found media agenda-setting on environmental issues to be stronger than on other issues.

Other longitudinal analyses have likewise found evidence of the influence of media coverage on the political and legislative forums. Thus, Jenner (2012) found that news photographs affect policy maker attention, but seem to have a more ambivalent impact on public attention to environmental issues. Dolšak and Houston (2013), in a study of legislative activity across a sample of US states during the period 1998–2006, found that newspaper coverage of climate change played an important role in setting the agenda for and influencing subnational policy processes.

As in agenda-setting research generally, determining the particular direction of influence between media coverage and public opinion with regard to environmental issues has generally proved challenging. While there is thus much evidence from longitudinal studies that the media can play a potentially powerful role in influencing the public opinion agenda, there is also, however, some evidence on some environmental issues of the reverse type of agenda-setting where public issue concerns appear to drive issue coverage in the news (Uscinski, 2009). This apparent contradiction, itself in part a product of examining agenda-interaction over extended periods of time, merely confirms what communication researchers have argued for some time, that media roles are best understood – not from a simple linear perspective – but from a much more dynamic and interactive perspective that recognises the dynamic and fluid nature of interactions between key 'meaning creating forums' (Gamson and Modigliani, 1989) in society.

In one of the most complex and sophisticated tests of agenda-setting on environmental issues to date, Liu *et al.* (2011) examine media and congressional attention to climate change over the extended period of 1969 to 2005. Their study casts valuable light on how problem indicators (e.g. quantitative statistical indicators), high-profile international events and science feedback regarding climate change influence media and political attention. Their findings confirm that the three sets of 'attention-grabbing' factors do 'indeed generally promote issue salience', albeit to different degrees and with different time-scales for media and political agendas. Indicative of the complexity of agenda-setting, they conclude that although their 'study sheds some light on where attention may come from and how attention level may be affected …, it certainly leaves us with even more questions for future agenda-setting research' (p. 415). Liu *et al.*'s study, with its sophisticated longitudinal design, then provides a salutary reminder of the challenges of mapping the highly complex dynamics and interactions of media, public, political, scientific, etc. agendas or meaning-creating forums.

Conclusion

While relatively rare in the first decades of environmental communication research, longitudinal research on media and public communication about the environment have – undoubtedly spurred on by technological and methodological developments in communication research – become considerably more common in the last couple of decades. More significantly, longitudinal studies have afforded significant advances in insights into media roles in the changing social and political values and approaches regarding the environment, as well as insights into the careers of environmental problems/issues and into the key drivers of ups and downs in media attention to the environment.

Longitudinal studies have provided macro-level evidence of the establishment in the 1960s and the steady consolidation since of what can broadly be called 'the environmental paradigm', and they have shown the long term changes in social outlooks and social constructions of the environment, from – in the last 60–70 years or so – a science/control/exploitation/resource-driven discourse to a 'nature-attuned' (Kroma and Flora, 2003)/sustainability/protection-oriented discourse.

At the meso and micro levels of analysis, longitudinal studies have drawn extensively on constructionist models of the issue-careers of environmental issues as social problems, notably as formulated by Anthony Downs in the early 1970s, as well as on the closely related frameworks of agenda-building and agenda-setting theory. Generally, studies drawing on the issue-attention cycle framework have shown this to be a productive framework for understanding the ups and downs of media, political and social attention to environmental issues, although it has also become clear that the careers of environmental issues rarely map onto Downs's five-stage issue-attention model in any neat linear fashion. One particularly important finding is the evidence that, once introduced, environmental issues – ups and downs and changing framings notwithstanding – have remained firmly established in the public sphere.

Building on the questions naturally raised by the issue-attention cycle framework, i.e. particularly with regard to key factors driving the cycle and interaction between issues as well as different forums, agenda-building/setting studies have provided significant insights into the dynamics of environmental issues in the public sphere. These have included insights into, *inter alia*, the impact of 'real-world' events on media and political attention, inter-issue dynamics, the role of key sources/voices/claims-makers in influencing the media agenda, changing framings of environmental issues (and associated questions about causes, responsibility and solutions) and, not least, interactions of the media agenda, the public opinion agenda and the political agenda.

Longitudinal agenda-setting studies have generally been able to demonstrate that especially with regard to environmental issues (compared with other issues such as crime or health policy) the news media agenda can exert considerable influence on the public and political agendas. But perhaps their chief contribution lies in their ample demonstration of the complexity and dynamic nature of how media, political and public agendas interact. Longitudinal studies, by their very nature of looking at extended periods of time, in particular benefit from the ability to observe how the careers of environmental issues in the public sphere are driven by multiple factors, intricate feedback loops or spirals, and indeed timing, rather than by simple linear cause–effect dualities. In this they have thus helped confirm that simple linear-model questions – once popular and still not uncommon in media and communication research – about the 'effects or influence' of news media are perhaps best characterised, to paraphrase the late prominent media scholar James Halloran, as 'asking the wrong questions'.

Note

1   Two of the defining criteria of comparative communication research are that 'The macro-level units … should be compared not only *cross-spatially* but also *cross-temporally* to capture processes of internally and externally motivated change' and '*Relationships between contexts and the objects of analysis should be specified*, as they inform the explanation of differences and similarities' (Esser and Hanitzsch, 2012: 7 – emphasis added).

Suggestions for further reading

Djerf-Pierre, M. (2013). Green metacycles of attention: Reassessing the attention cycles of environmental news reporting 1961–2010. *Public Understanding of Science*, *22*(4), 495–512.

This draws productively on and significantly advances Downs's issue-attention cycle framework, in an exemplary longitudinal analysis that produces major insights into the nature of cycles and meta-cycles in environmental news reporting.

Esser, F. and Hanitzsch, T. (2012). On the why and how of comparative inquiry in communication studies. In F. Esser and T. Hanitzsch (eds), *The Handbook of Comparative Communication Research* (pp. 3–23). London: Routledge.

Although making little reference to environmental communication research, this chapter sets out an extremely helpful framework for comparative research, including comparing longitudinal temporal change, and it highlights many of the difficulties associated with measuring valid indicators and interpreting their meaning.

Schmidt, A., Ivanova, A. and Schäfer, M. S. (2013). Media attention for climate change around the world: A comparative analysis of newspaper coverage in 27 countries. *Global Environmental Change: Human and Policy Dimensions*, *23*(5), 1233–1248.

This provides an interesting corrective to the absence of environmental communication studies in many of the recent overviews of advances in comparative communication research (including in Esser and Hanitzsch listed above). Schmidt *et al.* – attending to the key requirements of comparative communication research, that it be cross-national and cross-temporal – offer a comprehensive meta-analysis of longitudinal studies of climate change coverage in 27 countries. See also their specific focus on key drivers of media attention in:

Schäfer, M. S., Ivanova, A. and Schmidt, A. (2014). What drives media attention for climate change? Explaining issue attention in Australian, German and Indian print media from 1996 to 2010. *International Communication Gazette*, *76*(2), 152–176.

References

Ader, C. R. (1995). A longitudinal-study of agenda-setting for the issue of environmental-pollution. *Journalism and Mass Communication Quarterly*, *72*(2), 300–311.

Ahern, L., Bortree, D. S. and Smith, A. N. (2012). Key trends in environmental advertising across 30 years in National Geographic magazine. *Public Understanding of Science*, *22*(4), 479–494.

Bauer, M., Durant, J. and Gaskell, G. (eds). (1999). *Biotechnology in the Public Sphere: A European Source-Book*. London: The Science Museum.

Bengston, D. N., Fan, D. P., Reed, P. and Goldhor-Wilcock, A. (2009). Rapid issue tracking: A method for taking the pulse of the public discussion of environmental policy. *Environmental Communication-a Journal of Nature and Culture*, *3*(3), 367–385.

Beniger, J. R. (1978). Media content as social indicators: The Greenfield Index of agenda setting. *Communication Research*, *5*(4), 437–453.

Bowman, J. S. and Fuchs, T. (1981). Environmental coverage in the mass media: a longitudinal study. *International Journal of Environmental Studies*, *18*(1), 11–22.

Boykoff, M. T. (2007). Climate change and journalistic norms: A case-study of US mass-media coverage. *Geoforum*, *38*, 1190-1204.

Boykoff, M. T. (2008). Lost in translation? United States television news coverage of anthropogenic climate change, 1995–2004. *Climatic Change*, *86*, 1–11.

Boykoff, M. T. (2009). We speak for the trees: Media reporting on the environment. *Annual Review of Environment and Resources*, *34*, 431–457.

Brookes, S. K., Jordan, A. G., Kimber, R. H. and Richardson, J. J. (1976). The growth of the environment as a political issue in Britain. *British Journal of Political Science*, *6*, 245–255.

Brossard, D., Shanahan, J. and McComas, K. (2004). Are issue-cycles culturally constructed? A comparison of French and American coverage of global climate change. *Mass Communication and Society*, *7*(3), 359–377.

Condit, C. M., Achter, P. J., Lauer, I. and Sefcovic, E. (2002). The changing meanings of 'mutation': A contextualized study of public discourse. *Human Mutation*, *19*(1), 69–75.

Djerf-Pierre, M. (2012a). When attention drives attention: Issue dynamics in environmental news reporting over five decades. *European Journal of Communication*, *27*(3), 291–304.

Djerf-Pierre, M. (2012b). The crowding-out effect: Issue dynamics and attention to environmental issues in television news reporting over 30 years. *Journalism Studies*, *13*(4), 499–516.

Djerf-Pierre, M. (2013). Green metacycles of attention: Reassessing the attention cycles of environmental news reporting 1961–2010. *Public Understanding of Science*, *22*(4), 495–512.

Dolšak, N. and Houston, K. (2013). Newspaper coverage and climate change legislative activity across US states. *Global Policy*, Online publication: 16 November 2013. Available at: http://onlinelibrary.wiley.com/doi/10.1111/1758-5899.12097/abstract;jsessionid=9FDA2AA3563A57E4BA220D055F7766D0.f01t03 (accessed 21 May 2014).

Downs, A. (1972). Up and down with ecology – the issue attention cycle. *The Public Interest*, *28*(3), 38–50.

Dryzek, J. (1997). *The Politics of the Earth: Environmental Discourses*. Oxford: Oxford University Press.

Einsiedel, E. and Coughlan, E. (1993). The Canadian press and the environment: Reconstructing a social reality. In A. Hansen (ed.), *The Mass Media and Environmental Issues* (pp. 134–149). Leicester: Leicester University Press.

Escobar, M. P. and Demeritt, D. (2014). Flooding and the framing of risk in British broadsheets, 1985–2010. *Public Understanding of Science*, *23*(4), 454–471.

Esser, F. and Hanitzsch, T. (2012). On the why and how of comparative inquiry in communication studies. In F. Esser and T. Hanitzsch (eds), *The Handbook of Comparative Communication Research* (pp. 3–23). London: Routledge.

Fan, D. P. (1988). *Predictions of Public Opinion from the Mass Media: Computer Content Analysis and Mathematical Modelling*. New York: Greenwood Press.

Funkhouser, G. R. (1973). The issues of the sixties: An exploratory study in the dynamics of public opinion. *Public Opinion Quarterly*, *37*(1), 62–75.

Gamson, W. A. and Modigliani, A. (1989). Media discourse and public opinion on nuclear power: A constructionist approach. *American Journal of Sociology*, *95*(1), 1–37.

Hajer, M. (1995). *The Politics of Environmental Discourse: Ecological Modernization and the Policy Process*. Oxford: Clarendon Press.

Hansen, A. (1993). Introduction. In A. Hansen (ed.), *The Mass Media and Environmental Issues* (pp. xv–xxii). Leicester: Leicester University Press.

Hansen, A. (1994). Trends in Environmental Issues Coverage in the British National Press: Centre for Mass Communication Research, University of Leicester, and WBMG (Environmental Communications), London.

Hansen, A. (2010). *Environment, Media and Communication*. London: Routledge.

Hilgartner, S., and Bosk, C. L. (1988). The rise and fall of social problems: A public arenas model. *American Journal of Sociology*, *94*(1), 53–78.

Howlett, M. and Raglon, R. (1992). Constructing the environmental spectacle: Green advertisements and the greening of the corporate image. *Environmental History Review*, *16*(4), 53–68.

Janowitz, M. (1976). Content analysis and the study of sociopolitical change. *Journal of Communication*, *26*(4), 10–21.

Jenner, E. (2012). News photographs and environmental agenda setting. *Policy Studies Journal*, *40*, 274–301.

Kepplinger, H. M. and Habermeier, J. (1995). The impact of key events on the presentation of reality. *European Journal of Communication*, *10*(3), 371–390.

Kroma, M. M. and Flora, C. B. (2003). Greening pesticides: A historical analysis of the social construction of farm chemical advertisements. *Agriculture and Human Values*, *20*(1), 21–35.

Liu, X., Lindquist, E. and Vedlitz, A. (2011). Explaining media and congressional attention to global climate change, 1969–2005: An empirical test of agenda-setting theory. *Political Research Quarterly*, *64*(2), 405–419.

Mazur, A. (1981). Media coverage and public opinion on scientific controversies. *Journal of Communication*, *31*(2), 106–115.

McComas, K. and Shanahan, J. (1999). Telling stories about global climate change: Measuring the impact of narratives on issue cycles. *Communication Research*, *26*(1), 30–57.

McGaurr, L. and Lester, L. (2009). Complementary problems, competing risks: Climate change, nuclear energy, and the *Australian*. In T. Boyce and J. Lewis (eds), *Media and Climate Change* (pp. 174–185). Oxford: Peter Lang.

McGeachy, L. (1989). Trends in magazine coverage of environmental issues. *Journal of Environmental Education*, *20*, 6–13.

Meisner, M. S. and Takahashi, B. (2013). The nature of *Time*: how the covers of the world's most widely read weekly news magazine visualize environmental affairs. *Environmental Communication: A Journal of Nature and Culture*, *7*(2), 255–276.

Mikami, S., Takeshita, T., Kawabata, M., Sekiya, N., Nakada, M., Otani, N. and Takahashi, N. (2002). *Unsolved Conflict among Europe, Japan and USA on the Global Warming Issue: Analysis of the Longitudinal Trends in News Frame*. Paper presented at the IAMCR Conference, Barcelona, Spain.

Mikami, S., Takeshita, T., Nakada, M. and Kawabata, M. (1995). The media coverage and public awareness of environmental issues in Japan. *International Communication Gazette*, *54*(3), 209–226.

Mitman, G. (1999). *Reel Nature: America's Romance with Wildlife on Film*. Cambridge, MA: Harvard University Press.

Nisbet, M. C. and Huge, M. (2006). Attention cycles and frames in the plant biotechnology debate: Managing power and participation through the press/policy connection. *Harvard International Journal of Press-Politics*, *11*(2), 3–40.

Peeples, J. (2013). Imaging toxins. *Environmental Communication: A Journal of Nature and Culture*, *7*(2), 191–210.

Rutherford, P. (2000). *Endless Propaganda: The Advertising of Public Goods*. Toronto: University of Toronto Press.

Sampei, Y. and Aoyagi-Usui, M. (2009). Mass-media coverage, its influence on public awareness of climate-change issues, and implications for Japan's national campaign to reduce greenhouse gas emissions. *Global Environmental Change: Human and Policy Dimensions*, *19*(2), 203–212.

Schäfer, M. S., Ivanova, A. and Schmidt, A. (2014). What drives media attention for climate change? Explaining issue attention in Australian, German and Indian print media from 1996 to 2010. *International Communication Gazette*, *76*(2), 152–176.

Schmidt, A., Ivanova, A. and Schäfer, M. S. (2013). Media attention for climate change around the world: A comparative analysis of newspaper coverage in 27 countries. *Global Environmental Change: Human and Policy Dimensions*, *23*(5), 1233–1248.

Soroka, S. N. (2002). Issue attributes and agenda-setting by media, the public, and policymakers in Canada. *International Journal of Public Opinion Research*, *14*(3), 264–285.

Strodthoff, G. G., Hawkins, R. P. and Schoenfeld, A. C. (1985). Media roles in a social-movement - a model of ideology diffusion. *Journal of Communication*, *35*(2), 134–153.

Trumbo, C. (1995). Longitudinal modelling of public issues: an application of the agenda-setting process to the issue of global warming. *Journalism and Mass Communication Monographs*, *152*. Columbia, SC: Association for Education in Journalism and Mass Communication.

Trumbo, C. (1996). Constructing climate change: Claims and frames in US news coverage of an environmental issue. *Public Understanding of Science*, *5*(3), 269–283.

Uscinski, J. E. (2009). When does the public's issue agenda affect the media's issue agenda (and vice-versa)? Developing a framework for media–public influence. *Social Science Quarterly*, *90*(4), 796–815.

Walgrave, S., Soroka, S. and Nuytemans, M. (2008). The mass media's political agenda-setting power: A longitudinal analysis of media, Parliament, and government in Belgium (1993 to 2000). *Comparative Political Studies*, *41*(6), 814–836.

Wall, G. (1999). Science, nature, and the nature of things: An instance of Canadian environmental discourse, 1960–1994. *Canadian Journal of Sociology-Cahiers Canadiens De Sociologie*, *24*(1), 53–85.

Webb, T. J., Bengston, D. N. and Fan, D. P. (2008). Forest value orientations in Australia: An application of computer content analysis. *Environmental Management*, *41*, 52–63.

18

COMMUNICATING IN THE ANTHROPOCENE

The cultural politics of climate change news coverage around the world

Maxwell T. Boykoff, Marisa B. McNatt and Michael K. Goodman

Introduction

Over the past years, the number of *Reuters* stories about climate change has declined. This trend has been consistent with trends across other media outlets globally (see Figure 18.1) due largely to political economic trends of shrinking newsrooms and fewer specialist reporters covering climate stories with the same frequency as before. In 2010, the *Wall Street Journal* and the *Christian Science Monitor* closed their environmental blogs. Three years later, in January 2013, the *New York Times* dismantled its environment desk, assigning the reporters and editors to other departments, and discontinued its Green blog two months later. Yet, initially, *Reuters* had largely bucked those trends, continuing to employ top climate and environment reporters from around the globe, including Deborah Zabarenko (North America), Alister Doyle (Europe) and David Fogerty (Asia) who fed top media organizations with reporting comprised of a steady diet of climate and environment stories. So why this subsequent and precipitous drop in *Reuters's coverage of* climate change? In July 2013, David Fogerty—who left *Reuters* in late 2012—took to *The Baron* blog to explain why. He recounted that, after the appointment of Editor Paul Ingrassia in 2011, editorial decisions were made to deprioritize climate stories, and to shift these specialists to different beats. Fogerty, for example, was moved from the climate beat to instead cover issues around shipping in the Asian region. While climate stories had been already declining upon the appointment of Ingrassia, many argued that his revamping of the *Reuters* reporting priorities served to accelerate this drop.

Crucially, Fogerty and others asserted that Ingrassia's ideological and political leanings also played a detrimental part in continued coverage (Fogerty 2013). As such, in the summer of 2013, Reuters climate reporting faced deep levels of criticism and dismay from journalism colleagues and media critics and consumers. For example, journalist Alex Sobel Fitts at *Columbia Journalism Review* attributed variations in content and quantity to this new editorial turn (2013). And Max Greenberg at *Media Matters* framed this drop in coverage as a consequence of a "climate of fear" imposed by new contrarian editorial influences (2013).

Thus, while the specific situation with *Reuters* provides a worrisome glimpse into the contentious and high-stakes arena of global reporting on climate change in the twenty-first century, what it shows more generally is the way that environmental communication in the context of climate politics is thoroughly enmeshed in a combination of large-scale social, political and economic factors connected up with smaller-scale power-laden editorial decision-making, steeped in cultural economy and ideology.

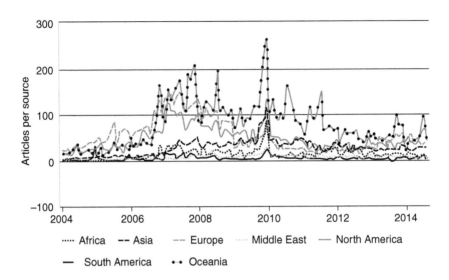

Figure 18.1 World newspaper coverage of climate change/global warming

Source: © 2014 Andrews, K., Wang, X., Nacu-Schmidt, A., McAllister, L., Gifford, L., Daly, M., Boykoff, M., and Boehnert, J. (2014). World Newspaper Coverage of Climate Change or Global Warming, 2004-2014. Center for Science and Technology Policy Research, Cooperative Institute for Research in Environmental Sciences, University of Colorado, Web. [Accessed November 11] http://sciencepolicy.colorado.edu/media_coverage.

Note: This figure tracks newspaper coverage of climate change or global warming in fifty newspapers across twenty countries around the world from January 2004 through October 2014. For monthly updates and the full list of sources go to http://sciencepolicy.colorado.edu/media_coverage.

Simultaneously, digital and social media are stepping into these spaces. This chapter explores key questions that arise regarding how the general decline in specialized reporting on climate change and rise in social media impacts overall quantity and quality of coverage as well as inputs into public awareness and engagement.

Most citizens around the world typically do not read peer-reviewed literature. Instead, to learn about climate change, people in the public arena turn to media communications—television, newspapers, radio, new and social media—to link formal science and policy with their everyday lives. Over the past decades, the dynamics of science and politics have clearly shaped media coverage of climate change. Yet, it is also worth noting and considering how 'news'—generated by mass media—has, in turn, shaped ongoing scientific and political considerations, deliberations and decisions. In other words, it is instructive to account for how mass media have influenced who has a say, when and how in the public arena.

"The media" around the world are actually much more heterogeneous and varied than at first glance. In their multiple dimensions, media are constituted by many institutions, processes and practices that together serve as "mediating" forces between communities such as science, policy and civil society. Media segments, articles, clips and pieces represent critical links between people's everyday realities and experiences and the ways in which these are discussed at a distance between science, policy and public actors. People throughout society rely upon media representations to help interpret and make sense of the many complexities relating to climate science and governance. Thus, media messages are critical inputs to what become public discourse on current climate challenges.

Yet, these media representations enter into an individual's pre-existing perceptions and perspectives, and are taken up or resisted in varied ways. For example, Wouter Poortinga and colleagues (2011) have found that people's enduring values and existing ideologies strongly influence their understandings and behaviors as they relate to climate change. Indeed, as Lorraine Whitmarsh put it in summarizing her research on climate contrarianism "attitudes to climate change are relatively entrenched and … information about the issue will be evaluated and used in diverse ways according to individuals' values and worldviews." She concluded:

> [S]imply providing climate change information is unlikely to be successful, as new information is often interpreted by people in line with their existing attitudes and worldviews … In other words, irrespective of how much information is provided, it is remarkably difficult to change attitudes that have become entrenched.
>
> (Whitmarsh 2011: 698)

These dynamic science–policy–media–public interactions have been spaces where claims-makers in the media have been changing (e.g. Baum and Groeling 2008; Fahy and Nisbet 2011), and traditional media outlets have faced newfound challenges (Boykoff and Yulsman 2013; Siles and Boczkowski 2012) while shifts to new/digital/social media tools have recalibrated who has a say and how these claims circulate (Baek et al. 2012; Cacciatore et al. 2012; Graham et al. 2013). Traditional and legacy media organizations themselves have worked to adapt to these changing conditions and researchers have increasingly sought to make sense of the shifts (e.g. Horan 2013; Nielsen 2012; Zhu and Dukes 2013) and their implications (e.g. Jacobson 2012) in various cultural, political, social and environmental contexts (e.g. Adams and Gwynnald 2013; Schuurman 2013).

In recent decades there has been significant expansion from traditional mass media into consumption of social and digital media. Essentially, in tandem with technological advances, this expansion in communications is seen to be a fundamental shift from broadcast, or "one-to-many" (often one-way) communications to "many-to-many" more interactive webs of communications (O'Neill and Boykoff 2010; van Dijk 2006). This movement has signaled substantive changes in how people access and interact with information and who has access.

Together, traditional/legacy and digital/social media spaces comprise a key part of what many now refer to as the "cultural politics of climate change": dynamic and contested spaces where various "actors" battle to shape public understanding and engagement (e.g. Boykoff and Goodman 2009). These are places where formal climate science, policy and politics operate at multiple scales, through multiple media forms and are dynamic as well as contested processes that shape how meaning is constructed and negotiated. In these spaces of the "everyday," cultural politics involve not only the discourses that gain traction in wider discourses, but also those that are absent (Derrida 1978). Contemplating climate considerations in this way helps to examine "how social and political framings are woven into both the formulation of scientific explanations of environmental problems, and the solutions proposed to reduce them" (Forsyth 2003: 1).

Media attention in the public sphere

Figure 18.1 appraises the trends in media coverage of climate change from 2004 into 2014 in fifty newspapers across the globe. This visual representation has provided an opportunity to assess and analyze further questions of *how* and *why* there were apparent ebbs and flows in coverage. For instance, notably 2009 ended with soaring media coverage of climate change around the world and numerous studies have sought to better understand events and developments during this time period (see Boykoff 2013 for example). At this time, climate news seemingly flooded the public arena and was dominated by the much-hyped and highly anticipated United Nations climate talks in Copenhagen, Denmark (COP15), along with news about the hacked emails of scientists from the University of East Anglia Climate Research Unit (referred to by some as "Climategate"). These events also linked to ongoing stories of energy security, sustainability, carbon markets and green economies that were unfolding during this time.

Across this nearly ten-year look, increases in each of the regions have not been symmetrical. For example, there were a relatively low number of stories on climate change or global warming in the regions of South America and Africa throughout this period. This points to a critical regional "information gap" in reporting on these issues, and relates to capacity issues and support for reporters in these regions and countries (developing and poorer regions/countries).

Tracking media treatment of climate change and global warming through intersecting *political*, *scientific* and *ecological/meteorological* climate themes provides a useful framework for analyses of content and context. Such accounting helps to demonstrate how news pieces should not be treated in isolation from one another; rather, they should be considered connected parts of larger political, economic, social, environmental and cultural conditions.

Moreover, patterns revealed in the mobilizations of journalistic norms internal to the news-generation process cohere with externally influenced dominant market-based and utilitarian approaches that consider the spectrum of possible mitigation and adaptation action on climate change. Robert Brulle has argued that an excessive mass media focus merely on the debaters and their claims, "works against the large-scale public engagement necessary to enact the far-reaching changes needed to meaningfully address global warming" (2010: 94). As such, examinations of the content of media treatment of climate change need to be considered within a context of larger political and social forces.

The cultural politics of climate change reside in many spaces and places, from workplaces to pubs and kitchen tables. "Actors" on this stage range from fellow citizens to climate scientists as well as business industry interests and environmental non-governmental organization (ENGO) activists. Over time, individuals, collectives, organizations, coalitions and interest groups have sought to access the power of mass media to influence architectures and processes of climate science, governance and public understanding through various media "frames" and "claims." Questions regarding "who speaks for the climate" involve considerations of how various perspectives—from climate scientists to business industry interest and ENGO activists—influence public discussions on climate change (Boykoff 2011). "Actors," "agents," or "operatives" in this theatre are ultimately all members of a collective public citizenry. However, differential access to media outlets across the globe are products of differences in power, and power saturates social, political, economic and institutional conditions undergirding mass media content production (Wynne 2008).

In the highly contested arena of climate science and governance, different actors have sought to access and utilize mass media sources in order to shape perceptions on various climate issues contingent on their perspectives and interests. For example, "contrarians," "skeptics" or "denialists" have had significant discursive traction in the U.S. public sphere over time (Leiserowitz et al. 2013), particularly by way of media representations (Boykoff 2013). Resistances to both diagnoses of the causes of climate change, and prognoses for international climate policy implementation, in the U.S. more specifically, have often been associated with the political right: the Republican Party and more particularly a right-wing faction called the "Tea Party" (Dunlap 2008). John Broder of the *New York Times* described this right-of-center U.S. political party stance as an "article of faith," and polling data have shown that "more than half of Tea Party supporters said that global warming would have no serious effect at any time in the future, while only 15 percent of other Americans share that view" (2010: A1). Moreover, while carbon-based industry interests have exerted considerable influence over U.S. climate policy, associated scientists and policy actors who have questioned the significance of human contributions—often dubbed "climate contrarians"—have been primarily housed in North American universities, think tanks and lobbying organizations (Dunlap 2013; McCright 2007). In particular, U.S.-based non-nation state organizations such as the "Heartland Institute" have held numerous meetings to promote contrarian views on climate science and policy (Boykoff and Olson 2013; Hoffman 2011).

Contributions to climate storytelling through news

Climate change is a complex and multifaceted issue that cuts to the heart of humans' relationship with the environment. The cultural politics of climate change are situated, power-laden, media-led and recursive in an ongoing battlefield of knowledge and interpretation (Boykoff et al. 2009). Mass media link these varied spaces together, as powerful and important interpreters of climate science and policy, translating what can often be alienating, jargon-laden information for the broadly construed public citizenry. Media workers and institutions powerfully shape and negotiate meaning, influencing how citizens make sense of and value the world.

In various cultural, political, social, economic and environmental contexts, journalists, producers and editors as well as scientists, policy makers and non-nation state actors must scrupulously and intently negotiate how climate is considered as a "problem" or a "threat." As part of this process, it has been demonstrated that media reports have often conflated the vast and varied terrain—from climate science to governance, from consensus to debate—as unified and universalized issues (Boykoff 2011). As a consequence, these representations can confuse rather than clarify: they can contribute to ongoing illusory, misleading and counterproductive debates within the public and policy communities on critical dimensions of the climate issue.

To the extent that media fuse distinct facets into climate *gestalt*—by way of "claims" as well as "claims makers"—collective public discourses, as well as deliberations over alternatives for climate action, have been poorly served. For example, although scientific experts have reached the consensus that humans contribute to climate change, there remains some disagreement among climate science experts as to the severity of climate change impacts and when and where climate impacts will occur (Schmidt and Wolfe 2008). Rosenberg et al. explain:

> Those that disagree that the problem [of anthropogenic climate change] is acute or in need of decisive action like to note points of disagreement among scientists to bolster their position that the science is unsure and not defined enough to use as a foundation for policy decisions.
>
> (2010: 311)

Media focusing on an area of climate change that contains scientific nuances and uncertainties, such as the degree to which an extreme weather event is the result of climate change, may result in a specious conclusion that more knowledge is needed before taking action on climate change. In another sense, a lack of media coverage on climate change solutions, or the idea that mere individual actions can make the requisite difference may also limit actions for climate change.

Regarding "claims makers," efforts to make sense of complex climate science and governance through media representations involves decisions regarding what are "experts" or "authorities" who speak for climate. This is particularly challenging when covering climate change, where indicators of climate change—such as sea level rise, temperature shifts, changing rainfall patterns—may be difficult to detect and systematically analyze (Andreadis and Smith 2007). Moreover, in the advent and increasingly widespread influence of new and social media (along with fewer "gatekeepers" in content generation), the identification of "expertise" can be more, rather than less, challenging. The abilities to quickly conduct a Google or Bing search for information is in one sense very liberating; yet, in another sense, this unfiltered access to complex information also intensifies possibilities of short-circuiting peer review processes (and determinations by "experts"), and can thereby do an "end-run around established scientific norms" (McCright and Dunlap 2003: 359). In other words, these developments have numerous and potentially paradoxical reverberations through ongoing public discourses on climate change.

There are many reasons why media accounts around the world routinely fail to provide greater nuance when covering various aspects of climate change. Central among them, the processes behind the building and the challenging of dominant discourses take place simultaneously at multiple scales. Large-scale social, political and economic factors influence everyday individual journalistic decisions, such as how to focus or contextualize a story with quick time to deadline. These issues intersect with processes such as journalistic norms and values (e.g. Boykoff 2011), citizen and digital journalism (e.g. O'Neill and Boykoff 2010), and letters to the editor (e.g. Young 2013) to further shape news narratives. Moreover, path dependence through histories of professionalized journalism, journalistic norms and values as well as power relations have shaped the production of news stories (Starr 2004). These dynamic and multiscale influences are interrelated and difficult to disentangle: media portrayals of climate change are infused with cultural, social, environmental and political economic elements, as well as how media professionals must mindfully navigate through hazardous terrain in order to fairly and accurately represent various dimensions of climate science and governance (Ward 2008).

Overall, media representations are derived through complex and non-linear relationships between scientists, policy actors and the public that is often mediated by journalists' news stories (Carvalho and Burgess 2005). In this, multi-scalar processes of power shape how mass media depict climate change. Processes involve an inevitable series of editorial choices to cover and report on certain events within a larger current of dynamic activities, and provide mechanisms for privileging certain interpretations and "ways of knowing" over others. Resulting images, texts and stories compete for attention and thus permeate interactions between science, policy, media and the public in varied ways. Furthermore, these interactions spill back onto ongoing media representations. Through these selection and feedback processes, mass media have given voice to climate itself by articulating aspects of the phenomenon in particular ways, via claims makers or authorized speakers. In other words, through the web of contextual and dynamic factors, the stream of events in our shared lives gets converted into finite news stories. Thus, constructions of meaning and discourse on climate change are derived through combined structural and agential components that are represented through mass media to the general public.

The rise of #climate news through digital and social media

Embedded in this dynamism is the burgeoning influence of digital and social media. With it comes numerous questions: does increased visibility of climate change in new/social media translate to improved communication or just more noise? Do these spaces provide opportunities for new forms of deliberative community regarding questions of climate mitigation and adaptation (e.g. Harlow and Harp 2013; Rogers 2004) and conduits to offline organizing and social movements (e.g. Jankowski 2006; Tufekci 2013)? Or has the content of this increased coverage shifted to polemics and arguments over measured analysis? In this democratized space of content production, do new/social media provide more space for contrarian views to circulate? And through its interactivity, does increased consumption through new/social media further fragment a public discourse on climate mitigation and adaptation, through information silos where members of the public can stick to sources that help support their already held views (e.g. Hestres 2013; Yang and Kahlor 2012)?

Sharon Dunwoody has cautioned to not view various modes of media production equally. As she puts it,

> because of their extensive reach and concomitant efficiencies of scale, mediated information channels such as television and newspapers have been the traditional channels of choice for information campaigns. But research on how individuals actually use mass media information suggests that these channels may be better for some persuasive purposes than for others.
>
> (quoted in Boykoff 2009: 2)

Furthermore, Cass Sunstein (2007) offers a similarly complicating—and also less than rosy—perspective: he warned of the likelihood of the "echo chamber" effect where this interactivity actually walls off users from one another by merely consuming news that meshes with their worldview and ideology.

Such considerations within these new media developments prompt us to reassess boundaries between who constitute "authorized" speakers (and who do not) in mass media as well as who are legitimate "claims-makers." These are consistently being interrogated, and challenged (Gieryn 1999; Loosen and Schmidt 2012). Anthony Leiserowitz has written that these arenas of claims-making and framing are "exercises in power ... Those with the power to define the terms of the debate strongly determine the outcomes" (2005: 149). These factors have produced mixed and varied impacts: journalist Alissa Quart (2010) has warned of dangers of mistaken (or convenient) reliance on "*fauxperts*" instead of "experts" while Boykoff (2013) and Boykoff and Olson (2013) have examined these dynamics as they relate to amplified media attention to "contrarian" views on various climate issues.

Conclusions

Connections between media information and policy decision making, perspectives and behavioral change are far from straightforward (Vainio and Paloniemi 2013). Coverage certainly does not determine engagement; rather, it shapes engagement *possibility* in quantity, quality, depth and effect (Boykoff 2008; Carvalho and Burgess 2005). So our explorations of media coverage of climate change around the world in this chapter seek to help readers better understand the dynamic web of influence that media play amidst many others that shape our attitudes, intentions, beliefs, perspectives and behaviors regarding climate change. As we have posited here, media representations—from news to entertainment, from broadcast to interactive and participatory—are critical links between people's perspectives and experiences, and the ways in which dimensions of climate change are discussed at a distance between science, policy and public actors.

The road from information acquisition via mass media to various forms of engagement and action is far from straightforward, and is filled with turns, potholes and intersections. This is a complex arena: mass media portrayals do not *simply* translate truths or truth-claims nor do they fill knowledge gaps for citizens and policy actors to make "the right choices." Moreover, media representations clearly do not dictate particular behavioral responses. For example, research has shown that fear-inducing and catastrophic tones in climate change stories can inspire feelings of paralysis through powerlessness and disbelief rather than motivation and engagement. In addition, O'Neill et al. (2013) found that imagery connected with climate change influences saliency (that climate change is important) and efficacy (that one can do something about climate change) in complex ways, in their study across the country contexts of Australia, the U.S. and United Kingdom. Among their results, they found that imagery of climate impacts promoted feelings of salience, but undermined self-efficacy, while imagery of energy futures imagery promoted efficacy. Overall, media portrayals continue to influence—in non-linear and dynamic ways—individual to community- and international-level perceptions of climate science and governance (Wilby 2008). In other words, mass media have constituted key interventions in shaping the variegated, politicized terrain within which people perceive, understand and engage with climate science and policy (Goodman and Boyd 2011; Krosnick et al. 2006).

Over time, many researchers and practitioners have (vigorously) debated the extent to which media representations and portrayals are potentially conduits to attitudinal and behavioral change (e.g. Dickinson et al. 2013). Nonetheless, as unparalleled forms of communication in the public arena, research into media representational practices remains vitally important in terms of how they influence a spectrum of possibilities for governance and decision making. As such, media messages—and language choices more broadly (Greenhill et al. 2013)—function as important interpreters of climate information in the public arena, and shape perceptions, attitudes, intentions, beliefs and behaviors related to climate change (Boykoff 2011; Hmielowski et al. 2013). Studies across many decades have documented that citizen-consumers access understanding about science and policy (and more specifically climate change) largely through media messages (e.g. Antilla 2010; O'Sullivan et al. 2003).

Furthermore, mass media comprise a community where climate science, policy and politics can readily be addressed, analyzed and discussed. The way that these issues are covered in media can have far-reaching consequences in terms of ongoing climate scientific inquiry as well as policy maker and public perceptions, understanding and potential engagement. In this contemporary environment, numerous "actors" compete in these media landscapes to influence decision making and policy prioritization at many scales of governance. Multitudinous ways of knowing—both challenged and supported through media depictions—shape ongoing discourses and imaginaries, circulating in various cultural and political contexts and scales. Furthermore, varying media representational practices contribute—amid a complex web of factors—to divergent perceptions, priorities and behaviors.

More media coverage of climate change—even supremely fair and accurate portrayals—is not a panacea. In fact, increased media attention to the issue often unearths more questions to be answered and *greater* scientific understanding actually can contribute to a *greater* supply of knowledge from which to develop and argue varying interpretations of that science (Sarewitz 2004). At best, media reporting helps address, analyze and discuss the issues, *but not answer them*. And dynamic interactions of multiple scales and dimensions of power critically contribute to how climate change is portrayed in the media. As we have detailed above, mass media representations arise through large-scale (or *macro*) relations, such as decision making in a capitalist or state-controlled political economy and individual-level (or *micro*) processes such as everyday journalistic and editorial practices.

The contemporary cultural politics of climate change thread through a multitude of rapidly expanding spaces. Within this, the media serve a vital role in communication processes between science, policy and the public. The influence of media representations as well as creative and participatory communications—nested in cultural politics more broadly—can be ignored or dismissed in shaping climate science and governance at our peril.

Further reading

Doyle, J. (2011). *Mediating Climate Change*. Aldershot: Ashgate Publishing.
This book confronts how nature and the environment have been problematically separated from humans and culture: interrogating how climate change becomes meaningful in our lives, Doyle explores how imagery shapes our understanding, and how climate mitigation efforts in particular relate to our food consumption choices, support for social movements and commitments to creative experimentation and engagement. In the interstices of climate science, culture and society, Doyle examines how mediation and visualization—as intensely values-laden processes—shape how we consider and respond to climate challenges.

Painter, J. (2013). *Climate Change in the Media: Reporting Risk and Uncertainty*. New York: I.B. Taurus & Co.
In this study, Painter quantitatively analyzes how risks and uncertainty have been portrayed in international media accounts, and how these feed into and bolster wider narratives on climate science, policy, culture and society. With a focus on articulations of climate science through the Intergovernmental Panel on Climate Change (IPCC) and melting Arctic ice, his team at the Reuters Institute for the Study of Journalism examine media representations in Australia, France, India, Norway, the United States and the United Kingdom. This work takes strides to improve connectivity between the processes and practices of journalism and climate science and decision making.

Crow, D. and Boykoff, M. T. (eds) (2014). *Culture, Politics, and Climate Change: How Information Shapes Our Common Future*. New York: Routledge.

Drawing from multiple disciplinary perspectives and research from both developed and developing worlds, this book presents an overview of the knowledge related to the current understanding of climate change politics and culture. The book includes a section focusing on Communication and Media, an analysis of the challenges and opportunities for establishing successful communication on climate change among scientists, the media, policy makers and activists, and identifies future needs and improvements in studies on climate change communication linking science, media, activism and policy.

Boykoff, M. T. (2011). *Who Speaks for the Climate?: Making Sense of Media Reporting on Climate Change*. New York: Cambridge University Press.

This contribution has sought to make sense of how media representations influence the many complexities relating to climate science and governance. The book provides a bridge between academic considerations and real world developments, helping students, academic researchers and interested members of the public make sense of media reporting on climate change as it explores "who speaks for the climate" and what effects this may have on the spectrum of possible responses to contemporary climate challenges.

References

Adams, P. C. and Gynnild, A. (2013) "Environmental messages in online media: The role of place," *Environmental Communication: A Journal of Nature and Culture* 7(1), 103, 113–130.

Andreadis, E. and Smith, J. (2007) "Beyond the ozone layer," *British Journalism Review* 18(1), 50–56.

Antilla, L. (2010) "Self-censorship and science: A geographical review of media coverage of climate tipping points," *Public Understanding of Science* 19(2), 240–256.

Baek, Y. M., Wojcieszak, M. and Delli Carpini, M. (2012) "Online versus face-to-face deliberations: Who? Why? What? What effects?" *New Media and Society* 14(3), 363–383.

Baum, M. A. and Groeling, T. (2008) "New media and the polarization of American political discourse," *Political Communication* 25(1), 345–365.

Boykoff, M. (2008) "The cultural politics of climate change discourse in UK tabloids," *Political Geography* 27(5), 2008, 549–569.

Boykoff, M. (2009) "A discernible human influence on the COP15? Considering the role of media in shaping ongoing climate science," *Copenhagen Climate Congress* Theme 6, Session 53.

Boykoff, M. (2011) *Who Speaks for Climate? Making Sense of Mass Media Reporting on Climate Change*, Cambridge: Cambridge University Press, 240 pp.

Boykoff, M. (2013) "Public Enemy no.1? Understanding media representations of outlier views on climate change," *American Behavioral Scientist* doi:10.1177/0002764213476846.

Boykoff, M. and Goodman, M. K. (2009) "Conspicuous redemption? Reflections on the promises and perils of the 'celebritization' of climate change," *Geoforum* 40, 395–406.

Boykoff, M. and Olson, S. (2013) "Understanding contrarians as a species of 'Charismatic Megafauna' in contemporary climate science-policy-public interactions," *Celebrity Studies journal* (special issue Editors M. K. Goodman and J. Littler).

Boykoff, M. and Yulsman, T. (2013) "Political economy, media and climate change: The sinews of modern life," *Wiley Interdisciplinary Reviews: Climate Change* 4(5), 359–371.

Boykoff, M., Goodman, M. K. and Curtis, I. (2009) "Cultural politics of climate change: Interactions in everyday spaces," in M. Boykoff (ed.), *The Politics of Climate Change: A Survey*, London: Routledge/Europa, pp. 136–154.

Broder, J. M. (2010) "Skepticism on climate change is article of faith for tea party", *The New York Times*, October 21, A1.

Brulle, R. J. (2010) "From environmental campaigns to advancing a public dialogue: Environmental communication for civic engagement," *Environmental Communication: A Journal of Nature and Culture* 4(1), 82–98.

Cacciatore, M. A., Anderson, A. A., Choi, D-H., Brossard, D., Scheufele, D. A., Liang, X., Ladwig, P. J., Xenos, M. and Dudo, A. (2012) "Coverage of emerging technologies: A comparison between print and online media," *New Media and Society* 14(6), 1039–1059.

Carvalho, A. and Burgess, J. (2005) "Cultural circuits of climate change in UK broadsheet newspapers, 1985–2003," *Risk Analysis* 25(6), 1457–1469.

Crow, D. and Boykoff, M. T. (eds) (2014). *Culture, Politics, and Climate Change: How Information Shapes Our Common Future*. New York: Routledge.

Derrida, J. (1978) "Structure, sign, and play in the discourse of the human sciences," in J. Derrida (ed.), *Writing and Difference*, Chicago, IL: University of Chicago Press, pp. 278–293.

Dickinson, J. L., Crain, R., Yalowitz, S. and Cherry, T. M. (2013) "How framing climate change influences citizen scientists' intentions to do something about it," *The Journal of Environmental Education* 44(3), 145–158.

Doyle, J. (2011). *Mediating Climate Change*. Surrey: Ashgate Publishing.

Dunlap, R. E. (2008) "Climate-change views: Republican–Democrat gaps extend," *Gallup*, May 29.

Dunlap, R. E. (2013) "Climate change skepticism and denial: An introduction," *American Behavioral Scientist* 57(6), 691–698.

Fahy, D. and Nisbet, M. C. (2011) "The science journalist online: Shifting roles and emerging practices," *Journalism* 12(7), 778–793.

Fitts, A. S. (2013) "Reuters global warming about face," *Columbia Journalism Review*, July 26. Retrieved August 6: www.cjr.org/the_observatory/reuterss_global_warming_about-.php?page=2 (accessed 2014).

Fogerty, D. (2013) "Climate change," *The Baron*, July 15. Retrieved August 6: www.thebaron.info/blog/files/38aa6 54bc00bdf2bac925993875a7b67-694.php (accessed 2014).

Forsyth, T. (2003) *Critical Political Ecology: The Politics of Environmental Science*, London: Routledge.

Gieryn, T. F. (1999) *Cultural Boundaries of Science: Credibility on the Line*, Chicago, IL: University of Chicago Press.

Goodman, M. and Boyd, E. (2011) "A social life for carbon? Commodification, markets and care," *The Geographical Journal* 177(2), 102–109.

Graham, M., Schroeder, R. and Taylor, G. (2013) "Re:search," *New Media and Society*, 12(3), 1–8.

Greenberg, M. (2013) "Reuters climate change coverage declined significantly after 'skeptic' editor joined," *Media Matters for America*, July 23. Retrieved August 14: http://mediamatters.org/blog/2013/07/23/reuters-climate-change-coverage-declined-signif/195015 (accessed 2014).

Greenhill, M., Leviston, Z., Leonard, R. and Walker, I. (2013) "Assessing climate change beliefs: Response effects of question wording and response alternatives," *Public Understanding of Science* 22(3), 1–19.

Harlow, S. and Harp, D. (2013) "Collective action on the web," *Information, Communication and Society*, 15(2), 196–216.

Hestres, L. E. (2013) "Preaching to the choir: Internet-mediated advocacy, issue public mobilization, and climate change," *New Media and Society* 1(1), 1–17.

Hmielowski, J. D., Feldman, L., Myers, T. A., Leiserowitz, A. and Maibach, E. (2013) "An attack on science? Media use, trust in scientists and perceptions of global warming," *Public Understanding of Science*, April 3, 1–18.

Hoffman, A. J. (2011) "Talking past each other? Cultural framing of skeptical and convinced logics in the climate change debate," *Organization and Environment*, 24(3), 3–33.

Horan, T. J. (2013) "'Soft' versus 'hard' news on microblogging networks," *Information, Communication and Society* 16(1): 43–60.

Jacobson, S. (2012) "Transcoding the news: an investigation into multimedia journalism published on nytimes.com 200-2008," *New Media and Society* 14(5): 867–885.

Jankowski, N. W. (2006) "Creating community with media: History, theories and scientific investigations," in L. A. Lievrouw and S. Livingstone (eds), *The Handbook of New Media*, Updated Student Edition, London/Thousand Oaks/New Delhi: Sage, pp. 55–74.

Krosnick, J. A., Holbrook, A. L., Lowe, L. and Visser, P. S. (2006) "The origins and consequences of democratic citizens' policy agendas: A study of popular concern about global warming," *Climatic Change* 77(1), 7–43.

Leiserowitz, A. A. (2005) "American risk perceptions: Is climate change dangerous?" *Risk Analysis* 25, 1433–1442.

Leiserowitz, A. A., Maibach, E., Roser-Renouf, C., Smith, N. and Dawson, E. (2013) "Climategate, public opinion and loss of trust," *American Behavioral Scientist* 57(6), 818–837.

Loosen, W. and Schmidt, J-H. (2012) "Re-discovering the audience," *Information, Communication & Society* 15(6), 867–887.

McCright, A. M. (2007) "Dealing with climate contrarians," in S. C. Moser and L. Dilling (eds), *Creating a Climate for Change: Communicating Climate Change and Facilitating Social Change*, Cambridge: Cambridge University Press, pp. 200–212.

McCright, A. M. and Dunlap, R. E. (2003) "Defeating Kyoto: The conservative movement's impact on U.S. climate change policy," *Social Problems* 50(3), 348–373.

Nielsen, R. K. (2012) "How newspapers began to blog," *Information, Communication and Society* 15(6), 959–968.

O'Neill, S. J. and Boykoff, M. T. (2010) "The role of new media in engaging the public with climate change," in L. Whitmarsh, S. J. O'Neill and I. Lorenzoni (eds), *Engaging the Public with Climate Change: Communication and Behaviour Change*, London: Earthscan, pp. 233–251.

O'Neill, S. J., Boykoff, M. T., Day, S. A. and Niemeyer, S. (2013) "On the use of imagery for climate change engagement," *Global Environmental Change* (23ª2), 413–421.

O'Sullivan, T., Dutton, B. and Rayne, P. (2003) *Studying the Media*, Bloomsbury, USA: Hodder Arnold.

Painter, J. (2013). *Climate Change in the Media: Reporting Risk and Uncertainty*. New York: I.B. Taurus & Co.

Poortinga, W., Spence, A., Whitmarsh, L., Capstick, S. and Pidgeon, N. F. (2011) "Uncertain climate: An investigation into public scepticism about anthropogenic climate change," *Global Environmental Change* 21(3), 1015–1024.

Quart, A. (2010) "The trouble with experts," *Columbia Journalism Review*, July/Aug, 17–18.

Rogers, R. (2004) *Information Politics on the Web*, Boston, MA: The MIT Press.

Rosenberg, S., Vedlitz, A., Cowman, D. F. and Zahran, S. (2010) "Climate change: A profile of US climate scientists' perspectives," *Climate Change*, 101(1), 311–329.

Sarewitz, D. (2004) "How science makes environmental controversies worse," *Environmental Science and Policy* 7: 385–403.

Schmidt, G. and Wolfe, J. (2008) *Climate Change: Picturing the Science*, New York: W. W. Norton & Company.

Schuurman, N. (2013) "Tweet me your talk: Geographical learning and knowledge production 2.0," *Professional Geographer* 65(3), 369–377.

Siles, I. and Boczkowski, P. J. (2012) "Making sense of the newspaper crisis: A critical assessment of existing research and an agenda for future work," *New Media and Society* 14(8): 1375–1394.

Starr, P. (2004) *The Creation of the Media: Political Origins of Modern Communications*, New York: Basic Books.

Sunstein, C. R. (2007) *Republic.com 2.0*, Princeton, NJ: Princeton University Press.

Tufekci, Z. (2013) "Not this one: Social movements, the attention economy, and microcelebrity networked activism," *American Behavioral Scientist* 57(7), 848–870.

Vainio, A. and Paloniemi, R. (2013) "Does belief matter in climate change action?" *Public Understanding of Science* 22(4), 382–395.

van Dijk, J. (2006) *The Network Society*, London: Sage.

Ward, B. (2008) *Communicating on Climate Change: An Essential Resource for Journalists, Scientists and Editors*, Providence, RI: Metcalf Institute for Marine and Environmental Reporting, University of Rhode Island Graduate School of Oceanography.

Whitmarsh, L. (2011) "Scepticism and uncertainty about climate change: Dimensions, determinants and change over time," *Global Environmental Change*, 21(2), 690–700.

Wilby, P. (2008) "In dangerous denial," *The Guardian*, June 30, 9.

Wynne, B. (2008) "Elephants in the rooms where publics encounter 'science'?" *Public Understanding of Science* 17, 21–33.

Yang, Z. J. and Kahlor, L. (2012) "What, me worry? The role of affect in information seeking and avoidance," *Science Communication* 35(2), 189–212.

Young, N. (2013) "Working the fringes: The role of letters to the editor in advancing non-standard media narratives about climate change," *Public Understanding of Science* 22(4), 443–459.

Zhu, Y. and Dukes, A. (2013) "The selective reporting of factual content by commercial media," University of Southern California working paper. Retrieved: https://server1.tepper.cmu.edu/seminars/docs/FactualContent_2013-04-18.pdf (accessed 2014).

19

CONTAINMENT AND REACH

The changing ecology of environmental communication

Libby Lester

In 2012, the Australian Government issued a white paper, *Australia in the Asian Century*. It mapped a route to national prosperity, capitalising on the growing middle classes of Asia and their increased access to consumer items, resources and destinations for holidays, work or study. The report acknowledged that more trade of raw materials, manufactured goods, ideas and people would create some challenges and risks, among them regional conflict as countries competed for limited essentials including water and minerals; increased pressure on Australia's resources and infrastructure; and environmental degradation that could hinder the country's capacity to meet demand. Media and communications played a significant role in this vision, largely as a tool of public diplomacy, promoting Australia's goals and shaping perceptions of 'individuals and groups in other countries' (2012: 264).

As the Government was releasing its white paper, Miranda Gibson had already surpassed the Australian record for tree-sitting and was about to clock up one year living on a platform 60 metres above ground in a remote forest in southern Tasmania. This was a landmark event, not because of the passage of time or the wild location. What made this action notable was the sophisticated use of new media technologies and associated practices that formed part of a comprehensive campaign to influence distant markets and publics, including in Europe and across Asia (Hutchins and Lester 2013). Gibson wrote a daily blog, detailing life in her 80-metre-tall eucalypt and the surrounding treetops; she ran cyber protests aimed at Japanese customers of Tasmanian timbers; she organised direct action protests that attracted significant news coverage; and she skyped into meetings with her local and international campaign colleagues, into interviews with journalists and students around the world, and into live television broadcasts. Gibson wanted to draw attention to the uncertainty surrounding the procurement of the controversial timber products. Were the products certified? Were they produced through sustainable practices? Was there social conflict at their source? Were the companies – the harvesters, manufacturers, the traders, the retailers – meeting their own corporate and social responsibility guidelines?

These questions were also at the forefront of 'roundtable' negotiations taking place two hours away in the capital city, Hobart, between leaders of the forest industry and the environment movement. Locked in a bitter conflict for decades, these two groups had come together secretly two years earlier to attempt to secure a sustainable future for the industry – struggling to maintain international markets and profits – and to agree on additional protection for hundreds of thousands of hectares of forests. The secrecy was based on the belief that media coverage would create or amplify conflict, and the talks therefore needed to proceed without media or public input until participants could achieve 'common ground' (Lester and Hutchins 2012a, 2012b). Mediated campaigns and protest aimed at overseas customers of the forest products were also perceived as a threat, and pressure was applied across the environment movement to put these on hold. Protest actions largely ceased; movement, industry and political websites were rarely updated; few media releases were issued; journalists were not briefed. The result was a new form of political invisibility, one that was sustained for several months, and one that kept journalists and the public distant from negotiations over an issue that had heavily impacted the island state's political, economic and cultural life for three decades.

Meanwhile, the Malaysian forestry company Ta Ann announced it was sacking 40 workers from its veneer mills in Tasmania. In the political and media storm that followed, it blamed 'persistent market attacks' by environmentalists for halving sales of its flooring products to Japan (Ta Ann Tasmania 2012). Industry lobbyists and Australian politicians described the campaign, which had drawn attention to anomalies in the 'eco' marketing of products, as 'terrorism', 'sabotage', 'economic vandalism' and 'blackmail'. Pressure mounted in the roundtable negotiations, and public diplomacy missions were launched; government representatives embarked on trade missions to Japan, China and Southeast Asia in an attempt to reassure customers and restore damaged markets (ABC Northern Tasmania 2012; Arndt 2012). Ta Ann Tasmania's executive director admitted, however, that these attempts to 'reassure the markets' were 'clearly not sufficient to change the opinion of the companies involved' (ABC Tasmania 2012).

And a final 'moment' from the same period: the newly elected conservative Australian government declared that environmental and consumer campaigns encouraging boycotts of products could be outlawed under the Consumer and Competition Act in order to protect Australia's international markets (Denholm 2013).

Conflict over landscapes, lifestyles and environments is a recurring feature of our time. The mediation of such conflict has long played an important role in public debate and democratic practice, and increasingly crosses cultural and geographic boundaries. It is now crucial to understand how changing media technologies, logics and practices impact on the communication of complex environmental issues and concerns, and in turn influence emerging forms of collective and transnational decision making. Evolving forms of exchange between claims-makers and decision makers, combined with major geopolitical, technological and social change, have produced a new environmental politics (Cottle and Lester 2011). It is a complex and multidimensional politics, creating networks of concern, strategic webs of influence and pressures on policy debates that have not been previously seen nor studied in detail. This is a politics that influences local, national and international environmental negotiations and outcomes, and has real consequences for communities, industries and landscapes.

The study of environmental media and communications has grown substantially in recent years, but often remains bounded by cultural, language and political borders, or – from methodological necessity – by specific media formats, platforms and isolated periods of publication or broadcast. Comparative research is increasingly providing broader contexts for understanding the production and reception of environmental coverage in different media and political systems (Lester and Cottle 2009) or across time (Lester 2007; Painter, 2013). The Australian 'moments', presented above, highlight why it is now important to identify and analyse environmental concerns as they are communicated across emerging online media and news networks and transnationally, as this is where new forms of politics are being produced (Cottle 2008; Beck 2009). These flows have complex and varied ramifications. Governments, industry and civil society can deploy media and communications in an effort to mobilise or dampen transnational intervention, and to build or obstruct international cooperation and governance structures (Berglez 2008; WBGU 2011; Ibold and Ireri 2012). Climate change negotiations in the last decade provide evidence that media representations and communication strategies are capable of both promoting and undermining local and international agendas (Cottle 2008; Painter 2011). Important questions related to power, transparency and voice (Sen 2009; Couldry 2010) are also raised when media is deployed as a form of public diplomacy and a tool of government and industry to promote short-term national and corporate interests and counter environmental campaigns carried by news and other less well-resourced media (Melissen 2005).

In thinking about the changing 'ecology' of environmental communication, it is helpful to consider the following three broad processes of change:

- *Emerging geopolitical pressures*: Australia, for example, has shifted from historically being one of the remotest countries in the world to being positioned within 10,000 km of a third of global economic output, rising to half by 2025 (Australian Government 2012: 6). Transport and communication costs have fallen. This might create a broad range of national opportunities, but it also places unprecedented pressures on natural landscapes and resources. Climate change, declining soil fertility, loss of species and ecosystems, biosecurity risks and stresses on water systems threaten long-term prosperity; economically, environmentally and socially (Hiscock 2012). It is also important to consider, as the Australian Government's white paper does, the implications for national security as individuals and non-state actors are empowered, and the capacity 'for groups in society to organise within countries and across national boundaries' grows (2012: 226).
- *Emerging media roles:* In the white paper vision, media are expected to help strengthen ties with the emerging and mature economies of Australia's neighbours. The capacity of various forms of media to destabilise these relationships is significantly underestimated. It is well established that media and communications have expanded the ability to communicate risk and of political messages to carry transnationally (Castells 2007); we now need to consider further how traditional media platforms and the internet can be strategically deployed across borders and are variously resourced to carry symbolic resonances and political impact in ways that will not always serve national or public interests (Thompson 2005).
- *Emerging political flows:* New technologies and media practices are converging with local and global environmental politics to produce sophisticated and powerful campaigns, alerting distant companies and consumers to the environmental provenance of the goods and materials they are being offered (Lester 2010; Lester and Cottle 2011). Distant supporters and activists join with those affected locally to resist development, end resource procurement and undermine growth strategies (della Porta and Tarrow 2005; Lester and Hutchins 2009). They demand a voice and forums for decision making about shared environmental futures (Fraser 2007).

In order to reveal features of this emerging communication ecology, we need to identify how conflict over environmental issues is enacted, carried, contained, understood and responded to within the context of national and transnational politics, trade and consumption. What is the relationship between news media and what Manual Castells calls the networks of 'mass self-communication' (2009)? What is the capacity of new forms of mediated public debate operating in increasingly networked and transnational contexts to lead to sound and just collective decision making? How might such debates lead to the development of policy and other governance frameworks to manage shared environmental futures on a local, national and international level? To understand the role played by evolving media and communications practices and technologies, we must also isolate the aids and obstacles to the transnational flow of mediated environmental communication, both via news and other means, and gauge how pressure groups, governments and industry seek to deploy various forms of media to influence environmental and related decision making across complex and diverse media and political systems. In short, we need to examine two oppositional yet symbiotic forms of mediated politics – the *politics of containment* and the *politics of reach*.

Politics of containment

Actors operating across the political spectrum must now be present in both 'legacy' media and Castells' networks of 'mass self-communication' (2009). Influence over public opinion and political agendas results from successfully finding connections between the two. Communicative 'bridges' are built when activists upload footage to YouTube, update blogs and vlogs, tweet links to journalists, and use Facebook to raise money for the purchase of advertising space in national newspapers and on commercial television. The same tools are available to governments and corporations that seek to engage citizens and consumers in ongoing 'conversations' in an effort to make effective decisions and secure favourable public relations (Davis 2009). Networked digital communications technologies both amplify the importance of these new forms of 'visibility' and render them more variegated and difficult to manage.

For John B. Thompson, visibility carries symbolic power, or an ability to control and/or affect political affairs (see especially 1995, 2000, 2005, 2011). While publicness can exist in physical spaces such as the town hall or city streets, its contemporary form is deeply interwoven with the practices, formats and structures of media. Thompson observes that achieving 'visibility through the media is to gain a kind of presence or recognition in the public space'. The obverse of this situation is thought to be the debilitating condition of invisibility, that confines 'one to obscurity – and, in the worst cases, can lead to a kind of death by neglect' (2005, p. 49). With the advent of the internet and other digital technologies, Thompson notices 'something new' in the conditions under which visibility is sought and delivered. Accepted boundaries between the private and public have been increasingly contested 'sites of struggle for information and symbolic content that threatens to escape the control of particular individuals' (2011, p. 68). For Thompson, the key to understanding this change is control. While he argues that political leaders have, since the advent of print, always struggled to control their visibility, it is now more difficult than ever (2005, p. 38). Control is also the key to conceptualising contemporary forms of privacy:

In its most fundamental sense, privacy has to do with the ability of individuals to exercise control over something. Usually this 'something' is understood as information: that is, privacy is the ability to control information about oneself, and to control how and to what extent that information is communicated to others (Thompson, 2011, p. 61).

Importantly in Thompson's thinking, it is once words and images enter the 'despatialized space' of contemporary media and communications networks that control becomes uncertain. Here, is 'a largely uncontrollable space' (2011, p. 61).

In relating these ideas to the changing ecology of environmental communication and the roles of news media and the internet, it is useful to identify the key sites of contest over visibility, publicness and control. The first and most easily identified 'struggle' over visibility and the containment of information occurs – as Thompson suggests – once words and images have entered media, where they produce meanings, promote particular interpretative frames and prompt counter responses. While structural inequalities are evident in many of the contests that take place at this point of mediation, the capacity of individuals and political actors to voice their concerns and issues rationally and strategically, or via accompanying symbolic and cultural resonances, can contribute to unexpected outcomes, even a 'fair fight' (Wolfsfeld 1997).

Within the realm of environmental politics and conflicts, where risks and concerns are often subsumed by the routines and structures of doing politics (Waisbord and Peruzzotti 2009), media communications prompted by events such as oil spills, nuclear accidents or public health concerns (Molotch and Lester 1975) are publicly negotiated via mediated response and counter-response. The debate is structured in part by the professional practices and organisational requirements of news media, including those that demand comment from a variety of sources within specific time and space constraints. When news media involvement produces perceived barriers for some political actors via its preference for 'elite' sources or enslavement to particular news values, digital media opportunities provide the means to circulate alternate and counter messages. Many political actors perceiving structural inequalities, and therefore a lack of control in mainstream news media, now seek to sustain mediated communications away from the news arena altogether. However, various research findings support Castells' claim that all political actors are now present in established commercial and alternative media spaces, and all are seeking to build bridges between the two (Castells 2009; Lester and Hutchins 2009; Hutchins and Lester 2011; Cottle and Lester 2011).

The second 'struggle' is positioned at the private–public interface, where messages are shaped and moulded in private, and then prepared and rehearsed for their journey into the public space. Understanding of this strategic process has improved markedly over recent decades as research has turned to source activity and the behind-the-scenes dynamics of media production (see, for example, Schlesinger 1990; Hansen 1993; Anderson 1997). Here, access to economic, organisational or cultural resources matter. These resources can include an understanding of newsroom practices, the capacity to produce news-ready events or messages, and physical support to update websites and social networking.

It is now normal for developers and resource extractive industries to deploy public and media relations in an attempt to control information as it travels through increasingly complex communication networks. However, it is protest by NGOs and other movement actors that is most readily associated with this site. Sea Shepherd and Greenpeace produce their own media-ready messages with symbolic plays on historic and cultural resonances and via for-the-camera events that bring the commitment of individual protesters to the fore (Hansen 2010). More poorly resourced groups organise tree sits or blockade gates to threatened landscapes, producing their own videos for YouTube and circulating media releases to newsrooms. The capacity for source control at this political site has increased as newsrooms cut resources, including staff available to cover geographically remote events, or alter newsroom practices such as removing once widespread restrictions on using activist-supplied footage (Lester 2010, 2011).

The third location for the 'struggle for visibility' could be described in terms of its opposite – invisibility, confinement. Keeping messages from entering media networks is recognised as a desirable option for industry and political professionals, and individuals, in certain circumstances. A number of ways to 'protect' confidentiality or reputation are routinely adopted, including legal threats and 'no-comment' responses to journalist queries. The commercial-in-confidence norm, Chatham House rule for negotiations, and 'behind closed doors' government decision making are accepted as legitimate and often essential steps for ensuring positive industrial and political outcomes within a democratic society (Ericson *et al.* 1989). Likewise, individuals and companies actively seek to protect reputations by gag orders, SLAPP lawsuits and careful public and media management (Davis 2003; Walters 2003). In environmental debate, industry and government seeking to avoid or silence opposition deploy these mechanisms.

Indeed, a defining role of the environmental movement to date has been to make visible or disclose to the 'affected' public these private, contained interactions and decisions, and in so doing to weaken industry and/or government control (Anderson 1997; Hansen 2010; Cox 2010). Here, the goals of environmental campaigners and news media converge. Disclosure of 'private' dealings that take place outside the norms of corporate or political behaviour allows news media to meet its self-promoted 'fourth estate' function and to work on behalf of the 'public interest'. Meanwhile, environmental movement power, built largely on its capacity to make the previously unseen visible, is built or sustained. Recent examples include Greenpeace Japan's exposure of whale meat sales, and the circulation by local environmental groups of visual evidence of healthy Tasmanian devil populations in forests earmarked for logging and mining. Yet, as the Australian 'moments' at the start of this chapter suggest, what happens when these political challengers who have deployed visibility as core practice and relied on public exposure for their political power retreat to private, contained spaces and corporate-like structures and rules for negotiation and publicity? How do news media respond? Such retreat not only leads us to question the relationship between media and movement, but complicates accepted understandings of political visibility, news media practices and the opportunities for 'voice' provided by digital technologies and fluid media-saturated contexts.

Politics of reach

Just as environmental politics is an important site to reveal the emerging tensions related to invisibility and the politics of containment, it is also useful in helping to understand the shifting dynamics of public debate, opinion and decision making in cosmopolitan, translocal and transnational contexts (Murphy 2011). Ulrich Beck's 'global risk society' and 'cosmopolitization' theses point to the essential role of media, publicity, staging and 'relations of definition' in exposing the way risks such as climate change and nuclear radioactivity work to form cosmopolitan communities that transcend the political boundaries of the traditional public sphere and 'imagined communities' (Beck 2011; see also 2006, 2009). These are communities dependent on the internet and 'worldwide communications and mobility processes, networks, forums of debate' for their existence (Beck 2011: 1353). Crucially for Beck, the 'reality' of risks 'can be dramatized or minimized, transformed or simply denied according to the norms that decide what is known and what is not' (2011: 1349).

> In this perspective, cosmopolitan events are highly mediatised, highly selective, highly variable, highly symbolic, local and global, national and international, material and communicative reflexive experiences and blows of fate. They transcend and efface all social boundaries and overturn the global order that holds sway in people's minds.
>
> (Beck 2011: 1349)

For a suggestion of how the mediated knowledge of risks creates cosmopolitan communities, and importantly how those communities might be recognised transnationally in determining environmental futures, it is worth revisiting Nancy Fraser's 2007 critique of Habermas's original public sphere thesis – and indeed of her own and others' earlier critiques that implicitly adopted Habermas's nation-focused viewpoint (see also Fraser 2014). In its widely understood form, a public sphere is both legitimate and effective when it a) provides an opportunity for all those affected to participate in public debate, b) provides a space for a diverse range of views to be put and importantly heard, and c) holds decision makers accountable through processes of publicity and the pressures of public opinion. In its Habermasian form, the 'public' is the citizenry of a nation state, equally contributing to opinion formation and being affected by decisions, who influence that state's laws and economic governance via debate carried out largely within the confines of a national and language-bounded media and political community. Here, a public sphere functions to check power and democratise governance. Thus, asks Fraser (2007: 8), what happens when these 'discursive spaces no longer correlate with sovereign states'? What happens when global institutions and agencies have greater control than states over economies, and economic decisions do not line up with the best interests of specific populations? What happens to national labour and environmental laws when markets 'tame' politics instead of politics taming markets, 'then how can citizens have any impact?' (Fraser 2007: 17).

Much public sphere theorising has also rested upon the presupposition of a single national language, and on the assumption that 'public opinion is conveyed through a national communications infrastructure, centered on print and broadcast' (Fraser 2007: 17):

> Given a field divided between corporate global media, restricted niche media and decentered Internet networks, how could critical public opinion possibly be generated on a large scale and mobilised as a political force? Given, too, the absence of even the sort of formal equality associated with common citizenship, how could those who comprise transnational media audiences deliberate together as peers?
>
> (Fraser 2007: 18)

In general then, Fraser asks, how can public opinion be considered *legitimate* or *efficacious* under current conditions when the 'who' of communication is a 'dispersed collection of interlocutors'; the 'what' of communication now stretches across a 'transnational community of risk'; the 'where' is 'decentralized cyberspace'; the 'how' encompasses a 'vast translinguistic nexus of disjoint and overlapping visual cultures'; and the addressee, once theorised as a sovereign state, is 'now an amorphous mix of public and private transnational powers that is neither easily identifiable nor rendered accountable' (Fraser 2007: 19).

As the 'moments' that began this chapter suggest, foregrounding the local as a dominant critical reality within environmental politics can help highlight emerging conditions of environmental communication in the context of transnational public debate and decision making. The local is where specific landscapes and resources provide minerals, fossil fuels, timber products or the locations for nuclear power plants, and communities and individuals carry the anxieties and lived realities of damaged environments. Localised threats and concerns coalesce symbolically into discourses of global risk, and discourses of global risk are synthesised for localised decision making. Staging and publicity are most successful when they represent both individual and shared risk (Lester and Cottle 2009), and see cosmopolitan communities of concern and opinion converge with those directly affected as they attempt to influence decision makers. In considering the interaction of the local within these transnational flows and cosmopolitan forces, a useful aspect is a 'translocal' one (Kraidy and Murphy 2008: 345), in which 'global communication processes can be understood by ethnographies of the local that nonetheless maintain the global as a counterpoint' (see also Hannerz 2003; della Porta and Piazza 2008).

But, of course, to 'get' at change, we also need to maintain a critical eye on the passage of time. What is actually new about this transnationalism, translocalism and cross-culturally enacted influence and empathy? As the history of civil unrest, social change and political movements shows, there is little revolutionary in the notion of 'speaking out' beyond a local and impacted community in order to provoke external interference. The US civil rights movement of the mid-twentieth century could only succeed by seeking the attention of the distant north. The fearful, disinterested or antagonistic news media of the south limited the movement's access to symbolic power at the physical site of its struggle: they needed to be bypassed in order to speak to the north. The movement was able to do so, as Jeffrey Alexander (2006) writes, by 'proving itself to be a master of the translating craft'. The main task for the movement was to relocate the particulars of the black movement in the south to the contexts and understandings of the north, that is, to translate to the distant. Highly symbolic, dramaturgical and emotional framing allowed protests to be recognised as legitimate by an American public removed from the site of conflict. Likewise, the fledgling contemporary environment movement in Australia could only win if it too successfully managed meanings as they travelled into distant locales. In Tasmania in the 1970s and 1980s, a conservative local news media and government needed to be side-stepped if the protest movement was to capture the attention of mainland voters and politicians, as well as international media and celebrities, who might support world heritage listing of the wilderness area and stop the damming of the Franklin River. Spectacular and emotional displays against a backdrop of wilderness offered symbolic touchstones for those distant from the river and the protest action. While the platforms, technologies and practices associated with speaking to the distant have evolved remarkably this century, we should not forget that communicative forms aimed at forcing political, social or environmental change have long been conditioned to speak to the distant.

Conclusion

This chapter has touched on key tensions and trends related to the changing ecology of environmental communication. It has suggested some ways to think about how and whether new media practices and technologies are impacting on environmental outcomes. How, it has asked, do these still emerging forms of media and communications change the way the environment and associated risks are made visible or contained? How are they transmitting and helping to enact new forms of transnational mediated politics? In focusing on two interrelated forces – the politics of containment and the politics of reach – it has raised questions about how campaigners, journalists, industry PR professionals and decision makers in various countries and contexts use news and new media to pressure, inform and act; and about the difficult task of identifying the environmental empathy, engagement and action that might result.

There is no easy way to reveal these changes, particularly when some are rendered deliberately invisible from the public (including media researchers) and others emerge locally to travel across national and language boundaries. Access to media and political actors in local settings is essential if we are to identify political communications flows in the context of vastly diverse political and media systems. The analysis of local media strategies and environmental outcomes in the context of varied political and economic systems, media literacy and technologies, and access to financial, human and symbolic resources needs also responds to Anders Hansen's call to reconnect not only the three major foci of communication research – production, content and impact – but to issues of power and inequality in the public sphere (Hansen 2011).

What we currently know is that a young woman up a tree with a laptop, mobile phone and internet connection can indeed send powerful messages a great distance. This might feel obvious to us now. The internet has meant that the practices and 'rules' surrounding secrecy, privacy, invisibility and containment have changed. In this new space of boundary-less flows and reach, we expect industry, government and self-proclaimed protectors of legitimate debate and information to have reduced opportunity to control and contain political and environmental communication. But we know that a more complicated story is unfolding. On the one hand, this story reflects the new complex and multidirectional flows of meanings, images and messages across the internet; on the other, it reflects the continued presence and viability of the same old restraints to the reach of information that have always existed when it comes to environmental messages and access to resources.

The internet may have expanded the capacity for messages to travel further. But – as the Australian 'moments' that began this chapter show – we are also witnessing attempts to control and contain those messages by deploying the well-worn practice of questioning the credibility and legitimacy of communications that are based on emotion, of what is 'polite' and 'civil'. The history of civil unrest tells us that emotionally and symbolically laden messages are the ones that travel best to distant 'northerners', engaging them and prompting them to act and intervene. And they are the ones that authorities and industry have always been most eager to contain. However, the other major change of recent decades – the recognition of global environmental risks, of being in this together – adds a new dimension to how these messages will carry or be contained, and we are only just beginning to witness the impact of this on environmental communications. Distant 'northerners' are no longer distant. They are, like us, affected by environmental harm, wherever it is occurring. This will clearly impact on how we speak to distant publics, and how they choose to speak back to us.

Acknowledgements

This chapter draws on research funded by the Australian Research Council 'Changing Landscapes: Online Media and Politics in an Age of Environmental Conflict' (DP1095173), and published in *Australian Journalism Review* (Lester and Hutchins 2012b) and *Media International Australia* (Lester 2014). Thank you to Janine Mikosza, Brett Hutchins and Simon Cottle for their assistance and participation in this research.

Further reading

Alexander, J. C. (2006) *The Civil Sphere*. Oxford: Oxford University Press.
How can political engagement and empathy be enacted from afar to produce social change?

Lester, L. and B. Hutchins (eds) (2013) *Environmental Conflict and the Media*. New York: Peter Lang.
Authors ask how conflicts emanate from and flow across multiple sites, regions and media platforms and examine the role of the media in helping to structure collective discussion, debate and decision making.

Nash, K. (2014) *Transnationalizing the Public Sphere*. Cambridge: Polity.
Fraser's attempt to reinvigorate public sphere theory by decoupling it from its national bounds is critically debated by authors from a range of disciplines.

Thompson, J. B. (2005) 'The new visibility'. *Theory, Culture and Society*, 22(6): 31–51.
What has changed about publicness and visibility with the advent of the Internet?

References

ABC Northern Tasmania (2012) 'Selling Tasmania to South East Asia', 20 February, accessed 16 November 2012 from www.abc.net.au/local/audio/2012/02/20/3435007.htm

ABC Tasmania (2012) 'Mornings', 16 November, accessed 16 November from http://blogs.abc.net.au/tasmania/2012/11/mornings-on-demand-131112.html?site=hobart&program=hobart_mornings

Alexander, J. C. (2006) *The Civil Sphere*. Oxford: Oxford University Press.

Anderson, A. (1997) *Media, Culture and the Environment*. London: Routledge.

Arndt, D. (2012) 'Trade mission could be a trip of mixed messages'. *The Examiner*, 20 February. Accessed 16 November 2012 from www.examiner.com.au/story/430835/trade-mission-could-be-a-trip-of-mixed-messages/.

Australian Government (2012) *Australia in the Asian Century*. Accessed 16 November 2012 from http://asiancentury.dpmc.gov.au/white-paper

Beck, U. (2006) *The Cosmopolitan Vision*. Cambridge: Polity Press.

Beck, U. (2009) *World at Risk*. Cambridge: Polity Press.

Beck, U. (2011) 'Cosmopolitanism as imagined communities of global risk'. *American Behavioral Scientist*, 55(10): 1346–1361.

Berglez, P. (2008) 'What is global journalism?' *Journalism Studies*, 9: 845–858.

Castells, M. (2007) 'Communication, power and counter-power in the network society'. *International Journal of Communication*, 1: 238–266.

Castells, M. (2009). *Communication Power*. Oxford: Oxford University Press.

Cottle, S. (2008). 'Reporting demonstrations: The changing media politics of dissent'. *Media, Culture and Society*, 30(6): 853–872.

Cottle, S. and L. Lester (eds) (2011) *Transnational Protests and the Media*. New York: Peter Lang.

Couldry N. (2010) *Why Voice Matters: Culture and Politics after Neoliberalism*. London: Sage.

Cox, R. (2010) *Environmental Communication and the Public Sphere*. Thousand Oaks, CA: Sage.

Davis, A. (2003) 'Public relations and news sources', pp. 27–42 in S. Cottle (ed.), *News, Public Relations and Power*. London: Sage.

Davis, A. (2009) 'Journalist–source relations, mediated reflexivity and the politics of politics'. *Journalism Studies*, 10: 204–219.

della Porta, D. and G. Piazza (2008) *Voices of the Valley, Voices of the Straits: How Protest Creates Communities*. New York: Berghahn Books.

della Porta, D. and S. Tarrow (eds) (2005) *Transnational Protest and Global Activism*. Lanham, MD: Rowman & Littlefield.

Denholm, M. (2013) 'Companies to get protection from activists' boycotts', *The Australian*, 23 September. Accessed 10 October 2013 from www.theaustralian.com.au/national-affairs/companies-to-get-protection-from-activists-boycotts/story-fn59niix-1226724817535

Ericson, R. V., P. M. Baranek and J. B. L. Chan (1989) *Negotiating Control: A Study of News Sources*. Milton Keynes: Open University Press.

Fraser, N. (2007) 'Transnationalizing the public sphere: On the legitimacy and efficacy of public opinion in a post-Westphalian world'. *Theory, Culture and Society*, 24(4): 7–30.

Fraser, N. (2014) 'Transnationalizing the public sphere: On the legitimacy and efficacy of public opinion in a post-Westphalian world', pp. 8–42 in K. Nash (ed.) *Transnationalizing the Public Sphere*. Cambridge: Polity.

Hannerz, U. (2003) 'Being there … and there … and there!: Reflections on multi-site ethnography'. *Ethnography*, 4(2): 201–216.

Hansen, A. (1993) 'Greenpeace and press coverage of environmental issues', pp. 150–178 in A. Hansen (ed.) *The Mass Media and Environmental Issues*. Leicester: Leicester University Press.

Hansen, A. (2010) *Environment, Media and Communication*. Abingdon, Routledge.

Hansen, A. (2011) 'Communication, media and environment: Towards reconnecting research on the production, content and social implications of environmental communication'. *International Communication Gazette*, 73(1–2): 7–25.

Hiscock, G. (2012) *Earth Wars: The Battle for Global Resources*. Singapore: John Wiley & Sons.

Hutchins, B. and L. Lester (2011) 'Politics, power and online protest in an age of environmental conflict', pp. 159–171 in S. Cottle and L. Lester (eds) *Transnational Protests and the Media*. New York: Peter Lang.

Hutchins, B. and L. Lester (2013) 'Tree-sitting in the network society', pp. 1–17 in L. Lester and B. Hutchins (eds) *Environmental Conflict and the Media*. New York: Peter Lang.

Ibold, H. and K. Ireri (2012) 'The chimera of international community: News narratives of global cooperation'. *International Journal of Communication*, 6: 2337–2358.

Kraidy, M. M. and P. D. Murphy (2008) 'Shifting Geertz: Toward a theory of translocalism in global communication studies'. *Communication Theory*, 18: 335–355.

Lester, L. (2007) *Giving Ground: Media and Environmental Conflict in Tasmania*. Hobart: Quintus Publishing.

Lester, L. (2010) *Media and Environment: Conflict, Politics and the News*. Cambridge: Polity.

Lester, L. (2011) 'Species of the month: Whaling, mediated visibility and the news'. *Environmental Communication: A Journal of Nature and Culture*, 5 (1): 124–139.

Lester, L. (2014) 'Transnational publics and environmental conflict in the Asian century'. *Media International Australia*, 150: 67–78.

Lester, L. and S. Cottle (2009) 'Visualizing climate change: Television news and ecological citizenship'. *International Journal of Communication*, 3: 17–26.

Lester, L. and S. Cottle (2011) 'Transnational protests and the media: Toward global civil society', pp. 287–292 in S. Cottle and L. Lester (eds) *Transnational Protests and the Media*. New York: Peter Lang,

Lester, L. and B. Hutchins (2009) 'Power games: Environmental protest, news media and the Internet'. *Media, Culture and Society*, 31(4): 579–596.

Lester, L. and B. Hutchins (2012a) 'The power of the unseen: Environmental conflict, the media and invisibility'. *Media, Culture and Society*, 34 (7): 832–846.

Lester, L. and B. Hutchins (2012b) 'Journalism, the environment and the new media politics of invisibility'. *Australian Journalism Review*, 34 (2): 19–32.

Lester, L. and B. Hutchins (eds) (2013) *Environmental Conflict and the Media*. New York: Peter Lang.

Melissen, J. (ed.) (2005) *The New Public Diplomacy: Soft Power in International Relations*. Basingstoke: Palgrave Macmillan.

Molotch, H. and Lester, M. (1975). 'Accidental news: The great oil spill as local occurrence and national event'. *American Journal of Sociology*, 81(2): 235–260.

Murphy, P. D. (2011) 'Putting the Earth into global media studies'. *Communication Theory*, 21: 217–238.

Painter, J. (2011) *Poles Apart: The International Reporting of Climate Skepticism*. Oxford: Reuters Institute for the Study of Journalism.

Painter, J. (2013) *Climate Change in the Media: Reporting Risk and Uncertainty*. Oxford: I.B. Taurus with Reuters Institute for the Study of Journalism.

Schlesinger, P. (1990) 'Rethinking the sociology of journalism: Source strategies and the limits of media-centrism', pp. 61–83 in M. Ferguson (ed.) *Public Communication: The New Imperatives*. London: Sage.

Sen, A. (2009) *The Idea of Justice*. London: Penguin Books.

Ta Ann Tasmania Pty Ltd (2012) 'Media release: Green market campaign costs jobs', accessed 29 August 2012 from www.taanntas.com.au/userfiles/Documents/20120213%20job%20losses%20due%20to%20green%20campaign.pdf

Thompson, J. B. (1995) *The Media and Modernity: A Social Theory of the Media*. Cambridge: Polity Press.

Thompson, J. B. (2000) *Political Scandal: Power and Visibility in the Media Age*. Cambridge: Polity.

Thompson, J. B. (2005) 'The new visibility'. *Theory, Culture and Society*, 22(6): 31–51.

Thompson, J. B. (2011) 'Shifting boundaries of public and private life'. *Theory, Culture and Society*, 28(4): 49–70.

Waisbord, S. and E. Peruzzotti (2009) 'The environmental story that wasn't: Advocacy, journalism and the asambleismo movement in Argentina'. *Media, Culture and Society*, 31(5): 691–709.

Walters, B. (2003). *Slapping on the Writs: Defamation, Developers and Community Activism*. Sydney: UNSW Press.

WBGU German Advisory Council on Global Change (2011) A Social Contract for Sustainability Factsheet No1/2011. Accessed 13 January 2013 from www.wbgu.de/fileadmin/templates/dateien/veroeffentlichungen/factsheets/fs2011-fs1/wbgu_fs1_2011_en.pdf

Wolfsfeld, G. (1997). *Media and Political Conflict: News from the Middle East*. Cambridge: Cambridge University Press.

20

REPRESENTATIONS OF THE ENVIRONMENT ON TELEVISION, AND THEIR EFFECTS

James Shanahan, Katherine McComas, and Mary Beth Deline

In 2013, the American public ranked climate change at the very bottom of 21 policy issues that they believed the President and Congress should deal with; another poll found that 2013 represented the lowest year for environmental concern since polling on the topic had begun over 20 years ago (Globescan, 2013; Pew Research Center, 2013). Television is an important influence on the development of attitude, and given the context of apparent indifference toward environmental issues, it is prudent to revisit research into portrayals of the environment on American commercial/entertainment television. Does what television says about the environment have anything to do with what we think about its problems and solutions?

This article reviews a research program that began in the 1990s (see McComas, Shanahan and Butler, 2001; Shanahan, 1996; Shanahan and McComas, 1997) that has been revisited only sporadically. The research was conceptualized in the late 1980s and reached a first period of fruition in the mid 1990s, as an extension of the Cultural Indicators project, which itself began in the 1960s. The Cultural Indicators project gathers data on how television represents the world, and uses "cultivation theory" to explore television's impacts on public perceptions (Morgan, Shanahan and Signorielli, 2009). Cultivation theory essentially states that heavier viewers of television will be more likely to hold conceptions of the world that are consistent with TV portrayals than lighter viewers (Besley and Shanahan, 2004; Good, 2009; Morgan and Shanahan, 2010). In revisiting and revising the work on environmental issues from a Cultural Indicators perspective, we first review the original findings and then discuss some new findings about cultivation theory and media attention in relation to the environment.

Cultural indicators research and the environment

Narratives are stories that embody a timed sequence of events (Jones and McBeth, 2010); the stories we are exposed to, over time, combine to create a cultural gestalt that both guides and reflects our norms, roles, and customs (Howard-Williams, 2011; Morgan and Shanahan, 2010). The idea that a message "system" (one based on stories) plays an important role in a culture—especially one that incorporates mass communication—was at the heart of the original Cultural Indicators research (Gerbner and Gross, 1976). The message system is what is consistent and repetitive within stories, and this consistency is what is important to the culture; television has normally been seen as the institution contributing the most to that system.

In the environmental sphere, the idea is that television story and message systems contribute to the Dominant Social Paradigm (DSP): "a system of shared beliefs 'upheld by the constant repetition of ideas that fit within it'" (Howard-Williams, 2011, p. 28, quoting Meadows, 1991, p. 74). The DSP is the system of norms and ideas that privileges material and economic growth over environmental sustainability. Television's focus on this growth was seen (perhaps somewhat simplistically, given what we now know) as maintaining and reproducing an anti-environmental DSP. Given that watching television is still America's preferred leisure activity, that most of the television consumed is entertainment oriented (Mutz and Nir, 2010), and that the amount of overall viewing that Americans engage in tends to increase (Morgan and Shanahan, 2010), continuing to examine TV entertainment programming's stories and cultural content related to the environment appears to be a worthwhile undertaking.

Findings concerning the overall amount of exposure to such cultural products and the effects of such exposure on policy interpretations are informed by cultivation theory. As noted above, the theory looks at how exposure contributes to cultural stability through shaping and maintaining the worldviews of viewers (Gerbner and Gross, 1976; Morgan and Shanahan, 2010). The theory was among the first to explicitly examine how media narratives affect the political environment, and it continues to be highly cited in studies of mass communication (Morgan and Shanahan, 2010; Mutz and Nir, 2010). Operationally, Cultural Indicators research consists of three types of analysis—"institutional process analysis" that examines message construction in light of media organizations' opportunities and constraints; message system investigations that examine cultural patterns in content; and cultivation analysis, which looks at the relationship between these institutional and message portrayals and the public's attitudes (Morgan and Shanahan, 2010).

Our research into portrayals of the environment was originally published in 1996 using data collected from 1991 and 1993; the last year that content was culled for analysis was 1997. The research was discontinued at that point (McComas et al., 2001; Shanahan, 1996). Since that time, remarkably little research has systematically documented how the environment is depicted on TV, even though there are quite a few studies that examine journalistic portrayals or individual types of programs. This begs the question: on TV, what stayed the same over that time period, or has anything changed?

Our early findings showed that environmental and natural images appeared rarely on prime-time TV (Shanahan, 1996). This finding was replicated again in subsequent analyses of television content from 1991–1995 and 1991–1997, where nature occurred as a "predominant theme" in only about 2 percent of programs in both time series (McComas et al., 2001; Shanahan and McComas, 1997). Both analyses also found separation between the "human" and "natural" domains on TV, with the majority of programs focusing on relationship or law/crime themes (McComas et al., 2001; Shanahan and McComas, 1997). Turning from nature themes to environmental "episodes," (actual instances of environmental issues being dealt with, within TV programs), we found that such episodes were largely absent from our sample (McComas et al., 2001; Shanahan and McComas, 1997). Findings from the last analysis of 1991–1997 (McComas et al., 2001) in particular found that environmental episodes most often referenced "general" issues or issues related to waste disposal, that the majority of the few environmental representations we found were "neutral" or "concerned" expressions by characters, and that whites were involved in the majority of the episodes where concerns were expressed (McComas et al., 2001).

Other content analyses of the environment on TV

There is only one other content analysis focusing on this area, to our knowledge. Howard-Williams (2011) examined both news and non-news programming in New Zealand for four television channels. This analysis found that nature as a theme was completely missing from over 60 percent of non-news programs, and that it occurred as a primary theme in less than 10 percent of the non-news programs; these are somewhat higher figures than our American studies. In addition, nature and scientific values themes tended to occur together, while (and this is similar to the American findings) natural and relationships/ sex themes were usually separate. In other words, television seems to conflate environmental and science themes, and separates them from dominant programming themes such as relationships, crime, and sex. The most frequently seen type of environmental episode was the same as in our American findings: "general nature." This was followed by (differently from our findings) "sustainability" and "green living." Environmental episodes were often situated to promote the purchase of "green" goods and services (Howard-Williams, 2011). It seems that New Zealand media offer a few more representations of the environment in their TV programs, but the structure of the themes is roughly congruent with American television.

Comparing 2012 to the 1990s

To update our findings from the 1990s, and given the paucity of other studies that have collected these sorts of data, we decided to update our own data collection with a recent sample. In 2012 we collected data on 78 programs. These were the same programs that were coded for the ongoing Cultural Indicators research project (see, for example, Morgan, Shanahan and Signorielli, 2012); they are prime-time programs from major broadcast networks (ABC, CBS, NBC, Fox, and the CW). The sample is structurally similar to the ones we analyzed in the 1990s, with the addition of programs from Fox and the CW (see Shanahan and McComas, 1999, for details).

Table 20.1 shows the frequency of themes that appear in this 2012 sample, comparing the frequencies to those we observed in the 1990s. As can be seen in the table, attention to environmental issues overall has remained at a similar level from the 1990s to the current sample. That is, most programs deal with environmental issues not at all; the remaining deal with the environment in minor or secondary ways. This is a confirmation and replication of what we saw in the 1990s. The total number of "episodes" (identifiable substories, plots, themes, or events within programs) that dealt with environmental content was 18. Thus, across 71 hours of programming, there were 18 identifiable episodes dealing with environmental issues. Most of these episodes lasted less than a minute. By our computation, about 28.5 minutes out of these 71 hours of programming dealt with environmental themes; this is somewhat similar to the estimate we gave for programs in the 1990s.

Most of the other attributes of the environmental themes are similar to what we observed in the 1990s. Environmental themes are often kept separate from the main themes of family, relationships, crime, and sex that dominate entertainment TV. While this may surprise few who are inveterate television watchers, it's worthwhile to note that the world of commercial television is much as Bill McKibben described it in the early 1990s: "TV, and the culture it anchors, masks and drowns out the subtle and vital information contact the real world once provided" (1992, pp. 22–23).

Table 20.1 Percentages of programs focusing on selected themes in the 1990s and in 2012

	"Nature" 90s	"Nature" 2012	"Personal relationships" 90s	"Personal relationships" 2012	"Family" 90s	"Family" 2012
Theme is absent	79.2	84.6	37.8	42.3	25.7	41
Theme is minor	13.1	3.8	18.8	10.3	19.8	5.1
Theme is secondary	5.7	6.4	21.8	11.5	22.4	7.7
Theme is primary	2	5.1	21.6	35.9	32.2	46.1

Source: authors.

Note: N of programs in the 90s = 510. N of programs in 2012 = 78.

In the absence of other datasets on this topic, and limiting our analysis here due to space constraints, the landscape of American network entertainment TV is still similar to the 1990s with respect to environmental issues. Of course, since that time other networks have appeared that take on environmental themes with more frequency, especially the science- and nature-oriented cable networks such as NatGeo, Treehugger, or Animal Planet. But even these networks are subject to the inexorable logic of narrative competition that drives them toward human interest and reality programming as much as any other network. Even though the environment has become an issue that won't go away, it has not pervaded the culture enough to infect the day-to-day entertainment programming at more than a minor level.

Cultivation of beliefs about the environment

As part of our early research on TV and the environment, we also looked at the cultivation of environmental attitudes by television viewing. These results, along with the content findings, were summarized in various publications (summarized in Shanahan and McComas, 1999). Overall, our findings were that heavy television viewers were less likely to show concern for the environment. These findings represented the first real attempts that were made to look at relationships between exposure to the message system and environmental attitudes, and they embodied a somewhat "orthodox" approach to the way that cultivation research is usually conducted. Our conclusion then was that the absence of the environment on television represented a "symbolic annihilation" that would lead to lower concern among heavy viewers.

We know now that the picture is somewhat more complex. After our original studies on cultivation of environmental concern, other investigators both broadened and sharpened the research. One study found that environmental attitudes guided television genre choices, predicting the use of television nature documentaries and news, but not fictional television (Holbert, Kwak and Shah, 2003). That is, attitudes form viewing preferences, as opposed to the traditional conceptualization where viewing "affects" attitude. A second study (Dahlstrom and Scheufele, 2010) found that overall TV exposure was related to concern about environmental risks, but when a measure of exposure *diversity* was introduced (which examines how many different types of content the person watches) it reduced the TV exposure-attitude relationship to non-significance. Such a finding could be important in an atmosphere with many new channels to appeal to those of an environmental bent.

Good (2007, 2009, 2013) shows that TV exposure is associated with "materialism" and that heavier viewers of television (including environmentalists) become less concerned about the environment, by virtue of this materialism. Her point is not that TV makes people less environmentalist "directly," it is their materialism that brings this about. She notes that three rationales are typically given for an environmental cultivation effect:

1. more television viewing means less time experiencing the outdoor environment;
2. "symbolic annihilation," as discussed above; and
3. television's focus on materialism, which she argues is necessarily a counter-value to environmental concern.

Good's findings are consistent with other empirical work that has shown the TV–materialism link (Shrum, Lee, Burroughs and Rindfleisch, 2011; Shrum and Lee, 2012).

Mainstreaming refers to similarities among heavy viewers of different socioeconomic groups who otherwise differ when they are light viewers (Cox, 2012; Good, 2009). Often the groups compared are ideological. For instance, when liberals and conservatives differ on an issue, it is often the case that *heavy viewing* liberals and conservatives are closer in outlook, and usually this closeness bends more toward the conservative viewpoint. This attitudinal closeness was termed "mainstreaming" by Gerbner and colleagues (1980) to reflect the convergence in outlooks that seems so common among heavy viewers. In our earlier analyses, we saw this phenomenon demonstrated in the 1993 General Social Survey (GSS) sample, where we looked at mainstreaming for the issue of willingness to make sacrifices for the environment (Shanahan and McComas, 1999, p. 138). This pattern still shows up in more recent GSS data. Figure 20.1 shows a characteristic mainstreaming pattern, in which liberal, moderate, and conservative groups converge toward greater unwillingness to sacrifice for the environment as their viewing increases.

Recent research might shed light on the mechanism behind this effect. It suggests that mainstreaming depends not on changing viewers' values but suppressing them and providing another framework in which to consider the values (Slater, Rouner and Long, 2006). In other words, if television narratives do not support viewer's values, then these values are likely to be suppressed; the more television one consumes perhaps the more cumulative the suppression is. This might especially be the case with fictional television that does not operate from the norm of journalistic balance, which requires coverage of multiple agendas (Mutz and Nir, p. 2010). Instead, a fictional narrative can promote a singular set of values.

Latter years of the GSS do not include as many of the original measures we used to assess cultivation effects, and there are relatively few other datasets that have been built to this purpose. Still, other investigators have added incrementally to our understanding of this issue. As Dahlstrom (2012) details, three main theories—"the transportation–imagery model (Green and Brock, 2000), extended elaboration likelihood model (Slater and Rouner, 2002), and entertainment overcoming resistance model (Moyer-Gusé, 2008),"—argue in part that empathy with characters reduces resistance to persuasion effects (p. 304). In particular, a connection to characters' circumstances and plights affects persuasion responses (Moyer-Gusé and Nabi, 2010). In addition, cohesion between a narrative worldview and one's own makes it more likely that audiences will receive and accept new messages (Jones and McBeth, 2010). Given that cultivation theory is essentially the study of mass narrative forms (Morgan and Shanahan, 2010), a focus on new findings from narrative research might lead to new insights into cultivation's mechanisms of effects.

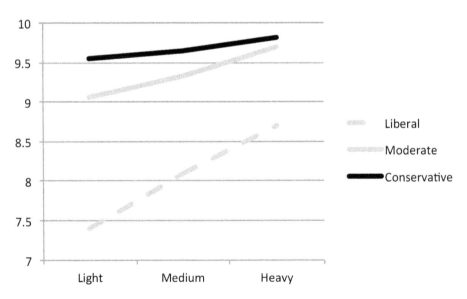

Figure 20.1 Mainstreaming of attitudes toward environmental sacrifice, 2010, in political subgroups (high scale values signify greater unwillingness to sacrifice)

Source: authors.

Media attention and the environment, other theories

Cultivation is not, of course, the only theory that looks at environmental media effects. Since the 1990s research in mass media and the environment has largely focused on print, in addition to television news coverage, with quite a bit of research focused on climate change (Boykoff, 2009). Examining media attention to environmental issues using issue-cycle and agenda-setting analyses has produced a number of relevant findings. Agenda setting refers to the role that the media take in defining topical issues for discussion among the public, while issue cycles refer to the up-and-down nature of attention paid to environmental issues (Djerf-Pierre, 2012; Downs, 1972; McComas and Shanahan, 1999; McCombs 2005).

Turning first to agenda setting, recent findings on this front show that the media have strong agenda-setting impacts on environmental issues such as climate change (Brulle, Carmichael and Jenkins, 2012). These studies focus on agenda setting via active information seeking through news consumption. New research shows how types of media are used for different purposes—active information seekers are prompted to find information while information scanning occurs by gleaning new information from habitually used media sources (Niederdeppe et al., 2007). In this case, those who undertake newspaper or TV news consumption might be motivated information seekers, while everyday monitoring of the media environment might result in most environmental information being obtained through television programming. Such a claim is exemplified in findings that the television and Internet, not newspapers, provided African American students with the majority of their environmental information (Lee, 2008).

Turning from agenda setting to issue cycles, Hansen (2011) argues that longitudinal studies are better at capturing how different meaning is made of issues over time. Other such work in the environmental arena has occurred. For example Djerf-Pierre (2012) turns to environmental news reporting in Sweden over a fifty-year period and identified four themes upon which environmental issues are grounded. These represent environmental "catastrophes, scandals, alarms and controversies" (2012, p. 505). Catastrophes are sudden events that trigger alarm, scandals are policy issues based on moral breaches, alarms represent scientific findings, and controversies represent the movement of an issue from the policy to the political arena (Djerf-Pierre, 2012). The dramatic nature of these frames seems to correlate with assertions that American news media's coverage of environmental issues fixates on spectacular rather than everyday representations of the environment (Weber and Stern 2011). Some of these frames also bear similarity to those found in other work specifically relating to climate change representations in newspapers, such as new scientific findings (alarms) or controversies among scientists (McComas and Shanahan, 1999).

★★★

Summarizing where we are with views on how television cultivates environmental concern, we see that the situation has not changed much since the 1990s. Though some environmental concepts have become buzzwords through the frequency of media mention (e.g., "climate"), there is not enough different about the institution of television and the ways that people watch it that would make it, overall, a net contributor to environmental concern. Television still seems to marginalize the environment as an issue, and its most avid viewers are concerned with other things. Though there is much hope that media attention and even hype can catalyze environmental concern, it is also useful to keep in mind that such hype could have exactly the opposite effect, possibly along the lines of the "narcotizing dysfunction" that is one of the oldest chestnuts of media effects theory (Lazarsfeld and Merton, 2000).

Further reading

Good, J. (2013). *Television and the Earth: Not a love story*. Black Point, CA: Fernwood Publishing.
A recent extension and update of some of the ideas from Shanahan and McComas (1997). The main finding is that television viewing is associated with higher levels of materialism, which mediates a relationship with lower levels of environmental concern.

McKibben, B. (1992). *The Age of Missing Information*. New York: Random House.
McKibben "analyzed" a whole day's worth of television content to see what it had to say about the environment, comparing it to a day spent in the woods. His conclusion is that television, for all of the "information" that it shows, misses a lot of the experience of being in nature.

Shanahan, J. and McComas, K. (1997). *Nature Stories: Representations of the environment and their effects*. Cresskill, NJ: Hampton Press.
An analysis of television viewing's relationship to the environment. Concludes that more viewing goes with less concern, and that the environment per se is not really seen that much on TV.

Shanahan, J. and Morgan, M. (1999). *Television and Its Viewers*. Cambridge: Cambridge University Press.
A review of cultivation analysis and cultural indicators research, the basis of Shanahan and McComas' original research and the topic of this chapter.

References

Besley, J. C. and Shanahan, J. (2004). Skepticism about media effects concerning the environment: Examining Lomborg's hypotheses. *Society and Natural Resources, 17*(10), 861–880.

Boykoff, M. T. (2009). We speak for the trees: Media reporting on the environment. *Annual Review of Environment and Resources, 34*(1), 431–457.

Brulle, R. J., Carmichael, J., and Jenkins, J. C. (2012). Shifting public opinion on climate change: An empirical assessment of factors influencing concern over climate change in the U.S., 2002–2010. *Climatic Change, 114*(2), 169–188.

Cox, R. (2012). *Environmental Communication and the Public Sphere*. Thousand Oaks, CA: Sage.

Dahlstrom, M. F. (2012). The persuasive influence of narrative causality: Psychological mechanism, strength in overcoming resistance, and persistence over time. *Media Psychology, 15*(3), 303–326.

Dahlstrom, M. F. and Scheufele, D. (2010). Diversity of television exposure and its association with the cultivation of concern for environmental risks. *Environmental Communication: A Journal of Nature and Culture, 4*(1), 54–65.

Djerf-Pierre, M. (2012). Green metacycles of attention: Reassessing the attention cycles of environmental news reporting 1961–2010. *Public Understanding of Science, 22*(4), 495–512.

Downs, A. (1972). Up and down with ecology: The "issue-attention cycle." *The Public Interest, 28*, 38–50.

Gerbner, G. and Gross, L. (1976). Living with television: The violence profile. *The Journal of Communication, 26*(2), 173–99.

Gerbner, G., Gross, L., Morgan, M., and Signorelli, N. (1980). The mainstreaming of America: Violence profile number 11. *Journal of Communication, 30*(3), 10–29.

Globescan. (2013). *Environmental Concern at Record Lows*. Obtained July 1, 2013 from www.globescan.com/commentary-and-analysis/press-releases/press-releases-2013/261-environmental-concerns-at-record-lows-global-poll.html

Good, J. (2007). Shop 'til we drop? Television, materialism and attitudes about the natural environment. *Mass Communication and Society, 10*(3), 365–383.

Good, J. (2009). The cultivation, mainstreaming, and cognitive processing of environmentalists watching television. *Environmental Communication: A Journal of Nature and Culture, 3*(3), 279–297.

Good, J. (2013). *Television and the Earth: Not a love story*. Black Point, NS: Fernwood Publishing.

Green, M. C., and Brock, T. C. (2000). The role of transportation in the persuasiveness of public narratives. *Journal of Personality and Social Psychology, 79*, 701–721.

Hansen, A. (2011). Communication, media and environment: Towards reconnecting research on the production, content and social implications of environmental communication. *International Communication Gazette, 73*(1–2), 7–25.

Holbert, R., Kwak, N. and Shah, D. (2003). Environmental concern, patterns of television viewing and pro-environmental behaviors: Integrating models of media consumption and effects. *Journal of Broadcasting and Electronic Media, 47*(2), 177–196.

Howard-Williams, R. (2011). Consumers, crazies and killer whales: The environment on New Zealand television. *International Communication Gazette, 73*(1–2), 27–43.

Jones, M. D. and McBeth, M. K. (2010). A narrative policy framework: Clear enough to be wrong? *Policy Studies Journal, 38*(2), 329–354.

Lazarsfeld, P. F. and Merton, R. K. (2000). Mass communication, popular taste and organized social action. In Paul Marris and Sue Thornham (eds), *Media Studies: A reader,* (pp. 18–30). New York: New York University Press.

Lee, E. B. (2008). Environmental attitudes and information sources among African American college students. *The Journal of Environmental Education, 40*(1), 29–43.

McComas, K. and Shanahan, J. (1999). Telling stories about global climate change: Measuring the impact of narratives on issue cycles. *Communication Research, 26*(1), 30–57.

McComas, K. A., Shanahan, J., and Butler, S. (2001). Environmental content in prime-time network TV's non-news entertainment and fictional programs. *Society and Natural Resources,* 37–41.

McCombs, M. (2005). A look at agenda-setting: Past, present and future. *Journalism Studies, 6*(4), 543–557.

McKibben, B. (1992). *The Age of Missing Information.* New York: Random House.

Meadows, D. (1991). Changing the world through the informationsphere. In: C. LaMay and E. Dennis (eds) *Mass Media and the Environment,* (pp. 67–79). Washington, DC: Island Press.

Morgan, M. and Shanahan, J. (2010). The state of cultivation. *Journal of Broadcasting and Electronic Media, 54*(2), 337–355.

Morgan, M., Shanahan, J., and Signorielli, N. (2009). Growing up with television: Cultivation processes. In J. Bryant and M. B. Oliver (eds), *Media Effects: Advances in theory and research* (pp. 34–49). New York: Routledge.

Morgan, M., Shanahan, J., and Signorielli, N. (eds) (2012). *Living with Television Now: Advances in Cultivation Theory and Research.* New York: Peter Lang.

Moyer-Gusé, E. (2008). Toward a theory of entertainment persuasion: Explaining the persuasive effects of entertainment-education messages. *Communication Theory,* 18, 407–425.

Moyer-Gusé, E. and Nabi, R. L. (2010). Explaining the effects of narrative in an entertainment television program: Overcoming resistance to persuasion. *Human Communication Research, 36*(1), 26–52.

Mutz, D. C. and Nir, L. (2010). Not necessarily the news: Does fictional television influence real-world policy preferences? *Mass Communication and Society, 13*(2), 196–217.

Niederdeppe, J., Hornik, R. C., Kelly, B. J., Frosch, D. L., Romantan, A., Stevens, R. S., Barg, F. K., et al. (2007). Examining the dimensions of cancer-related information seeking and scanning behavior. *Health Communication, 22*(2), 153–167.

Pew Research Center. (2013). *Climate Change: Key Data Points.* Retrieved July 1, 2013 from www.pewresearch.org/key-data-points/climate-change-key-data-points-from-pew-research/

Shanahan, J. (1996). Green but unseen: Marginalizing the environment on television. In M. Morgan and S. Leggett (eds) *Margin(s) and Mainstreams: Cultural Politics in the 90s* (pp. 176–193). Westport CT: Greenwood.

Shanahan, J. and McComas, K. (1997). Television's portrayal of the environment: 1991–1995. *Journalism and Mass Communication Quarterly, 74*(1), 147–159.

Shanahan, J. and McComas, K. (1999). *Nature Stories: Depictions of the Environment and their Effects.* Cresskill, NJ: Hampton Press.

Shrum, L. J. and Lee, J. (2012). The stories TV tells: How fictional TV narratives shape normative perceptions and personal values. In L. J. Shrum (ed.), *The Psychology of Entertainment Media: Blurring the lines between entertainment and persuasion.* New York: Routledge.

Shrum, L. J., Lee, J., Burroughs, J. E., and Rindfleisch, A. (2011). An online process model of second-order cultivation effects: How television cultivates materialism and its consequences for life satisfaction. *Human Communication Research, 37*(1), 34–57.

Slater, M. D., and Rouner, D. (2002). Entertainment-education and elaboration likelihood: Understanding the processing of narrative persuasion. *Communication Theory,* 12, 173–191.

Slater, M. D., Rouner, D., and Long, M. (2006). Television dramas and support for controversial public policies: Effects and mechanisms. *Journal of Communication, 56*(2), 235–252.

Weber, E. U. and Stern, P. C. (2011). Public understanding of climate change in the United States. *The American Psychologist, 66*(4), 315–28.

21

CARTOONS AND THE ENVIRONMENT

Anne Marie Todd

Mediated environmental messages are critical to public perception of environmental issues (Cantrill and Oravec 1996, Corbett 2006, Herndl and Brown 1996, Killingsworth and Palmer 1992, Myerson and Rydin 1996, Neuzil and Kovarik 1996). Media "provide us with the frames with which to assimilate and structure information" (Anderson 1997: 18). Increasingly, we perceive the world through screens: television and the internet provide news of environmental disasters, interpret science about climate change, and chronicle local environmental changes. For "non experts, understandings of environmental issues and policies are constructed from such mediated news reports, literature, or entertainment" (Meister and Japp 2002: 3). Environmental discourse holds significant influence on people's attitudes and perceptions about the world around them.

Popular culture plays an important role in how we communicate about our environment. Popular culture includes the entirety of social and cultural discourse, including "powerful modes of advertising, board games, newscasts, print news, cable television, greeting cards, film, and animated cartoons" (Meister and Japp 2002: 1). By definition, such discourse is important because it is appealing, and is thus a significant way people receive messages about important public issues such as the state of the environment. Visual and textual cues in entertainment media interpret humanity's place in the world. Language and images in popular culture "situate humans in relation to natural environments, create and maintain hierarchies of importance, reinforce extant values and beliefs, justify actions or inaction, suggest heroes and villains, create past contexts and future expectations" (Meister and Japp 2002: 4). Analysis of entertainment media provides a way of understanding how we encounter environmental communication in our daily leisure pursuits.

Cartoons are a beloved and enduring form of popular culture. Scholars in the fields of science, political science, sociology, and communication have studied the effects of cartoons on society. "The most common use of comic art is to entertain. That function is germane to cartoons even when they are employed for serious purposes—such as propagandistic, social consciousness raising (conscientization), educational, and developmental" (Lent 2008: 353).

The cartoon, whether printed comic or televisual animation, is a "literal and visual text [with] thematic, symbolic, and ideological material" (King 1994: 106). Cartoons produce meaning around a range of important social and political issues. Cartoons and animation are rich areas for communication scholarship concerned with how we construct and contest environmental issues. Cartoons offer significant environmental messages through character studies, comic corrective, and crisis response. Before contemplating cartoons as environmental discourse, communication researchers must first consider questions of medium and audience.

Cartoon media

When reading cartoons as communicative texts, it is important to consider questions of medium. Cartoons are part of an art form that includes both printed comics such as graphic novels, Sunday newspaper comics, editorial cartoons, and moving animation on film, television, and now the internet. Environmental issues are frequent subjects of editorial cartoons (Harris and Fromm 2008). Animation in film, television, and internet shorts offers a wide range of environmental perspectives: from feature-length film adaptation of *The Lorax* to the television series *The Simpsons* and *South Park* to internet videos parodying the 1990s television series *Captain Planet and the Planeteers.*

Scott McCloud defines comics as "juxtaposed pictorial and other images in deliberate sequence, intended to convey information and/or produce an aesthetic response in the reader" (McCloud 1994: 9). Robert Harvey argues the central characteristic of cartoons are the blending of visual and verbal content: "comics consist of pictorial narratives or expositions in which words … usually contribute to the meaning of the pictures and vice versa" (Harvey 2001: 76). The legal definition of cartoons is among the most helpful to understanding the social impact of cartoons: a "pictorial parody which by devices of caricature, analogy and ludicrous juxtaposition sharpens the public view of a contemporary event, folkway or political (or social) trend. It is normally humorous but may be positively savage" (Beutel 2001: para. 27). Cartoons are "a product of social interaction and interpretation" in that they reflect public debates and viewpoints (King 1994: 105).

Marshall McLuhan offers cartoons as an example of "cool media"—those that provide little information, requiring viewers to fill in details (McLuhan 2001: 22–23). Asa Berger notes that audiences "have to be more active in decoding the texts carried by cool media. They invite our participation" (Berger 2007: 32). Unlike photographs or live action video, cartoons are drawn, conjured from the imagination, and are not grounded in reality. Characters can take any form, action can take place anywhere in the world, and landscapes can change. Animation is "not bounded by the physical laws governing 3-D space … and take place in any geographic or historical time frame" (Alberti 2004: xiii). Animation creates the appearance of motion from still pictures and thus is by definition, an illusion.

The freedom that animation provides allows for distribution and appeal to a wide variety of audiences. Because of the simplicity of dubbing animation in multiple languages, animation "can transcend cultural boundaries and become a universal language, provide strong messages, and smooth over situations and make them appear as if they can happen anywhere" (Lent 2008: 376). Cartoons thus have a global appeal—voices are not tied to a human character whose appearance sets particular expectations about action, so dubbing allows freedom for reinterpretation. The same is true for analysis of cartoons—there are many more interpretations possible than with a photograph or live action film. Cartoons "can serve as agents for cultural critique" because they are "created with the greatest freedom from cultural constraints" (Bruce 2001: 243). Cartoon narratives are not bound to constraints of social or cultural norms and can thus do "forbidden and disruptive things" (Bruce 2001: 231). Cartoons can see the future, recreate the past, and offer alternative or mainstream readings. The cartoon is a "cultural text embedded with codes and representations that can be read from a number of perspectives, including the margins" (King 1994: 106).

Cartoons offer a lens to historical development of human perspectives. Historically, cartoons were part of television programming that played a key role in the framing of life (Spigel 1992). Today, environmental messages exist at the intersection of internet and television (Slawter 2008: 213). Questions of environmental communication must consider the "media matrix that constitutes our social milieu," and "[take] technology seriously" (DeLuca and Peeples 2002: 131). In his discussion of the cartoon parody of *The Matrix* aimed at revealing the horrors of factory farming, Dylan Wolfe writes,

the popularity of *The Meatrix* highlights the possibilities that new media technologies produce for environmentalists and other advocates. More than just a clever movie spoof, a humorous piece of cultural kitsch, *The Meatrix* is significant for its dynamic use of emerging technology, minimal production cost, and successful dissemination.

(Wolfe 2009: 319)

In considering cartoons as communication, researchers must take into account the changing mediascape in which cartoons are produced. As the number of screens that make entertainment available multiplies, cartoons gain broader reach and become more embedded in daily life.

Cartoon audiences

Cartoons initially gained popularity among younger audiences and attracted wide viewership on Saturday mornings (Bruce 2001), but primetime cartoons aimed at adults now produce some of the more pointed environmental messages (Stewart and Clark 2011). Cartoons remain an important element of popular culture for children, and play a large role in socializing children to environmental issues.

[This] can be easily demonstrated in a typical afternoon of children's television watching. Sesame Street, MTV, Nickelodeon, Barney and the Backyard Gang, all targeted to young viewers, regularly air environmental messages for children. Commercial cartoons have capitalized on environmental issues with characters such as the Toxic Crusaders and Captain Planet.

(King 1994: 105)

Cartoons can serve as edutainment, offering environmental lessons in visually appealing, memorable narratives for children.

Adult cartoons combine the "traditional children's medium, the cartoon, with the social and political content of prime-time programming" (Stewart and Clark 2011: 323). Such an "ambiguous cultural space allows producers and writers to … treat serious and even controversial issues under the cover of 'just being a cartoon'"(Alberti 2004: xiii). The medium of animation allows presentation of adult themes conveyed by silly characters through ridiculous storylines. While cartoons are not always realistic portrayals of life, they often reflect reality, offering relevant, critical commentary on environmental issues.

Cartoons have multiple audiences. Modern animated movies appeal to an adult audience with double entendres, political references, and double-edged humor. Adult cartoons such as *The Simpsons* and *South Park* are "often described as having multiple layers, including one aimed at the high cultural capital-possessing intellectual, but this 'higher' level is the level of its pastiche, not its parody" (Gray 2005: 235). Animation can offer social commentary with unrealistic or exaggerated portrayals that can reflect or interpret reality. *The Lorax*, a computer 3-D animated adaptation of the classic Dr. Seuss story demonstrates the cross-over appeal of narratives where "ideologies [are] given force in a visually-prominent format" (Wolfe 2008: 3). Cartoons present sophisticated environmental stories that appeal to both adults and children and are rich opportunities for scholarly inquiry.

Cartoon characters

Animated characters play an influential role in cartoon environmental discourse. Characters help form narratives about environmental issues, modeling how viewers might themselves approach a particular situation. Of course, by virtue of the medium, "cartoon characters are also allowed freedom from hypernormatic characterizations." Audiences do "not expect characters to suffer the consequences of their actions taken in prior episodes" (Stewart and Clark 2011: 323). *South Park*'s Kenny dies in nearly every episode and Bart Simpson will always be in fourth grade. The freedom from constraint that cartoon characters enjoy highlights the simplicity of their decision making and can help clarify the social significance of environmental action.

The classic comic book superhero is a familiar character that has significant sway over our imaginations (Bongco 2000, Lawrence and Jewett 2002). Smokey the Bear is a longheld cartoon role model designed to persuade people to remember to take action to prevent forest fires. Captain Planet is an environmental superhero summoned by five Planeteers to fight planetary destruction. The Green Ninja is a climate action superhero acting to reduce individuals' ecological footprints to fight global warming. These hero narratives proclaim that individuals have the power to solve the environmental crisis.

The superhero narrative is not without critique; some scholars argue that it appeals to an American myth of masculinity and whiteness (Lang and Trimble 1988, Palmer-Mehta and Hay 2005). For example, Captain Planet follows the universal archetype for superhero—a privileged, muscular, white male. Even though the Planeteers hail from five continents, Captain Planet holds the power, and sociologist Donna King concludes that "Captain Planet is clearly *not* a superhero 'every kid' can look up to" (King 1994: 110). Dylan Wolfe offers an alternative read of environmental characters, examining The Lorax as a "simple, colorful, charismatic prophet," which he argues, "provides a crucial element for the production of an environmental 'people'" (Wolfe 2008: 20). A prophet has believers, followers who may invigorate the spirit of the environmental movement. Whether prophet or superhero, human or nonhuman, cartoon characters provide exemplars of environmental action that offer opportunities for scholarly attention.

Popular culture positions nature as a resource, "emphasiz[ing] nature's 'use-value.' Simply, we consciously and unconsciously learn from popular culture the practice of consuming nature" (Meister and Japp 2002: 1). Advertising produces persuasive examples of popular culture that reflect "the environment's utility and benefit to humans." Such discourse "commodifies the natural world and attaches material value to non-material goods, treating natural resources as private and possessible, not public and intrinsic" (Corbett 2002: 143). In cartoon hero narratives about the environmental crisis, saving the planet requires, essentially, resource management to stem causes of pollution or stop environmental destruction (King 1994). In environmental popular culture discourse, Julia Corbett argues, nature is "merely a backdrop … for all but the most critical media consumers, the environment blends into the background." Nature's characteristics, the "qualities and features of the nonhuman world," are used to convey messages and sell products (Corbett 2006: 150). Cartoon depictions of natural features offer visions of the future. When Springfield's landfill grows too large, forcing the Simpsons (and all of Springfield) to move to the next town, the land lurches and gurgles, exaggerating the impact of waste disposal on the physical environment, highlighting what most people do not experience. Cartoons offer character-driven narratives that portray human perceptions of and attitudes toward the environment.

Comic corrective

Kenneth Burke introduces the concept of frames for understanding human experience (Burke 1937). For Burke, frames are the symbolic structures by which human beings impose order upon their experiences. Frames are "the more or less organized systems of meaning by which a thinking man gauges the historical situation and adopts a role with relation to it" (Burke 1937: 5). Frames are perspectives of interpretation and explain the order of human experience. "Comedy emphasizes the limitations on human knowledge, the lack of transparency of the world around us" (Forster 2002: 111). Comic frames offer a way to understand human imperfections and broader deficiencies in the social system.

Burke describes the comic frame as a rhetorical frame of acceptance that enables "people to be observers of themselves, while acting" (Burke 1937: 171). This self-reflective perspective provides a humane way of dealing with the destruction of the order of the status quo. Comic strategies are tools through which individuals can "point out the failings in the present system" (Powell 1995: 87). The comic frame sees "human antics as a comedy, albeit as a comedy ever on the verge of the most disastrous tragedy" (Burke 1937: iii). In response to imminent tragedy, the comic frame allows for human error; it provokes charitable self-reflection, which is conducive to social change because audiences can be sympathetic toward causes that they might be complicit in perpetuating. The comic frame "provides the charitable attitude toward people that is required for purposes of persuasion and co-operation" (Burke 1937: 166). Acceptance frames provide approval of the status quo, identifying with the system or perspective in question and thereby creating order.

The comic frame reveals attitudes that are intended to provoke charitable self-reflection (Powell 1995, Wills-Toker 2002). A. Cheree Carlson notes that this perspective can "free society by creating a consciousness of the system as a system, revealing its inherent weaknesses, and preparing an aware populace to deal with them" (Carlson 1986: 447). A comic perspective can lead to societal change through "reconstitution or re-education of the public audience, increasing the receptivity of the public to the marginalized, and increasing the resistance of public to the dominant institutional structures" (Madsen 1993: 174-175). Cartoons provide acceptance frames that highlight flaws in the human condition, and present a way for these to be rectified.

Adult cartoons demonstrate the comic frame as corrective. *The Simpsons* reveals that "the comic frame fosters more than an ironic self-awareness, but also constructs a position of semi-detachment, where one is able to reflect and comment on human foibles without guilt, shame, or other negative emotion" (Todd 2002: 66). The show's characters "display an overall disregard for the environment, are separated from nature, and often oppose nature" (Todd 2002: 66). Homer Simpson, the quintessential buffoon whose singular destruction of the environment is boundless, demonstrates the "potential clown in all human beings" (Carlson 1986: 448). Homer's choice to dump his pet pig's waste in the Springfield Pond is ridiculous and horrifying—highlighting the unsustainable disposal practices in humans. His ridicule of Lisa's choice to be a vegetarian demonstrates the pervasive meat-eating culture in America. The comic frame allows viewers to laugh at human behaviors, and while they may recognize a bit of themselves in Homer, they are not forced to judge that behavior.

The comic frame explains parody's pedagogical function because: "lessons rarely even feel like lessons … Jokes make us laugh, many viewers are likely to seek out parody, and few of us are likely to feel imposed upon in the way we might react to overtly didactic messages" (Gray 2005: 234). *South Park*'s four impious elementary-school children can offer a parody of environmentalism to frame a critique of environmental debates. Julie Stewart and Thomas Clark argue that *South Park*'s parody:

> simultaneously spoofs and reinforces many of the myths central to the American national character, parodies the condescension and intolerance of environmental advocates and their equally strident opponents. It employs a comic frame to ridicule extreme political behavior and language while promoting, through the children, an ethic of pragmatism and populism, with anti-elitism, anti-authority, and anti-hypocrisy themes.
>
> (Stewart and Clark 2011: 333)

South Park offers criticism of a range of environmental stakeholders. An episode entitled "Smug Alert" caricatures self-satisfied environmentalists while "Rainforest Schmainforest" parodies someone who holds a romantic view of nature without an understanding of the dangers and fragility of the ecosystem. Cartoons can take "environmentalist and environmental advocacy to absurdity" (Slawter 2008: 221). *South Park* offers a "'comic corrective' identifying in a seemingly children's medium, the cartoon, the characteristics of environmental advocacy rhetoric and that of its equally vociferous opponents that are most offensive to mainstream American audiences" (Stewart and Clark 2011: 333). Cartoons can exaggerate extremist views to present serious environmental debates in a new light.

Cartoons enfranchise viewers to learn without taking offense at the messages. "Parodic humor includes viewers by positing them on the knowing inside, rather than alienating them by positing them on the ignorant outside. Parody does not patronize us or talk down" (Gray 2005: 235). The inclusive framework of cartoons has a pedagogical function. "Parody goes down easy and so may be consumed more freely

[and] may inspire a more permanently critical disposition toward its targets" (Gray 2005: 235). Each of these shows present humans as comic fools, allowing viewers to consider their actions with ironic self-awareness. Cartoons present critical messages about wide-held beliefs or values that solicit reflection on the environmental impact of human society.

Crisis response

Environmental issues compel public communication, which defines how the public sphere responds to environmental issues (Cox 2011). Robert Cox's claim that environmental communication is a crisis discipline (Cox 2007, see also Cox 1982) prompted a debate about crisis as orientation, democratic process, and moral obligation (Schwarze 2007, Peterson et al. 2007). Cartooning responds to public crisis, for example, it is a common form of communication in development projects aimed at changing lifestyles in Asia and South America (Lent 2008: 361). Cartoon commentary on cultural and social norms can be considered a form of crisis response.

> Cartooning in the forms of animation, comic books, comic strips, and editorial cartoons, has been involved in public crises alert and/or relief campaigns worldwide. In most cases, cartoons have been used to raise social consciousness levels concerning dangerous or potentially dangerous threats to the public.
>
> (Lent 2008: 352)

Cartoons offer a form of risk communication, and thus have an educational purpose. Of course cartoons are not always educational, or scientifically accurate: images of science in cartoons and comics often distorted ideas that can sustain stereotypes by society (Vílchez-González and Palacios 2006). But cartoons "can be effective in pointing out risks and providing ways to manage them" (Lent 2008: 383). Through their comic form with humor and visual appeals, cartoons provide warnings that are palatable. Cartoons rely on "easy-to-understand bridging metaphors derived from the popular culture" to impart knowledge and foster "public understanding and concern"(King 1994: 297). Cartoons can portray environmental risks in ways that make viewers pay attention.

Cartoons are a way to talk about the apocalypse, a theme that pervades environmental discourse (Cox 1982, Killingsworth and Palmer 1992). Apocalyptic visions in the animated films *Wall.E* and *The Lorax* offer critiques of consumer lifestyles. Wall.E is a small robot living on Earth abandoned by humans. Visiting the space cruise ship holding humans who evacuated Earth he encounters a population immobilized by incessant media entertainment and processed food. In *The Lorax*, Thneed-Ville is a walled plastic city, where natural elements such as air and water have become commodified. Outside both of these encased population centers, the world is devastated at the hands of humans. These apocalyptic visions reveal the future that overconsumption of environmental resources will bring.

Animated environmental messages, particularly those aimed at children, typically offer a solution or moral to the story. Lisa Simpson is the "social conscience" of *The Simpsons*, which expresses its "ethical stance ... most explicitly through [her] words and deeds" (Turner 2004: 191). Cartoon characters that provide the foil to the comic fool can provide alternative ways of viewing and acting toward the earth. Whether animation, a form uninhibited by realism and laws of gravity, can offer realistic or authentic solutions is an open question. Donna King deems the environmental ethic promoted by Captain Planet "salve but little substance for avoiding further environmental disaster" (King 1994: 117). In many ways, cartoons reflect the challenges facing environmental communication of all sorts—how to address "complex, systemic problems that demand serious social, political and economic consideration and concern" (King 1994: 116). This is the question communication scholars must address when analyzing cartoons as crisis communication.

Conclusions

Comics, cartoons, animation—these visual art forms demonstrate potential to raise awareness and possibility for change. Through their appeal to broad audiences, and increased reach through distribution across global networks, cartoons can be considered a democratic art form (Chatterjee 2007, Maggio 2007). The considerations of audience and medium enhance the multiple layers of meaning in cartoons. Animated environmental messages construct and contest the environment through hero narratives, the comic frame, and apocalyptic visions. Research on cartoons as environmental communication will contribute to greater understanding of the possibility of popular culture to influence public environmental knowledge and behavior.

Further reading:

Alberti, J. 2004. *Leaving Springfield: The Simpsons and the Possibilities of Oppositional Culture*. Detroit, MI: Wayne State University Press.

Corbett, J. 2006. *Communicating Nature: How We Create and Understand Environmental Messages*. Washington, D.C.: Island Press.

Meister, M. and Japp, P. M. (eds). 2002. *Enviropop: Studies in Environmental Rhetoric and Popular Culture*. Westport, CT: Praeger Publishers.

References

Alberti, J. 2004. Introduction. *Leaving Springfield: The Simpsons and the Possibilities of Oppositional Culture*. Detroit, MI: Wayne State University Press.

Anderson, A. 1997. *Media, Culture and the Environment*. New Brunswick, NJ: Rutgers University Press.

Berger, A. A. 2007. *Media and Society*. Lanham, MD: Rowman & Littlefield.

Beutel, R. V. 2001. *Ross v. Beutel*. In: Court of Appeal of New Brunswick (ed.) NBCA 62. Document Number: 97/98/CA. http://canlii.ca/t/4vjd

Bongco, M. 2000. *Reading Comics: Language, Culture, and the Concept of the Superhero in Comic Books*. New York: Garland Publishing.

Bruce, D. R. 2001. Notes toward a rhetoric of animation: *The Road Runner* as Cultural Critique. *Critical Studies in Media Communication*, 18, 229–245.

Burke, K. 1937. *Attitudes Toward History*. Berkeley, CA: University of California Press.

Cantrill, J. and Oravec, C. 1996. *The Symbolic Earth: Discourse and Our Creation of the Environment*. Lexington, KY: The University Press of Kentucky.

Carlson, A. C. 1986. Ghandi and the comic frame: "Ad bellum purificandum." *Quarterly Journal of Speech*, 72, 446–455.

Chatterjee, S. 2007. Cartooning democracy: The images of R. K. Laxman. *PS: Political Science and Politics*, 303–306.

Corbett, J. 2002. A faint green sell: Advertising and the natural world. In: Meister, M. and Japp, P. M. (eds) *Enviropop: Studies in Environmental Rhetoric and Popular Culture* (pp. 141–160). Westport, CT: Praeger Publishers.

Corbett, J. 2006. *Communicating Nature: How We Create and Understand Environmental Messages*. Washington, D.C.: Island Press.

Cox, J. R. 1982. The die is cast: Topical and ontological dimensions of the *locus* of the irreparable. *Quarterly Journal of Speech*, 68, 227–239.

Cox, J. R. 2007. Nature's "crisis disciplines": Does environmental communication have an ethical duty? *Environmental Communication: A Journal of Nature and Culture*, 1, 5–20.

Cox, J. R. 2011. *Environmental Communication and the Public Sphere*. Thousand Oaks, CA: Sage Publications.

Deluca, K. and Peeples, J. 2002. From public sphere to public screen: Democracy, activism, and the "violence" of Seattle. *Critical Studies in Media Communication*, 19, 125–152.

Forster, S. 2002. Don McKay's comic anthropocentrism: Ecocentrism meets "Mr. Nature Poet." *Essays on Canadian Writing*, 107–135.

Gray, J. 2005. Television teaching: Parody, *The Simpsons*, and media literacy education. *Critical Studies in Media Communication*, 22, 223–238.

Harris, S. and Fromm, H. 2008. *There Goes the Neighborhood: Cartoons on the Environment*. Athens, GA: University of Georgia Press.

Harvey, R. 2001. Comedy at the juncture of word and image: The emergence of the modern magazine gag cartoon reveals the vital blend. In: Robin, V. and Gibbons, C. (eds) *The Language of Comics: Word and Image* (pp. 75–96). Jackson, MS: University Press of Mississippi.

Herndl, C. G. and Brown, S. C. 1996. *Green Culture: Environmental Rhetoric in Contemporary America*. Madison, WI: University of Wisconsin Press.

Killingsworth, M. J. and Palmer, J. S. 1992. *Ecospeak: Rhetoric and Environmental Politics in America*. Carbondale, IL: Southern Illinois University Press.

King, D. L. 1994. Captain Planet and the Planeteers: Kids, environmental crisis, and competing narratives of the new world order. *The Sociological Quarterly*, 35, 103–120.

Lang, J. S. and Trimble, P. 1988. Whatever happened to the man of tomorrow? An examination of the American monomyth and the comic book superhero. *The Journal of Popular Culture*, 22, 157–173.

Lawrence, J. S. and Jewett, R. 2002. *The Myth of the American Superhero*. Grand Rapids, MI: William B. Eerdmans Publishing Company.

Lent, J. A. 2008. Cartooning, public crises, and conscientization: A global perspective. *International Journal of Comic Art*, 10, 352–386.

McCloud, S. 1994. *Understanding Comics*. New York: HarperCollins.

McLuhan, M. 2001. *Understanding Media: The Extensions of Man*. London: Routledge Classics.

Madsen, A. J. 1993. The comic frame as corrective to bureaucratization: A dramatistic perspective on argumentation. *Argumentation and Advocacy*, 29, 164–166.

Maggio, J. 2007. Comics and cartoons: A democratic art-form. *PS: Political Science and Politics*, 20(2), 237–239.

Meister, M. and Japp, P. M. 2002. Introduction: A rationale for studying environmental rhetoric and popular culture. In: Meister, M. and Japp, P. M. (eds) *Enviropop: Studies in Environmental Rhetoric and Popular Culture* (pp. 1–12). Westport, Ct: Praeger Publishers.

Myerson, G. and Rydin, Y. 1996. *The Language of Environment: A New Rhetoric*. London: UCL Press.

Neuzil, M. and Kovarik, W. 1996. *Mass Media and Environmental Conflict: America's Green Crusades*. Thousand Oaks, CA: Sage Publications.

Palmer-Mehta, V. and Hay, K. 2005. A superhero for gays?: Gay masculinity and green lantern. *The Journal of American Culture*, 28, 390–404.

Peterson, N. M., Peterson, M. J., and Peterson, T. R. 2007. Environmental communication: Why this crisis discipline should facilitate environmental democracy. *Environmental Communication: A Journal of Nature and Culture*, 1, 74–86.

Powell, K. 1995. The Association of Southern Women for the Prevention of Lynching: Strategies of a movement in the comic frame. *Communication Quarterly*, 43, 86–99.

Schwarze, S. 2007. Environmental communication as a discipline of crisis. *Environmental Communication: A Journal of Nature and Culture*, 1, 87–98.

Slawter, L. D. 2008. TreeHuggerTV: Re-visualizing environmental activism in the post-network era. *Environmental Communication: A Journal of Nature and Culture*, 2, 212–228.

Spigel, L. 1992. *Make Room for TV*. Chicago, IL: University of Chicago Press.

Stewart, J. and Clark, T. 2011. Lessons from South Park: A comic corrective to environmental puritanism. *Environmental Communication: A Journal of Nature and Culture*, 5, 320–336.

Todd, A. M. 2002. Prime-time subversion: The environmental rhetoric of "The Simpsons." *Enviropop: Studies in Environmental Rhetoric and Popular Culture*. Westport, CT: Praeger Publishers.

Turner, C. 2004. *Planet Simpson*. London: Ebury Press.

Vílchez-González, J. M. and Palacios, F. J. P. 2006. Image of science in cartoons and its relationship with the image in comics. *Physics Education*, 41, 240–252.

Wills-Toker, C. 2002. Debating "what ought to be": The comic frame and public moral argument. *Western Journal of Communication*, 66, 53–84.

Wolfe, D. 2008. The ecological jeremiad, the American myth, and the vivid force of color in Dr. Seuss's *The Lorax*. *Environmental Communication: A Journal of Nature and Culture*, 2, 3–24.

Wolfe, D. 2009. The video rhizome: Taking technology seriously in *The Meatrix*. *Environmental Communication: A Journal of Nature and Culture*, 3, 317–334.

22

CINEMA, ECOLOGY AND ENVIRONMENT

Pat Brereton

In this overview chapter, I will outline some of the broad trajectories around ecocinema, drawing on familiar case studies such as *An Inconvenient Truth* and popular fiction films such as *The Day After Tomorrow*, *Avatar* and *Wall-E* among others, to help tease out why these films have become some of the most successful and influential eco-narratives ever produced.

Scott MacDonald argues that certain experimental films can promote an ecocentric sensibility and considers the 'fundamental job of ecocinema as a retraining of perception, as a way of offering an alternative to conventional media-spectatorship' (Rust *et al.*, 2013: 45). Such critics emphasize the benefits of 'slow cinema', which can encourage the retraining of perception as a necessary condition for greater ecological awareness. Paula Willoquet-Maricondi (2010) similarly suggests that only certain types of independent lyrical and activist documentaries may be thought of as ecocinema, because they are the most capable of inspiring progressive eco-political discourses and action among viewers. Both scholars are correct at one level, nonetheless it is much more beneficial to broadly accept that all types of film, from the excesses of a commercial Hollywood blockbuster, alongside the most rarefied and explicitly ecological art-house film, can consciously or unconsciously foreground ecological issues and help situate these concerns within the general public consciousness.

Introduction

[Ecocinema] overtly strives to inspire personal and political action on the part of viewers, stimulating our thinking so as to bring about concrete changes in the choices we make, daily and in the long run, as individuals and as societies, locally and globally.

(Willoquet-Maricondi, 2010: 45)

Focusing on art cinema, she further argues that a filmmakers' unique use of long takes and slow pacing can promote contemplation across ecological lines, and celebrates for example the Slovenian director Andrej Zdravic's *Riverglass: A Ballet in Four Seasons* (1997) for articulating such an aesthetic. Meanwhile, other art-house narratives such as Peter Hutton's *Study of a River* (1997) can be read as like 'being on a ship forced to slow down and allowed to take the time to look' (MacDonald, 1998: 252) and thereby really experience the environment; echoing for example Bella Tar's embodiment of slow cinema.

Such analysis tends to emphasize ecocinema as first and foremost a cognitive, rather than an affective or emotional experience. Cognitive estrangement is set up as the first step by which the desired state of environmental awareness might be attained. By all accounts this is a very narrow, even elitist framing, according to Rust *et al.*, who suggest that eco-film criticism's overarching purpose should *not* be to impose a political programme – much less predefined aesthetic practices – but to help create public spaces for debate and argument over the claims of the environment for a place in political life (2013: 3).

Popular fiction film, alongside more conventional preoccupations with nature/ecology in televisual documentaries and animation, remain an excellent forum to promote and at the same time help puncture any simple[istic] formulations around the complexities of dealing with environmental issues and debates. Scholarship has however to move beyond simply creating a robust definitional and textual based corpus of ecocinema and develop a body of clearly differentiated empirical evidence to help underpin many of the ecological assertions made in the literature, which could in turn help evaluate and measure future attitudinal changes; especially as ecocinema becomes more provocative and pervasive for its growing cross-media audiences.

Sean Cubitt goes so far as to suggest '[T]hough many films are predictably bound to the common ideologies of the day, including ideologies of nature, many are far richer in contradictions and more ethically, emotionally and intellectually satisfying than much of what passes for eco-politics' (Cubitt, 2005: 1). Cubitt further insinuates that while film critics remain preoccupied with the realist image, environmental science deals in effects that are often too vast, too slow, or too dispersed to be observed photographically. Consequently, in a seminal documentary such as *An Inconvenient Truth*, to be discussed later, there is a cinematic move towards rendering the world as visual data.

Meanwhile, Adrian Ivakhiv reconceives cinema as 'a machine that moves us along vectors that are affective, narrative and semiotic in nature and discloses worlds in which humanity, animality and territory are brought into relationship with each other' (Rust *et al.*, 2013: 6). In describing cinema's complex interactions, Ivakhiv embraces three ecologies of the earth-world; namely the material, the social and the perceptual. His subsequent book length study *Ecologies of the Moving Image: Cinema, Affect, Nature* (2013) has major implications for future eco-film scholarship, as it grapples with a growing corpus of cinematic reflections and theoretical analysis. I will illustrate some of these issues presently with a discussion of the much written about eco-blockbuster *The Day After Tomorrow*, among others.

At the same time, psychologists and film scholars frequently affirm how we cannot expect dramatic changes in worldviews as a result of simply watching a movie. Naturally also viewers tend to be attracted by the kind of films that fit their beliefs and probably eco-films may only be preaching to the converted. Yet mass audiences continue to watch movies precisely because of cinema's ability to reframe a wide range of perceptions. The late great film critic Roger Ebert of the *Chicago Sun-Times* often talked of the power of cinema to invoke empathy and allow audiences to step into another world and see reality from a totally different perspective. For ecocinema scholars, cinema enables audiences to recognize ways of seeing the world, other than through the narrow perspective of the anthropocentric gaze that ostensibly situates individual human desires at the centre of the moral universe.

While much theoretical analysis concentrates on the semiosis of the text, less attention has been paid to analysis of actual audiences' perceptions and interpretations, in order to empirically demonstrate these resultant affects and emotional connections.[1] Audience reception studies have in past decades emphasized the active, interpretative, critical, creative and sometimes resistant nature of engaging with the media. Nevertheless, while most film analysis makes claims about the audience, they seldom make this explicit (see Brereton, 2012).

Surprisingly, much of the literature in film, much less than in reception studies, remains somewhat abstract. For example, a major study such as Janet Staiger's *Perverse Spectators: The Practices of Film Reception* (2000) proposes the figure of the perverse spectator driven by 'affective and emotional experiences' (34) and pleasures. The perverse spectator, according to this thesis, engages film as 'an event' that reflects various layers of a multifaceted identity. In describing film viewing as an event, Staiger distinguishes between activities of watching (place, genre of text, social-mixing) and reception activities after the event (discussion, star imitation, production of new materials and initiating new viewers).[2] Applying such particular aspects of textual and reception studies to a wide corpus of ecocinema for instance will take time and much more adaptive scholarship to further explain the environmental pleasures for audiences, alongside nailing down the key behavioural triggers.

An Inconvenient Truth

The most cited example of ecocinema, at least from a documentary perspective, remains *An Inconvenient Truth* (2006) and succeeds not only because of its predictions and persuasive cognitive logics but also because of the deep eco-memories and emotional affect that it evokes. Gore's film – albeit directed by Davis Guggenheim – argues more powerfully for a widely held nostalgia for a better, cleaner world. Gore's very direct message gains rhetorical force, according to another ecocinema study (Murray and Heumann, 2009: 195), by foregrounding what can be defined as environmental nostalgia, coupled with its powerful emotional appeal.

This eco-documentary argues powerfully for sustainable environmental policies, by invoking both personal and universal ecological memories, as earlier evident in classic science fiction films such as *Silent Running* (1971), *Omega Man* (1971) and (even more closely entwined with Gore's narrative) *Soylent Green* (1973). *An Inconvenient Truth* opens with two scenes illustrating two historical memories of the world thirty years beforehand. One of those memories grows out of a meandering river that flowed near Al Gore's family farm:

> [A] river we see flowing clean and clear through a pristine green landscape. The year is 1973 and Al and wife Tipper float along in a canoe over gentle ripples of the Caney Fork River. Living nature is highlighted here by the river, the foliage that lines it and the fact that Tipper is close to giving birth to the Gores' first child.
>
> (Murray and Heumann, 2007: 1)

According to Murray and Heumann, such an affective and nostalgic eco-text helps to create a 'tipping point' in audience engagement and affords the legitimation of environmentalism as a primary ethical imperative.

Riding on the crest of this notion of a global tipping point, while developing a persuasive visual style which draws upon the scientific truth of climate change, actively inspired media producers to permeate their media landscape with images of global warming in diverse fictional films such as *11th Hour* (2007), *The Day the Earth Stood Still* (2008), *Quantum of Solace* (2008), *Wall-E* (2008), *Avatar* (2009), etc. All of these contemporary narratives and others also, according to Rust *et al.* (2013), Whitley (2012) and Kääpä (2013) draw on the inspiration and potency of this small-scale cautionary documentary tale.

Furthermore, at an aesthetic level, data visualization through an innovative usage of graphs, PowerPoint and other visualizing tools serve to reaffirm the documentary's originality. According to Cubitt, *An Inconvenient Truth* most effectively embraces cartography, numbers, graphics and simulations, which are also integral to the explicitly scientific discourse of climate change. Since global events such as climate change do not occur in humanly perceptible scales or time frames, they consequently demand forms of representation that can capture massive but at the same time relatively slow ecological change. Godfrey Reggio most notably pioneered the use of time-lapse photography in his eco-parables *Koyaanisqatsi* (1982) and *Powaqqatsi* (1988); techniques which in turn feed into evolving tools of representation that help to visualize such large time shifts and effectively capture themes central to a deep ecological agenda. *An Inconvenient Truth* foreshadows the 2012 eco-documentary *Chasing Ice*, for example, revealing the inside story of Climate Science through the stunning time lapse photography of James Balog. The visualization of climate change in such documentaries helps to overcome the tempo-spatial problems highlighted as one of the most challenging aspects of climate change communication. Balog's solution employs the use of photographic stills taken from the same vantage point and separated by years; thus presenting the unfolding ecological crisis before our very eyes, in breathtaking simplicity through the use of time lapse photography; like Kilimanjaro's vanishing snows (Cubitt in Rust *et al.*, 2013: 280).

Incidentally, there are surprisingly more dramatized instances of data visualization in *An Inconvenient Truth* than in *The Day After Tomorrow* which I now turn to, with its surfeit of science fiction fantasy. Certainly, attempts to create more realistic representations of climate change have become possible and even prophetic. Following actual footage from recent extreme weather events such as Hurricane Sandy and the extensive flooding around New York and its environs, such dramatic storylines speak to a less committed and more suspicious mass audience and probably have important long-term consequences, at least in bringing environmental concerns into mainstream public consciousness. Furthermore, while fictional narratives often remain exaggerated and scientifically untenable, especially within futuristic science fiction fantasies, nonetheless they still provoke a form of surface realism that is suffused through futuristic news broadcasts, for example, seeking to highlight the truth concerning various consequences of global warming.

Cinematic affect: *The Day After Tomorrow*

At the outset, we must recognize that climate change is not the only global environmental risk exploited by Hollywood in recent years: one calls to mind for example nuclear war in *Terminator 2: Judgement Day* (1991); deforestation in *FernGully: The Last Rainforest* (1992); bioterrorism in *28 Days Later* (2002); species extinction in *Earth* (2007); population growth in *Slumdog Millionaire* (2008); ecology and religion in *The Tree of Life* (2011) and *The Life of Pi* (2012); among many other environmental examples.

It makes sense for instance that the first fictional film to directly portray global warming was a post-apocalyptic science-fiction film set in the future. *Soylent Green* stars Charlton Heston and Edward G. Robinson as detectives on the case of a murdered food-industry executive. Through the late twentieth century, as the science of anthropogenic climate change became more conclusive, the energy industry and conservative think-tanks led a concerted effort across the mainstream media to frame the issue as a theoretical debate rather than a practical concern.

In a chapter titled 'Hollywood and Climate change' (Rust *et al.*, 2013), Rust makes an unsubstantiated claim that climate change films have primarily influenced a shift in American popular environmental discourse by translating the science of 'global warming'[3] into the vernacular of cinema. Released in 2004, *The Day After Tomorrow* earned more than 500 million dollars at the global box office and offers a window into what Stephen Rust terms, the 'cultural logic of ecology', epitomizing the pronounced shift in American popular discourse around the relationship between human beings and the earth that is taking shape in the early twenty-first century.

By the time global warming re-emerged in cinema during the late 1980s and early 1990s, 'a majority of scientists [had become] convinced that global warming was occurring' (Leiserowitz, 2003: 8–9). In this disaster-framed fictional world of *The Day After Tomorrow*, neither scientific consensus nor increased weather anomalies inspire the government or the public to begin mitigating global warming in time to avert disaster. The film's narrative suggests, only when 'Americans finally *see* climate change and *feel* its direct impact within the United States', will 'they accept responsibility for causing global warming and begin to take action in response to it' (Rust *et al.*, 2013: 198).

Meanwhile, contemporary film research has become more preoccupied with the power of emotional empathy and affect, as against more cerebral cognitive engagement. See for instance the work of scholars such as Greg Smith and Carl Plantinga (1999), Alexa Weik von Mossner (2014), Nöel Carroll, Murray Smith, and many others. Greg Smith's 'associative model' (2003) for instance, usefully accounts for how different aesthetic registers work together to construct a film's meaning for the viewer; a process that involves cognitive, emotional as well as affective aspects. Affect essentially is a visceral, bodily response to a film, whereas emotion also includes a cognitive element. According to this model, a narrative film usually works by establishing the viewer's emotional relationship with the protagonists' goals and actions, as well as through lower level, non-verbal affects that he calls 'moods', produced by stylistic elements including music, *mise en scène*, lighting and colour, etc.

Without voice-over commentary or 'talking heads', as seen in documentaries such as *An Inconvenient Truth*, a film works as much through audio-visual affect, as against the sort of cognitive affects identified by Willoquet-Maricondi and MacDonald and remains central to the eco-film (Rust *et al.*, 2013: 46). The significance of these models/interpretative frameworks becomes more apparent when understood in the context of the evolution of film studies, which has focused on teasing out form as opposed to content, while attempting to differentiate and analyse various aesthetic strengths; whereas for some scholars ecocinema simply seeks to speak to and foreground specific thematic manifestations of environmental concern. Essentially therefore the focus of such analysis involves striving to extract the particular variables that might promote a proactive engagement with the environment.

With its broad strokes eco-fictional diegesis and its evocation of an environmental creative imaginary, *The Day After Tomorrow* certainly has the potential to cue and prompt viewers into an active, conceptual and sensory consideration of the relationships between humans and their global environment. For instance there have been useful audience studies in Germany and America (see studies by Leiserowitz, 2004)[4] focused on reactions to the film, which highlight the varying power of the text to speak to mass audiences across cognitive and emotional protocols. Much more substantial longitudinal studies are needed, however, to test many of the assertions made in such pilot studies, using more extensive textually based eco-film investigation and this form of scholarship is ongoing (see Brereton and Hong, 2013).

Wall-E

Directed by Andrew Stanton, this cautionary animated satire on consumer culture for the modern world – pushing the implicit assertions of *An Inconvenient Truth* to its ultimate conclusion – has rightly received much praise for its engaging storyline; ostensibly set in 2700, long after the earth is smothered by waste and declared unlivable for humans. *Wall-E* bravely foregrounds a non-talking waste allocation (analogue-like) load-lifter – the last 'inhabitant' and robot on planet earth – who initially makes friends with a stray cockroach, before finding his true love EVE (Extra Terrestrial Vegetation Evaluator), a pristine (high tech digital) robot sent to earth to investigate whether humans could possibly return to their erstwhile 'Garden of Eden'.

Much later we find out 'Buy-n-Large', a business corporation, has been largely responsible for the waste explosion on the planet. Its CEO, a bland hypocrite called Shelby Forthright has, as Philip French asserts in his *Guardian* review from 20 July 2008, whisked away the human inhabitants for a cruise on the luxury starship Axiom, which has lasted for several centuries. Meanwhile, robots like Wall-E are left marooned back on earth to clean up the mess. For over 40 minutes this 'rusty metal box with ET's eyes' does nothing much but potter around his city space engulfed in filth, waste and flotsam from a dead planet. By all accounts, this is a long way from the frenetic action adventure of *The Day After Tomorrow*.

David Whitley provides a most useful eco-reading of the film and its 'mode of emotional identification that includes rampant anthropomorphizing' (2012: 3) and goes on to argue how the chief protagonist remains in love with the consumer culture that he so effectively critiques.

Meanwhile, like in *Terminator2* with its more advanced computer organism T1000, EVE also appears at first to be more suspect and less ethical in her actions, by zapping everything in her way. But soon both learn to appreciate each other for what they are. Eventually, during their strange courtship, Wall-E shows off a living organic green plant, which was locked away in a safe. As in *Logan's Run*, *Blade Runner* (1982), *Waterworld* (1995) and many other eco-science fiction narratives, organic vegetation is greatly prized in such a synthetic world. This miracle of natural photosynthesis in turn proves that the planet is again habitable and secures the empirical proof EVE was sent out to discover.

According to a conference paper by Bob Mellin (2009), *Wall-E* assumes that the apocalyptic warnings found in documentaries such as *An Inconvenient Truth* are valid, and as such, we can be comforted by the movie's claim that the environmentally degraded planet in *Wall-E* can be restored to the garden that it once was. Near the end of the closing credits of the film, we witness Wall-E and EVE, who have seemingly made an escape from the degraded city where they first met, holding hands within a lush green and pastoral landscape, reminiscent of a new Eden – like the original closing of *Blade Runner*, or as also visualized in *Silent Running*; recalling the 1964 touchstone work *The Machine in the Garden*, where Leo Marx advanced the now commonplace argument that pastoralism is foundational for the quintessential American experience, with the Anglo-colonizers originally perceiving North America as literally a new Eden. One wonders, however, whether contemporary audiences are satisfied in the same way by such historical forms of pastoralism, as suggested by Murray and Heumann (2009).

Either way, on a narrative level, the residents of twenty-eighth century earth do not find refuge from the ills of civilization in the countryside; probably because there is no longer a pristine countryside to escape to. Instead they have to travel to outer space in a spaceship that combines the splendours of shopping malls, alongside the convenience of conventional cruise ships. One would almost instinctively agree with Murray and Heumann's conclusion that *Wall-E*'s artificial environments are anathema to the restorative qualities of romantic pastorals,[5] a trope that continues to have echoes in a major recent eco-blockbuster *Avatar* (2009), to be discussed presently.

In *Finding Nemo* (2003) and *A Bug's Life* (1998), alongside *Wall-E*, nature and the environment take centre stage. While liberal audiences certainly find *Wall-E* provocative – drawn one supposes from the environmental message posed, with its blatant critique of over-consumption – conservative Christians apparently find the film alternatively fills a wholesome niche, by essentially valorizing deeply felt conservationist values. Such conservatives particularly detest litterbugs, according to a review by Charlotte Allen (13 July 2008) in the *Los Angeles Times*, alongside all forms of parasites, who expect others to clean up after them.

As recalled in my Pixar chapter in *Smart Cinema* (Brereton, 2012), the film is reminiscent of Roman times and the crude political strategy of using 'bread and circuses' to keep the masses satiated. This futuristic artificial society was similarly visualized in classic science fiction films such as *Logan's Run*, where the populace is controlled by pleasure and spectacle, with its inhabitants not required to make personal decisions, much less forage for food. *Wall-E* follows a similar path, with its more contemporary obese-looking animated humans, drip-fed on synthetic food and thereby becoming more supine and docile in their massive spaceship, having all their corporeal needs serviced by a mechanical under-class. In such an artificial futuristic age, the allegory insinuates, humans have lost the capacity to appreciate the importance of scarcity and striving for basic needs, alongside more normative evolutionary human desires around freedom to control one's destiny.

There was much debate over *Wall-E*'s intended ecological message and whether it went too far, or not far enough, towards suggesting any solutions for our waste problems. In any case the film effectively plays out a food consumption allegory around how unchecked appetites (alongside more controversial population explosion concerns) pose a major danger to the planet and its inhabitants. Within such allegorical storylines, science fiction in particular offers a cautionary glimpse into a dystopic future in which our insatiable hunger and general rapaciousness threaten to destroy the planet, eating away at our basic humanity, as cogently represented in earlier classics such as *Logan's Run* and *Soylent Green*.

Avatar: ecology and big business

The story centres around the evocative representations of the inhabitants of Pandora called Na'vi, who literally plug into the exotic, ecologically benign, and idyllic flora and fauna, rather than just naturally appearing at one with their habitat, as in so many representations from Hollywood generic fare. Adapting a classic narrative framework, Hollywood frequently situates its anti-heroes, like the Native American Indians, within a clearly established environment and portrays them as totally in tune with their habitat, as opposed to the colonizing and destructive agency of the white settlers (see for instance Bird, 1996; Kilpatrick, 1999; Elsaesser, 2011). As affirmed frequently in film scholarship, the western genre has been transformed into a contemporary form of science fiction spectacle, following similar generic and thematic tropes, and this trajectory is especially evidenced in this eco-blockbuster.

This hugely successful 3D spectacle, directed by James Cameron, follows the journey of Jake Sully (Sam Worthington), a former marine who is paralyzed during combat on Earth. His twin brother had been working as a scientist for the so-called Avatar program on Pandora – the well-named planet with the much sought after energy source, unsubtly called unobtainium. This scientific project constructed genetically engineered machine-human-Na'vi hybrids that enable the humans to control these avatars with their minds, while their own bodies sleep. An avatar, we discover, can only be controlled by a person who shares its unique genetic material; consequently when Jake's twin brother dies, he is asked to join the squad, being the only one who has the appropriate genes to control that particular avatar.

After the initial vicarious thrill of being able to freely run around this fantastically rich habitat in his new agile body, Sully faces a major ethical dilemma during his fantasy journey, in being forced to participate in the mechanistic and cosmic Manifest Destiny that will ultimately lead to the destruction of Pandora and the Na'vi culture, including its magnificent and exotic vegetation under which most of these precious deposits are situated, echoing current concerns around fracking and various forms of deep mining. Alternatively and more heroically of course, he could choose to embrace a radical conversion and reject his destructive predetermined agency.

In the guise of a tranquil and harmonious interaction with the unknowable otherness of the Na'vi, Sully becomes the audience's eyes, literally engaging with this alien but idealized eco-utopia. This narrative plays into long established generic discourses of several indigenous native cultures portrayed in westerns, alongside more benevolent eco-narratives such as Terrence Malick's primal American allegory *The New World* (2005). The setting up of this exotic spectacle, using the extreme violence perpetrated by humans in battle, can at an individual level also be viewed as a conflict between the un-recuperated male war-hero Sully and his love interest embodied by a native princess called Neytiri (Zoe Saldana).

The image of a Great Mother protecting the balance of life clearly draws on a pantheistic and deep ecological vision in which energy continuously flows through discrete bodies of organic life. During this and other sequences, Neytiri teaches Jake to behave and think as a Na'vi by 'going native' – a trope eulogized for instance in *Dances with Wolves* (see Brereton, 2005: 98–102). *Boston Globe* film critic Wesley Morris and others have mischievously renamed the film 'Dances with Blue People' (see McGowan, 2010: 3), to signal the obvious reference to this revisionist film. Sully records in his video diary that Neytiri is 'always going on about the flow of energy, the spirits of animals', and adds, 'I'm trying to understand this deep connection the people have for the forest.' Further referencing a deep ecological agenda, she talks about 'a network of energy that flows through all living things' and affirms how 'all energy is only borrowed and one day you have to give it back'.

Surprisingly, such an overtly explicit controlling ecological agenda was too much for some eco-critics who found the *mise en scène* of the iconic Home-Tree for example as simply too crude and too obvious from an ecological perspective and deduced that the story was trying too self-consciously to get its didactic message across, which in turn militated against its final achievement. For a more comprehensive analysis of various critical readings, see Thomas Elsaesser's (2011) comprehensive analysis, alongside contrasting readings from a wide range of scholars in Bron Taylor's (2013) reader dedicated to the film.

Paradoxically, however, in more 'serious' art-house ecocinema, as articulated by Willoquet-Maricondi and others discussed above, this direct strategy is not considered a drawback in any way. The old propaganda debate comes to mind, weighing up subtle semiotic and thematic concerns against the power of unambiguous didactic messages that project a high level of moral self-righteousness. Without seeming to adopt a patronizing tone; for general audiences not inured into the intricacies of a deep-ecological mindset, much less worried about the danger of essentializing ecological precepts within a gendered or ideological address, such easily digestible cinematic experience and powerful eco-visual correlatives are essential ingredients for effective mass communication. Hearing about the network connections that link back to the exotic Home-Tree, alongside relating how its non-humanoid inhabitants commune with, rather than abuse their habitat, remains allegorically potent for a whole generation of cinema goers and might even help to promote a contemporary form of ecocinematic literacy. The importance of such an ecological allegory can be concretely appreciated at one level, by how it banked 2.98 billion dollars within the first two years after its release; 73 per cent of which came from outside the USA.[6]

The director James Cameron's well publicized visit to the Amazon Basin in mid April 2010, after the film's global release, no doubt reflects his personal commitment and support for the rights of native peoples and their particular resistance to a proposed hydropower project on the Xingu River that would flood its indigenous Kayapó homelands. Speaking like a well-versed politician: '[W]e're here to listen to what you are saying, to hear your concerns and, because I am a film-maker, to share this with the outside world' (cited in www.blospot.ie). Cameron noted that the writing of *Avatar* was, at least in part, inspired by such diverse native peoples' struggle to protect their homelands. Before bidding farewell to the elders, as a committed deep ecologist, Cameron affirmed in a review in the *Guardian* by Tom Phillips; 'the rivers and the forests have a moral right to continue to exist as they have for thousands of years' (18 April 2010).

Cameron refused to back down apparently, according to journalist Gina Salamone (18 February 2010), when Fox studio executives suggested he leave some of 'the tree-hugging *Fern Gully* crap out of this movie' (see www.nydailynews.com). Betting that viewers would feel moral outrage at the Company's treatment of the Na'vi, Cameron's film successfully tapped into audiences' increasing awareness of global warming and some even suggest frustration over the wars in Iraq and Afghanistan. Paradoxically, however, the destruction of the Na'vi Home-Tree required Cameron to fully exploit the special effects and visceral pleasures of the blockbuster business model that is a hallmark of capitalist consumption.[7]

Nevertheless, many observers less worried about creating ecological awareness highlighted that *Avatar* simply pilfered aspects from films such as *The Emerald Forest* (1985), *Ferngully* (1992) or *Pocahontas* (1995). Others complained about Sully's 'white messiah' stereotyping (see www.fantasy-matters.com or Elsaesser, 2011) and suggested that the film recreated a 'noble savage' narrative, further playing into regressive colonial discourses.

Most recently, Carolyn Michelle *et al.* addresses these and other concerns in a very useful audience analysis titled; 'Understanding Variation in Audience Engagement and Response: An Application of the Composite Model to Receptions of *Avatar* (2009)' (2012). Such tentative research, however, is at a very early stage of development and much more broad-based evidence is needed to tease out and defend the film's ability to promote pro-environmental messages, around the sacred right of nature to protect itself.

Similarly, Lisa Sideris's (in Taylor, 2013) evocation of empathy in the film is most informative by the way she poses the question, alluded to earlier with regard to Hurricane Sandy and *The Day After Tomorrow*: does our constant bombardment with images of suffering, environmental disaster and injustice inure us from these important issues, leading to what experts call empathy or compassion fatigue? See for instance a recent review piece by Hugh Wilson titled 'Have you got Green Fatigue?' (*The Independent*, 29 November 2013, accessed through www.independent.co.uk). Maybe this is the case, yet I would like to believe that education for empathy, taking on board various caveats, can nevertheless be effectively used to promote a transformational mode of active environmental engagement and even proactive citizenship. Advocates for instance cite how children need to form an emotional, visceral bond with the natural world and with nonhuman forms of life (see for instance the *Biophilia* thesis by Wilson, 1984), before learning the dispiriting details of the environmental crisis (see Louv, 2010). The potential of eco-film in mobilizing debate using successful examples such as *Avatar*, can by all accounts support and promote this process of engagement. The use of a broad range of ecocinema as an educational tool is becoming a growing preoccupation in recent studies, as documented in Greg Garrard's edited (2012) volume, which includes a wide range of pedagogical approaches, drawing from scholarship across more explicitly defined eco-literature and film studies.

Without question, however, the most provocative critic of *Avatar* remains Slavoj Zizek, who winces over its 'politically correct' themes, supporting 'an array of brutal racist motifs; a paraplegic outcast from earth is good enough to get the hand of a beautiful local princess and to help the natives win the decisive battle'. Rather than promoting a proactive ecological message, the film, he claims, teaches us that the only choice the aborigines have is to be saved by the human beings or to be destroyed by them. In other words, 'they can choose either to be the victim of imperialist reality, or to play their allotted role in the white man's fantasy' (cited in Taylor, 2013: 4). In a YouTube video entitled 'Ecology as Religion', Zizek further denounces the film, calling it a mystifying ideology and encapsulating 'the new opium of the masses'. According to him, we need to love and embrace the real world, not an idealized ecological one (Zizek, 2010).

While sensitive to Zizek's fears, Bron Taylor in his final summation of the influence of the film, sticks his neck out, affirming: 'that there may well be a gestalt change in consciousness beginning to emerge' (2013: 4) following the films' release. While I would certainly caution against such unbridled optimism, echoing criticism of Rust's earlier endorsement of the power of *The Day After Tomorrow*, nonetheless one might at least accept the film's flawed potency and staying power in helping to extend a growing corpus of eco-films, alongside provoking a more robust form of eco-criticism. Such a popular mainstream film can, like others discussed in this chapter, be used as an effective short-hand and as an evolving template for popular ecological discussion around the exploitation of scarce natural resources, together with other prescient ecological concerns.

Conclusion

Many critics including Zizek and Ivakhiv discussed above worry that a strictly aesthetic or moralistic approach to ecocinema falls short of offering critics a sufficient toolkit for identifying and analysing the contradictions inherent in all films. While an underlying question for David Ingram is whether one film style, genre or taste culture is more effective than another in promoting some form of ecological understanding. We are some way from confidently addressing these and other theoretical debates and assumptions concerning the effectiveness or otherwise of so-called ecological cinema. Furthermore, we do not all agree on a robust definition of what constitutes ecocinema. While many scholars cited above affirm the power of documentary and art-house cinema to provoke an ecological agenda, this chapter has tried to illustrate how a wide range of fictional film can also be effectively read as promoting a more all-encompassing ecological agenda.

In particular, much progress has been made in cataloguing a growing corpus of ecocinema and developing effective multilayered textual analysis protocols for appreciating ecocinema. Rust *et al.* (2013), together with Willoquet-Maricondi (2010), Taylor (2013) and other readers in the publishing pipeline are creating a growing body of scholarship that will embed a broad range of strategies and theoretical models to assist in this growing area of study. Most importantly now, however, as mentioned in this chapter, extensive audience research and reception studies are badly needed to test and evaluate many of the hypotheses and assertions developed within the eco-literature. Incidentally, Thomas Elsaesser has provided a useful model for web analysis of film reception that could be applied also to ecocinema. He suggests four layers including:

- raw data culled from statistics (such as Google's 'Insight for Search');
- data gathering from users on (nationally, linguistically and regionally specific) blogs, list-servers, chats (including IMDb users' reviews);

- critics' taste and classification (e.g. external reviews listed on IMDb, Rotten Tomatoes, Metacritic, Factiva.com);
- scholars' conceptualizations and systematization (trans-national cinema, postcolonial studies, etc.) (Elsaesser and Buckland, 2013: 180).

By all accounts, arguments will only have force if we physically feel them. In other words if an argument fails to generate feelings, or does not tap into an affective range of public engagement, then it will probably not persuade. This is why the creative imaginary of fiction remains so important in mobilizing and framing public opinion and the extensive power of emotions remains of primary importance in affecting audiences. Such arguments only motivate when they induce feelings including satisfaction, pleasure, excitement, interest, anger or distress. If it generates no feelings at all, they are unlikely to be persuasive.

Margrethe Bruun Vaage for instance suggests in a *Nordicom Review* piece (2009: 159–178) that fiction film elicits self-reflection through self-focused role-taking, where spectators with the help of fiction clarify emotional experiences that have relevance in their own lives. She uses the untheorized notion of 'transportation' and 'transformation' to help appreciate what happens when strong emotional experiences in film viewing take place, drawing on personal and non-conventional associations and reflections. These useful terms, together with the well-used notion of the creative imaginary of film, could be more explicitly foregrounded and theorized in future examinations of ecocinema.

Furthermore, at the reception level, it would appear much depends on the predisposition of various audiences and their engagement with ecocinema. At one extreme, eco-narratives designed for more attuned or susceptible spectators might be felt almost like a religious experience, leaving one open to unconditional surrender and life changing values – echoing the comments by Willoquet-Maricondi for instance at the start of the chapter. While, for a majority audience, one supposes they might probably experience a temporary affect at least when consuming ecocinema. But to test or validate such assumptions, extensive longitudinal and cross-cultural audience analysis is badly needed.

As also insinuated in this chapter, eco-film scholarship's overarching purpose probably should *not* be used to impose a political programme and still less to propose a more 'efficient' communication of scientific truths to a waiting audience, as evidenced in documentaries such as *An Inconvenient Truth*; but more democratically can be actioned through a broad spectrum of eco-inspired fiction and documentary to raise political awareness and help create a public sphere for debate and argument over the claims of the environment, demanding a central place in political life. Mass audience fiction film certainly remains a persuasive medium to promote and highlight complexity around environmental debates.

We urgently need more research and evidence to evaluate the overall effects of the production, circulation and especially reception of what can be loosely categorized as ecocinema (see Lin, 2013 for a very useful survey of Taiwanese students). Audiences seem to have a strong craving for stories and narratives of all kinds and this probably is connected to a constitutional functioning of our brain where narratives across cultures always have been used to help make sense of our world. Audience research in recent years has suggested that there is no contradiction between experiencing films as entertaining with an escapist potential on the one hand, and alternatively as creating profound meaning and deep engagement upon the minds of the spectators on the other hand. But further research is needed to confirm the most appropriate questions to investigate and provide robust ways of developing an ecological approach to film study. David Whitley hits the right note when he affirms with regard to *Wall-E*: 'we clearly need fables of enchantment of this imaginative quality if we are to develop the kind of ethical generosity and grounded vision that are necessary (as the film puts it) to live fully, or even perhaps to survive' (2012: 159).

Notes

1   There has been much recent debate and investigation into whether film can really affect psychological mood and our explicit attitudes towards climate change in particular. One study for example showed clips of *An Inconvenient Truth* and found that the clips shown did affect emotion and some participants were more inclined to do something about the problem (Beattie, Sale and McGuire 2011).

2   I tend to instinctively endorse Oliver and Hartmann's perspective, when they argue that film viewing 'may have the potential to do much more than provide viewers with feelings of gratification, but may also serve as a means for instigating positive social change' (2010: 145).

3   It is interesting that most US commentators continue to use the term 'global warming' which is recognized as a very unscientific term and tends to also be a politically charged term which is both inaccurate and misleading, while also confusing public understanding of science.

4   'Research has also demonstrated that the experiential system can have powerful influences on risk perception, decision making and behaviour.' See reports in journal *Environment* 46(9) 22–37, 2004.

5   See proceedings paper from 2009 Science fiction film conference in Chicago, USA delivered by Bob Mellin 'White Flights and the Environmental Minstrel in *Wall-E*' Purdue University bmellin@pnc.edu

6   The figures would have been significantly higher had not the Chinese government apparently cut short the film's run, reportedly out of fears that it might encourage resistance to development projects and their resettlement schemes.

7   See Elsaesser's 2011 reading of Cameron's motivation. Incidentally, in *Hollywood Utopia* (2005), I examined Cameron's *Titanic* as potentially ecological in its premise/execution. His oeuvre has come a long way, however, towards explicitly foregrounding an ecological agenda.

Further reading

Ingram, David (2000) *Green Screen: Environmentalism and Hollywood Cinema*. Exeter University Press.

Rust, Stephen, Salma Monani and Sean Cubitt (eds.) (2013) *Ecocinema, Theory and Practice*. AFI Film Readers. New York and London: Routledge.

Willoquet-Maricondi, Paula (ed.) (2010) *Framing the World: Explorations in Ecocriticism and Film*. Charlottesville and London: University of Virginia Press.

References

Beattie, G., L. Sale, and L. McGuire (2011) 'An inconvenient truth? Can a film really affect psychological mood and our explicit attitudes towards climate change'? *Semiotica* 187, 105–125.

Bird, S. (ed.) (1996) *Dressed in Feathers: The Construction of the Indian in American Popular Culture*. Boulder, CO: Westview Press.

Brereton, Pat (2005) *Hollywood Utopia: Ecology in Contemporary American Cinema*. Bristol: Intellect Press.

Brereton, Pat (2012) *Smart Cinema: DVD Add-ons and New Audience Pleasures*. Basingstoke: Palgrave MacMillan.

Brereton, Pat and Pat Hong (2013) 'Irish eco-cinema and audiences'. *Interactions: Studies in Culture and Communications*. Special issue on ecocinema, editor Pietari Kääpä.

Cubitt, Sean (2005) *EcoMedia*. New York: Radopi.

Elsaesser, Thomas (2011) 'James Cameron's *Avatar*: Access for all'. *New Review of Film and Television Studies* 9(3), 247–264.

Elsaesser, Thomas and Warren Buckland (2013) 'The life cycle of *Slumdog Millionaire* on the web', in Ajay Gehlawat (ed.) *The Slumdog Phenomenon: A Critical Anthology*. London: Anthem Press, pp. 179–199.

Garrard, Greg (ed.) (2012) *Teaching Eco-criticism and Green Cultural Studies*. Basingstoke: Palgrave MacMillan.

Ingram, David (2000) *Green Screen: Environmentalism and Hollywood Cinema*. Exeter: Exeter University Press.

Ivakhiv, Adrian (2013) *Ecologies of the Moving Image: Cinema, Affect, Nature*. Waterloo, Ontario: Wilfred Laurier University Press.

Kääpä, Pietari (2013) *Ecology and Contemporary Nordic Cinema*. New York: Bloomsbury.

Kilpatrick, Neva Jacquelyn (1999) *Celluloid Indians: Native Americans and Film*. Second Edition. Lincoln, NE: University of Nebraska Press.

Leiserowitz, Anthony (2003) 'Global Warming in the American Mind'. University of Oregon. Environmental Science Dissertation

Leiserowitz, Anthony (2004) 'Before and After *The Day After Tomorrow*: A US study of Climate Change Risk Perception'. *Environment* 46(9), 22–37.

Lin, Sue-Jen (2013) 'Perceived impact of a documentary film: An investigation of the first-person effect and its implications for environmental issues'. *Science Communication* 35(6), 708–733.

Louv, Rich (2010) *Lost Child in the Woods: Saving Our Children from Natural Deficit Disorder*. London: Atlantic Books.

MacDonald, Scott (1998) *A Critical Cinema 3: Interviews with Independent Filmmakers*. Berkeley, CA: University of California Press.

McGowan, Todd (2010) 'Maternity divided: *Avatar* and the enjoyment of nature'. *Jumpcut: A Review of Contemporary Media* 52, 1.

Marx, Leo (1964) *The Machine in the Garden: Technology and the Pastoral Ideal in America*. New York: Oxford University Press [2000 edition].

Mellin, Bob (2009) 'White flight and the environmental minstrel in *Wall-E*'. *Proceedings from Science Fiction Film Conference, Chicago*, 30 October–2 November.

Mitchelle, Carolyn, Charles H. Davis and Florin Vladica (2012) 'Understanding variation in audience engagement and response: An application of the composite model to receptions of *Avatar* (2009)'. *The Communication Review* 15(2), 106–143.

Murray, Robin and Joseph Heumann (2007) 'Al Gore's *Inconvenient Truth* and its skeptics: A case of environmental nostalgia'. *Jumpcut: A Review of Contemporary Media*, 49, 1.

Murray, Robin and Joseph Heumann (2009) *Ecology and Popular Film: Cinema on the Edge*. Albany, NY: Suny Press.

Oliver, M. B. and T. Hartmann (2010) 'Exploring the role of meaningful experiences in users appreciation of "good movies"'. *Projections* 4(2), 128–150.

Plantinga, Carl and Greg M. Smith (eds) (1999) *Passionate Views: Film Cognition, and Emotion*. Baltimore: John Hopkins University Press.

Rust, Stephen, Salma Monani and Sean Cubitt (eds) (2013) *Ecocinema, Theory and Practice*. AFI Film Readers. New York and London: Routledge.

Smith, Greg (2003) *Film Structure and the Emotion System*. Cambridge, UK and New York: Cambridge University Press.

Staiger, J. (2000) *Perverse Spectators: The Practices of Film Reception*. New York: New York University.

Taylor, Bron (ed.) (2013) *Avatar and Nature Spirituality*. Waterloo, Ontario: Wilfrid Laurier University Press.

Vaage, M. B. (2009) 'Self reflection: Being beyond conventional fictional film engagement'. *Nordicom Review* 2, 159–178.

Weik von Mossner, Alexa (ed.) (2014) Moving Environments: Affect, Emotion, Ecology and Film. Waterloo, Ontario: Wilfrid Laurier University Press.

Whitley, David (2012) *The Idea of Nature in Disney Animation*. 2nd edn. Aldershot: Ashgate Press.

Willoquet-Maricondi, Paula (ed.) (2010) *Framing the World: Explorations in Ecocriticism and Film*. Charlottesville and London: University of Virginia Press.

Wilson, Edward D. (1984) *Biophilia: The Human Bond with other Species*. Cambridge, MA: Harvard University Press.

Zizek, Slavoj (2010) 'Return of the natives'. *New Statesman*, 4 March.

23

NATURE, ENVIRONMENT AND COMMERCIAL ADVERTISING[1]

Anders Hansen

Introduction

Nature imagery and ideas regarding 'the natural' form an important part of advertising and popular culture generally. While appeals to environmental protection and 'green' or 'sustainable' consumption wax and wane in public communication, nature imagery continues to figure prominently in commercial advertising and other media discourse. Advertising is perhaps particularly interesting in this regard because it effortlessly and seamlessly draws on culturally deep-seated and taken-for-granted meanings of nature and the natural, and reworks these in ways that promote consumption, particular worldviews and particular identities. Starting from the recognition that nature is socially, temporally and culturally constructed, this chapter explores the key ways in which commercial advertising uses and articulates ideas of nature. It reviews research that has examined the nature-discourses deployed in advertising content and their role in articulating and reinforcing public definitions of the environment, consumption and cultural identity. The chapter examines what research has shown about how advertising's uses and constructions of nature change over time and how advertising articulates and reworks deep-seated cultural categories and understandings of nature, the natural and the environment, and in doing so communicates important boundaries and public definitions of appropriate consumption and 'uses' of the natural environment.

Constructed nature and ideology

A discussion about the role of nature and appeals referencing 'natural-ness' in advertising perhaps needs to start by reiterating the simple recognition of the complex and historically changing meanings associated with 'nature'. Not only does nature figure prominently in our cultural vocabulary and in popular culture, it means different things, at different times to different people. Cultural critic Raymond Williams called nature 'perhaps the most complex word in the language' (Williams, 1983: 219). He noted how dominant cultural views of nature change significantly over time, although always drawn from a rich and ever-present reservoir of binary opposites, chief among which is culture versus nature. The way that cultural views of nature change historically not only shows that the meanings that we associate with nature are 'constructed'(rather than 'natural' or pre-given) but also provides an entry point for understanding the ideological uses of such constructions.

Contrary to its surface appearance of referencing something ontological, authentic and god-given, nature can be, and has been, used for lending authority to particular ideas, interests and political objectives.

Nature, as Evernden (1989: 164) argues, 'is used habitually to justify and legitimate the actions we wish to regard as normal, and the behaviour we choose to impose on each other'.

Thompson (1990: 66), in his comprehensive discussion of ideology, culture and media, refers to this as the 'strategy of naturalisation', and others, such as Fairclough (1989) and Stuart Hall (1982), have similarly noted the ideological character and centrality of a discourse of naturalisation in media and political rhetoric. Stuart Hall thus notes that media discourse is best characterised, 'not as naturalistic but as *naturalized*: not grounded in nature but producing nature as a sort of guarantee of its truth' (Hall, 1982: 75).

If nature-referencing and naturalisation are key rhetorical components of the way in which ideology is communicated, then the semiotic linking of a (romanticised) view of nature *with* a rural (idyllic) past *with* national identity has undoubtedly been one of the most potent ideological uses in the modern age. And nowhere more so than in advertising/marketing, where, as Soper (1995: 194) has argued, 'referencing of nature and the clichés of nationalist rhetoric have become the eco-lect of the advertising copywriter'.

The use of nature and referencing of the natural in advertising may be partly to do with ideological power (selling products or corporate images by invoking the qualities of goodness, purity, authenticity, genuineness, non-negotiability (Cronon, 1995)) and partly to do with the format constraints of the advertising genre. Thus, the time-constraints of the advertising format calls for the use of easily identifiable and recognisable shorthand symbols, or what Gamson and Modigliani (1989) aptly refer to as 'condensing symbols', and nature and the natural are perhaps among the most universally recognisable such symbols.

The chief ideological power of referencing of nature/the natural in advertising derives partly from the complexity and semantic flexibility of these referents, and partly from their predominant 'taken-for-granted' inconspicuousness. The ability of advertising to forge signification links that convey such key nature-related values as freshness and health onto cigarettes and smoking (Williamson, 1978) is perhaps one of the clearest examples of the semiotic flexibility and power of uses of nature in advertising. Nor has the potential power of nature-referencing been lost on corporate image advertisers keen to enhance connotations with environmentally responsible, ethical, sustainable behaviour.

Environment and nature in advertising and other media

Numerous studies have documented the considerable increase in news media reporting (and public concern) about environmental issues that happened in the latter half of the 1980s and very early 1990s (See Hansen, Chapter 17 and Boykoff *et al.*, Chapter 18 in this Handbook). A number of studies have, likewise, begun to reveal some of the characteristics and trends in representations of the environment, nature or environmental issues in advertising. Comparing a sample of American television advertisements from 1979 with a sample a decade on, 1989, Peterson (1991) thus found that explicit environmental or ecological messages were used in less than a tenth of advertisements, although slightly more frequently in 1989 than in 1979. Several studies from the mid 1990s (e.g. Iyer and Banerjee, 1993; Banerjee *et al.*, 1995; Kilbourne, 1995; Buckley and Vogt, 1996; Beder, 1997) indicate that the increased news and public interest in the environment, seen in the late 1980s, was also reflected in advertising, where so-called 'green advertising' or 'green marketing' became prominent. Much of this advertising latched on to general public concerns by labelling products as 'green' and 'eco-friendly' or by emphasising measures taken to reduce the potential impact that advertised products might have on the environment. The trend also extended to corporate image advertising stressing the environmental credentials of large companies, and to advertising more directly in the tradition of persuasive information/education campaigns in the form of government and local authority advertising campaigns designed to promote recycling initiatives or environmentally responsible behaviour (Rutherford, 2000).

Much of recent research on images/messages regarding the environment has focused on 'green' advertising/marketing and on 'green-washing' (see Cox, 2013, chapter 10 for an insightful discussion of green marketing and corporate campaigns), interpreted as the use of deceptive claims or disinformation

'regarding the environmental practices of a company or the environmental benefits of a product or service' (Baum, 2012: 423). In a study of print advertisements in leading news, business, environmental and
science magazines in the USA and UK in 2008, Lauren Baum (2012) found three-quarters of 'environmental' advertisements to contain one or more aspects of green-washing, and she argues that without
stricter regulations, a trend of increasing use of green or environmental appeals – and of green-washing
– in product and corporate image advertising is set to continue unchecked.

The overall trend indicated here is confirmed by a comprehensive study by Lee Ahern and colleagues
(Ahern *et al.*, 2012; Bortree *et al.*, 2013) of environmental messages in advertisements in the *National
Geographic* magazine over the three decades from 1979 to 2008. Ahern and colleagues show a significant
increase in corporate environmental responsibility communication over the three decades, as well as considerable changes in the framing of environmental messages. They particularly note an increasing emphasis
on promoting the environmentally positive actions ('doing more') of corporations rather than an emphasis
on conservation ('taking less from the earth').

In an exemplary earlier study Howlett and Raglon (1992) chart the changing uses of appeals to nature
and 'the natural' in advertising during the twentieth century. They show that, while product association
with nature and the natural changes little, 'environmental' corporate image advertisements gain momentum principally since the early 1970s, as companies start to 'portray themselves as nature's caretakers;
environmentally friendly, responsible, and caring', resulting, as they continue, 'in the use of more scenes
from nature … and an almost total elimination of the factory and machinery visuals which were standard
fare in the corporate image ads of the 1950s' (Howlett and Raglon, 1992: 55).

These and other studies (e.g. Plec and Pettenger, 2012; see also the collection of papers in the
2012 special issue of the *Journal of Advertising* on green advertising – volume 41/4 edited by Sheehan
and Atkinson) then confirm the continued and increasing reference to environment, nature and
environmental issues in advertising, the prominence of corporate voices in such advertising, and
the increasing use of green-washing frames that show products and corporations as environmentally
friendly and responsible.

While the prominence of explicit appeals to 'environmentally responsible' behaviour in advertising
goes up and down in ways seemingly not dissimilar to the ups and downs seen in news media attention
to environmental issues (see Chapter 17 'News coverage of the environment: a longitudinal perspective'
in this handbook), there is considerable evidence (e.g. Howlett and Raglon, 1992; Goldman and Papson,
1996; Rutherford, 2000; Corbett, 2002; Hansen, 2002) that references to nature and the natural have
been prominent throughout the twentieth century and continue to be so in the twenty-first century.
Hansen (2002), in a study of British television advertising, thus showed that 28 per cent of all advertising
contained nature referencing and/or imagery.

Goldman and Papson (1996) persuasively argue:

> From its inception, modern advertising used nature as a referent system from which to derive sig-
> nifiers for constructing signs. Nature's landscapes were used to signify experiences or qualities that
> urban-industrial life failed to provide. … Advertising suggested that civilisation's deficiencies could
> be ameliorated by consuming commodities that contained the essence of nature.
>
> (1996: 191)

Drawing on a number of historical analyses (including Marchand, 1985), they argue that nature was prominently used in advertising in the 1920s and 1930s in a nostalgic way that was itself a response to the economic
and the social-psychological crisis of the 1920s. Tracing the general trends in uses of nature in advertising
up through the twentieth century, Goldman and Papson note that '[t]he nostalgia for nature evident in the
advertising of the 1920s and 1930s gave way to the fetish of gadgetry' (p. 191) for the middle decades of the
twentieth century, and not until the 1970s did nature once again take a central position in advertising. While

nature is thus seen to have been prominent in advertising throughout the twentieth century, Goldman and Papson also point out that the genre known as 'green' advertising did not emerge until the 1980s.

Taking his point of departure in the now famous Crying Indian advertisements of the early 1970s, historian Kevin Armitage (2003) shows how (American) popular culture has long used the stereotype of the noble savage to epitomise and articulate a nostalgic view of unspoilt nature and of an idealised past of harmony between man and nature. Armitage persuasively argues that 'The fascination with nature and the primitive that marked turn-of-the-twentieth-century American culture was rooted in a larger ambivalence about modern life' (p. 73) – not unlike the disillusion with modernisation identified by Raymond Williams (1973) in the British context – but crucially, as Armitage argues, the fascination with nature and the primitive 'did not involve a rejection of civilisation, but rather an accommodation to modern life that was simultaneously nostalgic and progressive, secular yet spiritually vital' (pp. 73–74). Armitage shows that the idealised referencing of nature – through or with representations of the American Indian – in advertising was well under way towards the end of the nineteenth century.

In an exemplary analysis of advertisements for pesticides in agricultural magazines, spanning the half century from the 1940s to the 1990s, Kroma and Flora (2003) demonstrate the changing prominence of three different discourses: in the 1940s–1960s a 'science' discourse articulating the post-war faith in progress through science; in the 1970s–1980s, a 'control' (of nature/the environment) discourse drawing extensively from military/combat control metaphors; through, in the 1990s, a 'nature-attuned' discourse reflecting environmental sensibilities – concerns about sustainability, protection of and harmony with nature – emerging in the latter half of the twentieth century. They conclude that 'changing images reflect how the agricultural industry strategically repositions itself to sustain market and corporate profit by co-opting dominant cultural themes at specific historical moments in media advertising' (Kroma and Flora, 2003: 21).

Wernick's (1997) comparison of 1950s and 1990s advertising adds further confirmation of the changes in the ways in which nature imagery has been deployed in advertising. Where the 1950s advertisements celebrate 'the fruits of industrial civilisation' (p. 209), gadgetry, technology, science and progress, the 1990s adverts appeal to nostalgic ideas of nature and the past, a nature and a past that exist only in myth: 'The Good identified as the essence of its product is located in the past, not in the future. It … is something to be recovered rather than attained' (p. 210).

Drawing on the arguments presented by these authors, it seems then that a key difference in the uses of nature between advertising of the 1940s–1970s and advertising of the late twentieth century is one of perspective: the adverts of the middle part of the twentieth century in short *look forward*, with optimism even, to the progress and prosperity of the techno-scientific urban society, while the perspective of the late twentieth century is one of *looking back* – to recover a lost idyll, harmony, authenticity and identity of a (mythical) past. Wernick refers to the 'progress myth' of the 1950s advertising; others, notably Gamson and Modigliani (1989) in their study of the framing of nuclear power since the mid-twentieth century in popular culture and in public opinion, refer to this as the 'progress package' – a common and prominent frame in media and popular culture accounts involving the relationship between technology and nature, and valorising technological, economic and scientific progress above concerns for the environment or nature. Wilson (1992) likewise identifies this period as one in which the relationship with nature was one of domination and greed, where the urge to 'acquire and consume' (p. 14) far outpaced any hint of concern about the environment, limited resources or the protection of nature.

Rutherford (2000) introduces his own label for advertising celebrating the progress myth: 'Technopia'. In his historical sweep of what he broadly terms 'advocacy advertising', he implies a similar trend to that identified by Wernick, Goldman and Papson and others. He contrasts the 'Technopia' type of advertising – advertising that principally promotes a belief in the scientific and technological control and domination of nature as synonymous with progress and development – with what he terms 'Green Nightmare' advertising. 'Green Nightmare' advertising is advertising that stresses and calls public attention to the 'Dystopia'

– the destruction of nature, the environment and our entire habitat – resulting from the unchecked and wasteful production and consumption practices characteristic of late modernity.

In very general terms, Rutherford's analysis maps onto the time-line indicated above, namely with 'Technopia' advertising most prominent in the 1960s–1980s, and 'Green Nightmare' advertising prominent from the 1970s onward. If Rutherford's categories seem to overlap considerably, it is perhaps confirmation, not only of a diversification of discourses on nature, but of a public sphere marked increasingly by discursive competition over the framing and meaning of nature generally, and more specifically of the framing of science, technology and progress in relation to public conceptions of nature.

In summary, then, it would seem from the work of the authors discussed above that nature imagery has been a feature of advertising since at least the early part of the twentieth century. It is also clear that the particular deployment and constructions of nature in advertising and other media have, broadly speaking, oscillated between, at the one extreme, a progress-package-driven view of nature as a resource to be dominated, exploited and consumed, and, at the other extreme, a romanticised – and often retrospective – view of nature as the (divine) source and embodiment of authenticity, sanity and goodness, to be revered and protected (or to use a more current invocation: 'not to be tampered with').

Nature, nostalgia and social (race/class/gender) identity

Nature imagery in advertising of the late twentieth century is, as we have seen above, often deployed in relation to a retrospective look, a yearning for the 'idyllic past'. Nature imagery in this context is used to construct a mythical image of the past (including childhood) as a time of endless summers, sunny and orderly green landscapes, and, perhaps most importantly of all, as a time and place of community, belonging and well-defined identity. Several researchers have referred to this view as one of 'nostalgia' (e.g. Davis, 1979).

The nostalgic view of the past, as enacted through the use of nature imagery in advertising, is not merely a longing for a mythic past, but it is very much also a romanticised view of the past. In its use of nature imagery, it draws particularly on the romantic view of the countryside, the view constructed not least by the poets (e.g. William Wordsworth, Samuel Taylor Coleridge) and painters (e.g. John Constable, J. M. W. Turner) of the Romantic period. As Williams (1973) has pointed out, the growing cultural importance of a romanticised view of the countryside perhaps not surprisingly coincided with a period of immense social upheaval, urbanisation, migration to the cities and the rapid decline of a rural/agrarian economy.

Against the tremendous social, economic and political upheaval characterising much of the twentieth century, not least the first half, it seems perhaps hardly surprising that advertising should respond with romanticised images of a more natural, rural, countryside past, where identities seemed more firmly fixed, if only through 'knowing one's place' in the highly hierarchical structure of rural society. What is particularly ideological about this reconstructed past is the way in which the deeply hierarchical structures are either glossed over *or* romanticised and portrayed as indeed natural, desirable and harmonious.

The romanticised construction of nature and the uses of idyllic nature in advertising are then not just a matter of advertising responding to a public sense of alienation or a public search for identity. They are an ideological reconstruction in the sense that they naturalise, and sometimes even celebrate, a deeply stratified society. There are, in other words, important social class, race and gender dimensions to these uses of nature.

Phillips *et al.* (2001) in their analysis of the construction of rural/countryside/nature imagery in British rural television drama thus show that the dominant construction of a rural idyll goes further to 'also enact particular social identities, including, but not exclusively, those of class' and that the class identity enacted is predominantly a middle-class identity. Others (Thomas, 1995; Scutt and Bonnet, 1996) have similarly commented on the social class, race and gender dimensions of television and print media constructions of the countryside and nature.

In a similar vein, but with a particular focus on racial exclusion in American magazine advertising from 1984–2000, Martin (2004) points to the racial dimension to nature imagery: 'Advertisements taking place in the Great Outdoors or featuring models participating in wilderness leisure activities rarely include Black models, while advertisements featuring White models regularly make use of Great Outdoors settings and activities' (p. 513). He notes that the dominant view of nature/wilderness in advertising is a white Eurocentric view that finds little resonance among Black and Native American audiences.

Nature and national identity

A considerable body of literature has pointed to the links between particular constructions of nature and national identity. MacNaghten and Urry (1998), drawing on a broad range of work from geographers, sociologists and historians, note how every nation celebrates its particular nature. 'National natures' may not seem particularly 'constructed' where these bear a seemingly obvious relation to the particularly striking features of those natures (the Alps of Switzerland, the fjords of Norway, the forests and lakes of Finland, etc.). However, on closer historical scrutiny, it becomes clear that 'national natures' are indeed very much 'constructed'. This is made particularly clear in historian Simon Schama's insightful analysis of *Landscape and Memory* (Schama, 1995), in which he demonstrates the particular historicity and political role and construction of nature in the culture and politics of a range of nations (with examples ranging as widely as the 'forest' in German culture and history to 'wilderness' and national parks in the United States). Geographers, historians, sociologists and media researchers in Britain have commented on the close links forged, from the 1800s onwards, between national identity – particularly that of the English and Englishness – and a romantic view of nature.

Thus, since the late 1800s, the dominant image of Englishness in literature, art and popular culture generally has become one of equating Englishness with the countryside, the countryside as the true home of the English (seen as white and middle-class) and the essence of Englishness. However, as Thomas (2002) among others has noted, the 'association of national identity with a country's rural roots is not confined to Britain, and may be connected to the cultural homogenisation which is one of the outcomes of globalisation' (p. 34).

Following a similar line of argument, Creighton (1997), looking at domestic tourism and popular culture in Japan, demonstrates a renewed search for authentic Japanese identity as manifested in the increasing popularity of 'traditional' rural Japan. She describes the 'retro boom' – a looking back to the past and a search for authentic Japanese identity – experienced in Japan since the 1970s as a reaction to 'the perceived threat of cultural loss to which the processes of modernisation and Westernisation have subjected modern Japan' (p. 242). As in British advertising and popular culture, the 'place' of authentic national culture is seen as the countryside or traditional village.

For the alienated urban masses, the search for identity is supposedly answered through the travel, as promoted by tourism advertising, 'back' to the true time and place of Japanese culture and identity, the romanticised rustic countryside setting. But, as Creighton demonstrates, this journey is increasingly commodified in popular culture, department store displays and consumer goods, so that the busy urban dweller need never leave the city in order to buy into the retro boom construction of Japanese cultural identity.

As in the West, the achievement of advertising deploying this kind of nature imagery is to channel the yearning for authenticity or identity or the pure goodness of nature into consumption: purchasing the advertised product becomes a means of 'buying into' the identity or the authenticity ostensibly anchored in the idyllic rural past.

While, as indicated by MacNaghten and Urry (1998) and others, there are different 'national natures', it is perhaps testimony yet again to the semantic flexibility hinted at by Williams (1983), that some have implied a degree of global universality in 'nature imagery' and cultural constructions of nature. Howlett and Raglon (1992) thus argue that the attractiveness of nature imagery and symbolism to advertisers stems from the simple recognition that '[N]atural symbols and metaphors are among any culture's most easily understood ones' and they 'tend to be long-lived and their meanings widely accessible' (p. 61).

The similarities noted above between British and Japanese linking of national identity with a (romanticised) rural, idyllic, countryside past likewise suggest an element of culture-transcending universality. In a comparative study of cultural values in American and Japanese advertising, Barbara Mueller (1987) found that on the two nature-related dimensions investigated ('oneness with nature appeals' and 'manipulation of nature appeals') there was remarkably little difference between the advertising of the two countries:

> Themes emphasizing the goodness of nature are found in both Japanese and American advertisements. A subtle difference in the application of this appeal exists: U.S. advertisements in this category focus on natural as opposed to man-made goods, while the Japanese advertisements emphasize the individual's relationship with nature.
>
> (1987: 56)

In an interesting follow-up study twenty years later, Okazaki and Mueller (2008), comparing findings for 1978 advertising with 2005 advertising in Japan and the USA, found the 'oneness with nature appeal', identified as more characteristic of Eastern culture, continued to be prominent in Japanese advertising, but had dropped significantly in American advertising. By contrast, 'manipulation of nature appeals' had increased from a mere 0.7 per cent (1978) to 2.6 per cent (2005) in Japanese advertising, but were rarely deployed in US advertising of either period. The overall conclusion from the follow-up study was that advertising appeals in the two countries had become increasingly similar.

In her study of cultural values in Chinese and American television advertising, Carolyn Lin (2001) notes that previous studies have shown that 'advertisements in China are more likely than Western advertisements to use appeals of traditional values such as status and oneness with nature, whereas U.S. advertisements reflect such values as individualism and manipulation or control over nature' (pp. 86–87). Her own study likewise confirms that Chinese advertisements are more likely to use oneness with nature appeals than US advertising.

Cho *et al.* (1999) identified three categories of appeals involving nature in American and Korean advertising: Manipulation of Nature, Oneness with Nature and Subjugation to Nature. They found that only Oneness with Nature featured prominently in the advertising of both countries, while Manipulation of Nature and Subjugation to Nature were comparatively rarely used. While the differences between the two countries were not statistically significant, Cho *et al.* found 'non-significant directional support for the contention that oneness-with-nature is found more often in Korean commercials and that manipulation-of-nature is found more often in U.S. commercials' (p. 68). Another study of Korean advertising, in a comparison with Hong Kong advertising (Moon and Chan, 2005), found a greater – but still relatively small – use of appeals to the natural in Korean advertising than in Hong Kong advertising, but, as in the study by Cho *et al.*, the difference was not statistically significant.

There is, however, considerable evidence from elsewhere (e.g. Kellert, 1995) to suggest that the uses and interpretation of nature do indeed vary across different cultures, that nature is indeed not only culturally constructed but also culturally specific in its construction/interpretation. Any apparent similarity, across (Occidental and Oriental) cultures, in advertising and other popular culture constructions/uses of nature are thus more likely to be symptomatic of the increasing globalisation, Westernisation and homogenisation, characteristic of modern advertising trends, than of some universality of nature as a sign and metaphor.

A particularly interesting – but somewhat different – inflection of nature and national identity in advertising is the use of national stereotypes. Advertising, as Armitage (2003) has shown, has long articulated and exploited the popular culture stereotype of the American Indian as the idealized 'child of nature', epitomising a nostalgic anti-modern sentiment and a yearning for a lost harmony between man and nature. But this type of inflection also extends beyond peoples (the American Indian) to national stereotypes. A particularly potent example is the referencing of the Irish and Irishness in the global marketing of Irish Spring soap by the American Colgate-Palmolive company. Elbro (1983) and Negra (2001) both offer insightful analyses of the ways in which the Irish Spring advertisements draw on and reinforce the (stereotypical) linking between nature (pure, cleansing and untainted by modernity), the (idyllic) past, nation and national (Irish) identity. As in much other popular culture construction of national identity (see Creighton on Japan, referred to above), nostalgia plays a key role, in that the linking or association also implies that Ireland is a place where the (natural) qualities of the past can still be found, visited and consumed, or alternatively, bought into through consumption of the advertised product.

The Irish Spring soap advertisements have used for example idyllic images of the Irish countryside (winding lanes, hedges, fresh and green) and the Irish (jolly courting couples in rural attire) to associate the qualities of freshness, authenticity, genuineness, romance, etc. with the advertised product. The advertisements trade on and reinforce a nostalgic stereotypical image of Ireland as a 'non-industrialized paradise populated by simple country folk' (Negra, 2001: 86) and of Irishness as synonymous with honest, authentic, natural, uncomplicated, pure and romantic qualities. Condensing-symbols of nation and national identity linked to landscape and the natural environment are of course widely exploited in tourism advertising (see e.g. Urry, 1995 and 2001; Negra, 2001; Nelson, 2005; Clancy, 2011; Uggla and Olausson, 2013), and very likely increasingly so, as the popularity of 'nature-tourism' and 'eco-tourism' continue to rise.

Although there is a large and growing body of comparative research on the cultural values reflected in advertising across different cultures, only a small number of these have touched specifically on nature imagery or uses/constructions of nature in such advertising, and this is clearly a field ripe for further investigation. Research needs to compare the uses/constructions of nature and environment in television and other media advertising of several different countries. It further needs to examine how far, in a time of increasing globalisation (not least in marketing and advertising), constructions of nature are either increasingly universal, homogeneous (or Western perhaps) or alternatively, how far advertisers in their use of nature imagery seek to reflect and appeal to regionally, nationally or culturally specific understandings of nature.

Conclusion

Raymond Williams argued that 'nature is perhaps the most complex word in the language' – the review in this chapter of how nature has been used and appropriated in media discourse essentially confirms this. It testifies to the signifying flexibility of nature, and to the historically changing underlying views of nature. It is only by examining how discourses of nature change over time that we can begin to understand how they are used ideologically for promoting everything from national identity, nationalism, consumerism and corporate identity to framing and circumscribing what kinds of questions can and should be asked about the environment, environmental issues and (sustainable) development.

Invoking nature/the natural in advertising and other public discourse is a key rhetorical device of ideology in the sense that referencing something as 'of nature' or as 'natural' serves to hide what are essentially partisan arguments and interests and to invest them with moral or universal authority and legitimacy.

While explicit environmental messages and 'green' advertising seem to come and go, with some periods of prominence in advertising since the early 1970s, it is clear from the literature examined here that nature and appeals to the natural have been a significant part of advertising since at least as far back as the early 1900s.

The particular deployment and constructions of nature in advertising and other media have, broadly speaking, oscillated between, at the one extreme, a progress-package-driven view of nature as a resource to be dominated, exploited and consumed, and, at the other extreme, a romanticised – and often retrospective – view of nature as the (divine) source and embodiment of authenticity, sanity and goodness, to be revered and protected, 'not to be tampered with'.

On the continuum between these extremes lie, as we have seen, a wide range of 'constructions' of nature, including: nature as resource, good, authentic, idyllic, healthy, spiritual, enchanting, the 'home' of identity, fragile, a threat, a 'proving ground' for both human and product qualities, vengeful, etc. Several authors have indicated that the dominant construction of nature in advertising and popular culture of the late twentieth century is one that draws heavily on nature imagery of the Romantic period. It is also one that invokes a nostalgic view of the past, with implications for the public construction of social-class, gender, race and, not least, national identity.

There are contrasting – and possibly contradictory – indications from a number of studies regarding the extent to which advertising and other media constructions of nature are culturally or nationally specific. Cross-cultural comparisons of advertising show relatively small differences in the views/constructions of nature in Occidental and Oriental advertising, and indicate that traditional – and well documented – differences in views of nature may be subsumed under the homogenising influence of globalisation. But the evidence is very tentative and further research should focus specifically on the construction of nature to examine the extent to which the discourses of nature in advertising are culturally and nationally specific.

Note

1   This chapter builds on and extends my Chapter 6 Selling 'nature/the natural' in Hansen (2010).

Further reading

Howlett, M. and Raglon, R. (1992). Constructing the environmental spectacle: Green advertisements and the greening of the corporate image. *Environmental History Review, 16*(4), 53–68.
This offers an excellently longitudinal examination of changes in environmental/nature referencing and imagery in advertising of the twentieth century.

Rutherford, P. (2000). *Endless Propaganda: The Advertising of Public Goods.* Toronto: University of Toronto Press.
See particularly Part IV: Progress and Its Ills, pp. 179–227.

Williamson, J. (1978). *Decoding Advertisements: Ideology and Meaning in Advertising.* London: Marion Boyars.
Williamson's study remains one of the best and most insightful applications of semiotic analysis of advertising. On the uses of nature/naturalisation/the natural in advertising, see particularly Chapter Four: 'Cooking' Nature, and Chapter Five: Back to Nature, pp. 103–137.

References

Ahern, L., Bortree, D. S. and Smith, A. N. (2012). Key trends in environmental advertising across 30 years in *National Geographic* magazine. *Public Understanding of Science, 22*(4), 470–494.

Armitage, K. C. (2003). Commercial Indians: Authenticity, nature and industrial capitalism in advertising at the turn of the twentieth century. *Michigan Historical Review, 29*(2), 71–96.

Banerjee, S., Gulas, C. S. and Iyer, E. (1995). Shades of green: A multidimensional-analysis of environmental advertising. *Journal Of Advertising, 24*(2), 21–31.

Baum, L. M. (2012). It's not easy being green … or is it? A content analysis of environmental claims in magazine advertisements from the United States and United Kingdom. *Environmental Communication: A Journal of Nature and Culture, 6*(4), 423–440.

Beder, S. (1997). *Global Spin: The Corporate Assault on Environmentalism.* Totnes, Devon: Green Books.

Bortree, D. S., Ahern, L., Smith, A. N. and Dou, X. (2013). Framing environmental responsibility: 30 years of CSR messages in National Geographic Magazine. *Public Relations Review, 39*(5), 491–496.

Buckley, R. and Vogt, S. (1996). Fact and emotion in environmental advertising by government, industry and community groups. *Ambio, 25*(3), 214–215.

Cho, B., Kwon, U., Gentry, J. W., Jun, S. and Kropp, F. (1999). Cultural values reflected in theme and execution: A comparative study of US and Korean television commercials. *Journal of Advertising, 28*(4), 59–73.

Clancy, M. (2011). Re-presenting Ireland: tourism, branding and national identity in Ireland. *Journal of International Relations and Development, 14*(3), 281–308.

Corbett, J. B. (2002). A faint green sell: Advertising and the natural world. In M. Meister and P. M. Japp (eds), *Enviropop: Studies in Environmental Rhetoric and Popular Culture* (pp. 141–160). Westport, CT: Praeger/Greenwood Press.

Cox, R. (2013). *Environmental Communication and the Public Sphere* (3rd edn). London: Sage.

Creighton, M. (1997). Consuming rural Japan: The marketing of tradition and nostalgia in the Japanese travel industry. *Ethnology, 36*(3), 239–254.

Cronon, W. (ed.). (1995). *Uncommon Ground: Toward Reinventing Nature.* New York: Norton.

Davis, F. (1979). *Yearning for Yesterday: A Sociology of Nostalgia.* New York: The Free Press.

Elbro, C. (1983). *Det overtalende landskab: ideer om menneske og samfund i digternes og annonceindustriens naturskildringer i 1970'erne.* Copenhagen: C.A. Reitzel.

Evernden, N. (1989). Nature in industrial society. In I. Angus and S. Jhally (eds), *Cultural Politics in Contemporary America* (pp. 151–164). New York: Routledge.

Fairclough, N. (1989). *Language and Power.* London: Longman.

Gamson, W. A. and Modigliani, A. (1989). Media discourse and public opinion on nuclear power: A constructionist approach. *American Journal of Sociology, 95*(1), 1–37.

Goldman, R. and Papson, S. (1996). *Sign Wars: The Cluttered Landscape of Advertising.* New York: Guilford Press.

Hall, S. (1982). The rediscovery of 'ideology': Return of the repressed in media studies. In M. Gurevitch, T. Bennett, J. Curran and J. Woollacott (eds), *Culture, Society and the Media* (pp. 56–90). London: Methuen.

Hansen, A. (2002). Discourses of nature in advertising. *Communications, 27*(4), 499–511.

Hansen, A. (2010). *Environment, Media and Communication.* London: Routledge.

Howlett, M. and Raglon, R. (1992). Constructing the environmental spectacle: Green advertisements and the greening of the corporate image. *Environmental History Review, 16*(4), 53–68.

Iyer, E. and Banerjee, B. (1993). Anatomy of green advertising. *Advances In Consumer Research, 20*, 494–501.

Kellert, S. R. (1995). Concepts of nature east and west. In M. E. Soule and G. Lease (eds), *Reinventing Nature: Responses to Postmodern Deconstruction* (pp. 103–122). Washington, DC: Island Press.

Kilbourne, W. E. (1995). Green advertising: Salvation or oxymoron. *Journal Of Advertising, 24*(2), 7–19.

Kroma, M. M. and Flora, C. B. (2003). Greening pesticides: A historical analysis of the social construction of farm chemical advertisements. *Agriculture and Human Values, 20*(1), 21–35.

Lin, C. A. (2001). Cultural values reflected in Chinese and American television advertising. *Journal of Advertising, 30*(4), 83–94.

Macnaghten, P. and Urry, J. (1998). *Contested Natures.* London: Sage.

Marchand, R. (1985). *Advertising the American Dream: Making Way for Modernity, 1920–1940.* Berkeley, CA: University of California Press.

Martin, D. C. (2004). Apartheid in the great outdoors: American advertising and the reproduction of a racialized outdoor leisure identity. *Journal of Leisure Research, 36*(4), 513–535.

Moon, Y. S. and Chan, K. (2005). Advertising appeals and cultural values in television commercials: A comparison of Hong Kong and Korea. *International Marketing Review, 22*(1), 48–66.

Mueller, B. (1987). Reflections of culture: An analysis of Japanese and American advertising appeals. *Journal of Advertising Research, 27*(3), 51–59.

Negra, D. (2001). Consuming Ireland: Lucky charms cereal, Irish Spring soap and 1-800-SHAMROCK. *Cultural Studies, 15*(1), 76–97.

Nelson, V. (2005). Representation and images of people, place and nature in Grenada's tourism. *Geografiska Annaler Series B-Human Geography, 87B*(2), 131–143.

Okazaki, S. and Mueller, B. (2008). Evolution in the usage of localised appeals in Japanese and American print advertising. *International Journal of Advertising, 27*(5), 771–798.

Peterson, R. T. (1991). Physical-environment television advertisement themes – 1979 and 1989. *Journal of Business Ethics, 10*(3), 221–228.

Phillips, M., Fish, R. and Agg, J. (2001). Putting together ruralities: Towards a symbolic analysis of rurality in the British mass media. *Journal of Rural Studies, 17*(1), 1–27.

Plec, E. and Pettenger, M. (2012). Greenwashing consumption: The didactic framing of ExxonMobil's energy solutions. *Environmental Communication: A Journal of Nature and Culture, 6*(4), 459–476.

Rutherford, P. (2000). *Endless Propaganda: The Advertising of Public Goods*. Toronto: University of Toronto Press.

Schama, S. (1995). *Landscape and Memory*. London: HarperCollins.

Scutt, R. and Bonnet, A. (1996). *In Search of England: Popular Representations of Englishness and the English Countryside* (pp. 33). Newcastle upon Tyne: University of Newcastle upon Tyne, Department of Agricultural Economics and Food Marketing; Centre for Rural Economy.

Sheehan, K. and Atkinson, L. (2012). Special issue on green advertising: Revisiting green advertising and the reluctant consumer. *Journal of Advertising*, *41*(4), 5–7.

Soper, K. (1995). *What is Nature?* Oxford: Blackwell.

Thomas, L. (1995). In love with Inspector Morse: Feminist subculture and quality television. *Feminist Review*, 51, 1–25.

Thomas, L. (2002). *Fans, Feminisms and 'Quality' Media*. London: Routledge.

Thompson, J. B. (1990). *Ideology and Modern Culture: Critical Social Theory in the Era of Mass Communication*. Oxford: Polity Press.

Uggla, Y. and Olausson, U. (2013). The enrollment of nature in tourist information: Framing urban nature as 'the other'. *Environmental Communication: A Journal of Nature and Culture*, *7*(1), 97–112.

Urry, J. (1995). *Consuming Places*. London: Sage.

Urry, J. (2001). *The Tourist Gaze: Leisure and Travel in Contemporary Societies* (2nd ed.). London: Sage.

Wernick, A. (1997). Resort to nostalgia: Mountains, memories and myths of time. In M. Nava (ed.), *Buy This Book: Studies in Advertising and Consumption* (pp. 207–223). London: Routledge.

Williams, R. (1973). *The Country and the City*. London: Chatto & Windus.

Williams, R. (1983 [1976]). *Keywords: A Vocabulary of Culture and Society*. London: Flamingo/Fontana.

Williamson, J. (1978). *Decoding Advertisements: Ideology and Meaning in Advertising*. London: Marion Boyars.

Wilson, A. (1992). *The Culture of Nature: North American Landscape from Disney to the Exxon Valdez*. Cambridge, MA: Blackwell.

24

CELEBRITY CULTURE
AND ENVIRONMENT

Mark Meister

Introduction

The popularity of celebrity environmental endorsements is undeniable. From Alanis Morissette's campaigning against oil drilling in Alaska and Robert Redford's fundraising letters for the Natural Resources Defense Council to Leonardo DiCaprio's plea to "do something to save the ocean now" (Eilperin 2014: para. 9), the backing of musicians and actors, particularly, has been a staple of environmental causes. At the same time, the analysis of celebrity, celebrities, and celebrity culture has been a growing area of inquiry in the humanities and social sciences (Turner 2004, 2010). More recently, such inquiries have encompassed the relationship between celebrity and social activism, for the most part focusing attention on those already "famous" celebrities who, themselves, engage in such activism (McCurdy 2013). Celebrity studies scholar McCurdy (2013) has argued, however, that scholars should also focus on those individuals whose fame and celebrity resulted from their activism.

Focus on celebrity activism inevitably raises two other, converging variables—the mediated nature of celebrity, and the implication of celebrity within a wider capital/commodity culture. In his study of the "mediated visibility" of Paul Watson, anti-whaling activist and founder of the Sea Shepherd Conservation Society, for example, Lester (2011) argues that the celebrity of activists is also, inevitably, an important media practice. Such involvement of celebrity in environmental causes also engages the commodity-driven function of capitalism, which often invites public criticisms about engaging the system (capitalism) against which many environmentalists rail.

This chapter surveys recent research on celebrity environmental activism in contemporary society by locating such activism within what has been called an "epideictic" genre of civic rhetoric (below). I argue more specifically that the celebrity environmental advocate's facilitation of epideictic rhetoric within the confluence of both "mediated visibility" and capitalism is a plausible impression management strategy. Part one introduces the genre of epideictic rhetoric and profiles both the epideictic and cultural function of celebrity as context, particularly, for Hollywood celebrity-endorsed, environmental activism. Next, I review research addressing the cultural and rhetorical functions of celebrities and the ways environmental activism shapes their distinctive roles, from celebrity–conservationist and conservationist–celebrity, to politician–environmentalist.

In the latter half of the chapter, I turn to a neoliberal political critique of celebrity environmental activism, challenging whether celebrities, in fact, can be credible advocates for the environment within a mediated entertainment/capitalist culture. Finally, I discuss the implications of celebrity environmental activism as it relates to the epideictic function of corporate social responsibility, an area of research presently addressing many of the same issues relevant to environmental communication.

The epideictic and cultural function of celebrity

Critical/interpretive research in environmental communication has often focused on moral or ethical appeals, particularly in sustainability initiatives (Dannenberg, Hausman, Lawrence, and Powell 2012), appeals that are broadly associated with what classical scholars called "epideictic" rhetoric. The epideictic genre of rhetoric reflected ancient Greek and Roman practice of civic virtue, that is, as an expression of citizenship duties and the moral impetus of practical wisdom (*phronesis*), virtue (*arête*), and good will (*eunoia*) (Too 2001). In an epideictic version of the public sphere, citizens practice civic virtue through a genre of discourse "which shapes and cultivates the basic codes of value and belief by which a society or culture lives" (Walker 2000: 9). Epideictic rhetoric, then, is marked by "the experience of members of an audience who find that the speaker is saying exactly what needs to be said, who find that they are being caught up in a celebration of their vision of reality" (Sullivan 1993: 128).

In contemporary public discourse, and in environmental discourse, such citizen engagement and participation is often more vicarious than the orator in antiquity. (But then, early writers were not describing discourse occurring within a convergence of capitalism, media, Internet, celebrity, fame, and advocacy as a context for epideictic expression of societal values.) Arguably, civic virtue today is not articulated in the rhetoric of the concerned (individual) citizen, but is, instead, the imagined ideals of those holding significant cultural legitimacy. In such cases, celebrities may be said to function as surrogates, or impression managers, of the type of civic engagement described by classical sources. Such is made possible, as Sullivan (1993) argues, in that reputation, vision, authority, and construction of consubstantiality with the audience are characteristics of epideictic ethos.

Obviously important to significant cultural legitimacy, celebrity (and celebrities) are reliant on the influence and growth of mass media—"newspaper, magazine, television, the Internet, and such technologically sophisticated art forms as cinema and pop music" (Rockwell and Giles 2009: 179–180). As a consequence, celebrities have cultural influence by the mediated impressions they construct in the public's consciousness. Impression management, therefore, is part of a multifaceted and strategic set of actions aimed at shaping the perceptions of a wide range of outside parties (Brockington 2009; Merkl-Davies and Brennan 2011).

Theories of celebrity and their effects have often veered between two differing accounts of the role of celebrity in the public's consciousness. Brockington (2009) argues that structuralist theories of celebrity, on the one hand, are culturally significant, but are often problematic in celebrating economic or other structural conditions as more significant than democratic values. Structuralist theories of celebrity, Brockington argues, further disenfranchise intellectual and human diversity because mediated messages of celebrities often focus on difference. The primary implication of this type of celebrity engagement is that conservation "has become neoliberalized because of the desire of key neoliberal actors (states, corporations, international financial institutions) to use conservation to their own advantage" (Holmes 2011: 4).

In contrast, Brockington (2009) argues that post-structuralist theories of celebrity reject the primacy of economics in the celebrity phenomenon. More important is a celebrity's cultural politics, that is, celebrity as a system of representation, including politics, economics, circulation, and structures "within which the celebrity self resonates with the public sphere" (Holmes 2005: 10). Post-structuralist theories of celebrity focus on the topic of selfhood and identity in relation to Western ideals of individualism and in particular, how the ironies of the celebrity phenomenon prompt "self-conscious debates about the status of the self … as a site for the working through of discourses on the construction of identity" (Holmes 2005: 14).

In the post-structuralist vein, celebrity performance, circulation, construction, and mediation provide opportunities for self-reflection within the broader context of human identity. For example, Gamson

(1994) traces the roots of modern American celebrity to the Industrial Revolution and P. T. Barnum's understanding that "fame was not just the result of individual recognition but also the invention and manipulation of image" (Smith 2009: 222). Barnum's promotional techniques, early Hollywood history, and the engagement and evolution of the media, in light of post-structuralist theories of celebrity, embody American individualism via the mythology of the American Dream (Smith 2009). Complicated by language, commodification, economics, and signification, celebrities potentially embody the American Dream myth (and others) that reveal the ironies of individual and cultural life.

Celebrities are likewise socially constructed as both performers and individuals with private lives. The tension between public performer and private individual predicates celebrity status (Rojek 2001) and in this way common interest in celebrities is piqued (deCordova 1990). Thus, when access to celebrities' private lives is offered, mediated by the press, the public often perceives the performers' private lives are a "site of knowledge and truth" (deCordova 1990: 98).

In summary, structuralist theories of celebrity, including Marxism, linguistics, psychoanalysis, or behaviorism, tend to "displace the individual as a guarantor of discourse, but [instead] ... can be read as positing a real which is beneath or behind the surface represented by the individual as a discursive category" (Dyer 2011: 85). Thus, feature stories about celebrities, "which tell us that the star is *not* like he or she appears to be on the screen," Dyer (2011) explains, "serve to reinforce the authenticity of the star image as a whole" (p. 85). From a post-structuralist perspective, conversely, celebrity environmental activists occupy multiple discursive formations, in addition to their public performances and private lives.

Hollywood's nature? Celebrities and environmental advocacy

Celebrities and environmentalists have enjoyed a long-standing relationship beginning with the twentieth century inventions of TV and film that have invited structuralist analyses. As celebrity status evolved because of TV, film, and celebrity journalism, so too did the importance of social (environmental) issues and science journalism (Thrall et al. 2008). In fact, the evolution of celebrity journalism and science journalism, according to Cornog (2005), often overlapped, synthesizing "soft" with "hard" news, with celebrities increasingly appearing as key voices within the climate change debate; such voices provide a powerful news hook and potential mobilizing agent (Anderson 2011).

The synthesis of the celebrity environmentalist seemingly widens the audience or readership so important to journalism. In fact, *People*, *Variety*, and *Cosmopolitan* as well as *Time*, *Newsweek*, and the *HuffingPost* routinely report about climate change, paying attention to the increasing role that celebrities play within popular culture (Anderson 2011). Brad Pitt and Angelina Jolie, Robert Redford, Leonardo DeCaprio, and Matt Damon, for example, are often affiliated in such media with their environmental opposition to oil pipelines, fracking, rainforest depletion, and climate change.

While early coverage of climate change was dominated by scientific sources, now several actors and musicians are taking action against global warming (Boykoff and Goodman 2009). For example, Don Cheadle works as a spokesperson with the United Nations, and Leonardo DiCaprio, Edward Norton, and Natalie Portman all serve on the boards of a variety of environmental organizations, including Sting's Rainforest Foundation (Sullivan 2011). Moreover, actor–activists Elle Macpherson, James Olmos, and Ian Somerhalder are committed Sierra Club members engaged in its campaign to reduce dependence on coal-produced energy (Holmes 2011).

Celebrities' involvement with environmental causes, nevertheless, has also brought criticism of non-genuine advocacy or authenticity of purpose (and, ironically, revealing the epideictic function of "blaming"). Examples are replete: Former Vice-President Al Gore's film, *An Inconvenient Truth*, for example, raised public awareness about global warming, and while he also has spearheaded several clean-energy projects, Gore has received criticisms from neo-conservatives who have labeled him the world's first "carbon billionaire" (Chouliaraki 2012). And, while actor/producer Robert Redford is a 30-year board member of the Natural Resources Defense Council, he has reportedly received more public criticism for his environmental advocacy than for his films (Brown 2003). And again, in 2009, the hard-rock band, Pearl Jam, partnered with numerous organizations to offset the carbon emissions of the estimated 1 million fans driving to and from their concerts (Anderson 2011). They were, like many music groups that identify with environmentalism, also accused of "green-washing"[1] because of the large-scale environmental toll concerts and the recording industry have on the global environment (Anderson 2011).

British conservation scholar Dan Brockington's 2009 book, *Celebrity and the Environment* is perhaps the most systematic analysis of the relationship between celebrity and environmental advocacy. Brockington outlines five provocative ideas that, he argues, are central to understanding the relationship between fame, celebrity, and environmental activism. First, Brockington argues that environmental celebrities are flourishing because of the intersection of conservation and corporate capitalism. Second, these celebrities are not broad conservation supporters, but rather support those conservation issues that have garnered significant corporate backing. Third, celebrities support "safe" conservation issues that propose market solutions, and fourth, celebrity conservation relies on a mediated relationship with "wild nature." Fifth, Brockington argues that celebrity environmentalists do not try to save the world, but rather remake it by participating in global production, circulation, and consumption.

The roles and involvements of the celebrity environmentalist within capitalism are further crystalized by Brockington's (2009) discussion of three different categories of celebrity conservationism: the celebrity conservationist, the conservationist turned celebrity, and the celebrity environmental politician.

The most visible celebrity conservationists have been those who initially gained fame, then used that fame for advocacy purposes. During the 1990s and 2000s, such celebrity environmentalists brought much public attention to a variety of environmental issues. Then and now, such popular celebrities, not unlike the celebrity advocate discussed above, represent a variety of popular, political, and academic culture occupations, including, but not limited to, actors, musicians, athletes, politicians, and scholars.

Without question, the most popular environmental celebrities tend to be those who became visible within the popular culture industry itself. TV and film actors and personalities such as Harrison Ford, George Clooney, Leonardo DiCaprio, Angelina Jolie, Oprah Winfrey, James Cromwell, Ed Begley, Jr., Betty White, and Robert Redford, among others, for example, have all spoken on behalf of environmental causes. Working directly with conservation organizations, the Environmental Media Association, and the Earth Communications Office, movie and TV stars such as Cromwell, Redford, Edward Norton, Brad Pitt, Jane Fonda, Kevin Bacon, Tom Hanks, Cate Blanchett, Daryl Hannah, Pierce Bronsnan, Susan Sarandon, Tim Robbins, and Cameron Diaz, along with legendary musicians James Taylor and the Rolling Stones, have helped raise environmental awareness and money via serving as spokespersons for environmental causes (Yolen-Cohen 2003). Other ways have included writing environmental themes into movie and TV plotlines and by requiring the use of renewable building materials in the construction of motion picture and TV film (Brown 2003).[2]

The conservationist turned celebrity, on the other hand, is the focus of Graham Huggan's 2013 book *Nature's Saviours: Celebrity Conservationists in the Television Age*. Huggan explains how the 1960s nature programs of Jacques-Yves Cousteau, the writings of primatologist Diane Fossey (author of the 1983 novel and eventual film, *Gorillas in the Mist*), and the TV programming of naturalist Steve Irwin (the "Crocodile Hunter") and scientist–broadcaster–activist David Suzuki exemplify how conservationists garnered fame for themselves and for "wild nature." In this vein, both celebrity conservationists and the conservationist turned celebrity contribute arguments regarding the need for policy changes.

The third category of the celebrity environmentalist is exemplified by celebrity politicians. The most evident example is former U.S. Vice-President Al Gore. Gore's documentary, *An Inconvenient Truth*, highlighted the detrimental effects of global warming, earned him an Oscar, and contributed to his 2007 Nobel Peace Prize award. Similarly, the celebrity politician and past Governor of California, Arnold Schwarzenegger, used his movie action hero–politician fame to express his concerns about global warming.

Inevitably, celebrity relationships with environmental concerns, as I argued earlier, function epideictically as a result of their "mediated visibility" (Lester 2011), that is, they figure as discursive categories in popular media. News about celebrities and their environmental activism seemingly merge "soft" with "hard" news, as an impression management strategy that raises public attention for both. But celebrity environmental activists, as I've noted above, face criticism (Bashir, Lockwood, Chasteen, Nadolny, and Noyes 2013). Meyer and Gamson explained: "participation in a social movement means embracing identification with that movement, and for any person whose livelihood and status are tied to their relationship with a larger audience, such identification can be terribly risky" (1995: 189). This "blame" function of epideictic rhetoric is clearly evident in the neoliberal critique of celebrity environmental activists.

Neoliberal critiques of celebrity environmentalism

Regardless of the relationship—that is, as celebrity conservationist, conservationist turned celebrity, or celebrity politician—celebrity environmentalists are sometimes accused of a kind of hypocrisy, what Brockington (2008) and others have criticized as participating in, and/or promoting "neoliberalism" (Brockington 2008; Igoe and Brockington 2007; Sullivan 2011). Neoliberalism, McChesney argues, "posits that society works best when business runs things and there is as little possibility of government 'interference' with business as possible. In short, neoliberal democracy is one where the political sector controls little and debates even less" (1999: 6). The charge of hypocrisy is, at its basis, an accusation of inconsistency of celebrity endorsed environmentalism and the structural nature of neoliberal economic systems.

Superficially, such critics sometimes point to the fact that celebrities (generally) have larger incomes than members of the general public. Is Barbara Streisand, therefore, a hypocrite in that she donates nearly $20 million to her foundation which uses it to give grants to organizations working on environmental preservation (Balasubramanian 2011)? Or more seriously, the work of the Environmental Media Association (EMA)? The EMA is a nonprofit organization dedicated to harnessing the power of the entertainment industry and the media to educate the global public on environmental issues and motivate sustainable lifestyles. Members of the EMA board of directors include Ed Begley, Jr., Blythe Danner, Daryl Hannah, Norman Lear, Jeffrey Tambor, and Ted Turner. While such media serve these purposes, they, nevertheless, function as corporations for their shareholders. Are criticisms by EMA, therefore, of entrenched corporate environmental practices or its support of environmental reforms, credible?

Understanding the difference between celebrity endorsement and "celanthropy," or celebrity activism, is key in assessing celebrity credibility. As Keel and Nataraajan (2012) explain, celebrity endorsements are often associated with advertising and the celebrity receives compensation for the endorsement. As a result, self-reflection about individualism and the role of the government becomes problematic. Within this context, celanthropy, or individual celebrity activism, so this criticism goes, reinforces media representations of a disjointed world of emergencies, events, and crisis, yet does not call for specific government reform. Celanthropy thus underwrites what Rojek (2013) argues is a stateless solution, one that promotes the Big Citizen (as opposed to the Big Government) by creating socially conscious "projects" designed as media events that may, in fact, be no more than tokenistic fundraising. Rojek argues:

> Big Citizens raise cash and build awareness about inequality, injustice, hunger and corruption. But in failing to go beyond the level of Event consciousness, they play the practical role of being apologists for the system. In order to move beyond this, we need to stop equating Big Citizen celanthropy with automatic approval. We must turn more critically to examine how the Big Citizen soothes the psychology of western citizens and bolsters predatory, vested interests.

(2013: 13)

Rojek's challenge also potentially implicates the environmentalist or academic consultant similarly as Big Citizen celebrity wonks who run "with the crowd, courting acclaim for personal gratification and failing to address the real structural causes behind the ills of the world" (Rojek 2013: 4).

Closely related to celanthropy, in the neoliberal critique, is the close alliance of environmentalism and corporate sponsorship and/or their facilitation of environmental issues. In a sense, the corporation in contemporary society functions as a celebrity of sorts. Corporations concerned about sustainability, corporate responsibility, and corporate citizenship often rely on the same epideictic influence as celebrities, and, as such, represent significant cultural capital or influence (Cheney 1991), including influence of strategic decisions related to the environment (Patterson and Watkins-Allen 1997; Porter and Kramer 2006).

Corporate influence is often seen in the governance and/or close association with environmental organizations themselves. For example, Dorsey (2005) noted that members on the board of directors of many international wildlife conservation groups (e.g., NGOs) are chief executive officers of major corporations. The growing alliance between corporate social responsibility and wildlife conservation groups illustrates that representations of nature can and do function as brands; in this sense, the affiliation of corporate logos against the backdrop of wilderness or other protected areas facilitates a kind of celebrity environmentalism. And although these alliances exemplify neoliberalism, they also represent the influence of corporate-themed environmental interests associated with the wildlife and nature conservation interests of wealthy nations. As a consequence, the publics in such corporate/mediated societies frequently know nature through the representations of corporation-identified wildlife conservation.

The corporatization of environmentalism, therefore, may be subject, in the eyes of critics, to many of the same tensions or dilemmas as the celebrity environmental activist. Both the celebrity advocate and the corporate citizen remain vulnerable to the criticisms of hypocrisy or the Big Citizen as well as the seeming displacement of nature in celebrity events or impression management.

Impression management versus proximity to nature

In our highly mediated, corporate culture, publics seemingly rely more and more on celebrities to perform an epideictic role, that is, to articulate society's environmental moral and ethical beliefs. Ironically, such celebratory discourse, along with corporate impression management, also suggests such publics are farther removed from nature. The impression of proximity to nature is what matters, rather than the authenticity of the celebrity or corporate endorsement or activity. (See also Opel's discussion of "Cultural representations of the environment beyond mainstream media," in Chapter 25 of this Handbook.) Still, I argue, such publics (and environmental communication scholars) cannot disregard the performative significance and relevance of impressions created by celebrity and or corporate environmental endorsers in functioning in this regard, i.e., in "shap[ing] and cultivat[ing] the basic codes of value and belief by which a society or culture lives" (Walker 2000, p. 9). In an age where impression does matter, it remains useful to engage the variety and often contradictory practices of celebrity environmental activism.

For one thing, the goals of many environmentalists are not grounded in shared or specific places (Brockington 2008). Frankly, for those living in industrial capitalistic societies, encounters with environmental causes are often characterized by a lack of contact and interaction with the actual environments they reference. Thus, such society's epideictic rhetoric is marked by a physical distance (close or far proximity) from nature (Thompson and Cantrill 2013). As a result, environmental communication scholars, in considering the relationship between celebrity and environmental advocacy, should attend not simply to the neoliberal critique, but also to the variance and performative tensions among different expressions of environmental value.

Perhaps most importantly, I argue, is the relationships among celebrity environmental activism, corporate impression management, and what Brockington (2008) has called "ungrounded environmentalisms," or how we come to appreciate representations of nature as meaningful replacements for closeness to nature. A representation of nature facilitated by celebrity conservationists, conservationist celebrities, or the celebrity politician may, in fact, fulfill a psychological need to be close to nature—leaving an impression of a moral and or ethical commitment to nature. And, as argued above, corporate environmental disclosures also arguably function as celebrity environmental activists, even as they engage in impression management (Cho, Roberts, and Patten 2010), for example, as they promote (and/or publicize) sustainable corporate practices and citizenship (Brulle and Jenkins 2006).

As celebrity/corporate environmental engagement, endorsements, or impression management grows, so does the inevitable tension between such mediated epideictic rhetorics and what might be termed democratic or civic epideictic performances. Navigating such variances or tensions is deservedly an important task as environmental communication scholars explore the nature and communicative functions of celebrity and environmentalism in the "mediated visibility" of contemporary culture.

Notes

1   While some musicians and celebrities may be accused of green-washing, Daryl Hannah's commitment to environmental advocacy has led to jail time. She has organized and participated in environmental protests, including recently being arrested for protesting against TransCanada Corporation's Keystone XL pipeline.
2   More recently, *Showtime* aired in 2014 an eight-episode documentary on climate change, titled *Years of Living Dangerously*. Earlier, the *Hollywood Reporter* noted that producers James Cameron and Jerry Weintraub and actors Matt Damon, Don Cheadle, Alec Baldwin, Arnold Schwarzenegger, and Pulitzer Prize winners Thomas Friedman, and Nicholas Kristof intended to explore the human impact of climate change (Rose 2012).

Further reading

Brockington, D. (2009) *Celebrity and the environment: fame, wealth, and power in conservation*, London: Zed Books.
Celebrity and the Environment provides a detailed critical analysis of the celebrity industry's influence in shaping global environmental politics. The book not only identifies how fame, wealth, power, and politics shape the contemporary modern environmental movement, but also provides thought-provoking insight on what celebrity conservation might signify.

Gamson, J. (1994) *Claims to fame: celebrity in contemporary America*, Berkeley, CA: University of California Press.
Gamson draws upon historical, empirical, and critical traditions in media studies in his investigation of how entertainment celebrities are produced, shaped, and consumed in American culture.

Huggan, G. (2013) *Nature's saviours: celebrity conservationists in the television age*, New York: Routledge.
This is a comprehensive critique that details the roles that celebrity conservationists David Attenborough, Jacques-Yves Cousteau, Dian Fossey, David Suzuki, and Steve Irwin play in achieving conservation goals.

Rojek, C. (2001) *Celebrity*, London: Reaktion Books.
This book is an expansive collection of theories related to celebrity studies that helps explain the psychological need celebrities fulfill for the general public.

References

Anderson, A. (2011) "Sources, media, and modes of climate change communication: the role of celebrities," *Wiley Interdisciplinary Reviews: Climate Change* 2: 535–546.
Balasubramanian, R. (2011) "Grant guidelines 2011." Available at: www.barbarastreisand.com/us/guidelines (accessed September 27, 2013).
Bashir, N.Y., Lockwood, P., Chasteen, A.L., Nadolny, D., and Noyes, I. (2013) "The ironic impact of activists: negative stereotypes reduce social change influence," *European Journal of Social Psychology* 43: 614–626.

Boykoff, M.T. and Goodman, M.K. (2009) "Conspicuous redemption? Reflections on the promises and perils of the 'celebritization' of climate change," *Geoforum* 40: 395–406.

Brockington, D. (2008) "Powerful environmentalisms: conservation, celebrity, and capitalism," *Media, Culture, and Society* 304: 551–568.

Brockington, D. (2009) *Celebrity and the environment: fame, wealth, and power in conservation*, London: Zed Books.

Brown, J. (2003) "Green chic: when it comes to conscious living, industryites are leading the way—and doing it with style. A five-part look at going green at home and on the set," *VLife*. Available at: www.sorensenarchitects.com/pdfs/SA-VLife.pdf (accessed July 14, 2013).

Brulle, R. and Jenkins, J.C. (2006) "Spinning our way to sustainability," *Organization and Environment* 19: 82–87.

Cheney, G. (1991) *Rhetoric in organizational society: managing multiple identities*, Columbia, SC: University of South Carolina Press.

Cho, C.H., Roberts, R.W., and Patten, D.M. (2010) "The language of US corporate environmental disclosure," *Accounting, Organizations and Society* 35: 431–443.

Chouliaraki, L. (2012) "The theatricality of humanitarianism: a critique of celebrity advocacy." *Communication and Critical/Cultural Studies* 9: 1–21.

Cornog, E. (2005) "Let's blame the readers: is it possible to do great journalism if the public does not care?" *Columbia Journalism Review*. 43–49.

Dannenberg, C.J., Hausman, B.L., Lawrence, H.Y., and Powell, K.M. (2012) "The moral appeal of environmental discourses: the implication of ethical rhetorics," *Environmental Communication* 6: 212–232.

deCordova, R. (1990) *Picture personalities: the emergence of the star system in America*, Urbana, IL: University of Illinois Press.

Dorsey, M. (2005) "Conservation, collusion and capital," *Anthropology News* October: 45–46.

Dyer, R. (2011) *In the space of a song: the uses of song in film*, New York: Routledge.

Eilperin, J. (2014, June 17) "Obama, Leonardo DiCaprio vow efforts to protect the ocean," *Washington Post*. Available at: www.washingtonpost.com/blogs/post-politics/wp/2014/06/17/obama-leonardo-dicaprio-vow-efforts-to-protect-the-ocean/ (accessed June 28, 2014).

Gamson, J. (1994) *Claim to fame: celebrity in contemporary America*, Berkeley, CA: University of California Press.

Holmes, G. (2011) "Conservation's friends in high places: neoliberalism, networks, and the transnational conservation elite," *Global Environmental Politics* 11: 1–21.

Holmes, S. (2005) "Starring … Dyer?: re-visiting star studies and contemporary celebrity culture," *Westminster Papers in Communication and Culture* 2: 6–21.

Huggan, G. (2013). *Nature's saviours: celebrity conservationists in the television age*, New York: Routledge.

Igoe, J. and Brockington, D. (2007) "Neoliberal conservation: a brief introduction," *Conservation and Society* 5: 432–449.

Keel, A. and Nataraajan, R. (2012) "Celebrity endorsements and beyond: new avenues for celebrity branding," *Psychology and Marketing* 29: 690–703.

Lester, L. (2011) "Species of the month: anti-whaling, mediated visibility, and the news," *Environmental Communication* 5: 124–139.

McChesney, R.W. (1999) *Rich media, poor democracy: communication politics in dubious times*, Urbana, IL: University of Illinois Press.

McCurdy, P. (2013) "Conceptualising celebrity activists: the case of Tamsin Omond," *Celebrity Studies* 4: 311–324.

Merkl-Davies, D.M. and Brennan, N.M. (2011) "A conceptual framework of impression management: new insights from psychology, sociology and critical perspectives," *Accounting and Business Research* 41: 415–437.

Meyer, D. and Gamson, J. (1995) "The challenge of cultural elites: celebrities and social movements," *Sociological Inquiry* 65: 181–206.

Patterson, J.D. and Watkins-Allen, M. (1997) "Accounting for actions: how stakeholders respond to the strategic communication of environmental activist organizations," *Journal of Applied Communication Research* 25: 293–316.

Porter, M.E. and Kramer, M.R. (2006) "Strategy and society: the link between competitive advantage and corporate social responsibility," *Harvard Business Review* 84: 78–92.

Rockwell, D. and Giles, D.C. (2009) "Being a celebrity: a phenomenology of fame," *Journal of Phenomenological Psychology* 40: 178–210.

Rojek, C. (2001) *Celebrity*, London: Reaktion Books.

Rojek, C. (2013) "'Big citizen' celanthropy and its discontents," *International Journal of Cultural Studies* 0: 1–15.

Rose, L. (2012, December 3) "Showtime orders climate change series from James Cameron, Arnold Schwarzenegger, and Jerry Weintraub," *Hollywood Reporter*. Available at: www.hollywoodreporter.com/live-feed/showtime-orders-climate-change-series-396815 (accessed July 12, 2013).

Smith, G.D. (2009) "Love as redemption: the American dream myth and the celebrity biopic," *Journal of Communication Inquiry* 33: 222–238.

Sullivan, D. (1993) "The ethos of the epideictic encounter," *Philosophy and Rhetoric* 26(2): 113–133.

Sullivan, S. (2011) "Conservation is sexy! What makes this so, and what does this make? An engagement with *Celebrity and the Environment*," *Conservation and Society* 9: 334–345.

Thompson, J.L. and Cantrill, J.G. (2013) "The symbolic transformation of space," *Environmental Communication* 7: 1–3.

Thrall, T.A., Lollio-Fakhreddine, J., Berent, J., Donnelly, L., Herrin, W., Paquette, Z., Wenglinski, R., and Wyatt, A. (2008) "Star power: celebrity advocacy and the evolution of the public sphere," *The International Journal of Press/Politics* 13: 362–385.

Too, Y.L. (2001) Epideictic genre. In T. O. Sloane (ed.), *Encyclopedia of rhetoric*, pp. 251–257. New York: Oxford University Press.

Turner, G. (2004) *Understanding celebrity*, London: Reaktion Books.

Turner, G. (2010) "Approaching celebrity studies," *Celebrity Studies* 1: 11–20.

Walker, J. (2000) *Rhetoric and poetics in antiquity*, Oxford: Oxford University Press.

Yolen-Cohen, M. (2003) "Green glitterati: celebrities give props to the environment," *Sierra Magazine*. Available at: www.sierraclub.org/sierra/200307/media.asp (accessed July 16, 2013).

25

CULTURAL REPRESENTATIONS OF THE ENVIRONMENT BEYOND MAINSTREAM MEDIA

Andy Opel

While many scholars have written about the power of advertising, the appropriation of nature by market-ers, the journalistic constraints on environmental news stories and the power of language to shape and direct our understanding of the natural world, there has been significantly less attention to the incidental environmental images that occupy cultural crevices, background spaces and wallpaper that run through much of the industrialized world. These cultural threads show up in unexpected places and often echo the themes found in advertising, news media and popular culture—nature as commodity, nature as divine force, nature as Eden (Cronon, 1996).

From the pastoral, agrarian images on food packaging to super-saturated images on our computer screen savers, representations of nature pervade our visual world. In addition to news coverage, advertising, film and popular television, images of nature occur in incidental cultural spaces, reinforcing a hegemonic formation that is at one and the same time in love with the natural world and at war with the natural world. The con-tradiction between the ubiquity of images of nature and decreasing time actually spent in the natural world (Louv, 2005) has produced a condition where our daily lives are filled with pristine *images* of wildlife, plants and landscapes and at the same time we are experiencing the greatest extinction rate since the dinosaurs died off 65 million years ago.[1] The insatiable desire to consume pictures of nature results in a predominance of incidental images that reveals the enduring power of biophilia—that instinctive bond between humans and other species (Wilson, 1986)—and our deep connections to the natural world.

This chapter explores some of the previous work on incidental environmental representations, exam-ining previous scholarship on a variety of cultural artifacts including computer-mediated representations, games and virtual spaces. This overview of selected literature on non-mainstream representations of nature is followed by an analysis of computer wallpapers as a site of embodied nature discourse. The everyday, widespread familiarity of computer screensavers, the location where we focus so much of our gaze during our waking hours, reminds us of both the widespread consumer narratives that rely on the appropriation of nature *and* the enduring human desire to connect with the natural world.

Incidental environmental media

In his introduction to his oft-cited volume, *Uncommon Ground: Rethinking The Human Place in Nature* (1996), William Cronon unpacks a series of maps and signs, revealing the complex encoding of human/ nature relations embedded in the everyday objects that surround us. He distills eight themes that dominate the human/nature discourse. Of these, "nature as Eden" is the theme most embodied by the incidental cultural products that form the background of contemporary life in the industrialized North. Cronon describes this narrative as one that "projects(s) onto actual physical nature one of the most powerful and value-laden fables in the Western intellectual tradition" (p. 37). "Edenic narratives" then construct actual places as "perfect landscapes" that are outside of any need for restoration or protection and at the same time beyond reproach, so unquestioningly pristine as to make any thought of resource extraction or human development inconceivable. This pattern of edenic narrative in text and images permeates contemporary popular culture, cutting across media and context, present in the mundane and spectacular spaces of daily life, presenting a recurring theme in the environmental communication scholarship of cultural artifacts.

Previous work has documented the incidental and familiar ubiquitous presence of environmental imagery in obscure cultural locations. For example, Rehling (2002) examined Hallmark greeting cards, finding a predominance of sanitized, edenic images where "the possible dangers present in the wilderness remain obscured" (p. 24). The reproduction of idealized natural spaces creates an environment where "some kinds of wildlife and some types of geographical formations are more likely than others to seem worthy or deserving of preservation and protection" (p. 25). The implications of these images are argued to shape "our understandings and meanings of both nature and the human world, with consequences for both" (p. 27). While greeting cards are one small item in a sea of mediation, the accumulation of culturally constructed landscapes and life forms reinforce a consumer worldview that is increasingly attuned to the siren call of consumption and increasingly deaf to the subtle silence of the breeze.

Another example of unpacking a cultural artifact comes from an analysis of the cultural connotations of the board game *Monopoly™: The National Parks Edition* (Opel, 2002). *Monopoly™* is the best selling copyrighted game in history and this particular edition embodies the tensions of neoliberalism as the public parks are symbolically subjected to the forces of monopoly capitalism, where the "winner" takes control of all the parks and raises the "rent" to such astronomical rates so as to bankrupt the other players (citizens). In unpacking this familiar game, the theme of decontextualized nature emerges. The names and images of the national parks replace the familiar names such as Boardwalk or Marvin Gardens, yet there is no other accompanying information about the parks, wildlife, park funding or other data that might educate the players as they move about the board. This special edition of *Monopoly™* was found to "take an interest in the National Parks and the environment and steering that interest back into a capitalist consumer impulse" (p. 42). Again we see the theme of decontextualized nature woven into a familiar cultural product. The implications of this theme include a myriad of missed opportunities to harness the biophiliac impulse to build knowledge and understanding. Instead, the impulse to connect with nature is rewarded with a simulation that redirects audience attention back toward consumption.

In addition to two-dimensional representations of nature, scholars have also examined the physical spaces of re-created nature. Susan Davis turned her ethnographic gaze to the Sea World theme park in her 1997 book, *Spectacular Nature: Corporate Culture and the Sea World Experience*. Davis analyzes the spaces, performances and language of the park, revealing a complex web of marketing that works to reconfigure the citizen into a consumer and conveys the idea that "a visit to the nature theme park is a form of action on behalf of the environment" (p. 39). Davis argues that Sea World functions in a long tradition of presenting circus animals while reconfiguring this act of spectatorship into an expression of environmental protection and concern.

Davis positions Sea World as a "definitional project" that works to "model reality by defining what issues are open for consideration, what problems can be solved, and what concepts can be used in thinking issues through" (p. 238). The assumptions built into the physical space of the park are said to reinforce spectatorship and consumption while directing attention away from the systemic problems faced by cetaceans—whales, dolphins, porpoises and the like. In naturalizing a neoliberal worldview, the park is able to capitalize on the display of the ocean mammals while constraining the possibilities for collective action to address the health of marine ecosystems. Davis (1997) claims the park creates a narrative where:

extraction and pollution can never be connected to exploitation in the human world, to inequalities between classes, peoples and nations. At the same time that Sea World and its proliferating cousins erase a human history, they claim to be in and of themselves, a path to preservation, conservation, and environmental action.

(p. 238)

The park as cultural object then functions to limit how we think about environmental issues at the same time that it collapses consumption of the park experience into a form of environmental action.

Davis's work lays out familiar themes repeated throughout this literature—themes of constraining the realm of thinkable thoughts while profiting from the biophilic impulse. Although greeting cards, board games and theme parks may appear disconnected, in these instances they are united by a hegemonic worldview that is dominated by neoliberal marketplace discourse. The accumulation of these incidental cultural projects presents a fairly consistent message about human/nature relations, a message reinforced by advertising, film and news media.

Building on Davis's work, Opel and Smith analyzed the virtual theme parks built by the videogame *ZooTycoon* (2004). In the case of *ZooTycoon*, the beauty and detail of nature are reduced to banal icons whose primary purpose is to generate revenue for the virtual zoo. Players' sole measure of success is measured in dollars, resulting in a game that "encourages human expansion, monopolization of space, and creation of a capitalist place. It subjugates wild animals as menial laborers for our own entertainment and suggests that manipulating the environment any way possible to achieve this is acceptable" (p. 117). The game becomes a platform for enlisting nature, and human interest in the natural world in the service of capitalism and resource exploitation. Instead of providing in-depth information about animals, habitat and wildlife conservation, the game becomes a managerial chore where appeasing customers' desire for new attractions trumps the comprehension of animal needs and desires. Success is measured in dollars generated by your virtual park, positioning players in the role of a corporate manager as opposed to veterinarian, conservationist or biologist.

Continuing in the analysis of virtual nature, Clark (2011a, 2011b), examined the natural constructions found in the virtual world *Second Life*. What he found was a vast world where participants had gone to great length to re-create familiar environments, from pristine beaches to redwood forests to dense jungles. These virtual environments are said to depict tourist destinations, spaces for virtual human relaxation and meditation. The impulse to recreate virtual nature in the idyllic vision of unspoiled places is part of a long tradition of imaging nature as a place apart, a place devoid of traces of human intervention. The irony being that this virtual space is the embodiment of human construction, yet the players strive to re-create a hyper-reality that portrays the sort of untrammeled places that are disappearing in the real world. What is troubling about Clark's findings is that although *Second Life* is an open platform that allows users to build virtual worlds to their own specifications, users invariably revert to an anthropocentric, consumer perspective. "There is a continual repetition, through connotation, of the notion that Nature is a commodity for purchase and use by humans" (2011a, p. 58).

The accumulation of scholarship around nature and popular culture reveals a consistent pattern where representations of nature reinforce the idea of harnessing nature in the service of the maintenance and expansion of neoliberal economic growth and associated emphasis on consumer behavior over public policy solutions. Hansen and Machin (2013) reiterated this in their evaluation of environmental representations.

> Visual representations of the environment tend to be decontextualized and aestheticized in ways that enhance that flexible and versatile use across different genres of communication while also affording the basis for flexible new significations, as well as ones that are firmly anchored in culturally deep-seated/resonant discourses on nature and the environment … In all cases, representations appear to

favor individual responses to environmental problems rather than those that call for major structural changes in terms of the way in which we organize our societies and the resource greedy nature of capitalism.

(p. 157)

What the collection of scholarship reveals is the ever-expanding contours of what atmospheric chemist and Nobel laureate Paul Crutzen refers to as the Anthropocene, a new geologic era defined by the facts that "human-kind has caused mass extinctions of plant and animal species, polluted the oceans and altered the atmosphere, among other lasting impacts" (Stromberg, 2013). This pervasive condition is accompanied by a proliferation of nature representations that belie the reality of the Anthropocene yet affirm the strength of the human/nature connection.

Techno biophilia: computer wallpapers and virtual nature

As media convergence has blended the realms of film, television and the Internet, our screen time has become an increasingly complex site of work and play, education and entertainment. The result is that virtual representations of nature are far more ubiquitous than the unique virtual spaces such as *Second Life*. One does not have to actively choose to visit a website or play a video game to encounter proliferating images of nature. Instead these images come prepackaged with computers and the images appear when the computer goes into "sleep" mode. This nexus of image and naturalized techno-language (e.g. sleeping computers, cloud computing, viruses etc.) is a manifestation of what Sue Thomas calls "technobiophilia, the innate tendency to focus on life and lifelike processes as they appear in technology" (2013, p. 12). The following section explores Thomas's concept of technobiophilia as evidenced through the images of computer wallpapers.

Our homes and workspaces are increasingly populated by representations of nature. The places where we spend the majority of our time are dominated by recreations of nature, simulations that on the one hand sanitize and fetishize the natural world and on the other hand reveal an enduring impulse to be *in* nature, to return to nature, to escape the confines of our homes and offices. This tension is nowhere better witnessed than in the space where we increasingly look for so many of our waking hours—our computer screens.

Screensavers and computer wallpapers are overwhelmingly dominated by lush, saturated images of nature. Both Apple and Microsoft operating systems come bundled with numerous images that users may assign to their screens to appear when the computer goes into an energy saving mode. The result is that when users return to the computer, they are greeted by any number of beautiful images of streams, mountains, oceans, wildlife, plants, clouds and other depictions of natural beauty. This natural beauty then lures the user back to the online world of computer work/play/diversion/simulation, quickly vanishing at the touch of the "mouse" as the data from "the cloud" comes rushing back and nature literally vanishes.

For example, if we look at Microsoft's galleries of themes and desktop backgrounds, we see a predominance of natural images.[2] Of the 15 themes presented with sample images, ten are based in nature. From "Colors of Nature" to "Fantastic Flowers" to "Indian Wildlife," supersaturated images of plants, animals and landscapes are the vast majority of prepackaged images for Microsoft computer desktop images. Under the "Most Popular" category on this same website, only two of the 15 categories are clearly NOT nature based, the "Disney Infinity" category featuring animated images from Disney/Pixar films, and the "Grow" category featuring computer generated fractal patterns (that coincidentally mimic patterns found throughout the natural world).

Looking more closely at the sets of images that constitute this "nature" wallpaper for our computers, we see themes that echo Rehling's (2002) analysis of greeting cards, where sanitized beauty appears disconnected from modernity, almost as if from another planet, lush and untouched by the very processes and technologies that produce and reproduce the images. For example, in the theme, "Colors of Nature," Microsoft describes the eight-image collection as:

> Photographer Popkov Alexandr's sensitivity to the colors, creatures, and moods of nature is showcased in this free Windows theme. These stunning images capture lovely flowers and birds, a wide-eyed owl, an ocean storm, and a close-up of a delicate dragonfly, all for you to enjoy on your desktop.[3]

In one image from this Microsoft collection, "a bird" is presented, decontextualized and unnamed. We are provided with no reference to habitat, geographic range, population numbers, feeding habits and nesting or breeding requirements, nothing that would connect us to the image beyond the color and detail of the bird itself. Similarly, an image of a dragonfly is left nameless, unidentified. The absence of any defining information amplifies the otherworldly nature of the images, taking the plants and animals outside the biological taxonomy of classification and presenting them as textures and colors, light and shadow that adorn our networked machines and calls out to our enduring, deep links back to the natural world yet offers no other points of connection beyond the visual.

The "Colors of Nature" theme is then associated with four other related themes, "Fauna Dynamic" (25+ images), "Garden Life 2" (nine images), "Garden Seasons" (11 images) and "Insects Dynamic" (50+ images). Although both "Garden Life 2" and "Garden Seasons" follow the pattern of not providing any context for the images, the other two related themes did offer bits and pieces of context to *some* of their images. In the two "dynamic" themes—where RSS feeds directly update new pictures to your computer, 18 sample images are provided for each theme. In the "Fauna Dynamic" theme, of the 18 images, ten included short text identifying the animal, in some cases the location and in two cases, the Latin name for the animal. One image of a rhino is accompanied by the text, "Black rhinoceros (Diceros bicornis), Kenya."[4] Other than this label, the viewer is not provided any information about the current status of black rhinos in Kenya, what groups are supporting them, how to learn more, how to contribute to their protection or any other information related to this animal. Given the power of images to spark curiosity and connection, making these images "hot" with links to relevant information would be a small programming effort that would transform the eco-porn[5] into environmental education with the possibility of agency and activism.

The images in these collections themselves are not misrepresentations of nature or distortions per se, rather they exist within a system of display and de-contextualization that strips the images of connection to the world from which they sprang. In isolating spectacular plants, animals or landscapes, the images eclipse the human connection. In a representational tradition that goes back over a hundred years (Deluca and Peeples, 2000), evidence of human influence or presence is literally framed out of the images, creating idealized spaces that transcend time and place. This framing is part of the attraction, reproducing idealized conditions that redirect out attention away from the encroachments of anthropoids and allow us to gaze into a world without climate change, a world away from mass extinction, a world that embodies the diversity and bounty that sustained and allowed the rise of the Anthropocene.

A similar pattern dominates the images Apple bundles with its computers. The *Lion* operating system comes preloaded with seven folders of images with the "Nature" and "Plants" folders containing the most images. These images are a close parallel to those offered by Microsoft, with lush color and detail of plants, animals and landscapes all floating on the computer screen without any additional information. In the operating system *Mountain Lion*, Apple added over 40 "hidden" images broken into four themes (National Geographic, Cosmos, Arial and Nature Patterns), all based on nature and environmental images.[6] The collection of the "National Geographic" images reveals the consistent pattern of super-saturated color and decontextualized plants, animals and landscapes.

When we look at the recent "wallpapers" of the two dominant computer operating system manufacturers in the world, we see consistent attention to the natural world. Large mammals—whales, polar bears, panthers and lions—dominate, alongside birds, insects and landscapes undisturbed by the very processes that produce and display these images, the machines that occupy so much of our time and attention and keep our gaze fixed on the screen. The impulse to gaze at nature is presented as a lure back to the screen, back to work, indoors, away from the images that attracted our gaze in the first place.

This critical analysis of computer wallpapers contrasts with Thomas's (2013) attempt to find healing and redemption in our attraction to technobiophilia.

> So here we are, poised at a moment of crucial tension. Do we embrace cyberspace as part of the natural world, with all of its opportunities and flaws, or do we keep it at arm's length, as an unnatural guilty pleasure we should not really enjoy?
>
> (p. 187)

Thomas's answer to this question is that, "over and over again, cyberspace brings us back to the physical" (p. 188). Thomas argues that the longing and connections that are manifested through the images and language of computing strengthen and enhance our biophilic impulses, offering further opportunities to learn about our physical world by using the tools of the virtual world. Thomas manages to avoid the particulars of how hyperlinking might connect an image to an environmental organization or even a Wikipedia page, preferring instead to offer a broad hope that the images online touch something deeper that will begin a process of discovery in the viewer.

Discussion and conclusion

The easy and all too simplistic analysis of these computer images is to claim a corporate/economic appropriation of nature and redirection of an interest in the natural world back into a consumption act that contributes to the further destruction of that natural world. This is definitely part of what is going in and through these products and the myriad other products that enlist nature in their advertising and marketing campaigns. This is what filmmaker Robert Kenner (2008) calls "the veil"[7]—the intentional, commercial/corporate use of images that appeal to our attraction to a simple, clean environment unburdened by the demands of 7 billion humans on the planet.

In addition to the commercial appropriation of nature is a recognition of the human/nature connection that produces this insatiable desire to gaze at nature and connect with the natural world despite our lifestyles that include less and less time outside. This yearning is part of what psychologist Glenn Albrecht (2005; Albrecht et al., 2007) calls, "solastaglia" or "the pain experienced when there is recognition that the place where one resides and that one loves is under immediate assault … a form of homesickness one gets when one is still at 'home'"(Smith, 2010). Where Albrecht identifies the psychological stress that results from a changing physical environment, the concept can be expanded to include the cumulative knowledge of global changes taking place across the atmosphere, habitats and species diversity. The slow drip of environmental bad news is the context for these pristine images, idealized representations that work to combat the pain of recognition of the realities of the Anthropocene and all its implications.

When Mitchell (2005) asks, "What do Pictures Want?" he confronts the "power of idols over the human mind … their capacity for absorbing human desire and violence and projecting it back to us"(p. 27). In our case, we also want to flip this question back to, "What do humans want from pictures?" The sanitized, decontextualized images of nature that permeate our culture serve a purpose and that purpose is in part to salve the longing to be in nature, outside, away from consumer culture and in part as a palliative antidote to the steady drum beat of data pointing to environmental collapse that comes through news media and political debate. While these cultural objects reveal the tensions of human/nature relations in the Western world, they offer hope in their ubiquity. If individual or collective change is going to occur around climate change or mitigating the worst impacts of the anthropocene, this change will have to come in part from our biophilia inspired impulse to connect with the species that share fragments of our DNA. Fragments of this shared legacy are present in these images. It is up to us to turn away from the screen, step outside and see the real world so that we can begin to name the plants and animals around us as easily as we hum the jingles of the hucksters.

Notes

1 www.biologicaldiversity.org/programs/biodiversity/elements_of_biodiversity/extinction_crisis/ accessed 9/4/13
2 http://windows.microsoft.com/en-us/windows/themes accessed 9/4/13
3 http://windows.microsoft.com/en-us/windows/colors-nature-download-theme accessed 9/5/14
4 http://windows.microsoft.com/en-us/windows/fauna-dynamic-download-theme accessed 9/5/14. The image of the rhino can be accessed by scrolling through the sample images at the top of the page.
5 Mander, J. (1972) Ecopornography: One Year and Nearly a Billion Dollars Later, Advertising Owns Ecology, *Communication and Arts Magazine*, 14 (2), 45–56.
6 http://osxdaily.com/2012/09/07/40-gorgeous-secret-wallpapers-os-x-mountain-lion/ accessed 9/5/13
7 The chapter "The Veil" in the film, *Food Inc.*, 2008, Magnolia Pictures begins at 01:16:10 and continues to 01:24:14.

Further reading

Meister, M., and Japp, P.M. eds (2002). *Enviropop: Studies in Environmental Rhetoric and Popular Culture*. Westport, CT: Praeger Press.
This edited collection marks one of the first attempts to identify and analyze the environmental representations in non-traditional cultural crevices. The themes and patterns identified in these essays have been echoed and expanded throughout the literature in the years since this work first appeared.

Thomas, S. (2013). *Technobiophilia: Nature and Cyberspace*. London: Bloomsbury Academic.
Thomas interrogates a broad range of sites where nature and computers converge as she traces the history and possibilities of virtual nature. Thomas integrates a personal perspective into the analysis, avoiding the sometimes over-determined outcomes of political economy and cultural studies as she proposes possible positive outcomes of expanding virtual environments.

Parker, L.J. (2002). *Ecoculture: Environmental Messages in Music, Art and Literature*. Dubuque, IA: Kendall Hunt Publishing Co.
Aimed at advanced secondary or undergraduate students, this book offers a wide range of excellent exercises that help reveal the environmental messages found across a broad range of cultural products. The exercises and accessible reading make this a great introduction to environmental cultural analysis and applied cultural studies more broadly.

References

Albrecht, G. (2005). Solastalgia: A New Concept in Human Health and Identity. *Philosophy Activism Nature*, 3, 41–44.
Albrecht, G., Sartore, G-M., Connor, L., Higginbotham, N., Freeman, S., Kelly, B., Stain, H., Tonna, A., and Pollard, G. (2007). Solastalgia: The Distress Caused by Environmental Change. *Australasian Psychiatry*, 15 (1), S95–S98.
Clark, J. (2011a). The Environmental Semiotics of Virtual Worlds: Reading the "Splash Aquatics" Store in *Second Life*. *Graduate Journal of Social Science*, 8 (3), 47–64.
Clark, J. (2011b). Second Chances: Depictions of the Natural World in *Second Life*. In Enslinn, A. and Muse, E. (eds) *Creating Second Lives: Community, Identity and Spatiality as Constructions of the Virtual*. Oxford and New York: Routledge, pp. 145–168.
Cronon, W. (ed.). (1996). *Uncommon Ground: Rethinking the Human Place in Nature*. New York: Norton Press.

Davis, S. (1997). *Spectacular Nature: Corporate Culture and the Sea World Experience.* Berkeley, CA: University of California Press.

DeLuca, K, and Peeples, J. (2000). Imaging Nature: Watkins, Yosemite, and the Birth of Environmentalism. *Critical Studies in Media Communication,* 17, 241–260.

Hansen, A. and Machin, D. (2013). Researching Visual Environmental Communication. *Environmental Communication,* 7 (2), 151–168.

Kenner, R. (2008). *Food Inc.* (video recording). Magnolia Pictures: Los Angeles, CA.

Louv, R. (2005). *Last Child in the Woods: Saving Our Children from Nature Deficit Disorder.* Chapel Hill, NC: Algonquin Books.

Mitchell, J. T. (2005). *What do Pictures Want: The Lives and Loves of Images.* Chicago, IL: University of Chicago Press.

Opel, A. (2002). Monopoly™ The National Parks edition: Reading neo-liberal simulacra. In Meister, M. and Japp, P.M. (eds) *Enviropop: Studies in Environmental Rhetoric and Popular Culture.* Westport, CT: Praeger Press.

Opel, A. and Smith, J. (2004). ZooTycoon: Capitalism, Nature and the Pursuit of Happiness. *Ethics and the Environment,* 9(2), 103–120.

Rehling, D.L. (2002). When Hallmark Calls upon Nature: Images of Nature in Greeting Cards. In Meister, M. and Japp, P.M. (eds). *Enviropop: Studies in Environmental Rhetoric and Popular Culture.* Westport, CT: Praeger Press.

Smith, D. (Jan 27, 2010). Is There an Ecological Unconscious? *New York Times.* Available on-line at: www.nytimes.com/2010/01/31/magazine/31ecopsych-t.html?_r=1& (accessed September 21, 2013).

Stromberg, J. (2013). What is the Anthropocene and Are We in It? *Smithsonian Magazine,* January 2013. Available online at: www.smithsonianmag.com/science-nature/What-is-the-Anthropocene-and-Are-We-in-It-183828201.html#ixzz2gIyRg7UX (accessed September 19, 2013).

Thomas, S. (2013). *Technobiophilia: Nature and Cyberspace.* London: Bloomsbury.

Wilson, E.O. (1986). *Biophilia.* Cambridge: Harvard University Press.

## PART IV

Social and political implications of environmental communication

26

MAPPING MEDIA'S ROLE IN ENVIRONMENTAL THOUGHT AND ACTION

Susanna Priest

Introduction

Scholars know a good deal about how media messages influence people, even though these influences are generally less strong than is sometimes assumed. Arguably, the media's most-studied and best-established effect is that of "agenda setting" (McCombs and Shaw 1972). Put most simply, this means that the more attention the news media give to a particular issue, the more important media consumers think the issue might be. Conversely, if media do not cover an issue at all, many people simply would not know about it. This is especially the case for environmental issues, which can often be invisible; we cannot actually see many kinds of pollution such as heavy metals in a water supply, for example, and while the effects of climate change are all around us, we cannot directly see the climate changing or observe greenhouse gas concentrations without instrumentation. If symptoms of pollution such as brown skies or dirty water are not obvious, environmental problems can go unrecognized. Even if people are getting sick, it may be a mystery as to why. Receding ice, rising sea levels, desertification, vanishing species, and erosion of top-soil may be found in other parts of the world, but we will not necessarily be aware of it. Environmental reporting is a crucial component of society's response to these and many other environmental problems: If we do not know that the problems exist, we cannot motivate people to act to address them.

Most analysts do not attribute the recognition of problems solely to the media, however. The idea of "agenda building" (Cobb and Elder 1971; Lang and Lang 1981), as opposed to agenda *setting*, captures the idea that it is the collective actions of many societal institutions that together result in society's directing its attention to particular problems at a given point in time. Environmental and consumer advocacy organizations, conservation groups, corporations whose work affects the environment (whether in positive or negative ways), government agencies, universities and other research-oriented institutions, and a host of similar institutional players all influence one another—and contribute to our collective sentiment that a particular issue constitutes a "problem" that we need to address. Just as for other social problems such as poverty, racism, or crime, both the definition of the problem and the attribution of responsibility or blame should not be taken for granted. Arriving at such conclusions is the result of complex social processes. The media agenda itself is heavily influenced by the actions of other "players." Many times these actions take the form of what have been called "information subsidies" (Gandy 1982) that influence, in turn, the media agenda, such as a press release about an environmental issue. Information subsidies may also influence the "framing" or the definition of the problem contained in a news story or other account, not just its prominence.

301

The research literature offers many insightful case studies that look more closely at these relationships with respect to particular issues involving environment and risk. For example, the interaction between the Natural Resources Defense Council, the US Environmental Protection Agency, the apple industry, and the newspaper industry that created the initial public reaction against the use of the pesticide alar on apples in 1989 has been studied in detail by Friedman et al. (1996), who conclude that the coverage was often superficial, especially outside of apple-growing regions, and often failed to focus on actual health risk information. More recently, Nucci, Cuite, and Hallman (2009) studied the dissemination of information about *E. coli* contamination of spinach and similarly conclude that the diffusion of vital information about the outbreak was less than ideal. So, on the one hand, the media play a vital role in alerting people to the existence of environmental threats (in both of these cases, threats to the safety of the food supply, which surely would be expected to be a source of intense audience interest). However, on the other, the most important risk information may not be getting through.

The contemporary economic and technological transformation that the media are now undergoing has introduced new issues and is reshaping institutional structures. Economic pressures are reducing the workforce in traditional journalism, while also opening up new opportunities. While statistics are hard to come by, specialized journalists (environment, science, health) seem to be particularly threatened as news organizations lay off workers. Environmental reporters may be joining health and science reporters in being seen as an expensive luxury in newsrooms, meaning more and more stories about environment could be covered by general assignment reporters for whom all this is unfamiliar territory. And the new opportunities, which generally involve new Internet-based media, may be very good for democracy in terms of the proliferation of voices, but may also dilute the impact of informed voices, making it more difficult for information consumers to discern what points of view should be deemed legitimate and which "truths" are simply made up to fit someone's preconceptions—or to serve their interests. According to the Pew Research Center, newspaper newsroom employees have been cut by 30 percent since 2000, cable coverage of live events during the daytime has also fallen by 30 percent, and entities seeking to push information out to the public "have been more adept at using digital technical and social media to do so on their own, without any filter," as well as in getting their messages into traditional media (Pew Research Center 2014). This lack of filtering has been observed, for example, in the US presidential race in 2012, which was characterized by Pew as involving "more direct relaying of assertions … and less reporting."

In today's world, it is sometimes argued, anyone can be a journalist: bloggers, tweeters, activists, scientists, and even ordinary citizens can act as journalists. So can public relations people telling the story from the point of view of a particular stakeholder. The Internet hosts the viewpoints of all these groups and more, and as traditional journalism retreats, stakeholder-supported websites proliferate. In one highly visible effort, in 2009 research universities in the English-speaking world banded together to create a site designed to publicize their own news—including environmental news—directly to the public (Futurity 2014). On the one hand, this may result in more publicity for important research work; on the other, its governing board is made up of the communication officers of the participating universities, a group with an unambiguous stake in institutional promotion.

These trends—one economic, one technological, both eroding traditional journalistic practice—may bode well for broader participation in both journalism and policy. Yet observers of contemporary trends might well be concerned that it is too easy, in today's world, for individuals to avoid confrontation with viewpoints incompatible with their own while seeking out "maverick" perspectives that could reinforce their prejudices. People can live in their own realities, in other words, making consensus irrelevant. The practice of "objective" and "balanced" journalism takes on new meaning in this context. Will tomorrow's news consumers limit themselves to information that reinforces their points of view on environmental or other issues? What will be the consequences for democratic debate and consensus building on environmental policy? We do not yet know the answers.

Often missing from studies that focus on media messages about environment and science is a careful account of audiences and how they differ. We do not always fully understand what specific factors cause dynamics such as agenda setting or framing to influence certain people in certain ways. We do know that in order to be motivated to act, people need to recognize the existence of a problem, *and they also need to see themselves as part of the solution*. Media accounts need to make clear to people what they can do—to suggest a clear and realistic path toward action—and why they should do it. This inevitably involves ethical and value-based reasoning, as well as scientific arguments. Journalists, trained to be "objective" in covering other kinds of stories, often shy away from these dimensions. Most do not see motivating audience action to be part of their job description. In addition, audiences bring their own ideas into the equation; they are not simply passive consumers of media messages.

Varied audiences for media accounts of environmental risk

Historically, we've thought about the "mass" media as having effects on "mass" audiences—and individual differences have been understated and understudied. In fact, audiences for environmental information vary widely, especially within modern pluralistic societies that incorporate great social and cultural diversity. As the "mass" media have faded in importance compared to new media, often Internet based and including "social" media that reach much narrower audiences, yet (as with tweets) much more quickly, this old "mass" audience concept is fading as well. At the same time, the roles of the media are shifting as a result of the economic restructuring of the media industry, as well as the rapid proliferation and diffusion of newer media forms. Vastly more choices and channels are available as was the case just a few decades ago, arguably making each one less powerful individually. News audiences' use of digital sources and devices has continued to grow (Pew Research Center 2014).

Some news and entertainment organizations are seeking out new niches—for example, by developing informational websites (Tanner and Friedman 2011) or blogs to supplement traditional news stories. The venerable *New York Times* hosts so many affiliated blogs that it posts an online directory of them. Scientific magazines and even academic journals—notably *Nature*—have expanded online presences as well. Another strategy is political audience segmentation. Even in the United States, where the goal of journalistic neutrality has been a dominant norm for many decades, a number of news organizations are now seeking to attract and hold audiences by adopting more explicit political positions; for example, Fox News is known as a conservative voice, while the website *The Huffington Post* has a reputation for being liberal.

This shift away from thinking in terms of the old "mass" media has refocused our attention on the fact that there are many audiences, with different expectations and underlying beliefs that strongly influence their media choices and their receptivity to particular explanations. The Internet empowers people to seek out news that interests them. Audiences seek information from different sources, utilize different media, and interpret what they come across in different ways. These dynamics generally involve beliefs and values, not necessarily "facts." We are now in an era where an understanding of audience differences is more crucial than ever to effective media communication.

What do today's media audiences make of news about environmental risks? Audience members bring their own attitudes, expectations, values, and beliefs about both science and environment to their interpretation of media stories. Dunlap, one of the originators of the New Environmental Paradigm measurement scale, has argued that US culture is inherently resistant to an "ecological worldview" (2008). Long term, these expectations and beliefs are themselves shaped in part by media representations of reality in a process referred to as "cultivation." But these beliefs are also shaped by a host of other social and cultural factors. Popular reactions to the environmental and other dimensions of emerging technologies such as biotechnology and nanotechnology, for instance, are strikingly different (Priest 2012). Such differences appear to have deep cultural roots rather than being shaped by exposure to recent media accounts alone, although particular types of media information can undoubtedly resonate with different dimensions of these powerful, yet varied, underlying social and cultural factors.

Although media framing has sometimes been credited with (or blamed for having) enormous influence over public reactions, direct evidence of this is not easy to find. Even so, the twentieth century emergence of the environmental movement has certainly influenced twenty-first century audience receptivity to media claims about environmental issues. As indirect evidence of this cultural shift, some segments of the corporate world have undergone a shift to "green" advertising and public relations. Whether this trend can help to cultivate and reinforce a "greener" attitude toward consumer choices or is simply insincere "green-washing" is a matter of considerable debate. Dahl (2010) argues that the regulatory systems of many countries have been unequal to the "green-washing" challenge and that public confusion, environmental harm, and even risks to public health are the inevitable result. But, in the end, which interpretation is more convincing likely depends on whether those sending the messages are seen as trustworthy and sincere, on the one hand, or manipulative and driven only by profit, on the other—with the media playing only a supporting role. Different audience members will respond differently to the same message, in other words.

Issues of interpretation extend far beyond "green-washing," however. While both practitioners and scholars may be tempted to assume that individual reactions to issues involving science, including environmental science, can be predicted in a fairly straightforward way on the basis of scientific knowledge, the relationship between facts and attitudes is a complex one across a wide range of issues (Sturgis and Allum 2004). The old "deficit model" idea that teaching about facts will always change attitudes, while it continues to reappear, is gradually being discarded. Non-environmental examples may help illustrate this. On the science side, popular opinion about cases such as evolution and stem cell research provide examples. Religion and science are not always incompatible, but these well-known examples where they have clashed help to suggest the complexity of the relationship between beliefs and (factual) knowledge.

Some people who have deep religious objections to the idea of evolution (for example, in parts of the US South) seem to understand how scientists think that evolution works perfectly well—the issue is not a knowledge deficit. This may be difficult for others to grasp, but these individuals seem to score better on basic tests of science literacy that do not ask them if they "believe in" or "accept" evolution but rather ask them what it is that evolutionary theory claims (Rughinis 2011). However, surrounding cultural traditions, including the advice of respected leaders, tells them that this idea is incompatible with religious values. They understand the theory but do not accept it, in other words. In the case of stem cell research, the distinction between facts and values is even more clear. Positions against embryonic stem cell research are not based on misunderstanding the science, but of adopting the premise that it is unethical to destroy human embryos, regardless of the goal. It might be possible to persuade some of these individuals otherwise with messages based on the medical promise of the science, but it is not a misunderstanding of that science that led them to object in the first place.

These examples may seem far afield from environment, but they help us to understand how it is possible for some intelligent people, who may be generally knowledgeable and well educated, to reject the scientific evidence on (say) climate change. This issue has a different dynamic from rejecting evolution or stem cell research in that it does not generally have a religious basis; indeed, the tradition of stewardship embraced by many religious groups argues for environmental concern and protection, and environmental communicators recognize that religious values can represent an opportunity rather than a barrier. But accepting that the world is changing in unexpected and threatening ways is difficult enough to begin with, and if those forming a person's immediate social network reject the idea, that may dictate how climate change information is evaluated (Yang and Kahlor 2013). Of course, political and economic interests have also capitalized on the inevitable existence of uncertainty in scientific results by encouraging audiences to take the easy way out and deny the reality of anthropogenic climate change.

Here again, as for any complex issue, there are many shades and nuances of perspective, and it is useful to think about *audiences* rather than a single "mass" audience. Leiserowitz et al. (2013) have helped us to conceptualize how this works for climate change by organizing their analysis of national opinion survey results in terms of six different Americas—six different subsets of the population, or audiences, that have different orientations to climate change, as influenced by ideology, trust, demographics, and other factors. These range from the "alarmed" to the "dismissive." It is unlikely that the same message will reach them all.

In cases of public controversy over specific technologies with environmental implications, such as the genetic modification of food crops or the further development of nuclear power, it is clear that both proponents and opponents can be well informed about the scientific facts; rather, it is differences in values and differences in trust in the stakeholders and managers involved that are the more likely drivers of attitudes and opinions. Like the other examples above, while each of these controversies has distinct features, they help underscore the point that forming opinions about environment-related issues is not always closely linked to knowledge of the underlying science. Risk estimations always involve uncertainty and usually involve social values, not just scientific facts; risk perceptions all the more so. Is a risk to an endangered species such as the spotted owl more important than the risk to an associated industry, in this case logging? The answer is a matter of value-laden judgment, not something that can be resolved exclusively by science. Even so, the claim that people's livelihoods are at stake resonates widely and can create heart-felt resistance to environmental regulation in some cases, of which the spotted owl controversy provides an enduring example (see Andrew and Velasquez 1991).

Awareness that audiences or "publics" for news about the environment may respond in different ways might tempt communicators to adopt strategies that target these audience segments one by one, much as any modern marketer targets advertising, whether for products or politicians. However, this strategy can be counter-productive to the extent members of different audiences will inevitably be exposed to messages designed to influence someone else. For example, the message that adoption of alternative energy sources such as solar, wind, and water will bring more jobs to an area could be persuasive with an audience of business people enthusiastic about development but have quite a different effect with an audience of environmental activists whose highest priority is the preservation of the local natural landscape and wildlife.

Critical science literacy and the interpretation of news

Intelligently navigating today's media landscape, with its vast proliferation of voices and viewpoints and proportionately fewer authoritative anchor points to go by, will require new skills. We may all believe we know who is speaking from an informed perspective and who is not, but do we? Elsewhere (Priest 2013a, 2013b), I have introduced the idea of "critical science literacy" to refer to the skills that audience members today need to interpret and evaluate news and information about environment, science, technology, and health. This goes beyond knowledge of the facts—understanding the relationship between carbon and climate, knowing the importance of biodiversity, recognizing the shrinking availability of critical habitat, grasping how pollution affects the earth and the life that depends on it. It also means understanding science in relation to society. We know that scientific knowledge makes only a weak contribution to attitudes about science-related controversies, including environmental ones. Human values, including valuation of the environment, play the crucial role.

People also need to know quite a bit about how science works. This means understanding the full range of methodologies used in science, not just experiments but observation, description, theory-building, and modeling. It means understanding that science always involves uncertainty, but at its best provides us with the best available evidence. And it means understanding the nature of scientific expertise (Collins and Evans 2007) and something about how science works *socially*.

Science, including environmental science, is a social enterprise, and scientific claims are distilled through a highly social process, the result of which is what we commonly refer to as scientific consensus. Awareness of such crucial elements as peer review, the roles of scientific meetings and scientific societies, and the meaning of disciplinary expertise are all vital to understanding the nature of that consensus. All of these dimensions can be criticized—peer review lets through highly imperfect research, scientific societies may have too much influence over which truths receive appropriate recognition, and the most stunning conceptual breakthroughs often happen when disciplinary boundaries are breached rather than respected. But even in making such critiques, it is useful to start with some understanding of the sociology of science. Not all expertise is alike, for example.

Part of critical science literacy is recognizing that uncertainty is inevitable—while also understanding that this does not mean that scientific truth is entirely "up for grabs" or a matter of arbitrary opinion. Another part is understanding the role that human values play, if not in the actual conduct of science, in the choice of what problems are worth studying and what role science should play in the resolution of controversies that are rarely purely scientific in nature. Both of these dimensions are especially relevant to environmental choices that take place on the basis of science that inevitably carries a high degree of uncertainty. This sometimes results from the deep complexity of ecosystems but is also inherent in the nature of scientific inquiry. Our opinions and decisions about environmental issues inevitably involve value judgments, not just scientific ones.

Science exists *within society* and is not free of political and ideological influences, both those that take place within science and those that exist more broadly. Awareness of the political and ideological influences on science should not imply that science is simply a matter of belief, and yet a reality of the politics of science is that political elements cannot be ignored. A critical consumer of scientific claims should think to ask what ideological and political elements such claims incorporate. Whose political and economic interests does it serve? Climate change is again a key example; one recent study showed that climate change "counter-movement" organizations in the United States have a combined income of around $900 million, much of it from conservative foundations (Brulle 2014). But climate change is hardly the only place where funding determines the direction of research.

Sorting all of this out is asking a great deal of consumers of information about science generally, or about science-based environmental claims in particular. Yet this standard is profoundly different from the old standard of science literacy that used multiple choice tests to measure knowledge of a chosen set of scientific facts. Facts matter. But the world of scientific and environmental knowledge is expanding so rapidly that a fact-based assessment of literacy is impractical when what is really needed are the skills to navigate an uncertain landscape of competing claims. The possession of critical science literacy—the ability to apply critical thinking skills to scientific claims—is also more vital than ever. It is especially important in today's new media world in which claims-makers proliferate. We should celebrate the diversity of voices, but we must learn to make wise choices among them.

The shifting influences of media and newswork
Integrating risk theory, media theory, and audience theory

Social psychologists studying societal reactions to environmental and other types of risk have coined the term "social amplification of risk" to refer to the collective process through which entire societies react to risk-related information; this idea was presented in an early paper by Kasperson et al. (1988). This work is concerned both with the amplification or magnification of a risk, in which society may focus on and perhaps even exaggerate a particular risk, and with the attenuation or diminishing of a risk, in which society may downplay or ignore a particular risk, generally underestimating its importance. (Of course, whether public perception is magnifying an unimportant risk or downplaying a very important one is often itself a matter of opinion.) This approach, abbreviated SARF for "social amplification of risk framework," does not incorporate a fully developed explanation (that is, a true theory) about how some risks come to be amplified while others are attenuated; it is therefore more often described as a conceptual framework.

The media (meaning primarily the news media, although entertainment media also have influence) have been identified as one of many institutions that can serve to amplify or attenuate societal perceptions of risk, and it is consistent with our general understanding of media agenda setting that the media have the power to focus our attention on certain risks rather than others. SARF is also consistent with agenda building theory that asserts that the media work in concert with other institutions in doing so. While the media should no longer be conceptualized as "mass," any more than today's audiences can be conceptualized as "masses," SARF remains a useful way of thinking about the impact of newswork in the context of environmental controversies. For example, Bakir (2005) used this framework in a case study analysis of media's role in a battle between Greenpeace and Shell Oil over deep-sea disposal of an oil rig, a study that was explicitly designed to evaluate the SARF approach. While acknowledging and articulating the limitations of the framework (including its reliance on a static and linear view of communication), the author argues that SARF can accommodate attention to the roles of non-media actors, as well as more systematic analysis of how media institutions operate.

Journalists do not work in a vacuum; their choices of which stories to cover are usually responsive to what other actors do, particularly those actors who routinely act as media sources (such as politicians, corporate spokespeople, advocates and activists, and sometimes researchers). These actors contact journalists through press releases, press conferences, press kits, or simply by picking up the phone or sending a text, tweet, or email. Since the news media are highly influenced by these "information subsidies," we can identify the deliberate actions of concerned stakeholders on either side of a controversy as one strong explanation of which path is followed. A well-orchestrated public information campaign, whether by an environmental advocacy group or a corporation or industry defending itself against charges of environmental negligence, can certainly influence media accounts, both in terms of what stories are covered and how they are framed.

Whether these stories always or predictably influence public opinion in a particular way is a different matter, of course, one that depends on pre-existing values and attitudes, underlying cultural factors, trust, and perceptions of the credibility or legitimacy of particular points of view. But that is not to say that there are never effects, only that the effects are not uniform and may not always be entirely predictable, given their complexity.

This account is somewhat at odds with the ideology of journalism as "objective" reporting. In the United States, and to a large extent contemporary commercial journalism around the globe, "objectivity" is the norm. The term is in quotation marks because no account of a complex situation (whether in the media or elsewhere) can be truly "objective"; all of our knowledge, arguably even scientific knowledge, is produced by various forms of social consensus and is subject to revision. Yet media reports are often understood uncritically as reflecting truth and do disproportionately influence what is sometimes called the "social construction of reality" (Berger and Luckmann 1966, Tuchman 1978)—that is, how we as social beings come to understand the world from a collective point of view, including our understanding of which issues are important and which less so—including which risks matter.

An understanding of how information subsidies work should temper our blind faith in media objectivity. However, good journalists know how this works as well and conscientiously guard against being unduly influenced. Even so, we very often have no way to know about trends or problems that the media do not report, we tend to take seriously those that we know about, and we often accept journalism's definition of the situation. It is almost inevitable that media definitions of reality creep into our thinking as "truth," in other words.

Media "objectivity" and environmental reporting

According to Nelkin (1995), the whole idea of objectivity in journalism was originally borrowed from the idea of objectivity in science. Journalistic objectivity may also have economic roots in the desire to maximize audiences by offending no one, or at least as few as possible. However, in practice, objectivity in journalism generally means equitable attention to both sides of an issue. Derived from political reporting, in which many issues are reduced to a "left" and a "right" position, this is actually a much more simplistic proposition than it might appear, since most complex issues actually have more than two sides—even in politics. For environment, this creates a strong tendency for journalism to reduce issues into a pro-environment and a pro-industry position. This may be counter-productive to the extent it reinforces positions at either extreme, downplays possibilities for compromise, and even precludes finding and embracing solutions that are both good for the environment and good for (or at least acceptable to) industry. In other words, this practice likely increases polarization on environmental issues. Rather than being "objective," this approach may bias us against the search for workable "middle ground" solutions.

For science and environmental reporting in particular, there is also the important issue of what constitutes a legitimate opinion. Legitimacy on political or social issues may be defined somewhat differently, but for for science and environment, legitimacy is normally linked to claims of specialized expertise. Claiming that the world is flat would not be treated seriously by today's science journalists, or so one hopes, since it doesn't meet the standard for empirical verifiability that has been established for science. However, things are rarely this clear-cut. Claiming that climate change is not happening has very often been treated as a legitimate scientific point of view, unfortunately—one deserving of "equal attention" in news stories (Boykoff 2011). Covering such claims as simple differences of perspective provides an easy solution for journalists who don't want to be put in the position of deciding what science is legitimate and what is not. This is understandable, but even so it can be a potentially confusing and even destructive practice. Indeed, while the study of news credibility has a long and venerable history, the conceptually related study of how the news media confer legitimacy (or illegitimacy) on a particular point of view has hardly begun. Yet, in the history of every social movement, the issue of legitimacy inevitably appears—and occasionally takes over the stage entirely. In science, this often becomes a struggle over the right to claim expertise.

In the case of climate change in particular, ideologically driven minorities have benefited by this obligation to report "both" sides of the issue. Journalistic stories about climate change were for many years characterized by the inclusion of "skeptical" perspectives by what were presented as equally qualified scientists. The mainstream scientific community has made progress in communicating to journalists that this kind of so-called "skepticism" is not a legitimate scientific view on climate. Even so, this is difficult territory to navigate. The best journalists covering environment and science realize that by the very nature of scientific inquiry, it may turn out that a minority view is correct. Journalists are rightly cautious about privileging only mainstream views. Commonly accepted scientific truth can, as Kuhn (1962) famously pointed out, be overturned by new thinking.

The idea that truth is constantly subject to revision is an important foundation of science itself, including environmental science. For environment in particular, though, policy decisions must very often be made today on the basis of science that must remain uncertain until at least tomorrow. And on many issues—that the world is roughly round and that climate change exists, for example—a strong scientific consensus surrounds truth claims. The lesson learned should not be that journalists should never question scientific authority, but that they need to exercise wisdom and responsibility in conferring this authority indiscriminately.

New models and new strategies

Environmental journalism, like other forms of specialized journalism, faces profound challenges. Despite apparent and possibly permanent reductions in traditional media staffing—and even though democracy itself may be enhanced in today's new media world—information consumers will need a high level of sophistication to make sense of what they read, view, and hear regarding complex issues. The number of authoritative voices they can trust to guide them through this ever-more-complex maze seems to be shrinking, while the role of information subsidies is stronger than ever. Journalism tends to divide the world into opposed pairs of viewpoints, inviting polarization and discouraging consensus. Future news consumers may be increasingly tempted to seek out just the information that confirms their own views, potentially leading to further polarization despite the new opportunities for constructive debate that the Information Age affords. A better understanding of how science works, in the environmental or any other sphere, may help consumers to grasp, appreciate, and interpret alternative points of view wisely.

Environmental communication scholars and practitioners often see media framing as a key strategy for public opinion formation and attitude change. However, in a new media world in which the audience for media material is not only diverse but splintered, the idea of message framing represents a big challenge to both scholars and practitioners. It has always been true that the same message will resonate differently with different audience segments, one reason why generalizable framing effects have been hard to document. In today's "new media" world, this phenomenon of diverse audience responses can be exaggerated, in part because everyone has more choice about what messages to seek or heed. Some Internet messages "go viral," becoming immediately popular with some audiences—but unheard of to many others. Information subsidies may have increased, but in a decentralized media system, the impact of a particular message or message frame is far more difficult to predict or control. Quite possibly, more research on why certain messages get taken up by key target audiences (the "going viral" phenomenon) is needed.

The current trend in science communication more generally—including environmental and health communication—is the turn toward "public engagement" as a communication strategy. Instead of conceptualizing audiences as passive, waiting to have their thinking "framed" by whatever messages we choose to send their way, the thinking is that it is more important (and arguably more ethical) to encourage dialogue, discussion, and debate about the issues that confront us. Does a community, a state, or a nation want to encourage nuclear power generation, or is it too risky? Is biotechnology a threat to the environment, a means to reduce world hunger, or both—or, perhaps, neither? What kind of industry and economy is most environmentally sustainable? Surely, in an ideal world, citizens would come together to discuss and decide such things. In practice, however, there are some problems with this approach. Only a small segment of the population has the time and interest to "engage" on any given issue that confronts us. And if they did, we have no system that would accommodate this activity on a large scale, and no particular mechanism (in large, politically pluralistic, multicultural societies such as the United States, at least) for incorporating the outcomes of popular debate into policy decisions. Furthermore, the debates over US health care reform demonstrated that popular discussion is sometimes unpredictable and not always ideally constructive.

Therefore, media representations of issues and actors in the environmental arena, among others, will continue to be vitally important. For all that individuals have access to ever-increasing amounts of information and opinion on the Internet, media agenda setting still matters. "Mainstream" media are the root source of many of the topics of discussion in the "blogosphere" (Tremayne 2007). Citizen journalism, greatly facilitated by the Internet, even so has neither the resources nor the credibility of traditional news organizations. As institutions, traditional media have been slow to change—but they are changing, even proliferating, and certainly not going away entirely. A key question for both scholars and practitioners to address in the future, then, is whether (and by what means) media can continue to become more interactive and engaging without losing all of their traditional authority, quality, or credibility.

Further reading

Friedman, S. M., S. Dunwoody, and C. L. Rogers, eds (1999). *Communicating Uncertainty: Media Coverage of New and Controversial Science*. Mahwah, NJ: Lawrence Erlbaum Associates.

This book, which grew out of discussions at the 1996 meeting of the American Association for the Advancement of Science, attempts to make sense out of scientific uncertainty as presented in media accounts, including a number of environmental examples.

McKenzie-Mohr, Doug, Nancy R. Lee, P. Wesley Schultz, and Philip Kotler. (2012). *Social Marketing to Protect the Environment: What Works.* Thousand Oaks, CA: Sage.

This book introduces community-based social marketing as applied to environmental issues, extending our thinking about media messages to encompass not just news but also marketing messages – including those delivered on posters, stickers, signs, and other distinctive media. Strengths of the book include its use of many concrete cases (with graphical message illustrations) and its organization by issue type.

Pidgeon, Nick, Roger E. Kasperson, and Paul Slovic, eds (2003). *The Social Amplification of Risk.* Cambridge: Cambridge University Press.

This collection brings the original SARF idea up to date through extensive scholarly analysis of media's role in communicating risk, consequences for public perception and the emergence of controversy, and effects on politics and policy. It is not limited to environmental risk but usefully combines environmental with other examples that together illustrate the more general theory.

References

Andrew, C. and M. Velasquez. (1991). "Ethics and the Spotted Owl Controversy." *Issues in Ethics* 4(1). Marcula Center for Applied Ethics, Santa Clara University. Downloaded from: www.scu.edu/ethics/publications/iie/v4n1/homepage.html (accessed October 31, 2013).

Bakir, V. (2005). "Greenpeace v. Shell: Media Exploitation and the Social Amplification of Risk Framework (SARF)." *Journal of Risk Research* 8(7–8): 679–691.

Berger, P. L. and T. Luckmann, (1966). *The Social Construction of Reality: A Treatise in the Sociology of Knowledge.* Garden City, NY: Doubleday.

Boykoff, M. (2011). *Who Speaks for the Climate? Making Sense of Media Reporting on Climate Change.* Cambridge: Cambridge University Press.

Brulle, R. J. (2014). "Institutionalizing Delay: Foundation Funding and the Creation of U.S. Climate Change Counter-Movement Organizations." *Climatic Change* February, 122(4): 681–694.

Cobb, R. W. and C. Elder. (1971). "The Politics of Agenda-Building: An Alternative Perspective for Modern Democratic Theory." *Journal of Politics* 33(4): 892–915.

Collins, H. and R. Evans. (2007). *Rethinking Expertise.* Chicago, IL: University of Chicago Press.

Dahl, R. (2010). "Green Washing." *Environmental Health Perspectives* 118(6): A246–A252.

Dunlap, R. E. (2008). "The New Environmental Paradigm Scale: From Marginality to Worldwide Use." *Journal of Environmental Education* 40(1): 3–18.

Friedman, S., K. Villamil, R. A. Suriano, and B. P. Egolf. (1996). "Alar and Apples: Newspapers, Risk and Media Responsibility." *Public Understanding of Science* 5(1): 1–20.

Futurity. (2014). "About Futurity." Downloaded from www.futurity.org/about/ (accessed October 31, 2013).

Gandy, O. (1982). *Beyond Agenda Setting: Information Subsidies and Public Policy.* Norwood, NJ: Ablex Publishers.

Kasperson, R. E., O. Renn, P. Slovic, H. Brown, J. Emel, R. Goble, J. Kasperson, and S. Ratick. (1988). "The Social Amplification of Risk: A Conceptual Framework." *Risk Analysis* 8(2): 177–187.

Kuhn, T. (1962). *The Structure of Scientific Revolutions.* Chicago, IL: University of Chicago Press.

Lang, G.E. and K. Lang. (1981). "Watergate: An Exploration of the Agenda-Building Process." In G. C. Wilhoit and H. Deboch (eds). *Mass Communication Review Yearbook* (Vol. 2, pp. 447–468). Beverly Hills, CA: Sage.

Leiserowitz, A., E. Maibach, C. Roser-Renouf, G. Feinberg, and P. Howe. (2013). *Global Warming's Six Americas: September 2012.* New Haven, CT: Yale Project on Climate Change Communication. Downloaded from: http://environment.yale.edu/climate-communication/files/Six-Americas-September-2012.pdf (accessed October 31, 2013).

McCombs, M. E. and D. L. Shaw. (1972). "The Agenda Function of Mass Media." *The Public Opinion Quarterly* 36(2): 176–187.

Nelkin, D. (1995). *Selling Science: How the Press Covers Science and Technology.* Rev. ed. New York: W. H. Freeman.

Nucci, M., C. L. Cuite, and W. K. Hallman. (2009). "When Good Food Goes Bad: Television Network News and the Spinach Recall of 2006." *Science Communication* 31(2): 238–265.

Pew Research Center. (2014). "The State of the News Media 2013: An Annual Report on American Journalism. Overview." Downloaded from http://stateofthemedia.org/2013/overview-5/ (accessed October 31, 2013).

Priest, S. (2012). *Nanotechnology and the Public: Risk Perception and Risk Communication*. Boca Raton, FL: CRC Press (Taylor & Francis).

Priest, S. (2013a). "Can Strategic and Democratic Goals Coexist in Communicating Science? Nanotechnology as a Case Study in the Ethics of Science Communication and the Need for 'Critical' Science Literacy." In *Ethical Issues in Science Communication: A Symposium. Proceedings of a Symposium at Iowa State University, May 30–June 1, 2013*.

Priest, S. (2013b). "Critical Science Literacy: What Citizens and Journalists Need to Know to Make Sense of Science." *Bulletin of Science, Technology & Society* October–December, 33(5–6): 138–145.

Rughinis, C. (2011). "A Lucky Answer to a Fair Question: Conceptual, Methodological, and Moral Implications of Including Items on Human Evolution in Scientific Literacy Surveys." *Science Communication* 33(4): 501–532.

Sturgis, P. and N. Allum. (2004). "Science in Society: Reevaluating the Deficit Model of Public Attitudes." *Public Understanding of Science* 13(1): 55–74.

Tanner, A. and D. B. Friedman. (2011). "Authorship and Information Sourcing for Health News on Local TV Web Sites: An Exploratory Analysis." *Science Communication* 33(1): 3–27.

Tremayne, M., ed. (2007). *Blogging, Citizenship, and the Future of Media*. New York: Routledge.

Tuchman, G. (1978). *Making News: A Study in the Construction of Reality*. New York: Free Press.

Yang, J. and L. Kahlor. (2013). "What Me Worry? The Role of Affect in Information Seeking and Avoidance." *Science Communication* 35(2): 189–212.

27

AGENDA-SETTING WITH ENVIRONMENTAL ISSUES

Craig Trumbo and Se-Jin 'Sage' Kim

Introduction

Public issues exhibit a wide variety of characteristics within the various domains in which they are active—domains such as public opinion, policy attention and action, public discourse and news media, and others. The salience of an issue within a given domain is an important issue characteristic. And since these domains cannot operate in unique vacuums but rather interact and affect one another in complex ways, the manner in which the saliencies of public issues are shaped by interactions between and among these domains becomes an important concern. This is the focus of agenda-setting research, a good share of which has been directed toward understanding the forces that shape environmental issues.

We assume that readers have a good general familiarity with agenda setting overall. Thus we will turn directly to an examination of the studies that have applied agenda setting in the context of environmental issues. This material is organized to highlight the various environmental issues that have been examined, the disciplinary and international perspectives that have been involved, the approaches used, and the findings across the range of this body of work. Our hope is to provide an organized treatment of this work that may provide a roadmap to experienced researchers and a point of entry for those interested in joining this inquiry.

Approach

It is difficult to draw distinct boundaries around the body of work called agenda setting. To identify the studies to include in this review a broad approach was taken. We used the literature index *Communication and Mass Media Complete* in conjunction with *Social Science Abstracts* to identify all studies in which the term agenda setting was used in the title, the index subject heading, keywords, or the author-supplied keywords. We ended up with just over 400 articles, which is consistent with what other reviewers have recently observed. A keyword search proved to be impractical to identify the subset of articles involving environmental issues, so we manually inspected all articles. A review of the bibliographies pointed toward additional material, especially book chapters that were not provided by the index searches. We ended up with 51 items published from 1972–2012, accounting for about 12 percent of the agenda-setting literature. Any such effort will yield omissions, for which we accept responsibility and offer regrets.

Our approach to an examination of this literature began with the creation of notes/codes on the articles to categorize aspects of the specific environmental issues involved, the mode of theory, elements of design and methods, and the sources of data used. Items were placed in two broad categories, those in which one or more environmental issues were of central concern to the research (e.g., as much about deforestation as about agenda setting, 30 articles), and those in which one or more environmental issues were peripheral to the contextual thrust (e.g., deforestation one among a larger set of issues or the issues themselves subordinate to agenda setting per se, 21 articles). We then used *Google Scholar* to observe citation counts for each item (chapters and some journals were not indexed in *Web of Science*).

Descriptive findings

We present a set of descriptive results before turning to a discussion of the substantive component of the work reviewed. First, Table 27.1 reports the results of our general coding scheme. In terms of the specific environmental issues represented in the work we see a strong diversity in that the largest category is that which involves multiple issues. Of those 19 articles a dozen have as their focus multiple environmental issues, with the balance involving one or more environmental issues along with several other non-environmental issues.

With respect to the agendas involved in the studies, we see that the three most often observed elements in agenda-setting generally are also dominant in this subset of research: media, public, and policy (note that these do not sum as most articles feature more than one agenda). This pattern shows that the subset of agenda-setting studies that involve environmental issues is largely traditional in form. One likely departure from the main-stream work on agenda-setting is the proportion of work that examines the interaction of issues themselves. Otherwise, real-world factors and inter-media effects are less represented.

The disciplinary tradition of this body or work is largely divided between scholars working in the fields of mass communication and in political science. This is clearly indicated by our observation of citations within the articles, with one branch referring to the McCombs and Shaw tradition and the other referencing Cobb and Elder (and in both cases other related works).

Looking at the design and methods employed, we found it interesting to observe that a fairly even split exists between qualitative and quantitative approaches. In a study showing how the media affected the public's environmental agenda in Ghana, Kwansah-Aidoo (Kwansah-Aidoo, 2001) used qualitative interview and focus group methods and offers a strong argument for the importance of integrating qualitative approaches into agenda-setting studies. This has clearly been the case, although the qualitative papers are strongly dominated by the Cobb and Elder tradition, with only four qualitative papers referencing the McCombs and Shaw tradition.

As would be expected, the majority of studies (and both quantitative and qualitative) made use of longitudinal design. Within the quantitative studies, we see that there is a tendency for the longitudinal designs to be more recent, with none appearing before 1995. Few experiments have been done in agenda-setting. The directionality of the effects investigated is about split between those studies looking at a one-way model (e.g., media affecting public) and those with either a bi-directional effect between two agendas or a systems approach with multi-directional effects among multiple agendas (the latter dominated by longitudinal quantitative papers). Analytical approaches are split between the interpretative methods used in the qualitative studies and a variety of statistical methods used in the quantitative. More recent work has involved longitudinal systems approaches employing time-series analysis tools.

Sources of data have been diverse, with various approaches to measuring media content and policy attention represented. Surprisingly few studies included the use of public opinion survey data, with half relying on secondary analysis (e.g., Gallup's "most important problem") and half executing a unique survey data collection.

The final basic descriptive characteristic we will present before turning to a more substantive discussion is the temporal nature of this body of work. Figure 27.1 plots the timeline of the articles we review. It seems a reasonable interpretation to say that a couple of key works seeded during the early years of agenda

Table 27.1 Gross characteristics of articles examined

Issues	Agenda	Tradition	Empirical	Time	Direction	Data	Analysis
multiple (19)	media (29)	C&E (17)	quantitative (25)	longitudinal (25)	directional (18)	newspaper (26)	interpretive (24)
climate (14)	public (24)	M&S (16)	qualitative (23)	cross-sectional (14)	systems (16)	television (12)	TSA (8)
pollution/waste (5)	policy (23)	other (11)	mixed (3)	N.A. (9)	bi-directional (8)	government docs (11)	regression (7)
policy/regulation (4)	issue (10)	Downs (5)		experimental (3)	N.A. (9)	congressional record (11)	zero-order r (7)
env. movement (3)	real world (5)	Mazur (2)				survey MIP (9)	rank-order r (6)
water resources (2)	inter-media (2)					survey unique (7)	Granger (4)
nuclear energy (2)	other (2)					literature review (6)	cross-lagged r (2)
economics (2)						RGPL (4)	cross-tab (1)
						interviews (4)	ANCOVA (1)
						magazine (3)	t-test (1)
						science publications (3)	
						real world (3)	
						experiment (2)	
						focus groups (2)	
						web traffic (1)	
						press releases (1)	

Source: authors.

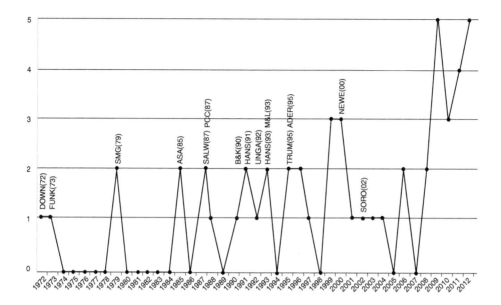

Figure 27.1 Articles published per year (top cited works indicated, see Table 27.2)

Source: authors.

setting—and at the height of the Earth Day influenced period of the environmental movement—set the stage for about a decade of low-level but sustained attention. The most recent decade-and-a-half has seen a significant increase in work in this area of inquiry, with fully half of the papers published since 2000.

Substantive findings

Turning to a more substantive discussion of this work, we first identify the most impactful articles. As described above, we used *Google Scholar* to obtain citation counts for all of the articles reviewed here. This obviously gives heightened scrutiny to older works, but nonetheless offers some utility. Table 27.2 identifies the top articles. As discussed above, we have used a somewhat broad delineation of the works to be considered here and some clearly fall outside of a "standard" definition of agenda-setting. Yet this handful of publications has exhibited significant influence over the interest and focus of agenda-setting researchers working in the context of the environment.

First among these is the 1972 paper by economist Anthony Downs (Downs, 1972) in which he postulates that major social issues exhibit a natural life course running from a pre-problem stage, to discovery, to realization of the cost of addressing the problem, through a decline of social interest and finally into a post-problem phase featuring an intermittent recurrence of interest, typically maintained at an average level higher than that in the pro-problem stage. The issue-attention cycle was clearly fuel for thinking about the environment within the framework of agenda-setting due to the theory's informal incorporation of a media effects perspective.

It was also at about this same period in time that agenda-setting was first elaborated, by McCombs and Shaw and shortly thereafter by Funkhouser. Neither of these papers had the environment explicitly as the focus of attention, but Funkhouser's somewhat more circumspect treatment of the media effect did include the environment as one of several contemporary issues analyzed. What remarkably happened within a couple of years of these two grounding papers was the publication of another report by

Table 27.2 Works cited greater than 100 times, representing top 30 percent

Cites	Abbrev.	Reference	C/P	Issue(S)	Findings
2035	DOWN(72)	Downs, A. (1972) Up and down with ecology: The "Issue-attention cycle." *Public Interest*, 38–50.	P	the environmental movement	proposes that social issues have a life cycle, with five specific phases
473	FUNK(73)	Funkhouser, G. R. (1973) The issues of the sixties: An exploratory study in the dynamics of public opinion. *Public Opinion Quarterly, 37*, 62.	C	environment one of 14 MIP issues	media exert limited effect on public, but both are disconnected from reality
312★	NEWE(00)	Newell, P. (2000) Climate of opinion: the agenda-setting role of the mass media (pp. 68–95). In: Newell, P. *Climate for change non-state actors and the global politics of the greenhouse.* Cambridge, UK: Cambridge Univ. Press.	C	climate change	book paints broad picture of dynamics, chapter shows media has greatest effect on policy early in the life of issue
285	SALW(87)	Salwen, M. B. (1987) Mass media issue dependency and agenda-setting. *Communication Research Reports, 4*, 26–31.	C	multiple environmental issues	personality factors affect degree to which people are susceptible to agenda setting
236★	SORO(02)	Soroka, S. N. (2002) *Agenda-setting dynamics in Canada.* Vancouver: UBC Press.	C	environment and seven other issues	demonstrates bi-directional effects among multiple agendas over attention cycles
231	UNGA(92)	Ungar, S. (1992) The rise and (relative) decline of global warming as a social problem. *The Sociological Quarterly, 33*, 483–501.	P	climate change	applies social arenas approach, shows how real world event triggered focus on climate, how cycle diminished
226	B&K(90)	Brosius, H.-B. and Kepplinger, H. M. (1990) The agenda-setting function of television news. Static and dynamic views. *Communication Research, 17*, 183–211.	P	environmental protection, energy supply, other non-env.	dynamic longitudinal analyses more appropriate to detect effect, found media-to-public effect
220	ADER(95)	Ader, C. R. (1995) A longitudinal study of agenda setting for the issue of environmental pollution. *Journalism and Mass Communication Quarterly, 72*, 300–311.	C	environmental pollution	media-to-public effect, also real world to public effect for waste recycling

Cites	Abbrev.	Reference	C/P	Issue(S)	Findings
208*	HANS(93)	Hansen, A. (ed.) (1993) *The Mass Media and Environmental Issues* (Part 1. Building the environmental agenda pp. 1–105). Leicester: Leicester University Press.	P	multiple environmental issues	collection of chapters demonstrate a wide variety of agenda-setting forces
204	HANS(91)	Hansen, A. (1991) The media and the social construction of the environment. *Media, Culture and Society, 13*, 443–458.	C	multiple environmental issues	literature review incorporating constructivist perspective
197	M&L(93)	Mazur, A. and Lee, J. (1993) Sounding the Global Alarm: Environmental Issues in the US National News. *Social Studies of Science, 23*, 681–720.	C	climate change, ozone, extinction, rain forest destruction	shows how multiple issues interact as each goes through an attention cycle
138	SMG(79)	Schoenfeld, A. C., Meier, R. F. and Griffin, R. J. (1979) Constructing a social problem: The press and the environment. *Social Problems, 27*, 38.	P	multiple environmental issues	insight into how the new media cover the environment, constructivist and environmental sociology perspectives
123	ASA(85)	Atwater, T., Salwen, M. B. and Anderson, R. B. (1985) Media Agenda-Setting With Environmental Issues. *Journalism Quarterly, 62*, 393–397.	C	multiple environmental issues	interpersonal communication shapes agenda-setting
106	PCC(87)	Protess, D. L., Cook, F. L. and Curtin, T. R. (1987) The impact of investigative reporting on public opinion and policymaking; targeting toxic waste. *Public Opinion Quarterly, 51*, 166–185.	C	toxic waste	media had minimal effect on public but strong effect on policy
106	TRUM(95)	Trumbo, C. (1995) Longitudinal modeling of public issues: An application of the agenda-setting process to the issue of global warming. *Journalism and Mass Communication Monographs, 52*, 1–57.	C	climate change	demonstrates bi-directional effects among multiple agendas over attention cycle, notably inter-media agenda setting

Note: C/P denotes Central/Peripheral, ★ denotes book.

Schoenfeld, Meyer, and Griffin (Schoenfeld et al., 1979). While certainly peripheral to agenda-setting, the authors brought to bear the contemporary literature on social problems from mainstream sociology and the side channel of environmental sociology. To this was added the exciting new perspectives being built on news making by authors. With their focus squarely on environmental issues, this remarkable paper brought forth important clarity to the complex factors working within the news media that would later be articulated as important mechanisms of social effect driving issue dynamics. Interestingly Schoenfeld (Schoenfeld, 1979) also published a paper that year describing how an environmental story failed to appear on the national agenda.

It was then in the mid 1980s that three papers were published that clearly set a focus on the environment within the rapidly growing area of agenda-setting research. Atwater, Salwen, and Anderson (Atwater et al., 1985b) offered arguably the first explicitly environmental treatment of agenda-setting. Looking at how multiple environmental issues were transferred from media to public agendas, the authors executed possibly the first primary data collection (survey) in the line of work. The study offered important insight as it illustrated (again among the first) how individual-level factors involving interpersonal communication shaped agenda-setting and highlighted the distinction between obtrusive and unobtrusive issues in their ability to exert an agenda-setting effect. Salwen also followed up on this project by publishing an additional analysis looking at the individual difference variable of media dependency as a factor that promotes an agenda-setting effect. Between these two papers Protess, Cook, and Curtin (Protess et al., 1987) published the results of their remarkable field experiment looking at the effect of the media on both the public and policy agenda as a consequence of the publication of a major report on toxic waste disposal in Chicago. The study did not show a strong effect on the public but was groundbreaking in this tradition of work in how it showed the media having a strong effect on the policy agenda, an interest in this area of research that continued to grow.

A more steady stream of influential works began appearing through the 1990s, each building on these earlier works vis-à-vis either scholarly follow-up or contextual inspiration, and all strongly influenced by the broader developments in agenda setting. At the start of this decade Brosius and Kepplinger (Brosius and Kepplinger, 1990) included energy supply and environmental protection among a total of 16 social issues represented in German television. Their study made use of newly emerging methods to examine longitudinal bi-directional effects and highlighted how the two environmental issues demonstrated media-to-public effects but not vice versa, attributing the differences in directional effect to the characteristics of the issues themselves. During this decade Anders Hansen made two critical contributions. The first was through his 1991 review and critique that brought together many of the earlier ideas on the role of the news media in shaping public issues, casting media effects (including agenda setting) within the more intellectually mature social constructivist framework (Hansen, 1991). This is arguably the first paper to provide a focused examination of the work reviewed herein up to this point (and of course does so in more detail). The second contribution was via an edited volume broadly examining how the mass media affect environmental issues (Hansen, 1993). The first part of the volume, "Building the environmental agenda," brings together five papers depicting how communicative practice, from environmental journalists to pressure groups to public relations interact to shape what the public comes to know as the issue of the environment. This collection is a solid example of the much more dynamic and nuanced view of media effects that was emerging at that time, and also offered a solid contextual emphasis on the environment as an issue.

During this period of time there were also two influential studies published from outside the formal boundaries of agenda setting, or for that matter mass communication, research. Sociologists Alan Mazur (with Jinling Lee) (Mazur and Lee, 1993) and Shelly Ungar (Ungar, 1992) each offered detailed examinations of the trajectory of important environmental problems as social issues and placed a solid focus on the force of the mass media. Both of these papers carried forth the heritage of earlier social constructivist work and social problems theory, but also blended in aspects of contemporary media effects theory. They both also featured a more sophisticated dynamics view of media effects, with influences flowing in multiple directions among multiple entities (and including real world events) and examined through time.

Probably one of the most enduring contributions of these papers, however, was that they overlapped on the issue of climate change, which was the sole focus of Ungar and one of four issues examined by Mazur and Lee. While climate change was rapidly becoming an obvious target for agenda-setting researchers, these two papers served as a strong foundation for that thread of work to rest on. The work published a few years later by Trumbo served to bring these two studies together, placing a square-on emphasis on climate change within the agenda-setting model (Trumbo, 1995). This look at the "agenda dynamics" of climate change demonstrated multiple effects, including reciprocal effects, longitudinally among media, public, policy, and scientific agendas.

The 1990s also includes one of the growing number of agenda-setting studies to model the effect of real world conditions. Christine Ader (Ader, 1995) used the *New York Times* and Gallup's Most Important Problem opinion series along with indicators of pollution levels to show that while a media-to-public effect was present, the real world of pollution did not directly influence either media attention or public concern. But with a narrower focus on waste recycling a real world to media effect was found.

As indicated above, citation counts are a strongly time-dependent metric. The articles among the strong growth of agenda-setting studies examining environmental issues in the last decade or so have yet had time to accumulate enough citations to break into the top 15 presented here, and arbitrarily cut at 100 citations. So we close this more detailed examination of the most influential works with two books. In his 2000 title *Climate for Change*, Newell (Newell, 2000) includes a chapter addressing how the mass media set the international agenda for climate change. His review of previous works, news reports, and other documents is of considerable, and unique value as it places the question of media effects in the broader context of how this issue of climate change has been shaped by international interests, most interestingly including non-state actors such as NGOs. In his book elaborating agenda-setting dynamics, Soroka (Soroka, 2002b) examines a set of eight public issues in Canada. This work, an expansion of the paper published the same year in the *International Journal of Public Opinion Research,* is peripheral with respect to our focus on agenda setting and environmental issues. But the work has quickly become very important in the broader arena of agenda-setting research. In his examination of the dynamics among multiple agendas—public, policy, media, real world—Soroka assimilates and strengthens the various approaches to time series analysis within agenda setting. The work demonstrates what would be appropriately labeled state-of-the art analytical methods for quantitative studies of agenda setting. Substantively, the study finds significant multi-directional effects among all of the agendas for the issue of the environment (as well as debt/deficit but not inflation), which Soroka attributes to the unobtrusive character of the environment as a public issue.

With these influential papers offered as a manner of an evolving story of the work on agenda setting and environmental issues, we now turn our attention more broadly to the full body of work. Our reading of these papers suggested three noteworthy topics: the role of international projects, an expansion of agenda-setting actors, and theoretical evolution. We address each of these in turn.

While the majority of the articles we review here have their focus in the U.S., a total of 19 are distributed around the world. Six of these studies were conducted in Canada. As discussed above, the work that Soroka did on issue dynamics focused on Canadian public issues (Soroka, 2002a). In methodologically relevant work, Howlett (Howlett, 1997) also looked at issue dynamics in a longitudinal approach, examining how policy agendas evolved for nuclear power and acid rain. Kamieniecki (Kamieniecki, 1991, 2000) similarly examines the evolution of the policy agenda, but with an aim toward theoretical testing, looking at forest practices from several agenda-building perspectives. And Lambright and Changnon (Lambright and Changnon, 1996) provide an interesting look at the issue of climate change within an urban context in Toronto. Another topic examined in the Canadian context is on the issue of water export by Bakenova (Bakenova, 2008), who shows that policy execution may differ markedly from what would be expected from the final agendas of various actors. Finally, Pralle (Pralle, 2006) executed a comparative study looking at how lawn pesticide regulation evolved differently between Canada and the U.S. The two countries diverged significantly due to matters of issue timing, actor mobilization, and changes in the venues in which the issue was emphasized.

A good number of papers have had a European focus as well, with studies done in Norway, Sweden, Denmark, The Netherlands, Germany, and Hungary. Those of note include two comparative studies. Green-Pedersen and Wolfe (Green-Pedersen and Wolfe, 2009) look at the policy issue of CO_2 emissions and compare agenda evolution between the U.S. and Denmark, concluding that the U.S. political system is more open to multiple influences giving rise to the generation of quick attention to new issues. But the parliamentary system prominent in Europe, which is slower to react initially, is more prone to sustaining an issue on the policy agenda. Berg (Berg, 1999) also offers an interesting comparative study, set in Hungary. Looking at the broad issue of environmental protection, Berg compares agenda setting and building under Communist and post-Communist governments. Interestingly, many of the processes of agenda setting were in operation well before the transition from Communism and Berg notes that despite the change in political structure the growth and effect of an environmental movement there was not significantly affected.

Other studies with an international focus have been set in the South Pacific Islands, Hong Kong, and Japan. Another location of significant note involves the work of Kwansah-Aidoo in Ghana (Kwansah-Aidoo, 2001, Kwansah-Aidoo, 2003). As mentioned above, this work is noteworthy, first, because of its strong advocacy for qualitative methods applied to the traditional agenda-setting model. In this study it is shown that clear evidence of agenda-setting effects (in this case the media affecting the public on environmental issues generally) can be observed through interviews and focus groups, allowing the theory to be examined in social and infrastructure contexts that might not be amenable to other data collection methods, such as phone surveys.

Possibly one of the most interesting aspects of this body of work resides in the rich variety of agenda-setting actors that have been evoked. Nearly a dozen unique agenda-setting actors, or forces are evoked in this work. Several can be grouped together, roughly, as they examine organizational actors. For example, Bakir (Bakir, 2006) provides an examination of how Shell Oil and Greenpeace competed against each other to set the policy agenda concerning the disposal of an oil drilling platform, while Davis and Hoffer (Davis and Hoffer, 2012) show how a similar competition for the policy agenda was played out between environmental and natural gas coalitions concerning fracking. The effect that non-governmental organizations (in Norway) and environmental movement organizations can have on the policy agenda is shown in two studies to be conditioned on the characteristics of the content of their messages (Johnson, 2008; Krovel, 2012). Other studies have shown how cooperating nations (in the South Pacific) can influence the international policy agenda (Shibuya, 1996), how party platforms and campaign rhetoric affect the public's agenda (Wattier and Tatalovich, 2000), and how alternate communication channels can play an agenda-setting role. In the latter, Hart and Leiserowitz (Hart and Leiserowitz, 2009) looked at the effect on the public agenda exerted by the release of the film *The Day After Tomorrow* (depicting climate change), doing so by tracking traffic changes on climate-related websites. Also, Jenner (Jenner, 2012) provides possibly the only study looking at how news photographs affect public and policy agendas, reporting that photographs depicting environmental problems have an effect on congressional committee attention but no effect on the public.

Three studies have evoked other unique influences. Asim and Todd (Asim and Todd, 2010) examine the effect of political ideology on the public's concern for climate change, showing that education and science literacy increase concern while conservatism decreases it. Interpersonal discussion was examined as having an influence over agenda setting by Atwater, Salwen and Anderson (Atwater et al., 1985a). In this study they observed through content analysis and phone survey data that individuals who were most engaged in interpersonal discussion of environmental issues demonstrated less of an agenda-setting effect.

And Clark (Clark, 2004) followed Kingdon's model to elaborate how the policy agenda was affected by the evolution of an idea (dam breaching).

Finally, it is worth emphasizing a recent paper that stands out on multiple points. Liu, Lindquist, and Vedlitz (Liu et al., 2011) present a very sophisticated study making use of time series analysis tools to look at climate change. They show how objective problem indicators, focusing events, and information feedback from the scientific community all increased issue salience on the congressional agenda.

Turning to our third area of emphasis in this work, we see that some (but not all) of the theoretical developments that have occurred in agenda setting over the years are present. Most prominently we see studies that have expanded the focus from one-way effects of a given agenda on another. These authors have taken a broader view and have turned attention toward the interactions among issues. Soroka's (2002a, 2002b) work discussed above is the most salient and recent example, and he makes the case that the key to achieving a more sophisticated and far looking form of agenda setting is to translate it into issue dynamics. The earliest work in this review that took such an approach was Mazur and Lee in the examination of how the issues of ozone depletion, climate change, rain forest destruction, and mass extinctions formed an interactive set of environmental issues active among multiple agendas (1993).

Between these two papers there are a number of others that have looked at issues themselves. For example, in a study set in Hong Kong, Chan (Chan, 1999) investigated how the issue of the environment was affected by other prominent national issues between 1983 and 1995, finding that it was essentially out competed for long-term attention. Flemming, Wood, and Bohte (Flemming et al., 1999) examine the macrodynamics of three issue areas—environmental rights, civil liberties, and civil issues—and compare their evolution across institutional (supreme court, congress, the president) and systemic (news media) agendas. They do not find support for unidirectional or non-interactive behavior among these agendas, and argue for a more holistic approach to agenda setting. In a somewhat similar vein, Uscinski (Uscinski, 2009) examines a comprehensive set of social issues from Gallup's Most Important Problem series and makes use of some 35,000 broadcast news segments from the Vanderbilt Television News Archive. The analysis was designed to investigate what issue types might be more likely to transfer salience from the media to the public or vice versa. While only a few environmental issues are included in this study (and it is thus peripheral in our scheme), it is found that issues that tend not to have spectacular events associated with them (including the environment) are those in which salience tends to be transferred from the public to the media and not the other way.

In a recent and quite remarkable set of publications, Monika Djerf-Pierre examines five decades (1961–2010) of environmental reporting in a single outlet, Swedish public service television (Djerf-Pierre, 2011, Djerf-Pierre, 2012a, Djerf-Pierre, 2012b). The study involves nearly 9,000 stories falling into 23 categories of environmental reporting. One core theme of her work goes all the way back to Anthony Down's issue-attention cycle. She builds a much more detailed, and quantitatively assessed, model of issue life cycles, which she terms metacycles. Four issue characteristics are found to drive metacycles: catastrophe, alarm, scandal, and controversy. Interestingly, it is also found that issues themselves do not necessarily compete against each other among metacycles. However, in an associated study that also made use of a previously published dataset (Asp, 2010), it is found that environmental issues are affected by issue fatigue and competition in which they are "crowded out" by economic and war stories.

Three other studies fall into the general area of individual effects in the broader scope of theoretical development in agenda setting. As described above, need for orientation and issue dependence have been found in other agenda-setting work to catalyze the propensity for a media-to-public agenda-setting effect. In a methodological work, Chernov, Valenzuela and McCombs (Chernov et al., 2011) experimentally tested extant and new versions of the need for an orientation scale using three issues, crime, drug abuse, and climate change. The scales were found to be equivalent. One of the early studies on issue dependency, by Salwen, found evidence for the effect using a survey looking at the suite of environmental issues in Lansing, Michigan (Salwen, 1987). And Lee tested whether issue salience or attribute salience was more

321

likely to influence priming (Lee, 2010), reporting that the priming effects were more likely influenced by second-level (focusing on issue attributes) agenda setting effects than first-level (focusing on issues or objects) agenda setting effects.

Conclusion

Agenda-setting has been a robust vehicle for the social scientific study of media and the environment over the last four decades. In terms of this body of work's contribution to the bigger picture on agenda-setting, we would argue that the major impact has been in the area of understanding longer-term and interactive cycles of attention. And along with that conceptual focus an emphasis is on the application of more sophisticated analytic tools for the examination of such cycles. It does seem clear that the environment presents a very fruitful domain in which to conduct socially relevant research. Some of the fundamental aspects of agenda setting make environmental issues a very good fit. Environmental issues can be quite unobtrusive, and thus amenable for this manner of media effect. Environmental issues are also intractable and inevitably play out over long periods of time, going through various cycles of attention and re-definition. Environmental issues are by their nature embedded in multiple agenda domains; not just policy, media, and public, but blogs, advertising and public relations campaigns, efforts by non-governmental organizations, websites, and many others. All of these domains in which environmental issues live present possibilities for agenda effects to be observed.

Further reading

These works are suggested to provide a glimpse into the future of agenda setting and a broader conceptualization of the practice and function of environmental communication.

Johnson, T. J. (ed.) (2013) *Agenda Setting in a 2.0 World: New agendas in communication*. New York: Taylor & Francis.
Lundgren, R. E. and McMakin, A. H. (2013) *Risk Communication: A handbook for communicating environmental, safety, and health risks*. New York: Wiley.
Hendry, J. (2010) *Communication and the Natural World*. State College, PA: Strata Pub Co.
Corbett, J. B. (2006) *Communicating Nature: How we create and understand environmental messages*. Washington, DC: Island Press.

References

Ader, C. R. (1995) A longitudinal study of agenda setting for the issue of environmental pollution. *Journalism and Mass Communication Quarterly*, 72, 300–311.
Asim, Z. and Todd, A. M. (2010) Evaluating the effects of ideology on public understanding of climate change science: How to improve communication across ideological divides? *Public Understanding of Science*, 19, 743–761.
Asp, K. (2010) Nordens plats i en medialiserad varld—frya nyhetshypoteser. In Torbjorn, B., Tufte, T., Weibull, L., and Helge, O. (eds) *Perspektiv fran forskningen om medier och kommunikation* (pp. 345–360). Gothenburg: JMG University of Gothenburg.
Atwater, T., Salwen, M. B., and Anderson, R. B. (1985a) Interpersonal discussion as a potential barrier to agenda-setting. *Newspaper Research Journal*, 6, 37–43.
Atwater, T., Salwen, M. B., and Anderson, R. B. (1985b) Media agenda-setting with environmental issues. *Journalism Quarterly*, 62, 393–397.
Bakenova, S. (2008) Making a policy problem of water export in Canada: 1960–2002. *Policy Studies Journal*, 36, 279–300.
Bakir, V. (2006) Policy agenda setting and risk communication: Greenpeace, Shell, and issues of trust. *Harvard International Journal of Press/Politics*, 11, 67–88.
Berg, M. M. (1999) Environmental protection and the Hungarian transition. *Social Science Journal*, 36, 227–250.
Brosius, H.-B. and Kepplinger, H. M. (1990) The agenda-setting function of television news: Static and dynamic views. *Communication Research*, 17, 183–211.

Chan, K. (1999) The media and environmental issues in Hong Kong 1983–95. *International Journal of Public Opinion Research*, 11, 135–151.

Chernov, G., Valenzuela, S., and McCombs, M. (2011) An experimental comparison of two perspectives on the concept of need for orientation in agenda-setting theory. *Journalism and Mass Communication Quarterly*, 88, 142–155.

Clark, B. T. (2004) Agenda setting and issue dynamics: Dam breaching on the Lower Snake River. *Society and Natural Resources*, 17, 599–609.

Davis, C. and Hoffer, K. (2012) Federalizing energy? Agenda change and the politics of fracking. *Policy Sciences*, 45, 221–241.

Djerf-Pierre, M. (2011) Green metacycles of attention: Reassessing the attention cycles of environmental news reporting 1961–2010. *Public Understanding Of Science (Bristol, England)*, 22, 495–512.

Djerf-Pierre, M. (2012a) The crowding-out effect. *Journalism Studies*, 13, 499–516.

Djerf-Pierre, M. (2012b) When attention drives attention: Issue dynamics in environmental news reporting over five decades. *European Journal of Communication*, 27, 291–304.

Downs, A. (1972) Up and down with ecology: The "Issue-attention cycle." *Public Interest*, 38–50.

Flemming, R. B., Wood, B. D., and Bohte, J. (1999) Attention to issues in a system of separated powers: The macrodynamics of American policy agendas. *Journal of Politics*, 61, 76–108.

Funkhouser, G. R. (1973). The issues of the sixties: An exploratory study in the dynamics of public opinion. *Public Opinion Quarterly*, 37(1), 62–75.

Green-Pedersen, C. and Wolfe, M. (2009) The institutionalization of environmental attention in the United States and Denmark: Multiple- versus single-venue systems. *Governance*, 22, 625–646.

Hansen, A. (1991) The media and the social construction of the environment. *Media, Culture and Society*, 13, 443–458.

Hansen, A. (ed.) (1993) *The Mass Media and Environmental Issues (Part 1. Building the environmental agenda pp. 1–105)*, Leicester: Leicester University Press.

Hart, P. S. and Leiserowitz, A. A. (2009) Finding the teachable moment: An analysis of information-seeking behavior on global warming related websites during the release of *The Day After Tomorrow*. *Environmental Communication*, 3, 355–366.

Howlett, M. (1997) Issue-attention and punctuated equilibria models reconsidered: An empirical examination of the dynamics of agenda-setting in Canada. *Canadian Journal of Political Science*, 30, 3.

Jenner, E. (2012) News photographs and environmental agenda setting. *Policy Studies Journal*, 40, 274–301.

Johnson, E. W. (2008) Social movement size, organizational diversity and the making of federal law. *Social Forces*, 86, 967–993.

Kamieniecki, S. (1991) Political mobilization, agenda building and international environmental policy. *Journal of International Affairs*, 44, 339–358.

Kamieniecki, S. (2000) Testing alternative theories of agenda setting: Forest policy change in British Columbia, Canada. *Policy Studies Journal*, 28, 176–189.

Krovel, R. (2012) Setting the agenda on environmental news in Norway. *Journalism Studies*, 13, 259–276.

Kwansah-Aidoo, K. (2001) The appeal of qualitative methods to traditional agenda-setting research. *Gazette: International Journal for Communication Studies*, 63, 521.

Kwansah-Aidoo, K. (2003) Events that matter: Specific incidents, media coverage, and agenda-setting in a Ghanaian context. *Canadian Journal of Communication*, 28, 43.

Lambright, W. H. and Changnon, S. A. (1996) Urban reactions to the global warming issue: Agenda setting in Toronto and Chicago. *Climatic Change*, 34, 463.

Lee, G. (2010) Who let priming out? Analysis of first- and second-level agenda setting effects on priming. *International Communication Gazette*, 72, 759–776.

Liu, X., Lindquist, E., and Vedlitz, A. (2011) Explaining media and congressional attention to global climate change, 1969–2005: An empirical test of agenda-setting theory. *Political Research Quarterly*, 64, 405–419.

Mazur, A. and Lee, J. (1993) Sounding the global alarm: Environmental issues in the US national news. *Social Studies of Science* (Sage Publications), 23, 681–720.

Newell, P. (2000) Climate of opinion: The agenda-setting role of the mass media. In Newell, P. (ed.) *Climate for change: Non-state actors and the global politics of the greenhouse* (pp. 68–95). Cambridge: Cambridge University Press.

Pralle, S. B. (2006) Timing and sequence in agenda-setting and policy change: A comparative study of lawn care pesticide politics in Canada and the US. *Journal of European Public Policy*, 13, 987–1005.

Protess, D. L., Cook, F. L., and Curtin, T. R. (1987) The impact of investigative reporting on public opinion and policymaking: Targeting toxic waste. *Public Opinion Quarterly*, 51, 166–185.

Salwen, M. B. (1987) Mass media issue dependency and agenda-setting. *Communication Research Reports*, 4, 26–31.

Schoenfeld, A. C., Meier, R. F., and Griffin, R. J. (1979) Constructing a social problem: The press and the environment. *Social Problems*, 27, 38.

Schoenfeld, C. (1979) The Press and NEPA. The Case of the Missing Agenda. *Journalism Quarterly*, 56, 577–585.

Shibuya, E. (1996) Roaring mice against the tide: The South Pacific islands and agenda-building on global warming. *Pacific Affairs*, 69, 541–555.

Soroka, S. (2002a) Issue attributes and agenda-setting by media, the public, and policymakers in Canada. *International Journal of Public Opinion Research*, 14, 264–285.

Soroka, S. N. (2002b) *Agenda-setting Dynamics in Canada*, Vancouver: UBC Press.

Trumbo, C. (1995) Longitudinal modeling of public issues: An application of the agenda-setting process to the issue of global warming. *Journalism and Mass Communication Monographs*, 1–57.

Ungar, S. (1992) The rise and (relative) decline of global warming as a social problem. *The Sociological Quarterly*, 33, 483–501.

Uscinski, J. E. (2009) When does the public's issue agenda affect the media's issue agenda (and vice-versa)? Developing a framework for media–public influence. *Social Science Quarterly*, 90, 796–815.

Wattier, M. J. and Tatalovich, R. (2000) Issue publics, mass publics, and agenda setting: Environmentalism and economics in presidential elections. *Environment and Planning C: Government and Policy*, 18, 115.

28

FRAMING, THE MEDIA, AND ENVIRONMENTAL COMMUNICATION

Matthew C. Nisbet and Todd P. Newman

Introduction

Framing as an area of research spans several scholarly disciplines. Frames as they appear in public debate and media coverage are interpretive storylines that set a specific train of thought in motion, communicating why an issue or decision matters, who or what might be responsible, and which political options or actions should be considered over others (Nisbet 2009a; Nisbet and Scheufele 2009). Framing an issue is also an important exercise in power (see Hansen 2011). By defining the terms of debate, groups and advocates can influence the amount of attention an issue receives, the arguments or considerations that are considered legitimate or out of bounds, and the voices who have standing to express their opinion or participate in decisions (Nisbet and Huge 2006; 2007).

There is no such thing as unframed information, and many readers of this chapter by way of their conversations, social media use or other interactions are already effective at framing their opinions and positions, whether using frames intentionally or intuitively. Framing, it should be noted, is not synonymous with placing a false spin on an issue, although some communicators do purposively distort evidence and facts. Rather, in an attempt to remain true to what is conventionally known about a complex topic, as a communication necessity, framing can be used to pare down information, giving greater weight to certain considerations and elements over others (Nisbet 2009b).

For these reasons, if scholars, professionals, and citizens are to effectively evaluate and/or participate in environmental debates, they will need to apply an understanding of framing as a cognitive, social, and political process. Members of the public rely on frames to make sense of and discuss complex environmental issues; journalists use frames to craft interesting and appealing news reports; policymakers apply frames to define policy options and reach decisions; and experts employ frames to simplify technical details and make them persuasive (Scheufele 1999; Nisbet 2009a).

In this regard, relative to the field of environmental communication, as Hansen (2011) has called for, research on framing provides an invaluable set of theoretical and methodological approaches for evaluating the factors that motivate social movements and protest; that influence the actions of political leaders and groups; that shape news coverage and patterns of attention; and that affect policy decisions and public opinion.

In this chapter, we begin by reviewing the role of framing at the macro-level as it relates to various social movement and advocacy strategies, the social construction of controversies such as those over nuclear energy, climate change, and food biotechnology, and the significance to political decision making. In doing so, we describe a generalizable typology of frames that can be applied to studying the social and political development of environmental debates and technology controversies as they play out in media coverage and public discourse. In the second section, we review research evaluating how the frames of reference found in media coverage and political debate resonate with and selectively activate individual mental models and cognitive "schema," thereby influencing public opinion. We also discuss examples of how this research is being applied to the design of effective public engagement campaigns on climate change. To conclude the chapter, we briefly review new directions for framing research in environmental communication, highlighting the relevance of work on cultural cognition and moral foundations theory; as well as "Big Data" methods for evaluating social media discussion and discourse.

Frame contests and environmental politics

In evaluating the factors influencing social movement strategies, media coverage, and political decisions, many framing scholars have followed the lead of sociologist William Gamson, adopting a "social constructivist" approach. According to this line of research, in order to make sense of political issues, citizens use as resources the frames available in media coverage, but integrate these packages with their own mental frames of reference forged by way of personal experience and conversations with others.

Media frames might help set the terms of the debate among political actors and the public, but rarely, if ever, do they exclusively determine public opinion. Instead, as part of a "frame contest," one interpretative package might gain influence because it resonates with popular culture or a series of events; fits with media routines or practices; and/or is heavily sponsored by powerful political actors (Gamson and Modigliani 1989; Gamson 1992; Price, Nir, and Capella 2005; Nisbet 2009a).

The framing of an environmental issue can also influence broader public attention while also shaping the "scope of participation" in a political debate (defined as the types and numbers of groups who are involved in policy making). In fact, across the history of many policy debates, power has turned on the ability to not only control attention to an issue within policy contexts or in the media, but also to simultaneously frame the nature of the problem and what should be done (Nisbet and Huge 2006; 2007; Hansen 2011).

If a group or coalition is favored by the status quo in environmental policy making, it is in their best interest to frame issues in highly technical, scientific, or legalistic ways and to downplay possible risks, since these interpretations tend to deflect wider news attention, and attract only narrow constituencies (Nisbet and Huge 2006; 2007, see also Schlichting 2013). Under these conditions, journalists lack the dramatic grist to produce coverage of an issue on an ongoing basis, meaning that overall news attention will remain low and sporadic (Nisbet and Huge 2006; 2007).

But, on the other hand, if a group or coalition is disadvantaged by the status quo in policy making, it is usually in their best interest to reframe an environmental issue in terms of dramatic risks/costs and in moral ways. These interpretations are more likely to shift decision making from regulatory arenas such as the Environmental Protection Agency (EPA) or the Food and Drug Administration (FDA) to overtly political contexts such as Congress or the White House, where arguments emphasizing dramatic risks and morality have more sway. Under these conditions, it becomes potentially easier to mobilize a more diverse coalition of groups to challenge the status quo, to generate widespread media coverage, and to influence broader public opinion (Nisbet and Huge 2006; 2007, see also Cox 2010; Hestres 2013).

Shifting the debate over food biotech

Consider the example of food biotechnology. Previous research has attempted to understand why the issue until recently has experienced relatively limited media and public attention in the U.S., especially in comparison to that in the United Kingdom and several European countries (Gaskell, Bauer, Durant, and Allum 1999). A major reason is that the biotech industry and scientists have been successful at limiting the scope of participation, as early policy decisions framed the issue around the technical aspects of scientific review and patenting rules. This ability to frame the terms of the debate and to limit the scope of attention and participation helped establish a virtual policy monopoly within regulatory policy arenas such as the FDA and EPA with very little attention from Congress or the White House or beyond the science and business beats at newspapers or other media outlets (Nisbet and Huge 2006; 2007).

Though increased media attention to plant biotechnology and more dramatic definitions of the issue have surfaced in recent years, challenging the status quo in U.S. regulation, the ability of the biotech industry and allies in early policy decisions to frame the debate around short-term environmental and health risks has led to lasting and powerful feedback effects. The early success of biotech proponents in defining the terms of the debate is attributable in part to minimal media coverage, which made precedent setting 1990s market approvals of genetically modified crops essentially 'non-decisions' for the wider public (Nisbet and Huge 2006; 2007).

This is in contrast to the UK and Europe, where from the beginning, there was a much wider scope of participation in policy decisions. The early inclusion of environmental, consumer, and labor groups, and the comparatively stronger framing of the issue in terms of transparency and public accountability, led to a very different European regulatory regime that took into account social and economic factors as well as the possibility of unknown future technical risks (Nisbet and Huge 2006; 2007; Listerman 2010).

Yet in the U.S. there are signs that the scope of participation and framing of the issue may be shifting. Critics of food biotechnology have helped expand and intensify public opposition to food biotechnology even as overall national news attention has remained low. Critics have done so by framing the issue in the context of parallel food system debates, broader consumer trends, and by taking advantage of the diffusion of online media strategies and favorable media outlets. These trends include public interest in localized economies and "buy local" efforts, as well as an idealization of "natural" and organic over conventional and industrial food production. Moreover, these trends have not only been popularized by advocates and social entrepreneurs, but also by way of the dramatic growth online in progressive media outlets such as *Mother Jones* or *Grist.org* and by documentary film campaigns such as *Food Inc.*, with articles or video excerpts widely shared and diffused by advocates via social media.

A generalizable typology of frames across environmental debates

For scholars and professionals analyzing debates over the environment, identifying the relevant frames as they appear and spread can be approached deductively. Drawing on previous work, studies usually work from a set of generalizable frames that appear to reoccur across policy debates and that tend to organize our thinking and conversations about the social implications of science, technology, the economy, and politics as they relate to the environment. Originally identified by Gamson and Modigliani (1989) in a study of nuclear energy, the typology of cultural frames was further adapted in studies of biotechnology in Europe and the United States (Durant, Bauer, and Gaskell 1998; Dahinden 2002; Nisbet and Lewenstein 2002), and in analysis of the debate over climate change (Nisbet 2009a). In these past studies, researchers have explored the frames available in the media via the qualitative and quantitative analysis of text-based media representations of environmental issues. More recently, scholars have also evaluated how visual images of climate change featured in media reports and advocacy strategies selectively define risks, and attribute responsibility and courses of action (Hansen and Machin 2008; O'Neill, Boykoff, Niemeyer, and Day 2013; O'Neill 2013; Meisner and Takahashi 2013; O'Neill and Smith 2014).

Even as researchers have shown that specific frames of reference about climate change in news coverage and political discourse differ by country and culture (see Dirikx and Gelders 2010; Gordon, Deines, and Havice 2010; Nerlich, Forsyth, and Clarke 2012; Takahashi and Meisner 2012; Zamith, Pinto, and Villar 2012; Dotson, Jacobson, Kaid, and Carlton 2012), and with respect to various energy policies or technological solutions (see Bickerstaff, Lorenzoni, Pidgeon, Poortinga, and Simmons 2008; Stephens, Rand, and Melnick 2009; Feldpausch-Parker et al. 2013), the findings of these studies tend to support a generalizable set of meanings that advocates, political leaders, and journalists tend to draw from across country setting and time.

These frames include:

- social progress: At stake is improving quality of life, or finding solutions to problems. Alternative interpretation is progress defined as living in harmony with nature instead of mastery, "sustainability," "balance," "quality of life";
- economic development/competitiveness: At issue is economic growth and investment, market benefits or risks; protecting local, national, or global competitiveness;
- morality/ethics: The issue is fundamentally a matter of right or wrong; respecting or crossing religious, ethical or "natural" limits, thresholds, or boundaries; and/or working towards justice for those who have been harmed;
- scientific/technical uncertainty: The issue or decision is a matter of expert understanding; what is known versus unknown; arguments either invoke or challenge expert consensus, call on the authority of "sound science," falsifiability, or peer-review to establish criteria for decisions;
- Pandora's box/runaway science/fatalism: Call for precaution in face of possible impacts or catastrophe. Defines problem or technology as out-of-control, a Frankenstein's monster, or as fatalistic, e.g., action is futile, the train has left the station, the path is chosen, no turning back;
- public accountability/governance: Is a decision or action in the public interest or serving private interests; emphasis on fairness, transparency, ownership, control including responsible use or abuse of expertise in decision making, e.g. "politicization";
- middle way/alternative path: An issue or decision is about finding a possible compromise position, or a third way between conflicting/polarized views or options;
- conflict/strategy: At stake is a broader power game among elites; emphasizing who's ahead or behind in winning debate, in public opinion polling, or political spending. Emphasis is on the battle of personalities; or groups; the tactics and strategies involved and how they will "play politically" (usually journalist-driven interpretation).

A few key details about this typology are worth noting. First, these frames serve as general organizing devices for public debate and should not be confused with specific policy positions. In other words, each frame can relate to pro, anti, and neutral arguments, though one set of advocates might more commonly activate one cultural schema over others. This distinction between frames and the valence of arguments becomes clearer after considering a few examples from the debate over food biotechnology.

In this debate, opponents of food biotechnology have defined the issue in terms of an idealized, pastoral vision of small-scale farms and the "natural," while also emphasizing fears of environmental contamination. These arguments frame food biotech in terms of social progress (specifically, living in harmony with nature rather than controlling it), relative to precaution in the face of scientific uncertainty; and relative to opening a Pandora's Box of "Frankenfoods" that are an "uncontrolled experiment" on nature and humans. Activists have also focused on the perceived inadequacy of regulation to ensure choice for farmers and consumers who have a "right to know" through labeling, emphasizing fairness, transparency, and equity. This line of argumentation frames food biotech in terms of public accountability, particularly science serving the public interest rather than the interests of biotech companies.

Industry and other proponents have countered by emphasizing via media reports and advertising the value of food biotechnology to meet world food demand in an era of climate change and growing population. This argument about food biotech frames the issue in terms of social progress, but emphasizes science and technology as a tool for mastering nature's adverse risks and as solving problems. Proponents have also strongly criticized anti-biotech activists for destroying crops and research installations, for promoting misinformation, and for generating unfounded public fears. This line of argumentation frames food biotech in terms of public accountability, emphasizing the pollution of science by the ideologically motivated actions of activists.

In each of the above examples, specific frames are also often efficiently translated and conveyed by way of "frame devices." These triggers of various interpretative packages include catch phrases (e.g. "Frankenfood," "right to know,"), metaphors (e.g. comparing food biotech to "playing God in the Garden" or "uncontrolled experiments," symbolic of runaway science), and visuals (e.g. an industry advertisement featuring an African farmer standing in an abundant field of crops, symbolic of social progress) (see also Hansen 2006; Maeseele 2010 for further discussion).

Framing effects on public opinion

With limited time and ability to process complex information, as we move through our daily lives trying to make sense of a constant flow of ambiguous signals, situations, and choices, we are heavily dependent on shifting cues that set the context for our perceptions. In this regard, both as a communication necessity and as a persuasion strategy, when experts, advocates, or journalists "frame" a complex environmental issue, they differentially emphasize specific cues relative to that complex issue, endowing certain dimensions with greater apparent relevance than they would have under an alternative frame (Scheufele 1999; Nisbet 2009a; Scheufele and Scheufele 2010). Depending on our own existing mental models about an issue and more generalizable schema about how the world should work, we are more inclined to pay attention to and accept some of these frames of reference over others, reinforcing and influencing our judgments and opinions.

For example, is climate change a grave environmental risk to animal species and ecosystems that requires regulation of industry to solve, or is it a public health threat to children and the elderly that requires government investment in clean energy technology? Both frames are essentially equivalent in accurately depicting the nature of the issue, though emphasizing different attributions about what is at stake and the possible courses of action. In the first context or "frame," the emphasis is on the risks to the environment, protecting nature from harm and the need to limit industry. In the second frame, the emphasis is on the risks to humans, protecting vulnerable people from harm, and the need to aid industry through government funding of technological innovation. Depending on our point of view and social outlook, we are likely to be more open to one of these accounts over the other in determining the relevance of climate change and what should be done.

As mental models and organizing devices for communication, frames set the context for perception and discussion by selectively activating different cognitive and affective schema (Marx et al. 2007). If frames are the software by which we navigate the complexity of risks and choices posed by a problem such as climate change, then cognitive and moral "schema" provide a deeper mental architecture, defining for us core concepts, such as the relationship between science and society, or the government and the economy. Schema can also be value constructs and moral intuitions that guide evaluations of personal behavior and societal choices, such as a desire to protect nature from harm, or to defend our communities from health threats. Once activated, schema provide short cuts for reaching an opinion about a complex topic, serving as a basis for inference, and operating as a mechanism for storing and retrieving information from memory (Scheufele and Scheufele 2010; Nisbet and Markowitz 2014).

In sum then, media and other discursive frames influence our judgments of complex policy debates when they are relevant—or "applicable"—to an individual's specific existing interpretive schema. Framing effects will vary in strength as a partial function of the fit between the schemas a frame suggests should be applied to an issue and the presence of those schemas within a particular audience (Scheufele and Tewksbury 2007).

In other words, media frames influence public perceptions of environmental problems by connecting the mental dots for the public. They suggest a connection between two concepts, issues, or things, such that after exposure to the framed message, audiences accept or are at least aware of the connection. Alternatively, if a frame draws connections that are not relevant to something a segment of the public already values or understands, then the message is likely to be ignored or to lack personal significance (Scheufele and Tewksbury 2007; Nisbet 2009a).

Framing and public engagement on climate change

Recent work in the U.S. has begun to apply the different streams of framing research reviewed so far – from the cultural and social to the cognitive – in order to better understand the types of investments that might motivate and enable increased public participation in decision making related to climate change. This includes efforts to protect and prepare communities against current climate change-related risks and to mitigate those risks over the long term. The work offers a useful model—and raises important implications—for thinking about the effects of framing on public opinion and the applications across environmental problems.

At its core, the research relies on identifying and mapping distinct "interpretative communities" of Americans, improving our understanding of why different segments of the public accept or reject certain arguments, risks, and dimensions of the climate debate (Leiserowitz 2007). An interpretative community is a group of individuals who share common risk perceptions about climate change, reflect shared schema, mental models, and frames of reference, and hold a common sociodemographic background. Not only do these interpretative communities share a common worldview, but the fragmented nature of the media system also helps reinforce, define, and shape a common shared outlook relative to climate change (see Roser-Renouf et al., Chapter 31, this volume, for additional discussion).

In a series of studies by Maibach, Nisbet and colleagues, we investigated how a diversity of Americans understand the health and security risks of climate change and how they react to information about climate change when it is framed in terms of these alternative dimensions. In this research funded by the Robert Wood Johnson Foundation, our goal was to inform the work of public health professionals, municipal managers and planners, and other trusted civic leaders as they seek to engage broader publics on the health and security risks posed by climate change (see Weathers, Maibach, and Nisbet 2013; Nisbet 2014 for overviews).

Framing climate change in terms of public health stresses climate change's potential to increase the incidence of infectious diseases, asthma, allergies, heat stroke, and other salient health problems, especially among the most vulnerable populations: the elderly and children. In the process, the public health frame makes climate change personally relevant to new audiences by connecting the issue to health problems that are already familiar and perceived as important. The frame also shifts the geographic location of impacts, replacing visuals of remote Arctic regions, animals, and peoples with more socially proximate neighbors and places across local communities and cities. Coverage at local television news outlets and specialized urban media is also generated (Nisbet 2009a; Weathers et al. 2013).

Efforts to protect and defend people and communities are also easily localized. State and municipal governments have greater control, responsibility, and authority over climate change adaptation-related policy actions. In addition, recruiting Americans to protect their neighbors and defend their communities against climate impacts naturally lends itself to forms of civic participation and community volunteering. In these cases, because of the localization of the issue and the non-political nature of participation, barriers related to polarization may be more easily overcome and a diversity of organizations can work on the issue without being labeled as "advocates," "activists," or "environmentalists." Moreover, once community members from differing political backgrounds join together to achieve a broadly inspiring goal such as protecting people and a local way of life, then the networks of trust and collaboration formed can be used to move this diverse segment toward cooperation in pursuit of national policy goals (Nisbet, Markowitz, and Kotcher 2012; Weathers et al. 2013).

To test these assumptions, in an initial study, we conducted in-depth interviews with 70 respondents from 29 states, recruiting subjects from six previously defined audience segments. These segments ranged in a continuum from those individuals deeply alarmed by climate change to those who were deeply dismissive of the problem. Across all six audience segments, individuals said that information about the health implications of climate change was both useful and compelling, particularly when locally focused mitigation and adaptation-related actions were paired with specific benefits to public health (Maibach, Nisbet, Baldwin, Akerlof, and Diao 2010).

In a follow up study, we conducted a nationally representative Web survey in which respondents from each of the six audience segments were randomly assigned to three different experimental conditions, allowing us to evaluate their emotional reactions to strategically framed messages about climate change. Though people in the various audience segments reacted differently to some of the messages, in general, framing climate change in terms of public health generated more hope and less anger than framed messages that defined climate change in terms of either national security or environmental threats. Somewhat surprisingly, our findings also indicated that the national security frame could "boomerang" among audience segments already doubtful or dismissive of the issue, eliciting unintended feelings of anger (Myers, Nisbet, Maibach, and Leiserowitz 2012).

In a third study, we examined how Americans perceived the risks posed by a major spike in fossil fuel energy prices. According to our analysis of national survey data, approximately half of American adults believe that our health is at risk from major shifts in fossil fuel prices and availability. Moreover, this belief was widely shared among people of different political ideologies and was strongly held even among individuals otherwise dismissive of climate change. Our findings suggest that many Americans would find relevant and useful communication efforts that emphasized energy resilience strategies that reduce demand for fossil fuels, thereby limiting greenhouse emissions and preparing communities for fuel shortages or price spikes. Examples include improving home heating and automobile fuel efficiency, increasing the availability and affordability of public transportation, and investing in government-sponsored research on cleaner, more efficient energy technologies (Nisbet, Maibach, and Leiserowitz 2011).

New directions for research

Two prominent lines of interdisciplinary research offer deductive typologies of schema that are likely to be generalizable across environment-related policy debates, shaping individual judgments and decisions as they are selectively activated by competing media and other discursive frames. Environmental communication scholars can benefit by integrating these important lines of research into assessments of media framing and their effects on public opinion, while also turning to more systematically analyzing how social and cognitive framing processes play out across online and social media.

Cultural cognition and moral intuitions

In research on "cultural cognition" led by Yale University's Dan Kahan and colleagues, individuals scoring high in terms of hierarchical and individualist values tend to reject the risks related to issues such as climate change, nuclear energy, and food biotechnology. Hierarchicalists view proposed regulations to limit such risks as threats to those they respect in power, to established order in society, and to status quo practices in the economy or their personal lives. Individualists, alternatively, view regulatory actions as unwise restrictions on markets, enterprise, and personal freedom. In contrast, for individuals scoring high in terms of egalitarian and communitarian values, such arguments for regulation align easily with more generalized views about the need to manage markets and industry in favor of the collective community and to protect the most vulnerable (Leiserowitz 2006; Kahan, Jenkins-Smith, and Braman 2010a).

Yet consider what happens when the frame of reference for these groups is shifted. In experiments, when hierarchicalists/individualists read that the solution to climate change was not regulations to limit emissions but investment in more nuclear power, their skepticism of expert statements relative to climate change decreased and their support for policy responses increased (Kahan et al. 2010a)

For environmental communication scholars, a major implication is that effective public engagement on climate change—no matter how effectively the science might be conveyed—will depend in part on the context set by proposed policy frameworks and technological solutions. Some actions such as tax incentives for nuclear energy, government support for clean energy research, or proposals to defend and protect local communities against climate change impacts are more likely to gain support from hierarchalists/individualists, segments of the public who lean toward strongly conservative in their political outlook. In this context, conservatives have less motivation to contest the risks identified by climate science since the proposed actions to deal with those risks affirm their schematic beliefs in the ability of human ingenuity to stretch environmental limits as well as maintain economic growth and the status quo in society.

Not only do political and technological solutions set the context and frame of reference for evaluating scientific advice about environmental risks and threats, but the perceived cultural similarity of those communicating about the problem also matters. This follows the longstanding finding that an individual's evaluation of the strength of an argument depends largely on how credible they find the source of that argument (Hovland and Weiss 1951; see also Druckman 2001).

For example, among hierarchalists/individualists, they are more likely to reject scientific information when it is conveyed by someone such as former Vice President Al Gore, whom they view as antagonistic and oppositional to their vision of a good society (Kahan et al. 2010a). On the other hand, they may be more open to the same scientific advice and policy recommendations when emphasized by a decorated military general. Similarly, egalitarians/communitarians who tend to lean toward liberal in their political outlook would be more likely to dismiss assurances about the safety of food biotechnology from an industry-employed scientist than from a university-based and publicly funded researcher or alternatively from a local farmer who has benefited from the technology.

In relation to Kahan's Cultural Cognition line of research, environmental communication scholars can usefully build on this work by examining how culturally consistent or antagonistic cues are embedded within different frames of reference on issues such as climate change or food biotechnology. Strong cultural affiliation with the identity of experts or advocates may enhance or mitigate the persuasiveness of a particular frame of reference (see Kahan, Braman, Cohen, Gastil, and Slovic 2010b). For example, in relation to testing a public health framed message about climate change, to what extent do the cultural background and identity of the featured experts and those portrayed as at risk play in enhancing or mitigating the frame's effects across audience segments? In a study testing the effects of a public health focused messaging strategy on climate change among residents of upstate New York, the more socially distant those at risk portrayed in the message (e.g. local state farmers versus European farmers), the more likely that Republicans were to dismiss the validity of the threat. In contrast, the social distance of the emphasized "victim" made little difference to the perceived risk perceptions of Democrats (Hart and Nisbet 2012).

Likewise, when evaluating frame contexts over climate change in media coverage, what are the culturally consistent and antagonistic cues differentially emphasized across conservative and liberal media? For example, how are frames of public accountability blaming conservatives for inaction on climate change as featured at the liberal cable network MSNBC further personalized and thereby made more resonant by referencing "deniers" such as Fox News, the libertarian Koch brothers, or U.S. Senator James Inhofe? Similarly, at Fox News, how are frames of reference emphasizing scientific uncertainty, economic costs, and public accountability further magnified and personalized by way of references, for example, to Al Gore, "Hollywood liberals," and the United Nations?

Moreover, in each case, what kinds of visual cues are also portrayed? Studies, for example, suggest that News Corp-owned news outlets tend to downplay the urgency of climate change by featuring imagery of political leaders; while in turn other news outlets play up the urgency of the issue by featuring iconic images of natural disasters and or melting glaciers and ice sheets (O'Neill and Smith 2014). Other studies could examine, for example, when discussing solutions to climate change, the impacts of different technological images on public attitudes, ranging from "soft energy path" technologies such as solar and wind to "hard energy path" technologies such as nuclear power or carbon capture and storage.

A second line of research provides clues on how framing relates to the morally relevant schema and intuitions that are strongly held across different segments of the public. "Part of what it means to be a partisan is that you have acquired the right set of intuitive reactions to hundreds of words and phrases," explains New York University social psychologist Jonathon Haidt (2012) in his best-selling book, *The Righteous Mind: Why Good People Are Divided by Politics and Religion*. In his research with several colleagues, Haidt draws on surveys of tens of thousands of individuals to develop and validate a typology of commonly held "moral foundations," schematic interpretations of right and wrong that in political debates can be triggered by competing frames as found in media coverage and public discourse:

- harm/care – concerns about the caring for and protection of others;
- fairness/cheating – concerns about treating others fairly, upholding justice;
- loyalty/betrayal– concerns about group membership; loyalty;
- authority/subversion – concerns about legitimacy, leadership, and tradition;
- liberty/oppression– concerns about personal freedom and control by others;
- sanctity/degradation – concerns about purity, sanctity, and contamination.

According to Haidt's research, liberals tend to communicate about issues in ways that mostly activate the moral foundations of harm/care, liberty/oppression, and fairness/cheating. This reflects their own schematic lens on how they make sense of society. Conservatives, in contrast, tend to focus more strongly on moral intuitions related to loyalty/betrayal, authority/subversion, sanctity/degradation, and liberty ("live free or die"), while liberals tend to emphasize the opposing oppression ("speaking truth to power"). Similarly, when they emphasize fairness, conservatives tend to focus on process and perceived merit rather than on equality of outcomes (Haidt 2012).

Consider how Haidt's typology of intuitive moral schema likely applies to climate change. Historically, advocates and their campaigns have framed the issue as an environmental problem that threatens ecosystems and wildlife, often in remote polar regions or other countries. This framing strategy activates the moral foundation of harm/care, though much of the focus is on harm to nature or the environment rather than humans, an interpretation uniquely persuasive to liberals (Feinberg and Willer 2013). Alternatively, in an effort to appeal more broadly to moderates and conservatives, environmental group campaigns in 2009 and 2010 supporting cap and trade legislation emphasized that a "cap=jobs" would "repower America." Though this emphasis promised economic benefits if legislation was passed, it did not activate a morally relevant schema for why we should act and why we have a responsibility to do so.

The emphasis on economic benefits in the context of the strong economic recession also turned the debate into "some economic benefits" as claimed by supporters of cap and trade versus "dramatic economic costs" as claimed by opponents, a balance that, given the economic context, favored the opposition. As a consequence, outside a committed base of environmentalists and progressive activists, during the cap and trade debate, most of the public lacked strong moral intuitions about climate change with appeals to participate lacking moral weight (see Nisbet et al. 2012 for analysis).

Environmental communication scholars can usefully build on and incorporate moral foundations research by examining a number of specific questions. First, how do these intuitive moral schema interact with and/or are activated by different framed messages? For example, does framing climate change as a public health problem have a broader appeal since it focuses on harm/care to humans rather than the environment? Similarly, in the debate over food biotechnology, how do different frames of reference promoted by environmentalists activate moral intuitions (and opposition to the technology) among liberals by activating moral intuitions related to the purity/sanctity of nature or by activating moral intuitions related to the fairness of food policy dictated by large corporations rather than consumers or small farmers?

The framing process across digital and social media

Environmental communication scholars are also still struggling to catch up to the rapid changes in our media system and what they might mean for understanding framing effects. For example, in complex policy debates such as those over climate change and food biotechnology, the editorial and business decisions at prestige news outlets have likely indirectly amplified differential risk perceptions across segments of the news audience. The *New York Times* and *Washington Post*, most notably, have cut back on news coverage of climate change and other science issues, letting go of many of their most experienced reporters, allowing advocacy-oriented media outlets and commentators to fill the information gap. As a consequence, careful reporting at outlets such as the *New York Times* on the technical details of science and policy have been replaced by morally framed interpretations from bloggers and advocacy journalists at other outlets that often dramatize and distort the risks related to these issues (Nisbet and Fahy in press).

Online news and commentary are also highly socially contextualized, passed along and preselected by peers and opinion leaders who are likely to share an individual's worldviews and political preferences (Nisbet and Kotcher 2009). Furthermore, individuals are also more likely to assign greater utility and perceived benefit to news items that contain strong social endorsements (Messing and Westwood 2012). Nonetheless, if an individual incidentally "bumps" into news about climate change or food biotechnology by way of Twitter, Facebook, or Google+, the news item is likely to be the subject of meta-commentary that frames the political and moral relevance of the information (Scheufele and Nisbet 2012).

Taking advantage of these self-reinforcing spirals (see Garrett 2009), advocacy groups devote considerable resources to flooding social media with politically favorable comments and purposively selected stories, anticipating that many news consumers may incidentally "bump" into these comments and stories by way of their social media networks. Yet, some experimental evidence suggests that across certain science issues, such as genetically modified foods, individuals may be more likely to pay attention to information online that challenges rather than supports their pre-existing schema (Jang 2013). Nonetheless, this trend suggests that in today's social and participatory news system, many news consumers are potentially exposed to multiple frames of reference when engaging with a single news item. Even before engaging with the framing featured in a news story, today's news consumer is potentially exposed to the frame emphasized in the blog post, Tweet, or Facebook feed that called their attention to the news story. If after reading the story, the individual decides to read the comment section, additional framing effects may occur (Scheufele and Nisbet 2012; Anderson, Brossard, Scheufele, Xenos, and Ladwig 2014).

Moreover, when individuals, prompted by a focusing event such as extreme weather or news of a major scientific report, do seek out more information about climate change or food biotechnology via Google and other search engines, further selectivity is likely to occur. In this case, for example, liberals might choose to search for information on "climate change" or "frankenfoods" and encounter one set of differentially framed search results, whereas a conservative searching for information on "global warming" or "genetically modified food" encounters an entirely different set of search results. Not only does word choice shape the information returned through Google, but also so does the past browsing and search history of the individual, adding an additional layer of selectivity and bias to the information encountered (Baram-Tsabari and Segev 2011; Brossard 2013; Li, Anderson, Brossard, and Scheufele 2014).

In assessing these trends, consumption patterns and possible effects, environmental communication scholars can learn from their peers among social and behavioral scientists as they experiment with "big data" analysis tools that can sample, capture, and code social media discussion of relevant topics. A next step in line with the research reviewed in this chapter is to analyze various social media statements and forms of expression by way of carefully developed and generalizable typologies of schema and frames. In this way, patterns of selectively framed discourse about a subject such as climate change or food biotechnology can be tracked in real time, by geographic location, and in relation to focusing events across online networks of groups and audience segments (Kirilenko and Stepchenkova 2014).

Environmental communication scholars, in this regard, can effectively work in tandem with behavioral scientists to design and run controlled experiments to test the effects of exposure to differentially framed social media conversations as they might be encountered online. This type of collaboration will likely benefit our overall understanding of the role of media framing in complex environmental debates while also fine-tuning the communication efforts of practitioners.

Further reading

Brossard, D. (2013). New media landscapes and the science information consumer. *Proceedings of the National Academy of Sciences, 110 (Supplement 3), 14096–14101.*
Overview and synthesis of relevant research in science communication and risk communication specific to online information seeking and consumption that is applicable and suggests important areas of research for environmental communication.

Feinberg, M. and Willer, R. (2013). The moral foundations of environmental attitudes. *Psychological Science*, 24: 56–62.
Introduction to moral foundations theory, its relevance to environmental debates, and models for testing linkages to framing and public opinion formation.

Hoffman, A. (2012, Fall). Climate science as culture war. *Stanford Social Innovation Review*, 31–27.
Introduction to the relevance of cultural cognition and audience segmentation research to understanding differences in U.S. public opinion about climate change with recommendations on how framing strategies can bridge perceptual divide.

Nisbet, M. C. (2009a). Communicating climate change: Why frames matter to public engagement. *Environment*, 51 (2), 514–518.
Detailed overview of generalizable typology of frames and their relevance to understanding political debates over climate change.

O'Neill, S. J. and Smith, N. (2014). Climate change and visual imagery. *Wiley Interdisciplinary Reviews: Climate Change*, 5(1), 73–87.
Synthesis and review of research on the visual portrayal and framing of climate change, differences across media outlets and countries, and possible influence on public opinion.

Price, V., Nir, L., and, Capella, J. N. (2005). Framing public discussion of gay civil unions. *Public Opinion Quarterly*, 69, 179–212.
A comprehensive review and synthesis of research on social constructivist approaches to framing as developed by sociologist William Gamson and the applicability to evaluating the relationship to individual-level opinion formation. Using the gay marriage debate as a test case, the study offers a useful framework for examining how media coverage reflects and shapes public discourse about climate change and other environmental issues.

Scheufele, D. A. (1999). Framing as a theory of media effects. *Journal of Communication*, 49(1), 103–122.
Synthesis of research and conceptualization of framing as a process that transcends "frame building" advocacy strategies, journalistic decision making, and audience schema and opinion formation.

References

Anderson, A. A., Brossard, D., Scheufele, D. A., Xenos, M. A., and Ladwig, P. (2014). The "nasty effect": Online incivility and risk perceptions of emerging technologies. *Journal of Computer-Mediated Communication*, 3, 373–387.

Baram-Tsabari, A. and Segev, E. (2011). Exploring new web-based tools to identify public interest in science. *Public Understanding of Science*, 20(1), 130–143.

Bickerstaff, K., Lorenzoni, I., Pidgeon, N. F., Poortinga, W., and Simmons, P. (2008). Reframing nuclear power in the UK energy debate: Nuclear power, climate change mitigation and radioactive waste. *Public Understanding of Science*, 17(2), 145–169.

Brossard, D. (2013). New media landscapes and the science information consumer. *Proceedings of the National Academy of Sciences, 110 (Supplement 3), 14096–14101.*

Cox, J. R. (2010). Beyond frames: Recovering the strategic in climate communication. *Environmental Communication*, 4(1), 122–133.

Dahinden, U. (2002). Biotechnology in Switzerland: Frames in a heated debate. *Science Communication*, 24: 184–197.

Dirikx, A. and Gelders, D. (2010). To frame is to explain: A deductive frame-analysis of Dutch and French climate change coverage during the annual UN Conferences of the Parties. *Public Understanding of Science*, 19(6), 732–742.

Dotson, D. M., Jacobson, S. K., Kaid, L. L., and Carlton, J. S. (2012). Media coverage of climate change in Chile: A content analysis of conservative and liberal newspapers. *Environmental Communication: A Journal of Nature and Culture*, 6(1), 64–81.

Druckman, J. N. (2001). On the limits of framing effects: Who can frame?. *Journal of Politics*, 63(4), 1041–1066.

Durant, J., Bauer, M. W., and Gaskell, G.(1998). *Biotechnology in the Public Sphere: A European Sourcebook*. Michigan, MI: Michigan State University Press.

Feinberg, M., and Willer, R. (2013). The moral foundations of environmental attitudes. *Psychological Science*, 24: 56–62.

Feldpausch-Parker, A. M., Ragland, C. J., Melnick, L. L., Chaudhry, R., Hall, D. M., Peterson, T. R., and Wilson, E. J. (2013). Spreading the news on carbon capture and storage: A state-level comparison of US media. *Environmental Communication: A Journal of Nature and Culture*, 7(3), 336–354.

Gamson, W. A. (1992). *Talking Politics*. New York: Cambridge University Press.

Gamson, W. A. and Modigliani, A. (1989). Media discourse and public opinion on nuclear power: A constructionist approach. *American Journal of Sociology*, 95: 1–37.

Garrett, R. K. (2009). Echo chambers online?: Politically motivated selective exposure among Internet news users. *Journal of Computer-Mediated Communication*, 14(2), 265–285.

Gaskell, G., Bauer, M. W., Durant, J., and Allum, N. C. (1999). Worlds apart? The reception of genetically modified foods in Europe and the US. *Science*, 285(5426), 384–387.

Gordon, J. C., Deines, T., and Havice, J. (2010). Global warming coverage in the media: Trends in a Mexico City Newspaper. *Science Communication*, 32(2), 143–170.

Haidt, J. (2012). *The Righteous Mind: Why Good People Are Divided by Politics and God*. New York: Pantheon Books, p. 58.

Hansen, A. (2006). Tampering with nature: "Nature" and the "natural" in media coverage of genetics and biotechnology. *Media, Culture and Society*, 28(6), 811–834.

Hansen, A. (2011). Communication, media and environment: Towards reconnecting research on the production, content and social implications of environmental communication. *International Communication Gazette*, 73(1–2), 7–25.

Hansen, A. and Machin, D. (2008). Visually branding the environment: Climate change as a marketing opportunity. *Discourse Studies*, 10(6), 777–794.

Hart, P. S. and Nisbet, E. C. (2012). Boomerang effects in science communication: How motivated reasoning and identity cues amplify opinion polarization about climate mitigation policies. *Communication Research*, 39, 701–723.

Hestres, L. E. (2013). Preaching to the choir: Internet-mediated advocacy, issue public mobilization, and climate change. *New Media and Society*, 2, 323–339.

Hovland, C. I. and Weiss, W. (1951). The influence of source credibility on communication effectiveness. *Public Opinion Quarterly*, 15(4), 635–650.

Jang, S. M. (2013). Seeking congruency or incongruency online? Examining selective exposure to four controversial science issues. *Science Communication*, 2, 143–167.

Kahan, D. M., Jenkins-Smith, H., and Braman, D. (2010a). Cultural cognition of scientific consensus. Cultural Cognition Project Working Paper No. 77. Yale University School of Law. Available at http://papers.ssrn.com/ sol3/papers.cfm?abstract_id¼1549444 (accessed August 10, 2014).

Kahan, D. M., Braman, D., Cohen, G. L., Gastil, J., and Slovic, P. (2010b). Who fears the HPV vaccine, who doesn't, and why? An experimental study of the mechanisms of cultural cognition. *Law and Human Behavior*, 34(6), 501.

Kirilenko, A. P. and Stepchenkova, S. O. (2014). Public microblogging on climate change: One year of Twitter worldwide. *Global Environmental Change*, 26, 171–182.

Leiserowitz, T. (2006). Climate change risk perception and policy preferences: The role of affect, imagery, and values. *Climatic Change*, 77, 45–77.

Leiserowitz, T. (2007). Communicating the risks of global warming: American risk perceptions, affective images and interpretive communities. In S. Moser and L. Dilling (eds), *Communication and Social Change: Strategies for Dealing with the Climate Crisis* (pp. 44–63). Cambridge: Cambridge University Press.

Li, N., Anderson, A. A., Brossard, D., and Scheufele, D. A. (2014). Channeling science information seekers' attention? A content analysis of top-ranked vs. lower-ranked sites in Google. *Journal of Computer-Mediated Communication*, 19, 562–575.

Listerman, T. (2010). Framing of science issues in opinion-leading news: International comparison of biotechnology issue coverage. *Public Understanding of Science*, 19(1), 5–15.

Maeseele, P. (2010). On neo-luddites led by ayatollahs: The frame matrix of the GM food debate in Northern Belgium. *Environmental Communication*, 4(3), 277–300.

Maibach, E., Nisbet, M. C., Baldwin, P., Akerlof, K., and Diao, G. (2010). Reframing climate change as a public health issue: An exploratory study of public reactions. *BMC Public Health*, 10, 299.

Marx, S. M, Weber, E. U., Orlove, B. S., Leiserowitz, A., and Krantz, D. H. (2007). Communication and mental processes: Experiential and analytic processing of uncertain climate information. *Global Environmental Change*, 17: 47–58.

Meisner, M. S. and Takahashi, B. (2013). The nature of time: How the covers of the world's most widely read weekly news magazine visualize environmental affairs. *Environmental Communication: A Journal of Nature and Culture*, 7(2), 255–276.

Messing, S. and Westwood, S. J. (2012). Selective exposure in the age of social media: Endorsements trump partisan source affiliation when selecting news online. *Communication Research*, 8, 1042–1063.

Myers, T., Nisbet, M. C., Maibach, E. W., and Leiserowitz, A. (2012). A public health frame arouses hopeful emotions about climate change. *Climatic Change Research Letters*, 1105–1121.

Nerlich, B., Forsyth, R., and Clarke, D. (2012). Climate in the news: How differences in media discourse between the US and UK reflect national priorities. *Environmental Communication: A Journal of Nature and Culture*, 6(1), 44–63.

Nisbet, M. C. (2009a). Communicating climate change: Why frames matter to public engagement. *Environment*, 51 (2), 514–518.

Nisbet, M. C. (2009b). The ethics of framing science. In B. Nerlich, B. Larson, and R. Elliott (eds). *Communicating Biological Sciences: Ethical and Metaphorical Dimensions* (pp. 51–74). London: Ashgate.

Nisbet, M. C. (2014). Engaging in science policy controversies: Insights for the U.S. debate over climate change. In M. Bucchi and B. Trench (eds) *Handbook of Public Communication of Science and Technology* (pp. 173–185). London: Routledge.

Nisbet, M. C. and Fahy, D. (in press). Why we need knowledge-based journalism in politicized science debates. *Annals of the American Academy of Political and Social Science*.

Nisbet, M. C. and Huge, M. (2006). Attention cycles and frames in the plant biotechnology debate: Managing power and participation through the press/policy connection. *Harvard International Journal of Press/Politics*, 11, 2, 3–40.

Nisbet, M. C. and Huge, M. (2007). Where do science policy debates come from? In D. Brossard, J. Shanahan, and C. Nesbitt (eds), *The Public, the Media, and Agricultural Biotechnology* (pp. 193–230). New York: CABI/Oxford University Press.

Nisbet, M. C. and Kotcher, J. (2009). A two step flow of influence? Opinion-leader campaigns on climate change. *Science Communication*, 30, 328–358.

Nisbet, M. C. and Lewenstein, B. V. (2002). Biotechnology and the American media: The policy process and the elite press, 1970 to 1999. *Science Communication*, 23(4), 359–391.

Nisbet, M. C. and Markowitz, E. (2014). Understanding public opinion in debates over biomedical research: Looking beyond partisanship to focus on beliefs about science and society. *PLoS ONE*, 9(2), e88473.

Nisbet, M. C. and Scheufele, D.A. (2009). What's next for science communication? Promising directions and lingering distractions. *American Journal of Botany*, 96(10), 1767–1778.

Nisbet, M. C., Maibach, E., and Leiserowitz, A. (2011). Framing peak petroleum as a public health problem: Audience research and participatory engagement. *American Journal of Public Health*, 101, 1620–1626.

Nisbet, M. C., Markowitz, E. M., and Kotcher, J. (2012). Winning the conversation: Framing and moral messaging in environmental campaigns. In L. Ahern and D. Bortree (eds), *Talking Green: Exploring Current Issues in Environmental Communication* (pp. 9–36). New York: Peter Lang.

O'Neill, S. J. (2013). Image matters: Climate change imagery in US, UK and Australian newspapers. *Geoforum*, 49, 10–19.

O'Neill, S. J. and Smith, N. (2014). Climate change and visual imagery. *Wiley Interdisciplinary Reviews: Climate Change*, 5(1), 73–87.

O'Neill, S. J., Boykoff, M., Niemeyer, S., and Day, S. A. (2013). On the use of imagery for climate change engagement. *Global Environmental Change*, 23(2), 413–421.

Price, V., Nir, L., and Capella, J. N. (2005). Framing public discussion of gay civil unions. *Public Opinion Quarterly*, 69: 179–212.

Scheufele, D. A. (1999). Framing as a theory of media effects. *Journal of Communication*, 49(1), 103–122.

Scheufele, D. A. and Nisbet, M. C. (2012). Online news and the demise of political disagreement. In C. Salmon (ed.), *Communication Yearbook 36* (pp. 45–53). New York: Routledge.

Scheufele, B. T. and Scheufele, D. A. (2010). Of spreading activation, applicability, and schemas: Conceptual distinctions and their operational implications for measuring frames and framing effects. In P. D'Angelo and J. A. Kuypers (eds), *Doing News Framing Analysis: Empirical and Theoretical Perspectives* (pp. 110–134). New York: Routledge.

Scheufele, D. A. and Tewksbury, D. (2007). Framing, agenda setting, and priming: The evolution of three media effects models. *Journal of Communication*, 57(1): 9–20.

Schlichting, I. (2013) Strategic framing of climate change by industry actors: A meta-analysis. *Environmental Communication*, 7(4), 493–511,

Stephens, J. C., Rand, G. M., and Melnick, L. L. (2009). Wind energy in US media: A comparative state-level analysis of a critical climate change mitigation technology. *Environmental Communication*, 3(2), 168–190.

Takahashi, B. and Meisner, M. (2013). Climate change in Peruvian newspapers: The role of foreign voices in a context of vulnerability. *Public Understanding of Science*, 22(4), 427–442.

Weathers, M., Maibach, E.W., and Nisbet, M.C. (2013). Conveying the human implications of climate change: Using audience research to inform the work of public health professionals. In D. Y. Kim, G. Kreps, and A. Singhal (eds), *Health Communication: Strategies for Developing Global Health Programs* (pp. 190–207). New York: Peter Lang.

Zamith, R., Pinto, J., and Villar, M. E. (2013). Constructing climate change in the Americas: An analysis of news coverage in US and South American newspapers. *Science Communication*, 35(3), 334–357.

29

ANALYSING PUBLIC PERCEPTIONS, UNDERSTANDING AND IMAGES OF ENVIRONMENTAL CHANGE

Lorraine Whitmarsh

Introduction

This chapter provides a critical introduction to research on public perceptions and images relating to environmental change and associated risks. The chapter focuses primarily on climate change, since this is where a considerable amount of research has focused in recent years, although research on landscape and ecosystem change is also touched upon. The chapter draws on quantitative and qualitative research within psychological and related domains from across several countries.

The next section reviews research on awareness and knowledge about environmental change; after this, research on attitudes and concern about environmental change is discussed. Building on this, the following section focuses on research on subjective imagery evoked by climate change. Research into the effects on public perceptions of climate change communication is then discussed. The final section concludes with implications of this research for communication efforts and future research.

Perceptions of environmental change: an overview

Initial research into public attitudes and responses to climate change was pioneered in the United States in the 1990s (e.g. Kempton, 1991), and today there remains a strong US concentration of expertise in this field. Yet, there also exists a growing literature on climate change perceptions within Europe, including the UK, as well as a small number of cross-national comparative studies (e.g. Bord *et al.*, 1998; Dunlap, 1998; Brechin, 2003; Lorenzoni and Pidgeon, 2006). A notable proportion of this literature has been commissioned by government departments or agencies (e.g. DEFRA, DfT, Scottish Executive) and other non-academic organizations (e.g. BBC, Energy Saving Trust), although there is also a growing academic literature relating to public perceptions and behavioural responses to climate change. Similarly, much of the research to date has been descriptive (e.g. trends in concern, themes in discourse) and atheoretical although, increasingly, relevant theories are being drawn on to explain and predict responses to environmental change. As discussed later, for example, identity and persuasion theories can help explain environmental denial and identify the elements of a persuasive message.

Literature relating to attitudes to particular climate change *impacts*, such as flooding and drought, has its roots in the hazard and risk perception studies of the 1960s (e.g. Kates, 1976) but has expanded to encompass new risks associated with climate change, such as sea-level rise and rapid climate change. Notably, though, certain risks (such as flooding) are more researched than others (such as heat stress), particularly in Europe, which has historically had more problems associated with precipitation than with a lack of it. The following sections focus on the key findings from this sizeable literature.

Research on attitudes to extant and prospective changes to ecosystems, landscapes and species is relatively limited (see Upham *et al.*, 2009 for a review). The existing research literature suggests that place attachment, environmental values, worldviews and trust in government and other actors (e.g. industry) strongly determine responses to environmental change, while scientific knowledge of biodiversity and environmental risks is limited.

Knowledge of climate change

Surveys show that awareness and self-reported knowledge about climate change has been rising over the last two decades (e.g. DEFRA, 2007). Since the early 2000s in the UK, for example, awareness of the terms 'climate change' or 'global warming' has been near universal (around 99 per cent; DEFRA, 2002, 2007; Whitmarsh, 2009). Most also claim they know 'a great deal' or 'a fair amount' about the issue in general (Whitmarsh *et al.*, 2011a), although in respect of more specific terms (e.g. carbon emissions, carbon offsetting), levels of awareness are lower (DEFRA, 2007; Whitmarsh *et al.*, 2011a). Most people also accept climate change is a human-caused problem (Hargreaves *et al.*, 2003; COI, 2008), and when prompted, most people can identify destruction of forests, carbon dioxide emissions, emissions from transport and emissions from power stations as contributors to climate change (e.g. DEFRA, 2002; COI, 2008). However, research also shows that individuals often make erroneous causal associations between climate change and other distinct issues (notably, ozone depletion; e.g. Kempton, 1991; Bostrom *et al.*, 1994; Eurobarometer, 2001).

When respondents are not provided with a checklist of possible causes, their understanding is shown to be lower (consistent with the recognized risk of acquiescence bias in survey research; Ray, 1990). For example, Whitmarsh (2009) found only 6 per cent explicitly identified CO_2/carbon emissions. Rather, when asked unprompted about climate change causes, the public tend to identify the more generic (and in effect secondary) cause 'air pollution' (Brechin, 2003; Whitmarsh, 2009). Pollution is also a more morally loaded term than 'fossil fuels' or 'carbon dioxide' (Kempton, 1991; Bord *et al.*, 1998; Hargreaves *et al.*, 2003; Whitmarsh, 2009). Indeed, qualitative research reveals that the public understands climate change as part of a broader set of social and environmental issues, such as industrialization, consumption and overpopulation (Darier and Schule, 1999; Wolf and Moser, 2011).

Similarly, when asked, unprompted, what the effects of climate change will be, UK (and to a lesser extent, US) publics most commonly identify changes in weather, including increased temperatures and rainfall (Dunlap, 1998; DEFRA, 2002; Whitmarsh, 2009). Qualitative studies indicate that there is a lack of distinction between weather and climate in lay understanding (Kempton, 1991; Bostrom *et al.*, 1994). This conceptual linking reflects the media coverage that tends to discuss climate change in the context of local weather-related stories, such as UK flooding (Hargreaves *et al.*, 2003; Gavin *et al.*, 2011). It is also likely due to the ways in which climate change, as a new concept, is made sense of in relation to familiar ideas and experiences (Ungar, 2000), a process known in social psychological literature as 'anchoring' (Breakwell, 1991; Whitmarsh *et al.*, 2011a).

A further feature of the public's understanding of the causes of climate change is a limited awareness of the relative contribution of different activities to climate change. In particular, there is an underestimation of the role of domestic energy use, meat eating/production, food miles and food waste (DEFRA, 2007; Whitmarsh *et al.*, 2011a). When shown a list of possible causes, only one in five people in England identify gas and electricity used in the home as a contributor to climate change (DEFRA, 2002). Unprompted, only 0.5 per cent mention domestic energy consumption as a cause (Whitmarsh, 2009).

Beyond this, there is also a general tendency to underestimate one's own contribution to causing climate change. People tend to identify causes of climate change with other people or countries, such as SUV drivers, firms, the US or China (Lorenzoni and Pidgeon, 2006; Whitmarsh, 2009), and with more 'distant' activities, namely industry and deforestation, rather than their own actions (Whitmarsh *et al.*,

2011a). This disconnection between individual actions and climate change is also reflected in British media coverage of the issue (Hargreaves *et al.*, 2003), but as discussed later, the underestimation of personal energy use in contributing to climate change may reflect a strategy of reducing cognitive dissonance by denying responsibility for tackling climate change.

Although most members of the public accept that climate change is, in part, a product of human activities, there remains a substantial minority who are sceptical (e.g. Whitmarsh, 2011; Smith and Leiserowitz, 2012). However, scepticism about rising temperatures ('trend scepticism') and the human causes ('attribution scepticism') are less common than doubts relating to the severity and risks of climate change ('impact scepticism'; Poortinga *et al.*, 2011). Scepticism may partly be a product of the media presentation of climate change as controversial and uncertain (e.g. Boykoff and Boykoff, 2004), and because the human causes of climate change are not self-evident. Yet, crucially, research highlights the interaction between values and scepticism: those with high pro-environmental values are less likely to be sceptical than those who value the environment less; and similarly, conservative political values are strongly associated with scepticism (e.g. Dunlap and McCright, 2008; Eurobarometer, 2009; Whitmarsh, 2011). As discussed later, this highlights the tendency to interpret information about climate change in relation to one's existing views of the world.

Many have argued that the public will need to have a reasonable understanding of climate change and their role in causing and responding to it, particularly if certain policies (carbon offsetting, personal carbon trading, etc.) are realized or expanded. Yet, how we define 'carbon literacy' is important: we should not assume the public is carbon illiterate simply because they do not know particular 'facts' or use technical language; more important is the contextual understanding and application of knowledge, which cannot be elicited through survey research (Whitmarsh *et al.*, 2011a). Yet, the evidence here is also not encouraging. Studies indicate there is low salience of climate change (and, indeed, energy and environment) in individuals' day-to-day choices and actions (Upham *et al.*, 2009; Whitmarsh, 2011). Growing awareness of environmental problems at an abstract or general level tends not to be translated into personally relevant cognitions or motivating attitudes. Most people (apart from the most environmentally conscious or most financially constrained) do not think about the energy they use in the home, and neither energy consumption (including embodied energy and potential future energy savings) nor environmental impact are typically considered when buying appliances (for which initial cost is the primary motivating factor) or when buying and preparing food or travelling (Upham *et al.*, 2009). As we now discuss, research on climate change perceptions indicates that climate change is a low salience issue.

Attitudes to climate change

Beyond awareness and knowledge, research has also explored how the public evaluates climate change (i.e. their attitudes), including levels of perceived risk and concern. Attitudes are dynamic, influenced by a range of factors, often ambivalent or uncertain, and frequently not predictive of behaviour (e.g. Haddock and Maio, 2012). Yet, they hold important functions for individuals, such as helping to organize knowledge, inform decisions, express identity and seek connections with others. Furthermore, the concept of attitudes is helpful in understanding how individuals interpret and respond differently to the same information, since pre-existing attitudes have been shown to bias perceptions and guide behaviour: people are more attentive to, and accepting of, attitude-consistent information and tend to ignore or reject dissonant information (e.g. Lord *et al.*, 1979). As we discuss later, this characteristic of attitudes highlights the heterogeneity of the public and helps explain the diverse effects of communication on environmental change issues.

In respect of climate change attitudes, most people have a negative affective (i.e. emotional) response to climate change (Lorenzoni and Pidgeon, 2006; see also next sub-section). For example, most people consider climate change to be a 'bad thing' rather than a 'good thing' (62 per cent versus 10 per cent, according to OST/MORI, 2004). Yet, while concern about climate change has tended to grow over the past two decades (e.g. Upham *et al.*, 2009), put in context, the issue is not a priority concern for most people, and may not even be one of the public's main environmental concerns (e.g. DEFRA, 2002). In the UK, for example, Pidgeon *et al.*'s (2008) survey found climate change/global warming was ranked the fifth most concerning environmental issue by Britons, after water pollution, air pollution, ozone depletion, deforestation and resource depletion. Furthermore, health, security and social issues feature higher in the public's concerns than environmental issues (Bord *et al.*, 2000; DEFRA, 2002; Poortinga and Pidgeon, 2003).

The low ranking of climate change as a concern reflects a widespread perception among the public that the issue is a spatially and temporally remote risk, affecting future generations and other countries. Gifford *et al.* (2009) describe this as 'temporal pessimism' (environmental quality will decrease over time) and 'spatial optimism' (environmental quality worsens as geographic distance increases), which is evident in most countries and may stem in part from self-esteem associated with 'place identity' (Proshansky, 1978) as well as from media coverage of distant impacts. While it is considered socially relevant, most individuals do not feel it poses a prominent personal threat (e.g. Bord *et al.*, 2000; Whitmarsh *et al.*, 2011a). For example, O'Neill and Nicholson-Cole (2009) asked UK participants to rate the severity of climate change risks for themselves, others in their community and others in their country, and found ratings increased significantly with distance; although recent work finds that climate change is seen as more proximal for some groups, particularly those in developing countries (e.g. Frank *et al.*, 2011).

There are a number of factors that explain why climate change is not widely conceptualized as a personal risk. First, according to the psychometric tradition in risk studies, risk perception is determined by whether risks are viewed as involuntary, catastrophic, dreaded, fatal, delayed, unknown, uncontrollable or inequitable (Slovic *et al.*, 1980; Etkin and Ho, 2007). Only some of these characteristics are perceived to apply to climate change. Poortinga and Pidgeon (2003) found Britons to be ambivalent about the potential for catastrophe arising from climate change and about the unfair distribution of risks on particular groups in society; but did characterize climate change as a moral issue, one with risks for future generations and something they did not have personal control over. Furthermore, most respondents agreed that climate change has 'unknown consequences'. Since dread is often associated with unfamiliar and 'unnatural' phenomena (e.g. Slovic *et al.*, 1980), the conceptual association of climate change and weather would further explain the low risk perception associated with it.

Second, the low ranking of climate change as a risk may be explained in respect of how it is learned about. Direct experience is more likely than communication to result in stronger, more confident, clearly focused and persistent attitudes, and in attitude-behaviour consistency (Fazio and Zanna, 1981). Similarly, direct experience and sensory evidence is often central to people's evaluation of environmental threats (e.g. Weber, 2010). According to the 'availability heuristic', the perceived likelihood of a risk increases if it has been experienced or can be readily imagined, so local risks are likely to seem more important than global risks (Slovic *et al.*, 1978; Burgess *et al.*, 1998). This represents a barrier to believing climate change poses a genuine threat: 'climate change' is a global and long-term phenomenon and so cannot be directly experienced or 'seen'. In fact, 'climate' technically refers to the average weather over 30 years, yet individuals' recall of past climates is distorted by anomalies and by cultural factors (e.g. Palutikof *et al.*, 2004); and while climate change risk is primarily detected through scientific data and computer models, human perception of climate change is filtered by various psychological and cultural factors (Weber, 2010). Several studies have explored whether related experience (e.g. of flooding, drought, temperature rise) shape climate change attitudes and findings are mixed. Whitmarsh (2008) found flooding experience did not predict climate change concern or behavioural response because the two issues were largely conceptually distinct for flood victims (see also Brody *et al.*, 2008; Dessai and Sims, 2010; Safi *et al.*, 2012). Yet, a recent study found flooding experience did influence climate change attitudes (Spence *et al.*, 2011), perhaps due to increased media coverage linking flooding to climate change (Gavin *et al.*, 2011).

Third, an important dimension of how risks are perceived and whether they are considered 'acceptable' is the balance between the costs and benefits associated with the risk issue (e.g. Slovic, 2000). Poortinga and Pidgeon (2003) found that the UK public rates the risks of climate change for themselves as moderate, and only somewhat higher than the benefits for themselves. The public may view the possible risks from climate change as being outweighed by the benefits associated with energy use (Lorenzoni and Pidgeon, 2006). Related to this, acknowledging the damage caused by one's personal energy consumption can result in cognitive dissonance. This refers to the uncomfortable psychological state experienced when aware of inconsistency in one's attitudes and behaviour (Festinger, 1957). For example, van der Pligt (1985) found that people who do not conserve energy, but are environmentally concerned, justify this dissonance by overestimating the prevalence of non-conservation, that is they claim non-conservation is a habit shared by other people. As discussed below, recent experimental evidence sheds further light on these dissonance processes.

In sum, the risks posed by climate change tend to be assumed to affect other people for a range of reasons relating to the characteristics of the issue itself (global, uncertain, complex, natural, linked to energy consumption) and to psychological and social factors (media communication, perceived credibility of communicators, denial and cognitive dissonance). The limited perception of climate change as a personal risk issue is one of the factors limiting behavioural response (Lorenzoni *et al.*, 2007).

Research also shows that the public is highly heterogeneous in their attitudes to climate change – both within and between nations. Internationally, there is considerable variation with sub-Saharan Africa and South American countries showing relatively high (and rising) concern, compared to most other countries (e.g. Gallup, 2011). Furthermore, US and UK segmentation studies expose distinct attitudinal clusters, ranging from the most concerned and active to the disinterested or active deniers, indicating multiple 'publics' in respect of climate change engagement (DEFRA, 2008; Leiserowitz *et al.*, 2012b). Belief in and concern about climate change varies according to a range of factors, such as gender, age and ethnicity (e.g. Upham *et al.*, 2009), although studies show values tend to be stronger predictors than demographic, knowledge or other factors. In the UK, Whitmarsh (2011) investigated whether climate scepticism is at heart a matter of ignorance (knowledge deficit) or divergent values; and found very clear evidence of the latter. Indeed, survey respondents' scores on a measure of scepticism did not significantly differ according to their self-reported knowledge about the issue or their qualifications (including scientific qualifications); whereas the strongest predictors were political affiliation and environmental values. This is consistent with other studies that show that sceptics score highly on tests of general scientific literacy (Kahan *et al.*, 2012) and that those with right-of-centre political beliefs and individualistic worldviews are more sceptical, at least within the UK and US (Dunlap and McCright, 2008; Eurobarometer, 2009; Poortinga *et al.*, 2011). In sum, there would appear to be an increasingly ideological component to climate change attitudes.

Latest experimental research sheds light on how worldviews shape climate change attitudes and the way in which climate change information is processed. Corner and colleagues (2012) measured participants' scepticism about climate change before and after reading two newspaper editorials that made opposing claims about the reality and seriousness of climate change (one sceptical and one stressing climate risks). Consistent with the well-established psychological phenomenon of biased assimilation (Lord *et al.*, 1979), they found that people who were less sceptical about climate change evaluated the convincingness and reliability of the editorials in a markedly different way to people who were more sceptical about climate change. Furthermore, partly consistent with prior evidence of attitude polarization, participants became more certain of their previous position after reading the editorials; although their attitudes did not polarize according to their measure of climate scepticism. Similar research conducted in the US demonstrates that the perceived scientific consensus of climate change is filtered through an ideological lens, such that those with right-of-centre political leanings do not tend to accept it, while left-of-centre voters do (Kahan *et al.*, 2010). Given that the public is exposed to diverse, uncertain and conflicting claims about climate change through mass and new media (e.g. Boyce and Lewis, 2009), this study shows that people tend to reinforce their existing views rather than process information in an open and unbiased way. This has clear implications for organizations endeavouring to persuade the public about the severity of climate change, as discussed below.

Other work sheds light on the mechanism underlying denial responses to climate change information. Previous work suggests denial can be a coping response to threatening information, including information about risks (Breakwell, 1983; Sherman *et al.*, 2000). Studies of climate change communication suggest information about climate change is perceived to be threatening (i.e. evokes fear and guilt; Ereaut and Segnit, 2006) and that information framed in this way tends to disengage publics (O'Neill and Nicholson-Cole, 2009; Feinberg and Willer, 2011). Climate scepticism may therefore be, at least in part, an identity protective response to threatening information (i.e. need for lifestyle change). Recent experimental work confirms that climate scepticism can indeed be a denial response to information threatening to lifestyle or identity. One study showed that, when provided with information about climate change that called for significant personal sacrifices (e.g. avoiding flying), participants were less convinced by the argument, compared to when they were asked to make only a small sacrifice (e.g. changing to low-energy light bulbs; Corner and Hahn, 2009). Similarly, Whitmarsh *et al.* (2011b) examined whether research participants responded differently to environmental change information if it was paired with a behaviour change message. In other words, they examined whether people would believe in climate change if it does not threaten their way of life/identity. They provided participants with one of two versions of a news article on a hypothetical environmental risk about water pollution from meat farming, which varied the extent of lifestyle changes (consumers drastically reducing meat consumption versus industry changes to livestock agriculture) required for mitigating the risks. They found that participants' prior attitudes to meat-eating interacted with the risk information: meat-eaters were more sceptical about the hypothesized risk (compared to those with less positive attitudes to eating meat) and perceived the article as less reliable and less convincing – but only in the 'behaviour change' condition (i.e. where they proposed meat rationing in order to mitigate the climate risks). They conclude that emphasizing the need for lifestyle changes may be interpreted as a threat to identity, disengaging the public and reducing the effectiveness of climate change communications.

Mental images of climate change

While elsewhere in this book climate change imagery used in communication is discussed, this section describes the mental imagery that is evoked when the term 'climate change' (or related terms) is discussed. Studies of the public's spontaneous associations with the issue show that it evokes associations that are typically negative but also spatially or socially distant. For example, one 2003 UK survey asked the public, unprompted, what they knew about 'climate change/global warming'[1] and found that impacts were more commonly mentioned than any other aspects of the issue (e.g. causes, process, information source; see Table 29.1). Possible solutions to climate change/global warming were not mentioned at all.

Table 29.1 Understanding of climate change and global warming as a whole (open-ended)

Responses given to 'What do you know about climate change/global warming?' (open-ended)	% of TOTAL survey respondents†	% of survey respondents by Questionnaire Version†		
		Climate change	Global warming	Sig.
# Temperature increase	23.6	16.2	30.1	p<.001
# Weather/seasons change	21.6	24.5	18.9	n.s.
# Melting icebergs/glaciers	19.9	13.7	25.3	p<.001
★ Ozone depletion/hole	19.9	13.7	25.3	p<.001

Responses given to 'What do you know about climate change/global warming?' (open-ended)	% of TOTAL survey respondents†	% of survey respondents by Questionnaire Version†		
		Climate change	Global warming	Sig.
+ Don't know much/anything	17.1	19.1	15.4	n.s.
+ Doubt about reality/causes	16.5	17.0	16.0	n.s.
« Pollution	11.9	6.9	16.3	p<.001
# Global impacts	10.0	8.7	11.2	n.s.
# Rising sea levels/land loss	9.7	8.7	10.6	n.s.
« Pollutants – other	9.2	3.2	14.4	p<.001
# Flooding	8.8	9.0	8.7	n.s.
« Carbon dioxide	7.8	4.7	10.6	p<.01
★ UV penetrating/reduced protection from sun	7.6	2.2	12.5	p<.001
# Impacts on climate	6.5	8.7	4.5	p<.05
« Natural variation in climate	6.3	7.9	4.8	n.s.
« Human caused (unspecified)	6.1	5.1	7.1	n.s.
# Summers hotter, winters wetter	5.3	10.1	1.0	p<.001
# Impacts already observed	5.3	7.2	3.5	p<.05
# Drought/less rainfall	5.3	7.6	3.2	p<.05
+ Contradictory views/debate	5.3	4.0	6.4	n.s.
+ Unsure/lack of knowledge	5.1	4.0	6.1	n.s.
★ Greenhouse effect	5.1	2.9	7.1	p<.05
Media	*4.6*	*5.8*	*3.5*	*n.s.*
★ Trapping of heat/gases; 'blanket' analogy	4.4	1.8	6.7	p<.01
« Deforestation	4.2	4.0	4.5	n.s.
« Natural causes – other	4.1	5.8	2.6	p<.05
« Greenhouse gases	3.9	1.4	6.1	p<.01
+ 'Climate change' differentiated from 'global warming'	3.9	5.1	2.9	n.s.
« Cars/vehicle emissions	3.7	2.2	5.1	n.s.
# Increased rainfall	3.7	5.4	2.2	p<.05
# Impacts – all other	18.0	19.1	17.0	n.s.
★ Process – all other	10.0	7.6	12.2	n.s.
« Causes – all other	9.2	5.1	12.8	p<.001
+ Uncertainty – all other	7.5	5.8	9.0	n.s.
Column total	311	274.2	343.6	

Source: Whitmarsh (2009).

Key:
« Causes   # Impacts   + Uncertainty   ★ Process   *Source of information*
† Respondents typically gave several responses, so column totals are greater than 100%.

Of the causes mentioned, most related to human activity, although fully 16.5 per cent expressed doubt about the reality or human causes of the issue (Whitmarsh, 2009). The most commonly mentioned impact of climate change was temperature increase (23.6 per cent of respondents), although this was significantly higher (30.1 per cent) among respondents of 'global warming' questionnaires. Change in weather/seasons (21.6 per cent) was the next most popular response overall, and the most commonly mentioned impact among respondents of 'climate change' questionnaires. Melting icebergs/glaciers, sea level rise, flooding, impacts on climate, drought/less rainfall and increased rainfall were also mentioned. Of the causes mentioned by respondents, by far the most common was pollution (11.9 per cent). Survey respondents also mentioned a number of other anthropogenic causes, including carbon dioxide, deforestation, greenhouse gases and vehicle emissions. Natural variation in climate was cited by only 6.3 per cent of respondents, and 4.1 per cent mentioned other natural causes.

US research on public imagery of climate change similarly shows that impacts of climate change are most commonly cited. Leiserowitz (2006) found that melting glaciers and polar ice were the most common response category, followed by generic associations with heat and rising temperatures, impacts on non-human nature, ozone depletion, disaster imagery, sea level rise and flooding, climatic change and climate scepticism. Analysis tracking this mental imagery over time shows that associations have tended to become more sceptical, while associations with melting ice and several other impacts declined, over the period 2002 to 2010 (Smith and Leiserowitz, 2012). Despite the increase in scepticism, most respondents' imagery was characterized by negative affect.

Other research has found sections of the public who are more optimistic (i.e. positive affect) about climate change. For example, when asked about the impacts of climate change for the UK, some people have identified beneficial impacts, such as wine growing (Upham *et al.*, 2009). The Scottish public is somewhat more positive about the impacts of climate change than the English, perhaps due to regional climate variation (Palutikof *et al.*, 2004). Poortinga and Pidgeon's (2003) UK study highlights that ambivalence (i.e. neutral affect) is common: a quarter of respondents considered climate change to be 'neither good nor bad', while the largest proportion (38 per cent) rate it as only 'fairly bad'; when asked to rate the acceptability of the risk from climate change, the largest proportion rated it as 'neither acceptable nor unacceptable'.

Across both the UK and US, then, it seems that climate change is seen most often in negative terms. However, the public rarely evokes personally relevant imagery; causes tend to be identified with other people and places, while impacts are understood in generic terms that would potentially affect all life, rather than local or human-specific impacts (e.g. Leiserowitz, 2006; Lorenzoni and Pidgeon, 2006). According to Whitmarsh's (2009) survey, 60 per cent of responses to an open-ended question about impacts related to generic impacts (e.g. changing weather, flooding, sea-level rise), compared to 11 per cent relating specifically to other organisms (e.g. impacts on wildlife, species extinction) and 19 per cent to humans (e.g. impacts on agriculture, spread of disease) – yet even here it is notable that respondents used virtually no personalized language. Research on how 'carbon' is conceptualized similarly finds it is seen in abstract, rather than concrete, terms (Whitmarsh *et al.*, 2011a). This highlights that climate change tends to be seen as a collective and distant rather than a personal or local problem.

Communicating climate change

The way in which environmental change is communicated fundamentally shapes public perceptions of and responses to the issue. Information source, content, media, audience and context interact to produce communication outcomes (e.g. Petty and Cacioppo, 1986). Since long-term global issues such as climate change cannot be directly experienced, trust and credibility of second-hand sources become central influences on whether the issue is perceived as a genuine and serious risk. In general, expertise, independence and familiarity are qualities that tend to be associated with credibility (e.g. Worcester, 2001). Surveys (Poortinga and Pidgeon, 2003; Whitmarsh, 2009) examining public trust in relation to particular sources of climate change information show that by far the most common source of information on climate change is mass media – television (92 per cent), newspapers (85 per cent) and radio (66 per cent) (Whitmarsh, 2009). However, media information only inspires a moderate amount of trust. Poortinga and Pidgeon (2003) found media was rated 2.7 on a 4-point scale (1 = not at all; 4 = a lot); while friends and family (4.1), environmental organizations (4.0), doctors (4.0) and scientists working for universities (3.9) are most trusted. Oil companies (2.3), car companies (2.4) and national government (2.7) are considered least trustworthy. The public's distrust of government as a source of climate change information clearly has implications for government efforts to raise awareness and change behaviour through information campaigns. In the context of other recent scientific and risk controversies (e.g. BSE, MMR), the public are more likely to distrust official risk information (Hargreaves *et al.*, 2003), as well as institutions' capacity to effectively deal with risks such as climate change.

As the most prevalent source of climate change information, media is thus a salient influence on public perceptions (Hargreaves *et al.*, 2003). As discussed elsewhere in this book, global media has played a major role in constructing climate change (e.g. Trumbo, 1996). Media has drawn public attention to scientific and political debate – and often controversy – about climate change (e.g. Hargreaves *et al.*, 2003; Corbett and Durfee, 2004; Boykoff and Boykoff, 2004). The media's focus on climate change in the context of social and political developments (e.g. climate change legislation, extreme weather impacts on communities) casts the issue as a concrete human narrative rather than an abstract science issue (Peters and Heinrichs, 2004), using shared social symbols, experiences and values to garner public interest (McComas and Shanahan, 1999; Hargreaves *et al.*, 2003). The media also offers visual 'evidence' of changing weather patterns and their impacts on humans and ecosystems (Darier and Schule, 1999; Smith and Joffe, 2009). However, Read *et al.* (1994) warn that public concern over climate change may only occur during periods of unusual or particularly hot weather, and wane at other times. This highlights the difficulty in communicating long-term global risks when people tend to define risks more locally and in terms of sensory evidence. At the same time, the media presentation of climate change (and other risk issues) as 'controversy' can strongly influence public perceptions about the reality and seriousness of an issue, and of the credibility of the scientific evidence and policy responses to it (Hargreaves *et al.*, 2003).

Recent research highlights the key role of the media in shifting public concern away from environmental issues towards other (e.g. economic) ones. Following a peak in public environmental concern in 2007, numerous studies show public concern about climate change has dropped, while doubt and scepticism have risen, in many regions (Pew Research Centre, 2010; Whitmarsh, 2011; Poortinga *et al.*, 2011; Gallup, 2011; Leiserowitz *et al.*, 2012a). For example, between 2003 and 2008, Whitmarsh (2011) found a sharp rise from 15 per cent to 29 per cent in the proportion of UK public agreeing that 'Claims that human activities are changing the climate are exaggerated'. Several explanations have been suggested for this widespread rise in doubt and apathy, including increased sceptic information campaigns and related media events (e.g. Climategate and Glaciergate in late 2009 and 2010), 'green fatigue' and reactance to media hype, reduced media coverage due to media 'issue cycles', politicization and public distrust, changing economic circumstances, (lack of) personal experience and short-term weather conditions, (threatened) values/lifestyles and cognitive dissonance (e.g. Whitmarsh, 2011; Feinberg and Willer, 2011; Pidgeon, 2012). Perhaps most convincingly, Brulle *et al.*'s (2012) analysis of US opinion data from 2002 to 2010 similarly show the economic recession, along with 'elite cues' (i.e. policy developments, politician statements) that together shape media coverage as having the most important influence on attitudes to climate change. On the other hand, qualitative research shows that, while the public has long been uncertain about the veracity or significance of certain media and political claims about environmental risk issues and cite barriers to taking action, they remain deeply concerned about environmental degradation as a result of human activities (e.g. Bulkeley, 2000; Lorenzoni *et al.*, 2007; Fisher *et al.*, 2012). The influence of contextual factors on concern points to the problem of relying on public concern at only one point in time, as this may reflect only a superficial or immediate response to topical news events or recent experience.

For communicators and policy makers interested in engaging the public with climate change, there is a growing body of evidence for how (and how not) to proceed. In respect of communicating climate change, the first clear message from research is that fear should be used with care, and perhaps avoided altogether (O'Neill and Nicholson-Cole, 2009; Feinberg and Willer, 2011). Unless concrete actions to reduce the danger are prescribed, audiences are likely to reduce their fear by other means, such as denying the risk (e.g. Maddux and Rogers, 1983). Framing information around positive messages of hope, rather than fear, seem to be far more likely to engage a range of audiences (e.g. Maibach *et al.*, 2010; Morton *et al.*, 2011; Bain *et al.*, 2012). Other message characteristics that seem to be important include tone of voice, with more 'open' and inclusive language being more persuasive (Haddad *et al.*, 2012); and terminology, with 'global warming' appearing to evoke more concern than 'climate change' (Whitmarsh, 2009).

Second, communication should be targeted to those most likely to be receptive; and in trying to change attitudes, the most receptive audiences are those who are ambivalent (Maio *et al.*, 1996). Hardened climate sceptics will perceive messages about climate change to be unreliable and unconvincing, and will reject them; while those with less entrenched views may be more open to considering the message (Corner *et al.*, 2012). Yet, it is certainly not impossible to engage with sceptical audiences; using alternative framings, such as national security, patriotism or public health, can be effective in achieving support for climate change or renewables policy, or individual energy saving behaviours (Feygina *et al.*, 2010; Maibach *et al.*, 2010; Hoffman, 2011). Bain *et al.* (2012), for example, found that framing policies to reduce greenhouse gases around impacts on interpersonal warmth and economic/technological progress shifted climate deniers' positions towards support for pro-environmental policies. Not all reframing towards non-climate benefits may be ultimately desirable, though, as appealing to economic self-interest may serve to erode intrinsic motivations, including environmental protection (e.g. Evans *et al.*, 2012). Audience understanding of the scientific process is also relevant: Rabinovich and Morton (2012) show that climate messages high in uncertainty are persuasive to those who see science as a process of debate, but not for those who see it as the pursuit of truth.

Third, communicator characteristics are important. The credentials of the communicator are likely to be critical for message persuasiveness, with audiences believing messages emanating from like-minded people, including politicians (Kahan *et al.*, 2010). For example, 47 per cent of Americans trust President Obama as a source of information about global warming, while only 21 per cent trust Mitt Romney on this issue (Leiserowitz *et al.*, 2012a). Scientists appear to be more widely trusted, with 74 per cent of US citizens claiming to trust climate scientists (Leiserowitz *et al.*, 2012a).

Of course, there are limits to what can be done with information. Public engagement with lifestyle change and broader societal change requires addressing the significant structural, social and financial barriers (Lorenzoni *et al.*, 2007; Whitmarsh *et al.*, 2010). Also, critically, political distrust and disengagement need to be overcome, while acceptable and effective climate policies imply a greater degree of public participation in decision making (e.g. Dietz and Stern, 2008).

Conclusion: future research

This chapter has highlighted the key themes and complexities in analysing public perceptions and designing effective communications on climate change and other environmental issues. The findings show widespread public awareness of climate change, though variable understandings that sometimes diverge with expert conceptions (notably in terms of lower perceived risk and framing in terms of broad societal discourses, e.g. pollution), and a strong ideological influence on attitudes. Studies of affective imagery of climate change indicate the issue evokes associations that are typically negative but also spatially or socially distant (e.g. polar regions). Information source, content, media, audience and context interact to produce communication outcomes, with media acting as a primary – but not always trusted – source of information, and prior attitudes providing a strong filter for information processing. Experimental evidence shows that positive messaging, inclusive language, targeting receptive or ambivalent audiences, and using like-minded or impartial sources are among the ways to effectively produce engaging climate change communications.

Further analysis should focus on understanding how public opinion about climate change is shaped, using time-series trend analysis methods used in recent US research (e.g. Brulle *et al.*, 2012) on other national or international public attitude datasets. More work is also needed to understand the efficacy of interventions designed to intentionally shape public attitudes. It has been widely observed that there is a deficit of rigorous evaluation studies on public attitude and behaviour change interventions, not only in respect of environmental issues, but more generally (e.g. House of Lords Select Committee on Science and Technology, 2011). While experimental studies to evaluate novel attitude-change methods are promising, more of this is needed if public funds to 'educate' the public are to be cost-effective and justified. Beyond this, more participatory approaches to involving the public in decision making are likely to help overcome public distrust and apathy and improve decision quality.

Note

1 The survey used 'climate change' in half the sample, and 'global warming' in the other, to allow for comparison of terminology (see Whitmarsh, 2009).

Further reading

Two recent review articles on climate change perceptions:

Pidgeon, N. F. (2012). Public understanding of, and attitudes to, climate change: UK and international perspectives and policy. *Climate Policy* 12(S1), S85–S106.
Weber, E. (2010). What shapes perceptions of climate change? *Wiley Interdisciplinary Reviews Climate Change* 1, 332–342.

An excellent review article on climate change communication:

Moser, S. (2010) Communicating climate change: History, challenges, process and future directions. *Wiley Interdisciplinary Reviews: Climate Change* 1(1), 31–53.

A range of international perspectives from academics and practitioners on how to communicate climate change and foster behavioural change among the public:

Whitmarsh, L., O'Neill, S. and Lorenzoni, I. (eds) (2010). *Engaging the Public with Climate Change: Behaviour Change and Communication*. London: Earthscan.

References

Bain, P. G., Hornsey, M. J., Bongiorno, R. and Jeffries, C. (2012). Promoting pro-environmental action in climate change deniers. *Nature Climate Change* 2, 600–603.
Bord, R. J., Fisher, A. and O'Connor, R. E. (1998). Public perceptions of global warming: United States and international perspectives. *Climate Research* 11, 75–84.
Bord, R. J., O'Connor, R. E. and Fisher, A. (2000). In what sense does the public need to understand global climate change? *Public Understanding of Science* 9, 205–218.
Bostrom, A., Morgan, M. G., Fischhoff, B. and Read, D. (1994). What do people know about global climate change? 1. Mental models. *Risk Analysis* 14, 959–970.
Boyce, T. and Lewis, J. (eds) (2009). *Climate Change and the Media*. New York: Peter Lang Publishing.
Boykoff, M. T. and Boykoff, J. M. (2004). Balance as bias: Global warming and the US prestige press. *Global Environmental Change* 14, 125–136.
Breakwell, G. M. (ed.) (1983). *Threatened Identities*. Chichester: John Wiley and Sons.
Breakwell, G. M. (1991). 'Social representations and social identity'. *Papers on Social Representations* 2, 198–217.

Brechin, S. R. (2003). Comparative public opinion and knowledge on global climatic change and the Kyoto Protocol: The U.S. versus the world? *International Journal of Sociology and Social Policy* 23(10), 106–134.

Brody, S. D., Zahran, S., Vedlitz, A. and Grover, H. (2008). Examining the relationship between physical vulnerability and public perceptions of global climate change in the United States. *Environment and Behavior* 40(1), 72–95.

Brulle, R. J., Carmichael, J. and Jenkins, J. C. (2012). Shifting public opinion on climate change: An empirical assessment of factors influencing concern over climate change in the U.S., 2002–2010. *Climatic Change* 114, 169–188.

Bulkeley, H. (2000). Common knowledge? Public understanding of climate change in Newcastle, Australia. *Public Understanding of Science* 9, 313–333.

Burgess, J., Harrison, C. and Filius, P. (1998). Environmental communication and the cultural politics of environmental citizenship. *Environment and Planning A* 30, 1445–1460.

Central Office of Information (COI) (2008). *Attitudes to Climate Change – Amongst Young People – Wave 2. Key Research Findings Prepared for Defra*. Defra, London. Available at: http://webarchive.nationalarchives.gov.uk/20130123162956/http://www.defra.gov.uk/environment/climatechange/uk/individual/attitudes/pdf/cc-youth-tracker-presentation.pdf (accessed 16 December 2014).

Corbett, J. B. and Durfee, J. L. (2004). Testing public (un)certainty of science: Media representations of global warming. *Science Communication* 26(2), 129–151.

Corner, A. and Hahn, U. (2009). Evaluating scientific arguments: Evidence, uncertainty and argument strength. *Journal of Experimental Psychology: Applied* 15, 199–212.

Corner, A. Whitmarsh, L. and Xenias, D. (2012). Uncertainty, scepticism and attitudes towards climate change: Biased assimilation and attitude polarisation. *Climatic Change* 114, 463–478.

Darier, E. and Schule, R. (1999). 'Think globally, act locally'? Climate change and public participation in Manchester and Frankfurt. *Local Environment* 4, 317–329.

DEFRA. (2002). *Survey of Public Attitudes to Quality of Life and to the Environment: 2001*. London: Department for Environment, Food and Rural Affairs.

DEFRA. (2007). *Survey of Public Attitudes and Behaviours toward the Environment: 2007*. London: Department for Environment, Food and Rural Affairs.

DEFRA. (2008). *A Framework for Pro-Environmental Behaviours*. London: Department for Environment, Food and Rural Affairs.

Dessai, S. and Sims, K. (2010). Public perception of drought and climate change in Southeast England. *Environmental Hazards* 9, 340–357.

Dietz, T. and Stern, P. C. (eds) (2008). *Public Participation in Environmental Assessment and Decision Making*. Washington, DC: National Academies Press.

Dunlap, R. (1998). Lay perceptions of global risk: Public views of global warming in cross-national context. International Sociology 13(4): 473–471.

Dunlap, R. and McCright, A. M. A. (2008). Widening gap: Republican and Democratic views on climate change. *Environment* 50, 26–35.

Ereaut, G. and Segnit, N. (2006). *Warm Words: How Are We Telling the Climate Story and How Can We Tell It Better?* London: IPPR.

Etkin, D. and Ho, E. (2007). Climate change: Perceptions and discourse of risk. *Journal of Risk Research* 10(5), 623–641.

Eurobarometer (2001). *Eurobarometer 55.2 – Europeans, Science and Technology*. Brussels, European Commission Directorate-General for Research. Available at: http://ec.europa.eu/research/press/2001/pr0612en-report.pdf (accessed 16 December 2014).

Eurobarometer (2009). *Special Eurobarometer 300. Europeans' Attitudes Towards Climate Change*. Brussels: European Commission.

Evans, L. Maio, G. R., Corner, A., Hodgetts, C. J., Ahmed, S. and Hahn, U. (2012). Self-interest and pro-environmental behaviour, *Nature Climate Change* 3, 122–125.

Fazio, R. H. and Zanna, M. P. (1981). Direct experience and attitude–behavior consistency. In L. Berkowitz (ed.) *Advances in Experimental Social Psychology, Vol. 14*. New York: Academic Press: 161–202.

Feinberg, M. and Willer, R. (2011). Apocalypse soon? Dire messages reduce belief in global warming by contradicting just-world beliefs. *Psychological Science* 22, 34–38.

Festinger, L. (1957). *A Theory of Cognitive Dissonance*. Stanford, CA: Stanford University Press.

Feygina, I., Jost, J. T. and Goldsmith, R. E. (2010). System justification, the denial of global warming, and the possibility of 'system- sanctioned change'. *Personality and Social Psychology Bulletin* 36(3), 326–338.

Fisher, A., Peters, V., Neebe, M., Vávra, J., Kriel, A., Lapka, M. and Megyesi, B. (2012). Climate change? No, wise resource use is the issue. *Environmental Policy and Governance* 22, 161–176.

Frank, E., Eakin, H. and López-Carr, D. (2011). Social identity, perception and motivation in adaptation to climate risk in the coffee sector of Chiapas, Mexico. *Global Environmental Change* 21, 66–76.

Gallup (2011). *Fewer Americans, Europeans View Global Warning as a Threat Worldwide.* Available at: www.gallup.com/poll/147203/Fewer-Americans-Europeans-View-Global-Warming-Threat.aspx#2 (accessed 8 October, 2012).

Gavin, N. T., Leonard-Milsom, L. and Montgomery, J. (2011). Climate change, flooding and the media in Britain. *Public Understanding of Science* 20, 422–438.

Gifford, R., Scannell, L., Kormos, C., Smolova, L., Biel, A., Boncu, S. *et al.* (2009). Temporal pessimism and spatial optimism in environmental assessments: An 18-nation study. *Journal of Environmental Psychology* 29, 1–12.

Haddad, H., Morton, T. and Rabinovich, A. (2012). The roles of 'tone of voice' and uncertainty when communicating climate change information. Presented at the Planet Under Pressure 2012 Conference, London.

Haddock, G. and Maio, G. R. (2012). *The Psychology of Attitudes.* London: Sage.

Hargreaves, I., Lewis, J. and Speers, T. (2003). *Towards a Better Map: Science, the Public and the Media.* London: Economic and Social Research Council.

Hoffman, A. (2011). Talking past each other? Cultural framing of skeptical and convinced logics in the climate change debate. *Organization and Environment* 24(1), 3–33.

House of Lords Select Committee on Science and Technology. (2011). *Second Report of Session 2010–12: Behaviour Change Report.* London: The Stationery Office. Available at: www.publications.parliament.uk (accessed 21 February 2012).

Kahan, D. M., Jenkins-Smith, H. and Braman, D. (2010). Cultural cognition of scientific consensus. *Journal of Risk Research* 14, 147–174.

Kahan, D., Peters, E., Wittlin, M., Slovic, P., Ouellette, L. L., Braman, D. and Mandel, G. (2012).The polarizing impact of science literacy and numeracy on perceived climate change risks. *Nature Climate Change* 2, 732–735.

Kates, R. (1976). Experiencing the environment as hazard. In H. Proshansky, W. Helson and L. Rivlin (eds) *Environmental Psychology*, 2nd edn. New York: Rinehart & Winston: 401–418.

Kempton, W. (1991). Lay perspectives on global climate change. *Global Environmental Change* 1, 183–208.

Leiserowitz, A. (2006). Climate change risk perception and policy preferences: The role of affect, imagery, and values. *Climatic Change* 77, 45–72.

Leiserowitz, A., Maibach, E. and Roser-Renouf, C. (2009). *Climate change in the American Mind: American's climate change beliefs, attitudes, policy preferences, and actions.* Yale Project on Climate Change, and Grorge Mason University Center for Climate Change Communication.

Leiserowitz, A., Maibach, E., Roser-Renouf, C., Feinberg, G. and Howe, P. (2012a). *Climate Change in the American Mind: Americans' Beliefs and Attitudes in September 2012.* New Haven, CT: Yale Project on Climate Change Communication, Yale University.

Leiserowitz, A., Maibach, E., Roser-Renouf, C., and Hmielowski, J. D. (2012b). *Global Warming's Six Americas, March 2012 and November 2011.* New Haven, CT: Yale Project on Climate Change Communication, Yale University.

Lord, C. G., Ross, L. and Lepper, M. R. (1979). Biased assimilation and attitude polarization: The effects of prior theories on subsequently considered evidence. *Journal of Personality and Social Psychology* 37 (11), 2098–2109.

Lorenzoni, I. and Pidgeon, N. F. (2006). Public views on climate change: European and USA perspectives. *Climatic Change* 77, 73–95.

Lorenzoni, I., Nicholson-Cole, S. and Whitmarsh, L. (2007). Barriers perceived to engaging with climate change among the UK public and their policy implications. *Global Environmental Change* 17, 445–459.

McComas, K. and Shanahan, J. (1999). Telling stories about global climate change: Measuring the impact of narratives on issue cycles. *Communication Research* 26(1), 30–57.

Maddux, J. and Rogers, R. (1983). Protection motivation and self-efficacy: A revised theory of fear appeals and attitude change. *Journal of Experimental Social Psychology* 19(5), 469–479.

Maibach, E. W., Nisbet, M., Baldwin, P., Akerlof, K. and Diao, G. (2010). Reframing climate change as a public health issue: An exploratory study of public reactions. *BMC Public Health* 10, 1–11.

Maio, G. R., Bell, D. W. and Esses, V. M. (1996). Ambivalence and persuasion: The processing of messages about immigrant groups. *Journal of Experimental Social Psychology* 32, 513–536.

Morton, T.A., Rabinovich, A., Marshall, D. and Bretschneider, P. (2011). The future that may (or may not) come: How framing changes responses to uncertainty in climate change communications. *Global Environmental Change* 21(1), 103–109.

ONeill, S. and Nicholson-Cole, S. (2009). 'Fear won't do it': Promoting positive engagement with climate change through visual and iconic representations. *Science Communication* 30, 355–379.

OST/MORI (2004). *Science in Society: Findings from Qualitative and Quantitative Research.* London: Conducted for the Office of Science and Technology, Department of Trade and Industry by MORI Social Research Institute.

Palutikof, J. P., Agnew, M. D. and Hoar, M. R. (2004). Public perceptions of unusually warm weather in the UK: impacts, responses and adaptations. *Climate Research* 26(1), 43–59.

Peters, H. P. and Heinrichs, H. (2004). Expertise for the public: The science–journalism interface in German discourse on global climate change. *Public Communication of Science and Technology Conference (PCST-8), Barcelona, Spain.*

Petty, R. E. and Cacioppo, J. T. (1986). *The Elaboration Likelihood Model of Persuasion.* New York: Academic Press.

Pew Research Center (2010). 22-Nation Pew Global Attitudes Survey. Available at: www.pewglobal.org/files/2010/06/Pew-Global-Attitudes-Spring-2010-Report-Topline.pdf (accessed 8 October 2012).

Pidgeon, N. (2012). Public understanding of, and attitudes to, climate change: UK and international perspectives and policy. *Climate Policy* 12(S1), S85–S106.

Pidgeon, N. F., Lorenzoni, I. and Poortinga, W. (2008). Climate change or nuclear power – No thanks! A quantitative study of public perceptions and risk framing in Britain. *Global Environmental Change: Human and Policy Dimensions* 18(1), 69–85.

Poortinga, W. and Pidgeon, N. F. (2003). Public Perceptions of Risk, Science and Governance. Norwich: UEA/MORI.

Poortinga, W., Spence, A., Whitmarsh, L., Capstick, S. and Pidgeon, N. (2011). Uncertain climate: An investigation into public scepticism about anthropogenic climate change. *Global Environmental Change* 21, 1015–1024.

Proshansky, H. M. (1978). The city and self-identity. *Journal of Environment and Behaviour* 10, 57–83.

Rabinovich, A. and Morton, T. (2012). Unquestioned answers or unanswered questions: Beliefs about science guide responses to uncertainty in climate change risk communication. *Risk Analysis* 32(6), 992–1002.

Ray, J. J. (1990). Acquiescence and problems with forced-choice scales. *Journal of Social Psychology* 130(3), 397–399.

Read, D. Bostrom, A., Morgan, M. G., Fischhoff, B. and Smuts, T. (1994). What do people know about global climate change? 2. Survey studies of educated laypeople. *Risk Analysis* 14 (6), 971–982.

Safi, A. S., Smith, Jr. W. J. and Liu, Z. (2012). Rural Nevada and climate change: Vulnerability, beliefs, and risk perception. *Risk Analysis* 32, 1041–1059.

Sherman, D. K., Nelson, L. D. and Steele, C. M. (2000) Do messages about health risks threaten the self? Increasing the acceptance of threatening health messages via self-affirmation. *Personality and Social Psychology Bulletin*, 26, 247–260.

Slovic, P. (2000). *The Perception of Risk.* London, Earthscan.

Slovic, P., Fischhoff, B. and Lichtenstein, S. (1978). Accident probabilities and seat belt usage: A psychological perspective. *Accident Analysis and Prevention* 10, 281–285.

Slovic, P., Fischhoff, B. and Lichtenstein, S. (1980). Facts and fears: Understanding perceived risk. In R. Schwing and W. A. Albers (eds) *Societal Risk Assessment.* New York: Plenum: 181–214.

Smith, N. and Joffe, H. (2009). Climate change in the British press: The role of the visual. *Journal of Risk Research* 12(5), 647–663.

Smith, N. and Leiserowitz, A. A. (2012). The rise of global warming skepticism: Exploring affective image associations in the United States over time. *Risk Analysis* 32(6), 1021–1032.

Spence, A., Poortinga, W., Butler, C. and Pidgeon, N. F. (2011). Perceptions of climate change and willingness to save energy related to flood experience. *Nature Climate Change* 1, 46–49.

Trumbo, C. W. (1996). Constructing climate change: Claims and frames in US news coverage of an environmental issue. *Public Understanding of Science* 5, 269–283.

Ungar, S. (2000). Knowledge, ignorance and the popular culture: Climate change versus the ozone hole. *Public Understanding of Science* 9, 297–312.

Upham, P., Whitmarsh, L., Poortinga, W., Purdam, K., Darnton, A., McLachlan, C. and Devine-Wright, P. (2009). Public attitudes to environmental change: A selective review of theory and practice. A research synthesis for the Living with Environmental Change Programme, Research Councils UK. Available at: www.lwec.org.uk (accessed 16 December 2014).

van der Pligt, J. (1985). Energy conservation: Two easy ways out. *Journal of Applied Social Psychology* 15(1), 3–15.

Weber, E. (2010). What shapes perceptions of climate change? *Wiley Interdisciplinary Reviews: Climate Change* 1, 332–342.

Whitmarsh, L. (2008). Are flood victims more concerned about climate change than other people? The role of direct experience in risk perception and behavioural response. *Journal of Risk Research* 11, 351–374.

Whitmarsh, L. (2009). What's in a name? Commonalities and differences in public understanding of 'climate change' and 'global warming'. *Public Understanding of Science* 18, 401–420.

Whitmarsh, L. (2011). Scepticism and uncertainty about climate change: dimensions, determinants and change over time. *Global Environmental Change* 21, 690–700.

Whitmarsh, L., O'Neill, S. and Lorenzoni, I. (eds) (2010). *Engaging the Public with Climate Change: Behaviour Change and Communication.* London: Earthscan.

Whitmarsh, L., Seyfang, G. and O'Neill, S. (2011a). Public engagement with carbon and climate change: To what extent is the public 'carbon capable'? *Global Environmental Change* 21, 56–65.

Whitmarsh L., Xenias, D. and Corner, A. (2011b). Responding to climate change information: Biased assimilation, identity protection and climate change communication. Presented at: *9th Biennial Conference on Environmental Psychology*, Eindhoven, Netherlands, 26–28 September 2011.

Wolf, J. and Moser, S. C. (2011). Individual understandings, perceptions, and engagement with climate change: Insights from in-depth studies across the world. *Wiley Interdisciplinary Reviews: Climate Change* 2(4), 547–569.

Worcester, R. M. (2001). Science and society: What scientists and the public can learn from each other. *Proceedings of the Royal Institution*, 71, Oxford: Oxford University Press.

30

PUBLICS, COMMUNICATION CAMPAIGNS, AND PERSUASIVE COMMUNICATION

Todd Norton and Natalie Grecu

Environmental communication campaigns (ECCs) are a common fixture in contemporary society. Daily we receive messages attempting to persuade our knowledge, perceptions, and/or behavior with regard to environmental issues. We are warned about the risks associated with toxins such as polychlorinated biphenyl (PCBs) (Agency, U. S. E. P., 2012), the emerging crises from global climate change (Leiserowitz et al., 2009), widespread loss of water resources due to unsustainable consumption and pollution (Agency, D. I., 2012), among other issues. So commonplace are these messages, we may forget that they often are part of highly orchestrated and persuasive campaigns by private, public, and non-profit organizations. In simple terms, ECCs provide information and persuasive messaging about environmental issues to various segments of the public.

And, there are good reasons for their prominence in contemporary society. Most environmental issues are rather complex problems both in basic science and public policy. PCBs, for example, involve over 200 congeners or specific biphenyl structures, are now measured in parts-per-quadrillion (a quadrillion being roughly equivalent of a single human hair to all hair on all humans in the world today (Ecology, 2013)), and essentially are pervasive in the environment including oil, hydraulic systems, some voltage regulators, fluorescent bulbs, adhesives, inks, and caulking. As such they contaminate fish, water, and humans (Agency, U. S. E. P., 2014). These and other factors make communication of important scientific information extraordinarily challenging yet absolutely critical.

Our objective in this chapter is to outline some of the principal developments and questions within ECCs. Given the space constraints, we do not attempt to fully summarize public information campaign literature, as such summaries are already available (Atkin and Rice, 2013; Day, 2000). Our hope is to integrate literatures in ways that are theoretically coherent and practically useful. In the remainder of this chapter we (1) discuss conceptual aspects and types of ECCs, (2) outline select methods for public segmentation processes, and (3) punctuate what we see as four especially fruitful 'frontiers' of campaigns research and practice. A point of caution: it is difficult to clearly delimit environmental campaigns from health or other campaign processes. This is due both to the nature of campaigns and the interdependence of environmental with other human problems. There is good reason, however, to parse out the analytical considerations of various environmental, health, and environmental-health issues. We briefly discuss some of these dynamics in the final section of the chapter.

Communication campaigns

Environmental communication campaigns emerged from a long and rich history both of theory and practice (Paisley and Atkin, 2013). The sub-discipline of public information campaigns closely parallels the communication discipline both in its integration of theory and practice, and in drawing together concepts from a generous range of allied disciplines. Researchers and practitioners alike range considerably in their approach to public information campaigns including enhancing understanding of issues, urging both policy adaptations and individual behavior responses in the face of specific hazards (Day, 2000).

Public information campaigns also developed through a number of distinct theoretical 'camps' and approaches including affect appeals, situational theory, mental models, issues management, trans-theoretical model, among many others (Atkin and Rice, 2013). Definitions and conceptualizations of campaigns differ among these various camps. For example, issues management focuses on the salience of an issue (e.g., energy policy) whereas fear appeals highlight the impact of specific message strategies on audiences and situation theory emphasizes clusters of perception and interest of various publics within society.

A related challenge, both practically and conceptually, is differentiating campaign efforts from non-campaign communication—e.g., chemical information on Materials and Safety Date Sheet versus a systematic attempt to analyze a population and make risks applicable to that specific publics' attributes. In a world constituted by ever-present communication, how do we meaningfully draw boundaries between campaigns' messaging and other types of communication: e.g., a website summarizing climate change data and a page advocating climate change action? It is not heuristic to assume that all communication is part of an orchestrated campaign.

To ground our discussion, we offer a definition. *Environmental communication campaigns (ECCs) are multi-platformed, two-way processes including both mediated and non-mediated messages strategically designed to inform and/or persuade targeted publics regarding the mutual impact between humans and the environment.* This definition is not and is not intended to be especially revolutionary. It is consistent with other similar definitions. Day (2000), for example, defines media campaigns as "Communications campaigns are varied, multifaceted, highly planned, and strategically assembled media symphonies designed to increase awareness, inform, or change behavior in target audiences" (p. 79). But this definition also differs from other approaches. We place emphasis on *multiple platforms* including both media outlets and events traditionally viewed as public participation (Senecah, 2004). In contrast, Day emphasized the *media symphonies* at the exclusion of community engagement and participatory processes. We thus draw attention to some aspects critical to ECCs.

First, ECCs must integrate from among available conceptualizations to create effective engagement and messaging. Utilization of different messaging strategies can increase the effectiveness of ECCs (Merry, 2010). Health interventions have long integrated campaign theories, such as Elaboration Likelihood Model and Prospect Theory for more effective change messaging (Jones, Sinclair, and Courneya, 2003). In a similar vein, Jones, Sinclair, Rhodes, and Courneya (2004) argued for integration of concepts from persuasion theories and the Theory of Planned Behavior. We need ECC research and practice to reflect similar integration of theories and concepts. We need to reach across sub-disciplines of environmental communication—civic dialogue (Toker, 2004), public participation (Nisbet, 2009), risk communication (Boyd, Jardine, and Driedger, 2009), and environmental rhetoric (Peeples, 2005)—and consider the possibilities in using these multiple platforms for ECCs (Brown et al., 2012).

Second, ECCs must integrate media and community engagement processes. Grunig and colleagues (Grunig, 1989b; Grunig and Grunig, 1992) crystallized this issue through four distinct types of public communication: one-way asymmetrical such as press releases and publicity, one-way symmetrical including public information meetings, two-way asymmetrical such as stakeholder relations with significant power imbalances, two-way symmetrical such as meaningful stakeholder and civic dialogue. Public relations long operated within asymmetrical modes of interaction with one-way information transfers. Increasingly, successful and effective public engagement necessitates an *interdependent, open-system* approach to information flow, in which information is exchanged among various organizations of knowledge and meaning, as well as responsibility to publics and other organizations, and conflict resolution through negotiation, communication, and compromise (Grunig and White, 1992, pp. 43–44). The effectiveness of media campaigns sans community engagement has come under continued scrutiny (Brown et al., 2012). As Grodzinska-Jurczak et al. (2006) found, "[t]raditional methods of promotion (including media campaigns, leaflet drops, newsletters, newspaper adverts, etc.) can only achieve a limited level of success in shifting public perception" (p. 187). This two-year public education program complemented information materials with home advisors trained on the principles of sustainable waste management who visited to increase recycling in Jaslo City, Poland. The result was a significant increase in both recycled wastes and in the number of participants joining the initiative.

Third and finally, our conceptualization raises the troubling question of what constitutes an environmental issue. The question is challenging precisely because of the close relationship environmental issues have with other aspects of people's lives. The food system, PCBs, nuclear and petroleum energy, waste water, ground and stream water, water contamination, and many other issues are simultaneously environmental and health concerns. In fact these issues matter primarily because they threaten human health (López-Marrero and Yarnal, 2010). But the issue is not simply a matter of using health frames to help people grasp environmental issues. An issue which is environmental to one public can fundamentally be defined as a health issue to a different public (López-Marrero and Yarnal, 2010). Furthermore, issues for people are often defined in substantially different ways based on perceptions of uncertainty (Norton, Sias, and Brown, 2011). We discuss these issues in the final section of the chapter.

Types of campaigns

There are a number of ways to parse campaigns. Communication campaigns can be viewed as a project management endeavor. Hallahan (2011) provided a sound outline for this approach involving situation analysis, audience, goals and objectives, action recommendations, communication strategy, tasks and tactics, budget, and evaluation plan. Atkin and Freimuth (2013) outlined the entire campaign process as an evaluation process. The authors summarized "Over the life of a campaign, evaluation research encompasses collection of information about audiences at the formative stage, followed by process evaluation to assess implementation as the campaign unfolds, and finally summative evaluation to track campaign impact" (p. 53).

Probably the most common means of distinguishing campaigns is by the desired impact of messaging. This distinction typically includes information/awareness and persuasive/behavior campaigns (Atkin and Rice, 2013). As this approach is well established, our objective here is to define and provide examples from ECCs.

Informing and instructing about environmental issues

Comprehension campaigns involve presentation of data to establish that an issue exists and measures to initiate appropriate processes on the public's part. As Atkin and Rice (2013) articulated, "informational messages that seek to create awareness or provide instruction play an important role" in empowering the target audience (p. 8). Campaigns of this type emerge mostly to fill an identified gap in environmental knowledge (Kollmuss and Agyeman, 2002), focusing less on a particular call to political action and more on the practical actions individuals can take to manage risks. Oceans Alert's campaign (Alert, 2012) for example provided basic information on PCBs:

Polychlorinated Biphenyls (PCB) are toxic chemicals that have pervaded the soil and waters. PCBs are mixtures of synthetic and organic chemicals with the same chemical structure that are either found as oily liquids or waxy solid ... They are most often exposed to the environment through leaky equipment, illegal dumping, waste oil from electrical equipment, and hazardous waste. The PCBs taint the soil and can run off into waters.

(Para. 1)

The focus here is entirely on a description of and sources of PCBs, their presence in the environment including hydrologic and human processes involved in their transfer. Absent is messaging about what the audience *ought* to do given problems with PCBs.

Instructional campaigns are very similar to informational campaigning but with an emphasis on processes, often to protect one's self. A PCB website from California's Office of Environmental Health Hazard Assessment (United States, 2007) provided a clear distinction of information and instructional campaigns. The top few paragraphs are information as they cover topics such as "where do PCBs come from?" and "how might I be exposed to PCBs?" The messaging on the latter paragraphs on the page "recommends that you clean and gut the fish you catch before cooking it" (para. 16).

As is evident in this passage, instructional, and even informational, messaging borders on behavior focused in an underlying or subtle urge to take action. With that, we turn our attention to behavior campaigns.

Preventing and promoting environmental behavior

Behavior change campaigns focus on characterizing individuals along a continuum of readiness for change (*Technical Guidance for Round 8 Global Fund HIV Proposals*, 2008; Ferrer et al., 2009). Many prevention campaigns utilize a Transtheoretical Model of Change (TMoC) model. This approach developed from transtheoretical therapy and modeling entailing preconditions for therapy, processes of change, content to be changed, and therapeutic relationship (Prochaska and DiClemente, 1982, p. 277). Research regarding individual social change provides significant insight into how people may facilitate change within a "stage" of more macro-level perspectives of social change (Prochaska et al., 1992).

Behavior campaigns are often distinguished as preventing or promoting behavior (Brulle, 2010). Atkin and Rice (2013) synthesized the thrust of behavior campaigns as offering "persuasion appeals emphasizing reasons why the audience should adopt the advocated action or avoid the proscribed behavior" (p. 9). However, this prevention/promotion distinction can also be problematic. Climate change campaigns often include information about the negative impacts of driving and the benefits of alternate transportation (Frank et al., 2006). But the easy dichotomy offers a place to begin.

Campaign scholars have extended TMoC and variations thereof to better understand the processes involved in prevention of environmental degradation at the individual level. Studies have included reduction of use of plastics (Chib et al., 2009) and Solar Water Disinfection (SODIS) Users (Kraemer and Mosler, 2011). Chib et al. (2009) utilized the Temporal Incentives Model of Social Influence based on Prochaska and DiClemente's TMoC and investigated the habits of youth regarding plastic waste management through the implementation of the "[minus]plastic" campaign. Findings indicated that Singaporean youth not only recognize the importance of responsible plastic management, but also adopted positive attitudes toward the adoption of such practices (p. 679). Similarly, Kraemer and Mosler (2011) employed the TMoC to examine behavior change processes to impel Zimbabweans from drinking harmful water to using a solar disinfection process for much safer water (p. 126). This longitudinal study using TMoC revealed distinct factors between groups, implicating the need to better inform campaign strategies to impel all user types to regularly use the solar water disinfection household treatment (p. 126).

Other campaigns have promoted physical health through activities such as commuter bicycling. These campaigns work to spotlight the health benefits of such activities, while conceptualizing environmental factors as components such as accessibility, aesthetic attributes, and convenience (de Geus et al., 2008, p. 698). Similarly, Duncan and Mummery (2005) utilized Geographic Information System (GIS)-derived measures in Queensland and found that aesthetics and safety were the most significant influences of physical activity while proximal footpaths increased the probability of recreational walking (p. 363). Furthermore, initiatives to promote walking advocate physical activity and the health benefits of doing so (Rhodes et al., 2009), while environmental factors, such as decreasing carbon emissions by walking instead of driving, are not a factor in these promotion campaigns.

Cautions about the informational/behavioral divide

Though the information/behavior divide makes for an easy dichotomy, it does not fully reflect the realities of campaign research and practice. In particular the dichotomy wrongly relies on and advances, even if implicitly, a distinction between informational and persuasive communication. This is significant first because established persuasion theory makes no such distinction (McGuire, 2001). As Salmon (1989) argued, "Although information campaigns can be seen as strategies in and of themselves (through persuasion and normative reeducation), they are often essential supplements to other change strategies as well" (p. 32). Similarly, Henry and Gordon (2003) found no clear distinction between information and change oriented public information campaigns subsequent to the 1990 Amendments to the Clean Air Act. While public information campaigns were utilized to raise awareness through, for example, air quality alerts, campaign communication likewise attempted to change behavior such as driving habits to reduce health hazards associated with poor air quality (see p. 45). Similarly Teisl, Noblet, and Rubin (2009) argued that a major element of the Maine Clean Car Campaign was providing information to consumers at car and truck dealerships through the use of "eco-labeling" in the form of Clean Car window stickers to encourage purchase of more environmentally sound cars. So we should be rather cautious in the use of categorization systems and the implications for communication.

The limitation is significant also because campaign messaging often is both informational and persuasive. This is especially true of the advocacy efforts in risk communication as messaging attempts to persuade the targeted audience to know certain other information and to see some views as more valid than others (Kinsella, 2004; Boyd et al., 2009). A campaign video by Plane Stupid (Klienman, 2009) highlights perfectly the practical salience. The video explicitly is intended to "confront people with the impact that short-haul flights have on the climate" and … "used Polar Bears because they're a well understood symbol of the effect that climate change is having on the natural world" (para 1). The video begins with the sound of a jet plane flying overhead and proceeds with polar bears falling from the sky to their death upon impact with the ground, buildings, and cars. The sometimes bloodied polar bears are followed with this information: "an average European flight produces over 400kg of greenhouse gases for every passenger … that's the weight of an adult polar bear" (at 0:42 of video). The very next frame shows the company's web address planestupid.com. Though dramatic, this campaign highlights the interweaving of informational and behavior messaging as it advocates the audience associate a human action (an airline flight) with a broader problem (climate change) and an associated environmental cost (tons of carbon in units of polar bears).

In the above section, we provided both a brief outline of types of campaigns and drew some attention to the problematic difference among these supposedly different types. In the next section, we turn our attention to the issue of segmentation.

Segmentation of publics

The process of analyzing shared and different characteristics to create 'target publics' or 'publics' is the theoretical domain and statistical process known as segmentation. Discussions about segmentation are complicated by the considerable range of theories and methods used in the analysis process. This is true both within the communication field (Grunig and Grunig, 1992) and among social science more broadly including, for example, economics (Barnett and Mahony, 2011). Such diversity of theorizing is useful on one hand as different social sciences and even different campaign processes surely have different purposes for segmenting publics. As Barnett and Mahony (2011) summarized, we lack a unifying theory or conceptualization of segmentation processes itself, lack coherent thinking regarding translating the market segmentation into public engagement, and are beset by a range of largely unanswered ethical questions.

Though differing in process, the overall purpose of segmentation remains fairly consistent—identifying meaningful collectives within a general population based upon analysis of similar and dissimilar attributes. And where segmentation lacks a unifying theory, it is enriched from nuanced analysis across a range of disciplines, theories, and variable sets from economic, demographic, psychographic, and perceptual variables such as income, political affiliation, central versus peripheral route of messaging, ecological paradigm, among many other specific variables. Though it is impossible to summarize all segmentation processes, we provide a brief outline of approaches most closely associated with communication research.

Situational Theory of Publics

The Situational Theory of Publics (Grunig, 1989a) emerged in the 1960s and developed through the 1980s. The situational model of publics assesses likelihood of action in response to a problem based upon the interaction among and co-variance of five variables. These include three independent variables of problem recognition, constraint recognition, and level of involvement, and two dependent variables of information-seeking and information processing (Grunig and Hunt, 1984). As Grunig and Hunt (1984) argued, "Communication behaviors of publics can be best understood by measuring how members of publics perceive situations in which they are affected by such organizational consequences as pollution, quality of products, hiring practices, or plants closing" (p. 148). Simply put, a public with high problem recognition, low constraint recognition and high involvement is more likely to act than a public with low problem recognition, high constraint recognition, and low involvement.

However, Situational Theory of Publics is not without its critics. In an effort to advance and refine the approach, Featherstone et al. (2009) used this issues-based model to study publics' understanding and perceptions of climate change (p. 214). Featherstone et al. (2009) contended that the Situational Model of Publics was developed to predict how various publics would respond to issue-based communication (p. 215). However the authors also argued for a stronger analytical means to identify those publics in the first place, to which the model may need further adjustment for climate change issues (p. 214).

Simply put, we need more rigorous ways to incorporate the qualitative aspects of what an issue means to various publics with the quantitative assessment of the aforementioned variables. In a somewhat different vein, in a study examining a segmented public's level of trust in government, Hong et al. (2012) argued that the situational approach may not be practical in long-term relationship building between government and specified publics because a government must segment particular publics upon the emergence of an issue or problem (p. 42). The authors proposed an international public segmentation model in order to identify both cross-cultural and culture-specific characteristics of public segments of the varying countries (p. 38). Additionally, Hallahan (2000b) argued that the importance of "inactive publics" has been largely overlooked (p. 499), proposing the Motivation, Ability, and Opportunity (MAO) model to address publics in general, as well as inactive publics (Hallahan, 2000a).

Elaboration Likelihood Model

The Elaboration Likelihood Model (ELM) constitutes a major theoretical mainstay in persuasion and public campaigns literatures for communication. Prior to the ELM (Petty and Cacioppo, 1981), persuasion methodologies were characterized by a variable-oriented approach, that focused primarily on the source, message, or receiver, unlike the ELM that focuses on the *processes* by which these characteristics of the message influence people to modify their attitudes (Booth-Butterfield and Welbourne, 2002, p. 156). The ELM explains how likely an individual will become involved in an advocacy based on how the individual processes information, either through a central or peripheral route. Whether an individual chooses a central or peripheral route is dependent on how important the message, or issue, is to that individual. Central route processing is characterized by persuasion that has likely occurred due to an individual's careful consideration of the credibility of the information in support of an advocacy, while peripheral route processing is characterized by persuasion that has likely occurred to do with a simple cue in the persuasion context that has transpired without scrutiny of the creditability of the information (Petty and Cacioppo, 1986, p. 3). Persuasion researchers have expanded on the ELM model, including Chaiken's (1980) Heuristic Systematic Model (HSM), Moyer-Gusé and Nabi's (2010) Extended-Elaboration Likelihood Model (E-ELM), and Kruglanski et al.'s (2006) Unimodel that incorporated both the ELM and HSM. However, an ECC in particular used individual-level cognitive behavioral factors to extend the ELM model and to examine orientations toward environmental issues, such as managers' orientation toward sustainable development in an organization that can then be adapted into a corporate strategy (Eberhardt-Toth and Wasieleski, 2013, p. 735). Eberhardt-Toth and Wasieleski (2013) applied Street et al.'s (2001) Cognitive Elaboration Model by examining "the degree of influence of two factors on the ethical sensitivity to sustainability: the moral maturity of the individual, and the perceived moral intensity of a sustainability issue" (p. 735).

Tailoring

Tailored messaging contrasts with traditional segmentation approaches by sending consistent feedback and relevant information to individuals based upon their behavior patterns. As Hawkins et al. (2008) summarized "Tailoring means creating communications in which information about a given individual is used to determine what specific content he or she will receive, the contexts or frames surrounding the content, by whom it will be presented and even through what channels it will be delivered" (p. 454). While targeting approaches to environmental initiatives continues to gain popularity, especially through social marketing platforms, social marketing alone is inefficient in building support for more aggressive climate change policy and intervention initiatives (Corner and Randall, 2011, p. 1005). In a review of household energy consumption interventions, Abrahamse et al. (2005) found that while information increased knowledge, it did not necessarily change behaviors, while rewards more effectively encouraged conservation, but effects were short-lived, in addition to the uncertainty of whether or not effects were maintained over an extended period of time (p. 273). However, feedback, a tailoring approach to environmental campaign strategies, was proven more successful, particularly when provided frequently to households (p. 273).

Influenced by the emergence of computer-mediated communication, interactive information technology, and new opportunities for message design and delivery (David, Henry, Srivastava, Orcena, and Thrush, 2012, p. 2), tailoring addresses different individuals' unique needs, interests, and concerns (Kreuter, Strecher, and Glassman, 1999, p. 276). Utilizing a multitude of possible methods, tailoring describes the process of creating individualized communications for a particular receiver, with the expectation that such tailoring will result in greater intended effects of the communications (Hawkins et al., 2008, p. 454). As argued by Rimer and Kreuter (2006), the effects of tailoring aligns itself with the ELM (Petty and Cacioppo, 1981), in that both approaches suggest how personally relevant information can increase an individual's motivation to process the information, resulting in behavior change (p. S188) and greater receptiveness to a persuasive effort (Lustria, Cortese, Noar, and Glueckauf, 2009, pp. 156–157). Other researchers have operationalized tailoring strategies in addition to other frameworks such as Stage of Change (SOC) (David et al., 2012) and the congruency effect (Updegraff, Sherman, Luyster, and Mann, 2007).

Tailoring provides opportunities to segment populations in such a way as to promote environmentally conscious behaviors, utilizing a multitude of approaches such as combinations of tailored information, goal setting, and tailored feedback. For example, Abrahamse, Steg, Vlek, and Rothengatter (2007) employed the combination of tailored information and feedback, along with goal setting through the use of an Internet tool in order to encourage households to reduce both direct (gas, electricity, and fuel) and indirect (deep-seated in consumer goods production, transportation, and distribution) energy use (p. 265). Other research has focused on more organizationally focused initiatives, such as the reduction of oil pollution of wastewater at workshop garages, in which Daamen, Staats, and Wilke (2001) found that tailored messages resulted in more accurate knowledge and pro-environmental behavior, while the nontailored messages were as ineffective as no message (p. 230).

Summary

These three theoretical approaches—Situation Theory of Publics, Elaboration Likelihood Model, and Tailored Messaging—provide a sketch of the very different ways public segmentation has been approached in ECC research. These few pages though do highlight the ongoing commitment and diversity of ongoing research in campaigns. We now turn our attention to a critical review and consider what the future of environmental campaigns research will likely include.

Future campaign research trajectories

A substantial amount of theoretical and practical work has been undertaken in public information campaigns for environment, health, and associated issues over the past 20 years. But much work remains. In this section we briefly outline some principal, future research considerations in environmental campaigns. Our commentary and recommendations are aimed at campaigns generally and not intended to address all issues within a specific or a set of theoretical approaches. These considerations include the imperative for interdisciplinary approaches to communication campaign development and implementation, a systematic process for analyzing spatial variability of social and cultural factors, the cautious use of sub-discipline campaign processes and theoretical perspectives to inform various campaigns, in which we address potential obstacles associated with transplanting campaign strategies, and the further examination of "issues" identification in the segmentation stage of campaign development.

First, communication campaigns need to become far more interdisciplinary in scope. In particular, environmental campaign scholars need more in depth and sustained engagement with counterparts in environmental sciences and studies. The campaigns sub-discipline has tended to ground itself at the intersection of communication and psychology. But the challenges faced in environmental crises need to engage policy studies (Roe, Teisl, Rong, and Levy, 2001), adaptive capacity (Paveglio, Carroll, and Jakes, 2010), systems dynamic modeling (Langsdale, Beall, Carmichael, Cohen, and Forster, 2007), in tandem with the ongoing work in media analysis (Zhao, 2009) and message design (Merry, 2010). It would be exceptional to integrate systems dynamic modeling (Chen, QI, Zhou, Li, and Xiao, 2004) with segmentation processes (Grunig, 1997), for example, to understand not just the differences and similarities among various user groups but to take the additional step to better understand how the various pieces of the system are integrating and interacting. It is not just about understanding the variance in perceptions among groups but understanding how all of the parts of that system interact to make the system run as it does. This sort of systems-level thinking and analysis would greatly inform the messages needed to communicate with diverse but mutually relevant publics.

Second and by way of extending the interdisciplinary approach to campaigns, we need a much stronger mechanism for analyzing spatial variability of both social and cultural factors that impact peoples' behaviors as well as the spatial variability of messaging impacts. Environmental campaigns work to inform and persuade audiences of environmental issues (Cox, 2013). These campaigns frequently fail due to public misconceptions of how individual behaviors are related to the issue in terms of spatial scale. Many environmental and health communication campaigns are targeted at mass audiences, without attention to important economic, psychological, or social differences among population segments. Such programs have had limited success, particularly for issues that do not have immediate personal consequences or where the impacts of individuals' behaviors are not readily apparent to them (Moser, 2010). People may not be motivated to process messages, or they may rely on their own personal experience as a more valid source of information (Zehr, 2000). To remedy this analytical and practical limitation, segmentation processes must incorporate spatial modeling techniques and spatially explicit research endeavors.

Subsequently, our campaign strategies must also adopt far more rigorous efforts to deliver spatially variable yet consistent messaging to promote sustainable resource management. Given the social variability of a wide range of resource consumption and environmental behaviors, it only makes good sense to test the effectiveness of incorporating spatial variance into message design. Though conceptually articulated (Coles et al., 2011), we understand little about the most appropriate spatial scale to target and the relative impacts of consistent versus variable messaging across these various scales. Are different spatial scales more or less appropriate for different types of environmental resource issues (e.g., water at the census block scale as opposed to wildfire adaptation at the community scale and climate change at the regional scale)? Similarly, questions about message efficacy and boomerang effects become exponentially more challenging when spatial scales figure into our modeling efforts.

Third, we need to utilize multiple theoretical perspectives but be cautious about merely transplanting campaign variable analytics and outcomes from one sub-discipline to another. Especially relevant are the findings from health and environmental-health campaigns to environmental campaign contexts. The fundamental issue in question is that of social scale. Many of the findings emerging from health-oriented studies are able to assume a tight feedback cycle between the focal problem (cancer), the focal behavior in question (smoking), and messages for individuals within the target population to change the behavior. The bulk of health campaigns assume a reasonably clear connection between the benefits of cessation of unhealthful behaviors (e.g., stop smoking → reduction of cancer rates) or increase of healthful ones (e.g., better diet and exercise → heart health). Important here is that the benefits from behavior change as reaped directly by the person who changed their behavior.

Environmental issues, in general, do not have such a clear connection between problem, behavior change, and subsequent benefits. In fact, many changes in environmental behavior likely cost the person in some way. Stopping environmentally negative or starting environmentally positive behaviors make our world more sustainable to be sure. But those benefits are to society generally (and potentially to future generations). Little direct benefit and often direct costs are realized by the person and publics we are attempting to change.

An additional complication from importing concepts and transplanting results from other areas concerns the distinction among types of campaigns. As ECCs are complex and spatially/socially scaled differently than health campaigns, an ECC may simultaneously be both preventative and promotional. For example, the prevention of driving may also entail promoting alternative energy sources; therefore, should the ECC collapse such distinctions between prevention and promotion in a campaign? This link is likely grounded in the extractive/consumptive realities of human existence on the planet. Environmental issues are often about choices among consumptive options such that promotion of one option is, effectively, a preventive campaign about its alternative—biking to automobile driving, locally grown meats or vegetables versus corporate-processed varieties, and so forth. Through an interdisciplinary lens, environmental communication becomes complex when considering, for example, policy dynamics and the difficulty of knowing what policies will make sense because of the uncertainty regarding climate change, but also the difficulty of knowing how to modify behaviors in light of speculations (Jamieson, 1990). Therefore, past public relations campaign categorization, such as promotion and preventative classifications, do not capture the complex dynamics faced by environmental communication campaigners.

Fourth and lastly, there is a significant difference in the management of issues among "publics" versus organizational issues management among publics. Therefore, we call for the further examination of "issues" identification in the segmentation stage of campaign development. While organizational issues management and public relations approaches to campaign development may overlap in assessment and identification of publics, the deductive or inductive methods of identifying an issue are quite distinct. In reviewing public relations campaign literatures, the prevalent motivational framework in campaign development (Atkin and Salmon, 2010), consists of stages including planning, implementation, and evaluation measures (Solomon and Cardillo, 1985; Atkin and Salmon, 2010). However, prior to campaign development, a segmentation method to identify key audience(s) is essential for an effective campaign. As noted by Grunig and Hunt (1984), "[c]ommunication behaviors of publics can be best understood by measuring how members of publics perceive situations" (p. 148). Furthermore, "[t]he publics that develop around problems or issues differ in the extent to which they are aware of the problem and the extent to which they do something about the problem" (p. 147). This conceptualization of publics and issues varies significantly from organizational issues management, which emphasizes the importance of identifying key issues early in the strategic management program (Heath, 1997; Heath and Palenchar, 2009). By examining the distinction between "issues" processes in the segmentation stage of campaign development, we can better understand how place-based perceptions may be better understood through a public relations perspective on issues, versus other approaches to segmentation and issues identification.

Conclusion

In this chapter we attempted to synthesize ECC literature to provide the reader not just a summary of literature but also principal points of intersection and critical questions that can guide the ongoing work of researchers and campaigners alike. This involved outlining (1) conceptual aspects and types of ECCs, (2) methods for public segmentation processes, and (3) four fruitful 'frontiers' of campaigns research and practice. The challenges faced by ECC campaigners are in many ways different than other areas of public campaigns so we need to borrow concepts and methods from these areas but not without important revisions.

Further reading

Atkin, C. K. and Rice, R. E. (2013). Theory and principles of public communication campaigns. In R. E. Rice and C. K. Atkin (eds), *Public communication campaigns* (pp. 3–20). Los Angeles, CA: Sage.
One of the most comprehensive introductions to communication campaign theories and issues. The authors provide a summary of principal theories of public information (e.g., agenda setting and situational theory), discuss segmentation and campaign design issues, as well as message design, source, and channel considerations.

Botan, C. H. and Taylor, M. (2004). Public relations: State of the field. *Journal of Communication, 54*(4), 645–661.
Provides an extensive review of theoretical frameworks of public relations scholarship, including the significance of the transition from a functional to cocreatonal perspective in public relations practice and research.

Brulle, R. J. (2010). From environmental campaigns to advancing the public dialog: Environmental communication for civic engagement. *Environmental Communication: A Journal of Nature and Culture, 4*(1), 82–98.
Argues that cognitive approaches to public information and message design will not bring about any large-scale mobilization or sustained change, and stresses the important role participatory structures play in collective decision making processes; suggests an alternate approach focused on cultivating civic dialog.

Featherstone, H., Weitkamp, E., Ling, K., and Burnet, F. (2009). Defining issue-based publics for public engagement: Climate change as a case study. *Public Understanding of Science*, *18*, 214–228.

Critiques established persuasive communication theories by highlighting the limitation of testing message impact on predetermined and segmented publics, arguing that what is needed is more careful consideration of the science involved in segmenting people into various groups—aka publics—in the first place.

Hallahan, K. (2000). Enhancing motivation, ability, and opportunity to process public relations messages. *Public Relations Review*, *26*, 463–480.

Extends previous theorizing to approach the challenging issue of inactive publics; articulates the theory of Motivation, Ability, and Opportunity (MAO) including specific message design techniques such as issue relevance.

References

Abrahamse, W., Steg, L., Vlek, C., and Rothengatter, T. (2005). A review of intervention studies aimed at household energy conservation. *Journal of Environmental Psychology*, *25*, 273–291.

Abrahamse, W., Steg, L., Vlek, C., and Rothengatter, T. (2007). The effect of tailored information, goal setting, and tailored feedback on household energy use, energy-related behaviors, and behavioral antecedents. *Journal of Environmental Psychology*, *27*, 265–276.

Agency, D. I. (2012). *Global Water Security: Intelligence Community Assessment* (D. I. Agency, Trans.). Washington DC: Defense Intelligence Agency.

Agency, U. S. E. P. (2012). Health effects of PCBs. Retrieved from www.epa.gov/osw/hazard/tsd/pcbs/pubs/effects.htm (accessed June 12, 2013).

Agency, U. S. E. P. (2014). Basic information about polychlorinated biphenyls (PCBs) in drinking water. Retrieved from http://water.epa.gov/drink/contaminants/basicinformation/polychlorinated-biphenyls.cfm#area (accessed June 12, 2013).

Alert, O. (2012). PCB Information. Retrieved from www.oceansalert.org/pcbinfo.html (accessed June 14, 2013).

Atkin, C. K. and Freimuth, V. (2013). Guidelines for evaluating research in campaign design. In R. E. Rice and C. K. Atkin (eds), *Public communication campaigns* (pp. 53–68). Los Angeles, CA: Sage.

Atkin, C. K. and Rice, R. E. (2013). Theory and principles of public communication campaigns. In R. E. Rice and C. K. Atkin (eds), *Public communication campaigns* (pp. 3–20). Los Angeles, CA: Sage.

Atkin, C. and Salmon, C.T. (2010). Communication campaigns. In C. Berger, M. Roloff, and D. Roskos-Ewoldsen (eds), *The handbook of communication science* (pp. 419–435). Thousand Oaks, CA: Sage Publications.

Barnett, C. and Mahony, N. (2011). Segmenting publics. *National Co-ordinating Centre for Coordinating Public Engagement*, 69. Bristol: University of Bristol and University of the West of England.

Booth-Butterfield, S. and Wellbourne, J. (2002). The elaboration likelihood model: Its impacts on persuasion theory and research. In J. P. Dillard and M. Pfau (eds), *The persuasion handbook: Developments in theory and practice* (pp. 155–173). Thousand Oaks, CA: Sage Publications.

Boyd, A. D., Jardine, C. G., and Driedger, S. M. (2009). Canadian media representations of mad cow disease. *Journal of Toxicology and Environmental Health, Part A*, *72*, 1096–1105.

Brown, D. R., Soares, J., Epping, J. M., Lankford, T. J., Wallace, J. S., Hopkins, D., Buchanan, L. R., and Orleans, C. T. (2012). Stand-alone mass media campaigns to increase physical activity: A community guide updated review. *American Journal of Preventive Medicine*, *43*, 551–561.

Brulle, R. J. (2010). From environmental campaigns to advancing the public dialog: Environmental communication for civic engagement. *Environmental Communication: A Journal of Nature and Culture*, *4*, 82–98.

Chaiken, S. (1980). Heuristic versus systematic information processing in the use of source versus message cues in persuasion. *Journal of Personality and Social Psychology*, *39*, 752–766.

Chen, Y.-F., QI, J., Zhou, J.-X., Li, Y.-P., and Xiao, J. (2004). Dynamic modeling of a man-land system in response to environmental catastrophe. *Human and Ecological Risk Assessment*, *10*, 579–593.

Chib, A., Chiew, H. J., Kumar, C., Choon, L. G., and Ale, K. (2009). [minus]plastic: influencing pro-environmental attitudes among Singaporean youth. *Environmental Education Research*, *15*, 679–696.

Coles, A., Eosco, G., Norton, T., Ruiz, J., Tate, E., and Weathers, M. (2011). Mapeo del conocimiento tradicional sobre las amenazas. *Gestion y Ambiente*, *14*, 45–57.

Corner, A. and Randall, A. (2011). Selling climate change? The limitations of social marketing as a strategy for climate change public engagement. *Global Environmental Change*, *21*, 1005–1014.

Cox, R. (2013). *Environmental communication and the public sphere.* Thousand Oaks, CA: Sage Publications.

Daamen, D. D. L., Staats, H., and Wilke, H. A. M. (2001). Improving environmental behavior in companies: The effectiveness of tailored versus nontailored interventions. *Environment and Behavior, 33,* 229–248.

David, P., Henry, A., Srivastava, J., Orcena, J., and Thrush, J. (2012). Reactance to a tailored multimedia intervention encouraging teachers to promote cover-the-cough. *Journal of Health Communication, 17,* 915–928.

Day, B. (2000). Media campaigns. In B. A. Day and M. C. Monroe (eds), *Environmental education and communication for a sustainable world* (pp. 79–84). Washington DC: USAID.

de Gues, B., De, B. I., Jannes, C., and Meeusen, R. (2008). Psychosocial and environmental factors associated with cycling for transport among a working population. *Health Education Research, 23,* 697–708.

Duncan, M. and Mummery, K. (2005). Psychosocial and environmental factors associated with physical activity among city dwellers in regional Queensland. *Preventive Medicine, 40,* 363–372.

Eberhardt-Toth, E. and Wasieleski, D. M. (2013). A cognitive elaboration model of sustainability decision making: Investigating financial managers' orientation toward environmental issues. *Journal of Business Ethics, 117,* 735–751.

Ecology, W. D. O. (2013). What does that number mean? Supplemental information for ecology's water quality policy forum. Retrieved from www.ecy.wa.gov/programs/wq/swqs/HelpfulTerminologyAnalogies.pdf (accessed September 4, 2013).

Featherstone, H., Weitkamp, E., Ling, K., and Burnet, F. (2009). Defining issue-based publics for public engagement: Climate change as a case study. *Public Understanding of Science, 18,* 214–228.

Ferrer, R. A., Amico, K. R., Bryan, A., Fisher, W. A., Cornman, D. H., Kiene, S. M., and Fisher, J. D. (2009). Accuracy of the stages of change algorithm: Sexual risk reported in the maintenance stage of change. *Prevention Science: the Official Journal of the Society for Prevention Research, 10,* 13–21.

Frank, L. D., Sallis, J. F., Conway, T. L., Chapman, J. E., Saelens, B. E., and Bachman, W. (2006). Many pathways from land use to health. *Journal of the American Planning Association, 72,* 1.

Grodzinska-Jurczak, M., Tomal, P., Tarabula-Fiertak, M., Nieszporek, K., and Read, A. D. (2006). Effects of an educational campaign on public environmental attitudes and behaviour in Poland. *Resources, Conservation and Recycling, 46,* 182–197.

Grunig, J. E. (1989a). Publics, audiences, and market segments: Segmentation principles for campaigns. In C. T. Salmon (ed.), *Information campaigns: Balancing social values and social change* (pp. 199–228). Newbury Park, CA: Sage Publications.

Grunig, J. E. (1989b). Symmetrical presuppositions as a framework for public relations theory. In C. Botan and V. T. Hazelton (eds), *Public relations theory* (pp. 17–44). Hillsdale, NJ: Lawrence Erlbaum Associates.

Grunig, J. E. (1997). A situational theory of publics. In D. Moss, T. MacManus, and D. Vercic (eds) *Public relations research: An international perspective* (pp. 3–48). London: International Thomson Business Press.

Grunig, J. E., and Grunig, L. A. (1992). Models of public relations and communication. In J. E. Grunig (ed.), *Excellence in public relations and communication management* (pp. 285–326). Hillsdale, NJ: Lawrence Erlbaum Associates.

Grunig, J. E. and Hunt, T. (1984). *Managing public relations.* New York: Holt, Rinehart & Winston.

Grunig, J. E. and White, J. (1992). The effect of worldviews on public relations theory and practice. In J. E. Grunig and IABC Research Foundation (eds), *Excellence in public relations and communication management* (pp. 31–64). Hillsdale, NJ: Lawrence Erlbaum Associates.

Hallahan, K. (2000a). Enhancing motivation, ability, and opportunity to process public relations messages. *Public Relations Review, 26,* 463–480.

Hallahan, K. (2000b). Inactive publics: The forgotten publics in public relations. *Public Relations Review, 26,* 499–515.

Hallahan, K. (2011). Organizing a communications campaign/program—At-a-glance. Retrieved from http://lamar. colostate.edu/~pr/organizer (accessed September 4, 2013).

Hawkins, R. P., Kreuter, M., Resnicow, K., Fishbein, M., and Dijkstra, A. (2008). Understanding tailoring in communicating about health. *Health Education Research, 23,* 454–466.

Heath, R. L. (1997). *Strategic issues management: Organizations and public policy challenges.* Thousand Oaks, CA: Sage Publications.

Heath, R. L. and Palenchar, M. J. (2009). *Strategic issues management: Organizations and public policy challenges.* Los Angeles, CA: Sage Publications.

Henry, G. and Gordon, C. (2003). Driving less for better air: Impacts of a public information campaign. *Journal of Policy Analysis and Management, 22*(1), 45–63.

Hong, H., Park, H., Lee, Y., and Park, J. (2012). Public segmentation and government–public relationship building: A cluster analysis of publics in the United States and 19 European countries. *Journal of Public Relations Research, 24,* 37–68.

Jamieson, D. (1990). Managing the future: Public policy, scientific uncertainty, and global warming. In D. Scherer (ed.), *Upstream/Downstream: Issues in Environmental Ethics* (pp. 67–89). Philadelphia, PA: Temple University Press.

Jones, L. W., Sinclair, R. C., and Courneya, K. S. (2003). The effects of source credibility and message framing on exercise intentions, behaviors, and attitudes: An integration of the elaboration likelihood model and prospect theory. *Journal of Applied Social Psychology*, *33*(1), 179–196.

Jones, L. W., Sinclair, R. C., Rhodes, R. E., and Courneya, K. S. (2004). Promoting exercise behaviour: An integration of persuasion theories and the theory of planned behaviour. *British Journal of Health Psychology*, *9*(4), 505–521.

Kinsella, W. J. (2004). Public expertise: A foundation for citizen participation in energy and environmental decisions. In S. P. Dopoe, J. W. Delicath, and M.-F. A. Elsenbeer (eds), *Communication and public participation in environmental decision making* (pp. 83–98). Albany, NY: State University of New York Press.

Klienman, D. (Producer). (2009, 9/15/2013). Polar bears. [Video] Retrieved from www.youtube.com/watch?v=opk1bmjTFqo (accessed January 10, 2013).

Kollmuss, A. and Agyeman, J. (2002). Mind the gap: Why do people act environmentally and what are the barriers to pro-environmental behavior? *Environmental Education Research*, *8*(3), 239–260.

Kraemer, S. and Mosler, H.-J. (2011). Factors from the transtheoretical model differentiating between solar water disinfection (SODIS) user groups. *Journal of Health Psychology*, *16*, 126–136.

Kreuter, M. W., Strecher, V. J., and Glassman, B. (1999). One size does not fit all: The case for tailoring print materials. *Annals of Behavioral Medicine*, *21*, 276–283.

Kruglanski, A., Chen, X., Pierro, A., Mannetti, L., Erb, H.P., and Spiegel, S. (2006). Persuasion according to the unimodel: Implications for cancer communication. *Journal of Communication*, *56*, 105–122.

Langsdale, S., Beall, A., Carmichael, J., Cohen, S., and Forster, C. (2007). An exploration of water resources futures under climate change using system dynamics modeling. *Integrated Assessment*, *7*(1), 52–79.

Leiserowitz, A., Maibach, E., and Roser-Renouf, C. (2009). *Climate Change in the American Mind: American's Climate Change Beliefs, Attifudes, Policy Preferences, and Actions.* Yale Project on Climate Change, and Grorge Mason University Center for Climate Change Communication.

López-Marrero, T. and Yarnal, B., (2010). Putting adaptive capacity into the context of people's lives: a case study of two flood-prone communities in Puerto Rico. *Natural Hazards*, *52*, 277–297.

Lustria, M. L. A., Cortese, J., Noar, S. M. and Glueckauf, R. L. (2009). Computer-tailored health interventions delivered over the web: Review and analysis of key components. *Patient Education and Counseling*, *74*, 156–173.

McGuire, W. (2001). Theoretical foundations of campaigns. In R. E. Rice and C. K. Atkin (eds), *Public communication campaigns*. Thousand Oaks, CA: Sage.

Merry, M. K. (2010). Emotional appeals in environmental group communications. *American Politics Research*, *38*(5), 862–889.

Moser, S. C. (2010). Communicating climate change: History, challenges, process and future directions. *Wiley Interdisciplinary Reviews: Climate Change*, *1*(1), 31–53.

Moyer-Gusé, E. and Nabi, R. L. (2010). Explaining the effects of narrative in an entertainment television program: Overcoming resistance to persuasion. *Human Communication Research*, *36*, 26–52.

Nisbet, M. C. (2009). Communicating climate change: Why frames matter for public engagement. *Environment: Science and Policy for Sustainable Development*, *51*(2), 12–23.

Norton, T., Sias, P., and Brown, S. (2011). Experiencing and managing uncertainty about climate change. *Journal of Applied Communication Research*, *39*(3), 290–309.

Paisley, W. and Atkin, C. K. (2013). Public communication campaigns: The American experience. In R. E. Rice and C. K. Atkin (eds), *Public communication campaigns* (4th ed.). Los Angeles, CA: Sage.

Paveglio, T. B., Carroll, M. S., and Jakes, P. J. (2010). Alternatives to evacuation during wildland fire: Exploring adaptive capacity in one Idaho community. *Environmental Hazards*, *9*, 379–394.

Peeples, J. (2005). Aggressive mimicry: The rhetoric of wise use and the environmental movement. In S. L. Senecah (ed.), *The environmental communication yearbook* (Vol. 2, pp. 1–17). Mahwah, NJ: Lawrence Erlbaum Associates.

Petty, R. E., and Cacioppo, J. T. (1981). *Attitudes and persuasion: Classic and contemporary approaches.* Dubuque, IA: Wm. C. Brown.

Petty, R. E. and Cacioppo, J. T. (1986). *Communication and persuasion: Central and peripheral routes to attitude change.* New York: Springer-Verlag.

Prochaska, J. O. and DiClemente, C. C. (1982). Transtheoretical therapy: Toward a more integrative model of change. *Psychotherapy: Theory, Research and Practice*, *19*, 276–288.

Prochaska, J. O., DiClemente, C. C., and Norcross, J. C. (1992). In search of how people change: Applications to addictive behaviors. *American Psychologist*, *47*(9), 1102–1112.

Rhodes, R., Blanchard, C., Courneya, K., and Plotnikoff, R. (2009). Identifying belief-based targets for the promotion of leisure-time walking. *Health Education and Behavior*, *36*, 2, 381–393.

Rimer, B. and Kreuter, M. (2006). Advancing tailored health communication: A persuasion and message effects perspective. *Journal of Communication, 56,* 184–201.

Roe, B., Teisl, M. F., Rong, H., and Levy, A. S. (2001). Characteristics of successful labeling policies: Experimental evidence from price and environmental disclosure for deregulated electricity services. *Journal of Consumer Affairs, 35*(1), 1–26.

Salmon, C. T. (1989). Campaigns for social "improvement": An overview of values, rationales, and impacts. In C. T. Salmon (ed.), *Information campaigns: Balancing social values and social change* (pp. 19–53). Newbury Park, CA: Sage Publications.

Senecah, S. L. (2004). The trinity of voice: The role of practical theory in planning and evaluating the effectiveness of environmental participatory processes. In S. P. Depoe, J. W. Delicath, and M.-F. A. Elsenbeer (eds), *Communication and public participation in environmental decision making* (pp. 35–58). Albany, NY: State University of New York Press.

Solomon, D. S. and Cardillo, B. A. (1985). The elements and process of communication campaigns. In T. V. Dijk (ed.), *Discourse and communication: New approaches to the analysis of mass media discourse and communication* (pp. 60–68). Berlin: Walter de Gruyter & Co.

Street, M., Douglas, S., Geiger, S., and Martinko, M. (2001). The impact of cognitive expenditure on the ethical decision-making process: The cognitive elaboration model. *Organizational Behavior and Human Decision Processes, 86,* 256–277.

Technical Guidance for Round 8 Global Fund HIV Proposals (2008). Prevention: Social change communication. Working document.

Teisl, M. F., Noblet, C. L., and Rubin, J. (2009). Can environmental promotion backfire? Evidence from the vehicle market. *Social Marketing Quarterly, 15,* 2–32.

Toker, C. W. (2004). Public participation or stakeholder frustration: An analysis of consensus-based participation in the Georgia Ports Authority's stakeholder evaluation group. In S. P. Dopoe, J. W. Delicath, and M.-F. A. Elsenbeer (eds), *Communication and public participation in environmental decision making* (pp. 175–200). Albany, NY: State University of New York Press.

United States. (2007). Evaluation of results from EPA's schools chemical cleanout campaign: An examination of approaches and factors that impacted the results of programs funded by EPA's RCC schools chemical cleanout campaign in FY 2004. Washington, DC: U.S. Environmental Protection Agency.

Updegraff, J. A., Sherman, D. K., Luyster, F. S., and Mann, T. L. (2007). The effects of message quality and congruency on perceptions of tailored health communications. *Journal of Experimental Social Psychology, 43,* 249–257.

Zehr, S. (2000). Public representations of scientific uncertainty about global climate change. *Public Understanding of Science, 9*(2), 85–103.

Zhao, X. (2009). Media use and global warming perceptions. *Communication Research, 36,* 698–723.

31

ENGAGING DIVERSE AUDIENCES WITH CLIMATE CHANGE

Message strategies for Global Warming's Six Americas

*Connie Roser-Renouf, Neil Stenhouse, Justin Rolfe-Redding,
Edward Maibach, and Anthony Leiserowitz*

Introduction

Global climate change is an environmental threat of the gravest magnitude, recognized by virtually the entire climate science community. Among Americans, however, it remains a divisive issue, with large segments of the population discounting or ignoring the threat. To build public understanding and issue engagement, climate change communicators must recognize and respond to these varied points-of-view: Messages are unlikely to be effective if a diverse population is treated as a homogeneous mass.

In this chapter, we discuss climate change communication strategies in light of the information-processing propensities of *Global Warming's Six Americas*—six unique audience segments that perceive and respond to the issue in distinct ways. The Six Americas range across a spectrum of concern and issue engagement, from the *Alarmed*, who are very concerned about the issue and support aggressive action to reduce it, to the *Dismissive*, who do not believe it is real or a problem (Figure 31.1). Between these two extremes are four other groups—the *Concerned, Cautious, Disengaged,* and *Doubtful*—that vary in their issue concern, certainty, behavioral responses, and policy support.

The groups were first identified in a 2008 nationally representative survey using latent class analysis, and have been tracked since then with bi-annual surveys. They are strongly associated with a range of characteristics, including climate and energy beliefs and policy preferences; political ideology and party identification; cultural values; political efficacy; and consumer and political behavior (see Maibach et al., 2009, 2011; and reports posted on our websites: http://climatechange.gmu.edu and http://environment.yale.edu/climate-communication/). Developed to assist climate change communicators, the segmentation is being used by government agencies, informal science educators, and media organizations.

To date, publications describing the segments have been largely descriptive, detailing the beliefs, behaviors, and characteristics of each group. There are, however, two theoretical dimensions underlying the Six Americas—*attitudinal valence* and *issue involvement* (Figure 31.2)—which link the segmentation to well-developed literatures on persuasion, information-processing, science and risk communication, and opinion leadership, suggesting a wealth of communication strategies for reaching and influencing the Six Americas.

Attitudinal valence is defined here as *the inclination to accept or reject the science of climate change*, and is assessed with several key beliefs: Climate change is occurring; it is harmful; humans are causing it; humans

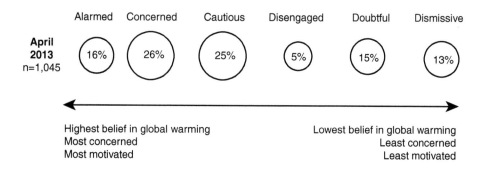

Figure 31.1 Global Warming's Six Americas

Source: Yale/George Mason University.

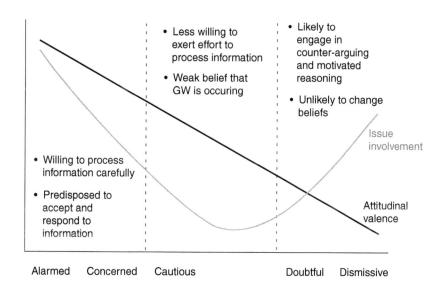

Figure 31.2 Information-processing propensities among the Six Americas

Source: Yale/George Mason University.

can reduce it; and scientists agree on its reality and human causes. These beliefs have been shown to predict support for national action on the issue and for mitigation policies, as well as political activism (Krosnick et al., 2006; Ding et al., 2011; Lewandowsky et al., 2013; Roser-Renouf et al., 2014).

Issue involvement refers to *cognitive and affective issue engagement*, and is assessed in terms of the amount of thought devoted to the issue and attitudinal certainty. Both the *Alarmed* and *Dismissive* think about the issue and are certain of their opinions, but the *Alarmed* are likely to accept all the key beliefs and are predisposed to accept messages that are consistent with the science, while the *Dismissive* reject the key beliefs and are predisposed to reject and counter-argue these same messages.

The remaining four segments, currently comprising about 70 percent of the U.S. population, have lower issue involvement and greater uncertainty regarding the reality, dangers, and causes of climate change; they differ, however, in their levels of uncertainty, predispositions to accept or reject climate science, cultural values (Figure 31.3), media use, attention paid to information about global warming (Figure 31.4), and, to a smaller extent, demographics. All of these differences have implications for the types of information the groups are interested in learning (Figure 31.5), the communication channels most likely to reach them, and the communication strategies that are most likely to influence them.

Our purpose in this chapter is to describe each segment very briefly. We aggregate the segments into three higher-order categories based on their level of issue involvement and their propensity to accept or reject climate information, and then discuss several theoretical perspectives that suggest how we may best communicate with the groups. Readers who desire more extensive descriptions of the segments may find these in the initial report identifying the six groups (Maibach et al., 2009), and in the subsequent reports posted on our websites that track the segments over time.

None of the theoretical perspectives discussed below are of exclusive relevance to a single segment; they are, however, more relevant to particular segments, as will be discussed. And little of what follows has been tested with climate change messages and audiences to date; rather, we are inferring from research with other issues what is likely to be effective. The theory discussions are extremely brief, doing justice to none of the perspectives discussed: In light of space limitations, we must simply point readers in the direction of literature of likely relevance to climate communications.

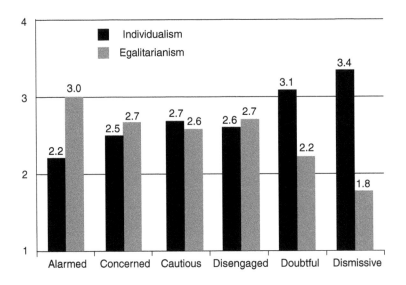

Figure 31.3 Cultural values of the Six Americas

Source: Yale/George Mason; estimates based on surveys in 2008, 2010, and 2011; unweighted n=5,127.

Note: Figure shows mean egalitarianism and individualism on 4-pt. scales.

Note: The Six Americas differ in the weight they ascribe to egalitarian values—i.e., equal opportunity, a more equal distribution of wealth, and protection for vulnerable minorities and the poor—as opposed to individualistic values— i.e., freedom from government intervention in the lives of individuals and in business.

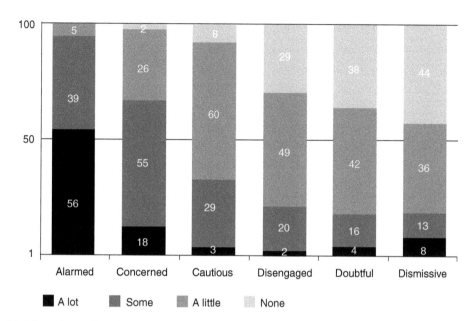

Figure 31.4 Attention paid to global warming information

Source: Yale/George Mason, April 2012, unweighted n=1,008.

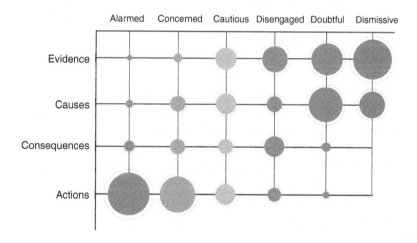

Figure 31.5 Nature of the one question respondents would most like to pose to a climate scientist

Source: Yale/George Mason University, May 2011; unweighted n=1,010; Figure credit: Ian Barin.

Note: The Six Americas are interested in learning different types of information about global warming, with the unconcerned segments most interested in information about the evidence for and causes of global warming, the concerned segments interested in information about action to mitigate climate change, and the uninvolved segments varying widely in their questions.

Publics with high involvement and positive climate change attitudes

The Alarmed

Key beliefs and issue involvement

The *Alarmed* show very high levels on measures of the five key beliefs: Almost all are certain that global warming is happening, believe their own family is at risk, and perceive future generations to be at risk; three-quarters or more believe that global warming is human-caused, recognize that most scientists think that global warming is happening, believe that people in the U.S. are being harmed now, and see global warming as potentially solvable. They are highly involved with the issue—much more so than even the *Concerned*: 63 percent report having thought a lot about global warming, a proportion more than four times as large as that of any other segment. For the *Alarmed*, global warming is a real, worrisome, and urgent threat (Figure 31.6).

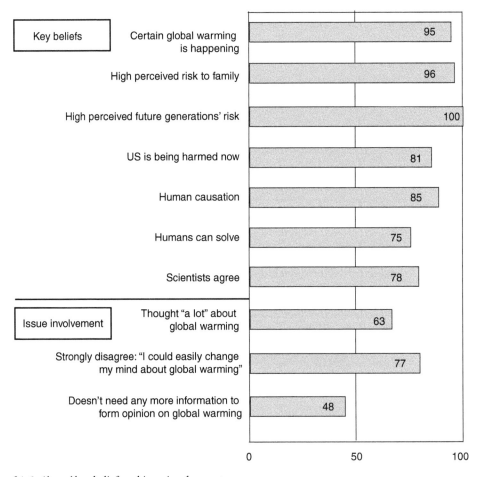

Figure 31.6 *Alarmed* key beliefs and issue involvement

Source: Yale/George Mason University, April 2013; n=1,045.

Note: See Appendix for item descriptions.

Characteristics

The *Alarmed* have a higher proportion of liberals and Democrats than any other segment. About half identify as liberal, compared to about 30 percent of the *Concerned*, and a quarter of all Americans. The *Alarmed* are the most egalitarianism segment, and the least individualistic.

The *Alarmed* are not homogenous, however: About half do not identify as liberal or as Democrats. And although the *Alarmed* are more educated than the national average—close to 40 percent have a bachelor's degree or higher, compared with just under 30 percent nationally—on other demographic variables, the *Alarmed* are not substantially different from national averages.

Informational needs and media use

Since the *Alarmed* are already convinced of the reality, danger, and human-caused nature of climate change, they are most likely to report an interest in finding out what kind of actions can be taken to reduce it, either by the U.S. or personally. They are very attentive to global warming news, compared to the other segments: 55 percent report paying "a lot" of attention to news stories about global warming, more than four times as high a proportion as any other segment. Almost 80 percent of the *Alarmed* follow environmental news, compared to a national average of 38 percent, and over half say they pay "a lot" of attention to global warming information. They are more likely to closely follow news about politics, science, and technology than any other segment except the *Dismissive*.

The Concerned

Key beliefs and issue involvement

On many measures, the *Concerned* are midway between the *Alarmed* and the less-engaged middle segments. The *Concerned* are less likely than the *Alarmed* to espouse some of the key beliefs on the issue, such as certainty that global warming is happening and feeling that their own family is at risk. They're less likely than the *Alarmed* to agree that global warming is human-caused or that future generations are at risk. They are, however, still much higher than all segments other than the *Alarmed* on all key beliefs. Perhaps the largest difference between the *Concerned* and *Alarmed* is the proportion reporting high levels of involvement with climate change: Only 13 percent of the *Concerned* report having thought "a lot" about climate change, compared to 63 percent of the *Alarmed*; and only 18 percent agree that they do not need more information to form a firm opinion about climate change, compared to 48 percent of the *Alarmed* (Figure 31.7).

Characteristics

The *Concerned* are less skewed to the left than the *Alarmed*, with the proportion of people reporting liberal ideology and Democratic party allegiance only slightly higher than national averages. They value egalitarianism over individualism, but are closer to the national averages than the *Alarmed*. Demographic distributions of the *Concerned*—gender, ethnicity, education, age, and income—are all close to national averages.

Informational needs and media use

Like the *Alarmed*, the *Concerned* say they'd most like to learn what the U.S. and they themselves can do to reduce global warming, but they are more likely than the *Alarmed* to want to know whether global warming is happening, and how experts know it is happening. Although close to three-quarters report paying at least "some" attention to information about global warming, the proportion paying "a lot" of

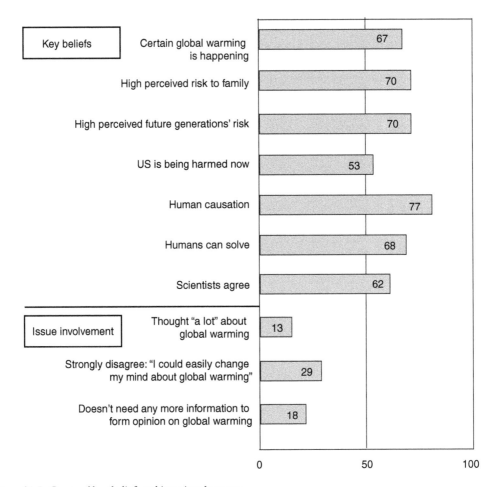

Figure 31.7 Concerned key beliefs and issue involvement

Source: Yale/George Mason University, April 2013; n=1,045.

Note: See Appendix for item descriptions.

attention (18 percent) is much lower than among the *Alarmed* (56 percent). Their other media use habits are quite similar to national averages, except that they are more likely to follow environmental news.

High involvement communication strategies

The challenge with the high involvement segments is motivating them to take action, particularly political action and opinion leadership: even among the *Alarmed*, political actions are not the norm; e.g., less than one-third have contacted an elected official about global warming over the past year.

Dual-processing theories such as the Elaboration Likelihood Model suggest that high-involvement audiences such as the *Alarmed* and *Concerned* will be receptive to messages with a great deal of information and complexity, including relatively high-level science and policy content (Petty et al., 2009); these messages may be delivered via print media, which require greater processing effort. Because messages to these audiences will likely be processed effortfully, message content is more likely to be remembered, and behavioral changes are more likely (Petty et al., 2009). A caveat is that it becomes more important to use strong, logi-cally sound arguments for action, since weaker arguments are more likely to be detected, and may lead to a potentially *lower* level of behavior change than if no message had been received (Petty et al., 2009).

Both the *Alarmed* and *Concerned* are most interested in learning about solutions to climate change—actions they can take and the U.S. can take to mitigate the effects. They are already strongly convinced of the reality and danger of climate change, so strong arguments on these topics aren't needed; they need instead well-reasoned information on solutions that are both feasible and effective.

In seeking to influence these groups, communicators may wish to focus on building efficacy to com-plement risk perceptions as a motivator for action. The *Alarmed* and *Concerned* already have high levels of concern about climate change, but lower levels of efficacy with regard to solving it: While majorities agree that humans *could* reduce climate change, the proportions who believe their own actions make "some" or "a lot" of difference in reducing their emissions have decreased over the past five years by 13 percentage points among the *Alarmed* (from 68 percent to 55 percent), and by 23 percentage points among the *Concerned* (from 61 percent to 38 percent).

Several forms of efficacy are relevant for climate change: Response efficacy—the belief that responses to the threat will be effective in reducing it; self-efficacy—the belief that one is capable of taking these actions; and collective efficacy—the belief that one's group is capable of acting effectively together (Bandura, 1986). Much evidence suggests that people who feel both threatened and capable of taking effective action to reduce the threat are more likely to take action (Witte and Allen, 2000), and a recent meta-analysis supported the idea that threatening information only promotes behavior change when effi-cacy is also high (Peters et al., 2013). The *Alarmed* and *Concerned* already feel threatened however, so, messages emphasizing the ability of individuals or groups to effectively fight climate change are likely to be most effective with these groups. The messages must, however, be convincing, or they may boomerang, lessening both confidence in solvability and behavior change.

An additional strategy for consideration with the *Alarmed* is tapping their potential to act as opinion leaders. This is actually a strategy for reaching the less involved middle segments that are more likely to be influenced interpersonally than through the mass media. But it entails a campaign objective for com-munications with the *Alarmed*—i.e., activating their opinion leadership potential.

Scholars have suggested using a "two-step flow" model of communication on climate change: Rather than trying to communicate with all citizens directly, climate communicators might instead promote opinion leadership among the *Alarmed*, encouraging them to discuss the issue with friends and family more frequently (Nisbet and Kotcher, 2009). Targeting those *Alarmed* who are already opinion leaders—i.e., people who are well-connected socially and who frequently give advice or have their advice sought out by those they are connected to—would be particularly desirable. These people may then proceed to use personal influence within their social networks to create a larger overall effect than if the communica-tor had tried to reach the same audience directly. One scholar has shown evidence that the ideal opinion leader is one who both sets a normative behavioral example and explicitly communicates about why behavior change is a good idea (Venkatraman, 1989). Our surveys of the *Alarmed* and *Concerned* show that in addition to being more likely to talk about global warming, they are more likely to engage in behaviors designed to reduce carbon emissions, making them good candidates for this type of leadership.

Low involvement publics

The Cautious

Key beliefs and issue involvement

The *Cautious* are weak on all key beliefs and have low issue involvement. They're more likely to believe climate change is occurring than not, but only one in five is certain; four out of five believe future generations are at risk, but half as many perceive their own family to be at risk. They have given very little thought to the topic, and only 5 percent are very certain of their opinions. Global warming is far from *Cautious* minds—a problem for people in the future (Figure 31.8).

Characteristics

The *Cautious* are in some ways the least distinctive segment: Their levels of egalitarianism and individualism, and their party identification and political ideology track population means closely. They're slightly less likely than average to have a college degree, but follow population distributions on other demographics, including ethnicity, gender, income, and age.

They show signs of a higher-than-average tendency to social comparisons, however—i.e., they are significantly more likely to say that having a home as well-equipped and furnished as their peers is important to them, and that they follow the latest fashion trends. They're also more likely than the *Alarmed* or *Dismissive* to say they prefer brands and products that make them feel accepted by others.

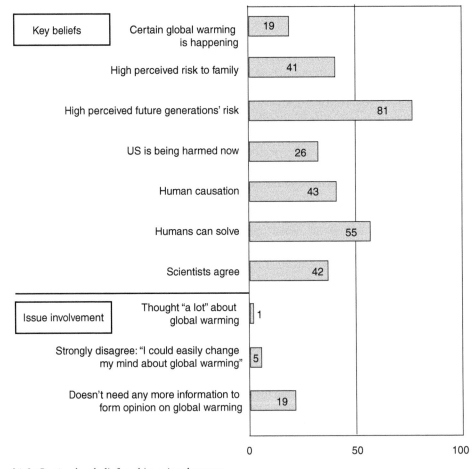

Figure 31.8 Cautious key beliefs and issue involvement

Source: Yale/George Mason University, April 2013; n=1,045.

Note: See Appendix for item descriptions.

Informational needs and media use

The top questions that the *Cautious* have about climate change are how scientists know it is caused by humans, and whether it is really happening. They're unlikely to encounter the answers, however—close to 70 percent say they pay little or no attention to global warming information.

While they report average levels of media use, they pay less-than-average attention to news, and have the lowest attention to environmental news of all six segments. Hence, reaching them through informational channels may be challenging.

The Disengaged

Key beliefs and issue involvement

The *Disengaged* are the group that has given the issue of global warming the least amount of thought. On questions with a "don't know" response option, they overwhelmingly choose this response—e.g., in April 2013, 88 percent said they don't know how great the threat is to their family, and 98 percent said they don't know how great the threat is to future generations. Only 6 percent are certain that global warming is happening, and only one in 20 are very certain of their opinions. If pressed, however, they are inclined to believe that global warming is somewhat dangerous: When no "don't know" response option is offered, their risk perceptions tend to be slightly higher than the risk perceptions of the *Cautious*—e.g., 32 percent of the *Disengaged* say Americans are being harmed now, as compared to of the 26 percent of the *Cautious* (Figure 31.9).

Characteristics

The *Disengaged* have lower socio-economic status than other segments: They are least likely of the segments to have graduated from college, and they have the lowest incomes. About 60 percent are women, and a quarter are African American. They're more likely than other segments to be retired, disabled, and renters.

They tend to be moderate Democrats who are politically inactive. Close to a quarter have no party identification, saying they are uninterested in politics; the group has the lowest proportion of registered voters. Their levels of egalitarianism and individualism are about equal and similar to national norms, but they are higher than average in biblical literalism and in rejection of the theory of evolution—beliefs associated with a rejection of climate science.

Informational needs and media use

The *Disengaged* do not follow political news very closely and while they say they need more information on global warming, they are unlikely to seek it. They are high television viewers, watching more entertainment programming than any other group, but less news and public affairs. They pay the smallest amount of attention to national politics of the six segments, and close to 80 percent say they pay little or no attention to global warming information.

Low involvement communication strategies

Reaching and engaging audiences that are uninterested in an issue begins with the recognition that no matter how important we believe our message to be, audience members are unlikely to pay attention if understanding the content requires cognitive effort—hence, we must turn to methods that are not effortful. These include message strategies that:

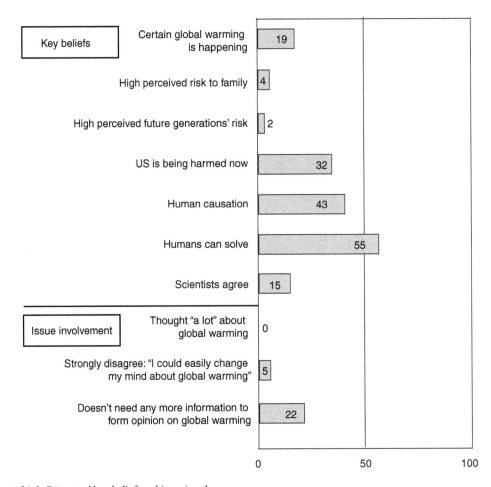

Figure 31.9 Disengaged key beliefs and issue involvement

Source: Yale/George Mason University, April 2013; n=1,045.

Note: See Appendix for item descriptions.

- require only peripheral/heuristic information processing, e.g., humor, and attractive or highly credible sources;
- promote positive social norms by suggesting that climate-friendly behaviors are popular, respected, and common;
- show rather than tell what is happening, thereby triggering automatic information-processing;
- personalize the threat by showing impacts on places that are physically close or emotionally significant (such as national parks), and on people with whom the audience can identify;
- generate involvement through the use of narratives.

These communication strategies apply to all segments, in that we are all influenced by social norms, we become emotionally engaged with compelling narratives, are drawn to attractive sources, and process visual information effortlessly and instantly. They are, however, particularly applicable to the *Cautious* and *Disengaged* because these groups lack the willingness to pay attention that characterizes involved segments.

Barriers communicators face with low involvement audiences are motivation and ability, two prerequisites for deep information processing: Three-quarters of the *Disengaged* and 44 percent of the *Cautious* say they have difficulty understanding global warming news; over half of the *Disengaged* and more than a third of the *Cautious* say they don't like to read or hear about the topic (Table 31.1). Note, however, that these barriers exist across all six segments, with close to a quarter of the *Alarmed* saying they have difficulty understanding, and majorities of the *Disengaged*, *Doubtful* and *Dismissive* saying they don't want to read or hear about the issue. Either barrier can be sufficient to halt information processing, and the challenge for communicators is to create content that will draw audiences in and be simple to understand.

While the use of attractive, credible sources and humorous messages may generate the short-term persuasion typical of peripheral/heuristic message processing, such effects tend to be short-term and unstable; hence, communicators may wish to employ additional strategies in reaching the *Cautious* and *Disengaged*.

Because neither segment attends to global warming information or news, narratives may be a more effective way of reaching them—particularly the *Disengaged*, with their high use of entertainment programming. Narratives foster involvement with a story and characters, and prior issue involvement is unnecessary for drawing the audience's attention. Memory of narrative content tends to be high, allowing educational content to be conveyed, and studies find that the persuasive effects of fiction can be as high as for non-fiction if the individual has become absorbed in the story (Green and Brock, 2000). An empathic response to story characters fosters acceptance of their values and beliefs, at least in the short term, and some evidence suggests that absorption decreases counter-arguing and increases message acceptance (Slater and Rouner, 2002).

Another strategy that may be effective with low involvement audiences is the promotion of positive social norms that can influence both attitudes and behaviors (Cialdini and Trost, 1998). Studies show that low issue involvement is associated with greater normative influence (Petty and Brinol, 2012), and the *Cautious* may be particularly good targets for this approach in light of their higher-than-average sensitivity to social appearances.

Social influence occurs at a largely unconscious level through our observation of the actions of others around us (descriptive norms), but also through learning what those we respect and care for expect us to do (injunctive norms). Environmental communicators unwittingly use descriptive norms to promote behaviors they wish to extinguish by stating how prevalent undesirable behaviors are. Instead, to the extent that it's possible to do so honestly, messages should suggest that desirable views and actions are widespread, growing in popularity, and characteristic of admired individuals; maintaining consistency between descriptive and injunctive norms is an important component of effective norm messaging: This is popular *and* it's socially approved (Cialdini, 2003).

Table 31.1 Ability and motivation barriers

	Alarmed (%)	Concerned (%)	Cautious (%)	Disengaged (%)	Doubtful (%)	Dismissive (%)
"I have difficulty understanding news reports about global warming."	23	39	44	77	35	19
"In general, I don't like to read or hear anything about global warming."	10	28	37	59	57	72

Source: Yale/George Mason University, June 2011; n=1,043.

Note: Cells show the proportions that agree with each statement.

Involved publics with negative climate change attitudes

The Doubtful

Key beliefs and issue involvement

The *Doubtful* have similar levels of issue involvement to the *Concerned*, but low acceptance of the key beliefs. While 40 percent are certain global warming is occurring, they view it as a low risk and take a dim view of the notions that humans have caused climate change or can solve it; few think that scientists agree that climate change is happening. They are more involved in the issue than the middle segments, however, and even though they do not actively think a lot about climate change on a daily basis, they are moderately certain of their views, with three-quarters very certain of their opinions, and close to half not needing any new information to make up their minds. The *Doubtful* have concluded that climate change is not an important issue, but are not strident in their views (Figure 31.10).

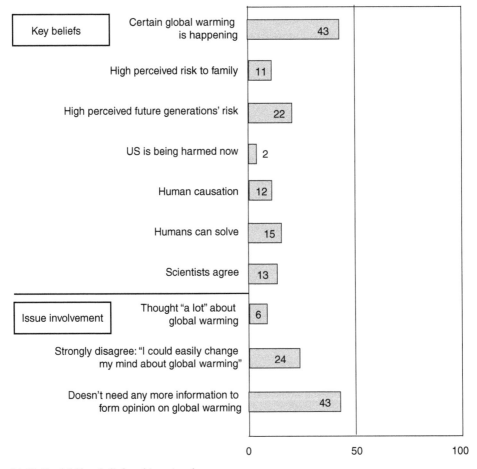

Figure 31.10 Doubtful key beliefs and issue involvement

Source: Yale/George Mason University, April 2013; n=1,045.

Note: See Appendix for item descriptions.

Characteristics

The *Doubtful* tend to be politically conservative; only around one in ten is a liberal. Party identification skews Republican, as do cultural values: Among the segments, the *Doubtful* are the second lowest in their level of egalitarianism, and second highest in their levels of individualism. While they are slightly more likely to be white and male than the national average, their income, age, and education do not substantially differ from the rest of the country.

Informational needs and media use

The *Doubtful* would most like to know how scientists know that climate change is real—the proportion who would ask this question is twice the national average. They are unlikely to attend to the topic, with only 3 percent saying they pay a lot of attention to global warming information. Few follow environmental news, but they do pay an average amount of attention to news about politics, science, and technology.

The Dismissive

Key beliefs and issue involvement

The *Dismissive* are the most certain that climate change is not happening and are highly confident in their views. They are an inverse of the *Alarmed*—strong issue partisans, but with a diametrically opposed position. *Dismissives* are very unlikely to hold any key beliefs about climate change. None believe that climate change is harming the U.S. now. While on several measures (belief certainty and risk perceptions) the *Disengaged* are nominally lower, this is due to their high proportion of "don't know" responses, whereas the positions taken by *Dismissives* indicate a firmer rejection of climate science. This is further evinced in *Dismissives*' high levels of issue involvement. While climate change is a greater presence in the everyday thoughts of the *Alarmed*, *Dismissives* are the most likely of any segment to say that they do not need any more information to make up their mind on the topic (Figure 31.11).

Characteristics

More than 70 percent of the *Dismissives* are conservative. Over half identify as Republicans, with only 3 percent Democrats, and their cultural values are the least egalitarian and the most individualistic of any segment. Demographically, they are more likely to be male and white than the national average. Their socio-economic status is also notably higher, with greater educational attainment, a lower-than-average proportion with less than a high school degree, and the highest income of any of the Six Americas.

Informational needs and media use

The question *Dismissives* would most like answered is how climate scientists know that climate change is real, and they are very unlikely to ask about anything else. They do, however, follow climate change news and information. Whereas the *Doubtful* are largely uninterested, the *Dismissives* are the second most likely to say that they pay "a lot" of attention to global warming, second only to the *Alarmed* (by a wide margin of 22 percent to 57 percent). And whereas virtually all of the *Alarmed* pay at least some attention to global warming, a substantial portion of the *Dismissives*—40 percent—choose to ignore information on the topic.

The *Dismissives* pay average attention to news on the environment, above average attention to science and technology news, and are the segment most likely to follow politics. Unlike other segments (including the *Doubtful*), the *Dismissives* are unlikely to trust scientists on climate change.

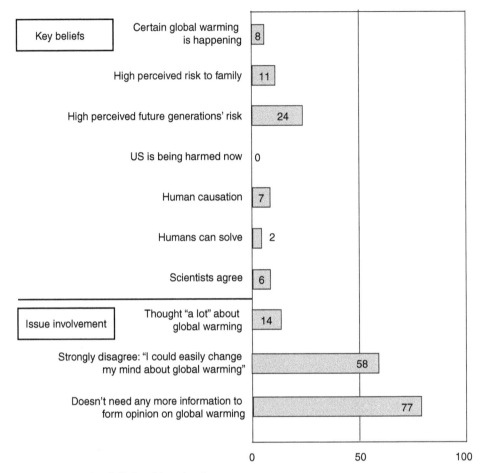

Figure 31.11 Dismissive key beliefs and issue involvement

Source: Yale/George Mason University, April 2013; n=1,045.

Note: See Appendix for item descriptions.

Counter-attitudinal communication strategies

Research suggests that reaching counter-attitudinal audiences such as the *Doubtful* and *Dismissive* may be achieved by adopting approaches that access the "back doors of the mind." A key challenge with the *Doubtful* and *Dismissive* segments is not merely to *inform* or *confirm* existing beliefs that climate change is real and problematic (as with the *Cautious* and *Disengaged*), but more fundamentally to *persuade* the *Doubtful* and *Dismissive* that their beliefs about the issue are incorrect. Yet, because the views of these segments are opposed to efforts aimed at addressing climate change, direct communication is likely to trigger counter-arguing, rather than persuasion, in a process of motivated reasoning (Kunda, 1990).

The *Dismissives'* high issue involvement means that their beliefs and attitudes are strongly held and well established; they will critically scrutinize messages about climate change, rejecting the science, while uncritically accepting information questioning climate change in a process of biased assimilation. The *Doubtful* are midway between that characterization and the low-involvement segments. Though skeptical, the *Doubtful* hold their attitudes and beliefs about climate change less fervently, spend less time and energy thinking about climate, and are less likely to have the motivation to closely scrutinize climate change communication.

Emphasizing scientific agreement on the reality of climate change may help the *Doubtful* become less skeptical, though this message is unlikely to work with the *Dismissive*, who are distrustful of climate scientists. Whereas the *Doubtful* are quizzical and potentially persuadable, the *Dismissive* are certain about their views and likely to resist efforts to change their minds (Taber and Lodge, 2006).

A prime risk of directly engaging with the *Dismissive* is that any mention of climate change may result in a "boomerang effect" (Hart and Nisbet, 2012) in which an attempt at persuasion results in attitude change in the opposite direction than desired, due to counter-arguments generated by the message recipient.

An important aspect of the "back door of the mind" approach is appreciating the underlying motivational structure beneath expressions of skepticism about climate change and mitigation proposals. Research on the cultural cognition of risk suggests that individuals develop their understanding of societal issues with reference to their underlying cultural values (Kahan and Braman, 2006). Climate change poses a threat to the values of individualism and respect for established order that mark political conservatism in the United States (Kahan et al., 2011; Kahan et al., 2012), values that are strongly held by the *Doubtful* and *Dismissive*. This implies that communicating about climate change in more value-congruent ways may increase engagement. One experiment indicates that trust in a fictive climate change scientific expert increased among those with individualistic and hierarchical values when that expert advocated nuclear power (as opposed to government regulation) as a policy solution (Kahan et al., 2011).

Although communicating with these segments may be difficult, several approaches may be of use: Pointing out concrete ways in which people have personally experienced climate change may be effective with the *Doubtful*, as research has found that personal experience of climate change leads people to become convinced of its reality (Myers et al., 2013). Health frames have been shown to have some resonance with these segments, and language choice may also be important: "Climate change" is a less polarizing term than "global warming" for Republicans and *Dismissives*, and they are more likely to accept its reality (Schuldt et al., 2011).

Discussion

While theory and prior research can guide us on communication strategies appropriate for publics with different beliefs and issue involvement, real-world communication presents us with audiences containing multiple publics. This challenge may be addressed in several ways:

1. Examination of the channels most used by particular segments permits targeting to some extent: The *Alarmed* are unlikely to listen to Rush Limbaugh, or the *Dismissive* to view Jon Stewart. Building opinion leadership among the *Alarmed* may be best accomplished through specially focused channels, such as environmental magazines, email newsletters, and social media postings by environmental, scientific, and social action organizations. A strategy employed by a number of organizations is to ask those who have signed a petition or made an online donation to repost the original request they received on Facebook or to email it to their friends and families, encouraging them to act as opinion leaders, fostering interpersonal (although mediated) communication, and broadening the original message's impact.
2. Reaching the middle segments is likely to require the use of channels that have a broad, mass audience. Low involvement strategies are most likely to be effective in these channels, as they have demonstrated efficacy across audiences.

3. Messages should be layered, including both efficacy-building and threat content. The low involvement publics need to be taught the danger posed by climate change, but placing too much emphasis on the threat may lead to defensive avoidance and despair among the *Alarmed* and *Concerned*, who already understand the threat and are fearful. It has sometimes been suggested that threat information should be dropped altogether—that the audience has heard enough about the threat and positive, efficacy-building messages are sufficient. A recent meta-analysis finds, however, that both risk perceptions and efficacy beliefs are necessary to motivate action (Peters et al., 2013).

There remains a gulf between the communication strategies we have suggested and the actual crafting of effective messages. Communicators are often advised, for example, to frame messages in ways that are consistent with the values and beliefs of the audience. A recent effort to influence *Dismissives* using a national security frame backfired, however: Although national security is prized among the *Dismissive*, a short essay describing the national security threat posed by global warming resulted in anger, rather than persuasion (Myers et al., 2012). Subsequent surveys have found that *Dismissives* simply do not believe this to be the case, and the essay is likely to have fostered counter-arguing, resulting in boomerang effects. A public health frame was more effective, across segments.

The time window within which we can act to prevent the most severe impacts of climate change is steadily shrinking; scientists in a range of disciplines have shown the impacts that are occurring and that will occur in the absence of action. It's now up to social scientists to find the communication tools to spur people to action. In light of the urgency, studies on effective climate communication should be topping our field's research agenda.

Appendix

Measures of key beliefs and issue involvement

Figures 31.6 through 31.11 show the proportions of respondents with:

1. *certainty that global warming is happening*: "Extremely sure" or "very sure" global warming is happening;
2. *Risk perceptions*:
   - *amount of harm*: Their families and future generations will be harmed "a great deal" or "a moderate amount";
   - *timing of harm:* People in U.S. are being harmed now;
3. *human causation*: "Assuming global warming is happening," it is caused mostly by human activities;
4. *solvability*: Humans could reduce global warming, but it's unclear whether we will *or* we can reduce global warming and will;
5. *scientific agreement*: Most scientists think global warming is happening;
6. *prior thought*: Say they have thought "a lot" about global warming before today;
7. *opinion certainty*: Strongly disagree with statement: "I could easily change my mind about global warming";
8. *need for information*: Say they do not need any more information to form a firm opinion on global warming.

Further reading

Maibach, E., Leiserowitz, A., Roser-Renouf, C., and Mertz, C. K. (2011). Identifying like-minded audiences for climate change public engagement campaigns: An audience segmentation analysis and tool development. *PLoS ONE, 6*(3), e17571.

Maibach, E., Roser-Renouf, C., and Leiserowitz, A. (2009). *Global Warning's Six Americas 2009: An Audience Segmentation Analysis*. Available at: http://climatechange.gmu.edu

Myers, T., Nisbet, M., Maibach, E., and Leiserowitz, A. (2012). A public health frame arouses hopeful emotions about climate change. Climatic Change, *113*, 1105–1112.

References

Bandura, A. (1986). *Social Foundations of Thought and Action: A Social Cognitive Theory*. Englewood Cliffs, NJ: Prentice Hall.

Cialdini, R. (2003). Crafting normative messages to protect the environment. *Current Directions in Psychological Research*, *12*(4), 105–109.

Cialdini, R. B. and Trost, M. R. (1998). Social influence: Social norms, conformity and compliance. In Gilbert, D. T., Fiske, S. T., and Lindzey, G. (eds) *The Handbook of Social Psychology*, Vols. 1 and 2 (4th edn). New York: McGraw-Hill, 151–192.

Ding, D., Maibach, E., Zhao, X., Roser-Renouf, C., and Leiserowitz, A. (2011). Support for climate policy and societal action are linked to perceptions about scientific agreement. *Nature Climate Change*, *1*, 462–466.

Green, M. and Brock, T. (2000). The role of transportation in the persuasiveness of public narratives. *Journal of Personality and Social Psychology*, *79*, 701–721.

Hart, S. and Nisbet, E. (2012). Boomerang effects in science communication: How motivated reasoning and identity cues amplify opinion polarization about climate mitigation policies. *Communication Research*, *39*(6), 701–723.

Kahan, D. and Braman, D., (2006). Cultural cognition and public policy. *Yale Law and Policy Review*, *24*, 149.

Kahan, D., Jenkins-Smith, H., and Braman, D. (2011). Cultural cognition of scientific consensus. *Journal of Risk Research*, *14*, 147–174.

Kahan, D., Peters, E., Wittlin, M., Slovic, P., Ouellette, L. L., Braman, D., and Mandel, G. (2012). The polarizing impact of science literacy and numeracy on perceived climate change risks. *Nature Climate Change*, *2*(10), 732–735.

Krosnick, J. A., Holbrook, A. L., Lowe, L., and Visser, P. S. (2006). The origins and consequences of democratic citizen's policy agenda: A study of popular concern about global warming. *Climatic Change*, 77, 7–43.

Kunda, Z. (1990). The case for motivated reasoning. *Psychological Bulletin*, *108*, 480–498.

Lewandowsky, S., Gignac, G., and Vaughan, S. (2013). The pivotal role of perceived scientific consensus in acceptance of science. *Nature Climate Change*, *3*, 399–404.

Maibach, E., Leiserowitz, A., Roser-Renouf, C., and Mertz, C. K. (2011). Identifying like-minded audiences for climate change public engagement campaigns: An audience segmentation analysis and tool development. *PLoS ONE*, *6*(3), e17571.

Maibach, E., Roser-Renouf, C., and Leiserowitz, A. (2009). *Global Warning's Six Americas 2009: An Audience Segmentation Analysis*. Available at: http://climatechange.gmu.edu (accessed 23 January 2015).

Myers, T., Maibach, E., Roser-Renouf, C., Akerlof, K., and Leiserowitz, A. (2013). Personal experience or belief in the reality of global warming: Which comes first? *Nature Climate Change*, *3*, 343–347.

Myers, T., Nisbet, M., Maibach, E., and Leiserowitz, A. (2012). A public health frame arouses hopeful emotions about climate change. *Climatic Change*, *113*, 1105–1112.

Nisbet, M. and Kotcher, J. (2009). A two-step flow of influence? Opinion-leader campaigns on climate change. *Science Communication*, *30*(3), 328–354.

Peters, G., Ruiter, R., and Kok, G. (2013). Threatening communication: A critical re-analysis and a revised meta-analytic test of fear appeal theory. *Health Psychology Review*, 7 (sup1), S8–S31.

Petty, R. and Brinol, P. (2012). A multi-process approach to social influence. In Kendrick, D. R., Goldstein, L. J, and Braver, S. L. (eds). *Six Degrees of Social Influence: Science, Application, and the Psychology of Robert Cialdini*. New York: Oxford University Press, 49–58.

Petty, R., Brinol, P., and Priester, J. (2009). Implications of the elaboration likelihood model of persuasion. In Bryant, J. and Oliver, M. B. (eds), *Media Effects: Advances in Theory and Research*. New York: Taylor & Francis, 125–164.

Roser-Renouf, C., Maibach, E., Leiserowitz, A., and Zhao, X. (2014). The genesis of climate change activism: From key beliefs to political action. *Climatic Change*, *125*(2), 163–178.

Schuldt, J., Konrath, S., and Schwarz, N. (2011). "Global warming" or "climate change"? Whether the planet is warming depends on question wording. *Public Opinion Quarterly*, *75*(1), 115–124.

Slater, M. and Rouner, D. (2002). Entertainment-education and elaboration likelihood: Understanding the processing of narrative persuasion. *Communication Theory*, *12*(2), 173–191.

Taber, C. and Lodge, M. (2006). Motivated skepticism in the evaluation of political beliefs. *American Journal of Political Science*, *5*(3), 755–769.

Venkatraman, M. (1989). Opinion leaders, adopters, and communicative adopters: A role analysis. *Psychology and Marketing*, 6(1), 51–68.

Witte, K. and Allen, M. (2000). A meta-analysis of fear appeals: Implications for effective public health campaigns. *Health Education and Behavior*, 27(5), 591–615.

PART V

Conclusions

Future trajectories of environment and communication

32

BEYOND THE POST-POLITICAL ZEITGEIST

Pieter Maeseele

Introduction

This chapter is the product of an invitation to write a focused and incisive contribution on interesting new perspectives and emphases for potential future trajectories in the field of environment and communication. The perspectives and emphases I will advance in this chapter, however, are not consistent with what is generally associated with 'new', an overvalued label that immerses twenty-first century public and academic discourse, and communication and media studies in specific (Murdock 2004). To the contrary, this chapter calls for reinvigorating classic sociological concerns and approaches in research on environment and communication, which have lost their glamour during the last decades (Hansen 2011). Instead of calling for the analysis of specific understudied or 'new' media outlets or communication practices (such as online representations, user generated content, persuasive communication, games, multimedia platforms, etc.), technologies (synthetic biology, nanobiotechnology, etc.) or environmental risks (fracking, fine dust particles, etc.), it calls for a reorientation of research aims and questions towards the social roles of media in liberal democratic societies and the relationship between media(ted) discourses, power and democratic politics.

More specifically, an analytical framework will be put forward that allows conclusions to be drawn on the contribution of public discourse(s) to facilitating democratic debate and citizenship, and as a result, on how to communicate more effectively from the perspective of democratic politics. A defining characteristic of this analytical framework, i.e. the risk conflicts-perspective, is its *politicization* of research in the field of environment and communication: its respective conceptual, methodological and empirical choices are aimed at accommodating research designs to function as spaces for conflict and dissent to be expressed and registered.

To this end, it is urgent to integrate literature from political theory on agonistic democracy and post-politics. I argue in this chapter that public and academic discourse on the environment is deeply characterized by the post-political zeitgeist that has swept through Western societies these last decades, with consensus and de-politicization as fundamental logics. In that respect, the existing work of an emergent Belgian school is broadened and deepened by applying its insights to the field of environment and communication.

In the first section of this chapter, this post-political zeitgeist is characterized in general, before going deeper into existing work on the environment in specific. From this reading, three conditions are identified for developing a research agenda regarding the analysis of the contribution of public discourse to democratic debate and citizenship. In the second section, the extent to which academic discourse[1] in the field of environment and communication conforms to these conditions is investigated, by evaluating existing communication models and research literature on public discourse regarding climate change, respectively. In the third section, the risk conflicts-perspective is introduced, by focusing on its conceptual, methodological and empirical implications.

The post-political zeitgeist

The literature on post-politics has been developed mainly by authors of Belgian origin. While political theorist Mouffe has been one of the leading theoretical figures in articulating the post-political condition, others such as Swyngedouw, Goeminne, Kenis and Lievens have recently applied this work to the environment.

The post-political condition

The post-political zeitgeist refers to how the post-Cold War period has generally been approached in public and academic discourse as the arrival of a 'post-ideological' era, characterized by the belief in a universal rational consensus, with experts reconciling conflicting interests and values through impartial procedures and technical knowledge (Fukuyama 1992, Giddens 1994). A particular school of political philosophers (Mouffe 2005, Rancière 1998, Žižek 1999), however, criticizes this conceptualization as embodying not a 'post-ideological', but a 'post-political' or 'post-democratic', condition, characterized by the de-politicizing nature of the politics of consensus. The post-political or post-democratic condition refers to how the essence of democratic politics, i.e. the confrontation of hegemonic political projects, is abandoned in favour of a *de-politicized* technocratic management of social, economic and ecological matters within the framework of an *inevitable* hegemonic neoliberal project and global market forces.

In this process of *de-politicization*, the political is transformed from a matter of ideological contestation to a matter of administration, where decision making is not a question of political position but of expert knowledge. As a result, technocratic decision making and market considerations come to substitute properly political debate and the notion of (political) conflict is reduced to political bickering: the political is only addressed at the ontic or empirical level and not at the ontological level. As a result, a democratic struggle between alternative (e.g. economic, technological or socio-ecological) futures beyond the existing socio-political status quo, and more specifically the continuing neoliberalization of all social spheres, is foreclosed.

Mouffe in particular (2005) has argued how antagonism and conflict are constitutive, not only of the social condition, but more importantly, of *democratic politics*. She argues that any form of *consensus* is always based on acts of exclusion, entailing the naturalization of particular power relations. The politics of consensus turn anyone who disagrees with the consensus into a fundamentalist, traditionalist or blind radical, through a *moralization* and *rationalization* of politics. This implies that the construction of the us/ them opposition in political categories constitutive of democratic politics, is, respectively, replaced by the moral categories of 'good' versus 'evil', or neutralized by striving for a consensus reached by 'rational' argumentation between 'rational' experts. These processes eliminate not only the existence of fundamental (political) oppositions, but also of (political) adversaries by turning these into enemies of the consensus. Consequently, the dichotomy politicization versus de-politicization serves as a framework for revealing strategies of in- and exclusion, and more specifically, how processes of de-politicization separate *legitimate*, *responsible* actors and demands from *illegitimate, irresponsible* actors and demands, excluding the latter from democratic debate. Furthermore, this dichotomy also serves as a framework for revealing the hegemonic constitution of society, and more specifically, the role of power relations in the construction of particular forms of objectivity.

To move from a politics of consensus to democratic politics, or from de-politicization to politicization, implies rethinking the properly political and re-establishing the horizon of democratic politics by organizing spaces for the agonistic confrontation of competing alternative futures and fundamental politico-ideological conflict, in academic as well as public discourse. In other words: by the creation of a genuine political space of disagreement overall.

The post-political environmental consensus

During the last decade, an emerging literature has been concerned with identifying symptoms of the post-political condition within discourses on the environment.

Environmental geographer Swyngedouw (2007, 2010) has been a leading figure in this regard, arguing how discourses on sustainability, nature and the environment have served as a key arena for the configuration, entrenchment and consolidation of the post-political condition. He primarily takes aim at the singular view of Nature as a harmonious equilibrium underlying existing discourses on sustainability. This (predefined consensual) concept of Nature turns sustainability into an empty signifier and reduces the politics of sustainability to a negotiation about the technomanagerial fixes at our disposal to 'save' nature from current 'unsustainable' paths: in other words, to retrofit nature to an apparently benign former status quo. By concealing the competing imaginations mobilized by various social actors, this idea of nature precludes democratic debate about the kind of nature we would like to inhabit and how this can be achieved, since it inhibits the articulation of alternative socio-ecological futures beyond the liberal–capitalist order. Swyngedouw calls for turning the question of sustainability (and nature and environment) into a question of democracy, by creating spaces for the recognition *and* cultivation of conflict about the naming and trajectories of competing socio-environmental futures.

When it comes to climate change discourse in specific, Swyngedouw (2010) takes aim at its consensual presentation and mainstreaming as the struggle to stabilize rising CO_2 concentrations in terms of a global humanitarian cause. He argues how this consensual framing is sustained by apocalyptic imaginaries and ecologies of fear, a particular science-politics short-circuiting procedure and the reification and commodification of CO_2. In these processes, scientific expertise is put forward as the only legitimate foundation for policy making, which is narrowed down to an issue of rationality claims. Furthermore, the framing of climate change in terms of a struggle of 'us' versus 'CO_2' represents climate change as a universalizing and socially homogenizing threat to humanity and externalizes and objectifies CO_2 as the enemy. Since this disavows social conflicts and antagonisms, obfuscates structural inequalities and eliminates any space for dissent, these de-politicizing processes preclude democratic debate since climate change is disassociated from alternative political programmes or socio-ecological futures from which to choose, while constructed as remarkably fit for technomanagerial machinery.

Physicist and Science and Technology Studies (STS) scholar Goeminne (2010, 2012) elaborates Swyngedouw's arguments by focusing on how the consensual focus on the scientifically registered level of CO_2 emissions in UN climate politics[2] sustains a techno-scientific and market-oriented framing of climate change that naturalizes the neoliberal foundations of the Western economic development paradigm and conceals the political question of which society we want to live in. In this scientization of environmental discourse (and resultantly, policy), Goeminne identifies two accumulative levels in which particular forms of objectivity are constructed in terms of separating internalities from externalities, determining who and what is to be taken into account.

First, in addition to narrowing the potential range of dispute to 'controversies between believers and non-believers … regarding the validity of the answers science provides' (2010: 212), this scientization brings in its wake the nature/society (ontologically) and fact/value (epistemologically) dichotomies. This results in the creation of a discursive space in which the (epistemic) superiority of rational decision making (legitimized by the authoritarian status of science and the efficiency of technological developments) is put against the (epistemically vacuous) inferiority of political judgement. The claim for rational decision making in 'consensual' climate policy making therefore functions as an exclusionary mechanism for anyone questioning the (neoliberal) alliance between science and policy.

Second, these dichotomies can only be overcome by starting from a constructivist rather than representational account of science (which is based on the alleged universal and non-exclusive character of scientific knowledge), since the recognition of science as a situationist, compositionist practice reveals the political dimension of scientific representation, in terms of the separation of internalities from externalities: in other words, dividing between what is taken into account and what is not in the construction of scientific knowledge. Goemmine thus calls for shifting the focus from 'a dispute over matters of fact in terms of true and false to a struggle for matters of concern in terms of internalities and externalities' (2012: 212), since this opens the discursive space for a political struggle over what to be concerned about. Only then, the conditions are fulfilled to revive the environment, and the climate in specific, 'as a matter of genuine political concern that is open to struggle and contestation between alternative visions of society, in this way constituting an essential component of social change' (213).

Elaborating these analyses, political ecologist Kenis and philosopher Lievens (Kenis and Lievens 2015) have recently identified a discursive shift with the rise of the Green Economy as a hegemonic project. Responding to what they identify as 'the quantum leap' in ecological awareness during this last decade, following Al Gore's *An Inconvenient Truth* (2006) and the hype surrounding the climate summit in Copenhagen (2009), this project calls for the mobilization of market mechanisms and capitalism's innovative nature for making the transition to sustainability, based on (i) technomanagerial innovation,[3] (ii) sustainable entrepreneurship (i.e. corporate social responsibility), and (iii) sustainable consumption (i.e. individual behaviour change). While the articulation and hegemonization of this project has initially been led by international institutions, governments, corporations, think tanks and banks, its success is based on (a discourse of all-round) collaboration of environmental NGOs and green parties. The Green Economy project aims to incorporate environmental protest and turn it into a new regime for capital accumulation, therefore Kenis and Lievens equate green economy to green capitalism, and link it to earlier discourses of ecological modernization, transition management and people planet profit. The authors argue how the success of this project depends on the all-round collaboration between former antagonists against a common enemy (CO_2), which in turn depends on active processes of de-politicization, which, they argue, the ecological issue lends itself more easily to for the following reasons: (i) the existing post-political context, (ii) the lack of a clear emancipatory actor (sustaining discourses of all-round cooperation, dialogue and consensus), (iii) the technical complexity of the problem and proposed solutions such as emissions trading, (iv) the conservative nature of the singular view of nature, and (v) a specific framing of climate change, primarily emphasizing the urgency, scale and nature of the threat, that favours pragmatic short-term solutions.

But more importantly, this discourse of all-round collaboration is to a significant degree sustained by the articulation of an antagonistic political relation towards those groups that deny the existence of a scientific consensus on the nature of climate change. With the aim of delegitimizing any fundamental government regulations or public interventions, this coalition of fossil fuel capitalists and climate skeptics contests and politicizes the epistemic level, resulting in a scientific non-debate replacing a proper political debate. However, what this latter coalition as well as the adherents of the Green Economy project have in common is an instrumentalization of science, reducing policy making to a matter of rationality claims, with profound de-politicization as a result: the reduction of climate change to an epistemic debate conceals the political positions underlying scientific positions and precludes a democratic debate between alternative futures beyond the existing socio-political status quo. Therefore Kenis and Lievens argue how the re-politicization of the environment and the climate will involve a twofold struggle: (i) the recognition of politico-ideological conflict as legitimate by opening up spaces of contestation and dissent, and (ii) the articulation of alternative futures to the Green Economy project.

In addition to these Belgian authors, British political theorist Machin published a book in the summer of 2013 in which she argues that only a radical democratic approach will allow the politics of climate change to be reinvigorated and produce the collective action needed to address climate change. She takes aim at what she puts forward as four dominant approaches for failing to facilitate any conclusive decision making, which she attributes to their common goal of inclusive agreement and rational discussion: (i) the techno-economic approach, based on technology and markets, (ii) the ethical-individual approach, based on developing good conscience, (iii) the green republican approach, based on the assertion of a common good by responsible citizens, and (iv) the deliberative democratic approach, based on the transcendence of disagreement through rational discussion and mutual understanding. On the other hand, she argues how only the celebration and encouragement of disagreement results in a real choice between real alternatives, fostering democratic debate and citizenship, and creating the conditions for collective as well as decisive action. In that respect, she calls for the recognition of 'nature' and 'climate' as *political* categories constructed within particular sociocultural imaginaries. A discursive construction of climate change in terms of an exclusionary scientific consensus on the other hand impedes democratic citizenship, since it encourages either political apathy by alienating people from owning the issue or polarization between acceptance and denial.

The return of the political

These authors have identified the following characteristics of the post-political condition in discourses on the environment: first, the de-politicizing nature of the politics of consensus precludes debate on the meaning(s) of the environment and resultantly on the articulation of alternative (environmental, technological, etc.) futures, by concealing the competing imaginations mobilized by social actors. This naturalizes the existing socio-political status quo and neoliberal foundations of the Western economic development paradigm in specific, and reduces the politics of the environment to a negotiation about potential technomanagerial fixes within this framework. Second, a rationalization of politics is sustained by a focus on the epistemic level and the assumption that the politics of the environment is a matter of translating a scientific consensus into a political consensus, thereby reducing policy making to a matter of rationality claims. Third, a moralization of politics is sustained by framing the environmental question as a global humanitarian threat, based on processes of universalization and social homogenization and the externalization and objectification of the problem. These processes of rationalization and moralization are deeply characterized by mechanisms of exclusion, since those actors and demands that either disagree with the scientific consensus or with framing climate change as an epistemic matter or a global humanitarian threat are stigmatized as enemies of the consensus. Fourth, by foreclosing the space for politico-ideological conflict, these processes result in precluding a democratic debate between alternative futures and in stifling democratic citizenship, since people are turned into passive spectators and not active participants in the articulation and shaping of alternative futures.

If we translate this work into a research agenda for the field of environment and communication, the central question becomes: to what extent do we find public discourse facilitating or impeding democratic debate and citizenship regarding environment and communication? To be able to provide an answer to this question, three conditions need to be fulfilled: (i) the recognition of an underlying politico-ideological conflict to each and every social issue, including controversies on the environment, (ii) the creation of a discursive space to reveal the nature and extent of politico-ideological struggle and dissent, and (iii) the identification within this space of those discursive strategies that aim at its foreclosure (de-politicization) or cultivation (politicization) and of related ideological preferences. In other words, to avoid the post-political trap, a politicization of academic discourse is called for. In what follows, we investigate the extent to which any of these conditions are fulfilled in existing communication models in the field of environment and communication and in the existing research literature of public discourse on climate change.

Environment and communication

Analytical perspectives

When talking about environment and communication, and media(ted) discourses on the environment in specific, we need to engage with the literature on media and science, since science is inevitably involved in discourses on the environment. I have argued before how this literature can be characterized in terms of two analytical perspectives that are based on different underlying communication models (Maeseele 2013a).

First, there is the science- and media-centric traditional approach, which is heavily influenced by the adjacent fields of science popularization, science communication and risk communication. This approach characterizes the relationship between science and society as a *problem* of communication and information: its conceptual and methodological focus is the efficiency of the *transmission* process between the 'scientific realm' on the one hand and (a presumably non-scientific) society on the other (Farago 1976, Krieghbaum 1967). As a result, research questions are reduced to measuring the level of accuracy in media reports or to measuring public knowledge of individual scientific facts. Since any dislocation in the relationship between media and environment is attributed to an inadequate transmission of scientific information and since scientific information as such is not problematized, while 'media' and 'the public' (or society in general) are, this is a media- and science-centric approach. Finally, it equates a lack of environmental progress to overcoming resistance towards specific scientific facts, either from special interest groups, professional mediators (e.g. journalists) or citizens. The solution, as always, is expected to be found in more and better science diffusion, media coverage and public understanding. Although many scholars in the field (mainly sociologists, historians of science and media sociologists) have been proclaiming the end of this conceptualization for two decades for its reliance on outdated communication and sociology of science models (Bucchi 1996, Meyer 2006), it remains the prevailing way of thinking: it displays a remarkable tenacity in official circles and public campaigns and is still prevalent in publications by science organizations and public institutions (Royal Society 2006), advocacy groups (Hartz and Chapell 1997) and science communication scholars (Willems 2003).

These last two decades, a more sociological approach has increasingly made its way into the academic field, with which the meaning-making practices in public discourse regarding the environment have become the central point of concern (Anderson *et al.* 2005, Nisbet and Huge 2006). Here, the conceptual and methodological focus is how public discourse functions as a site of contestation over different representations put forward by various (heterogeneous and unequal) social actors, not in the least over science's representations. This approach allows a much broader range of research questions, focusing either on understanding (i) how scientific claims are represented in public discourse and by whom, (ii) how this relates to issues of access to media and social debate in general, and (iii) how these discourses are interpreted and used by various audiences. In other words, the production, representation and audience reception levels become equally important levels to analyse the circulation, (re-)production and (re-)configuration of meanings about environmental issues. Furthermore, the mediatization of science itself, i.e. the professionalization of marketing practices, public relations and image management within science organizations, also becomes an object of critical concern.

A vital difference, however, between both approaches is their radically divergent interpretation of the relationship between science and society, and science and media in specific. Starting from a representational account of science, the traditional model is de-politicizing by definition: based on the assumption of a (predefined) scientific consensus, the *scientific* demands of *responsible* actors are distinguished from the (epistemically vacuous) *illegitimate* demands of *irresponsible* actors, shifting the site of struggle from the political level to the epistemic level in terms of dichotomies such as science versus anti-science, religion, ideology, politics, emotions, fear, etc. Furthermore, it serves to reify science and scientific authority in terms of a discourse of 'sound science' that is used as a rhetorical tool to isolate unacknowledged (and unaccountable) value-laden assumptions and material interests in competing demands. In so doing, the traditional model is concerned with strategies to safeguard institutionalized conceptions of environmental risk by distinguishing these from 'false' manifestations instead of 'alternative' ones, starting from an assumption of communication practices as the (in)efficient communication of predefined matters. As a result, the traditional model forecloses the discursive space for politico-ideological conflict by delegitimizing its potential in advance.[4] On the other hand, by starting from an assumption of communication practices as indefinite articulations of meaning, the sociological approach allows the discursive space for politico-ideological conflict to be created. Ultimately, while both interpretations represent a struggle between two politico-ideological models for interpreting the relationship between science and society, only the latter creates the discursive space for approaching public discourse as a site of politico-ideological conflict, while the aim of the former is exactly to delegitimize its acknowledgement. In the end, the politico-ideological struggle starts in the first instance in academic discourse on science and society, and science and media in specific, where the potential value of public discourse is determined, while only in the second instance in public discourse itself, by starting from those academic starting points that allow the ideological nature of public discourse to be revealed.

However, the existing research literature appears to suggest that the sociological approach's potential in this respect is dependent on specific conditions: in a recent review of the research literature (Maeseele 2013a), the institution of science was found to successfully adapt to the mediatization of society, to the extent of a relatively effective control of its public image in media(ted) discourses: for instance, this last decade, science reporting is found to be characterized by a largely affirmative, sometimes even hyperbolic nature that is often the result of the success of science PR. Furthermore, a high degree of satisfaction is found among scientists regarding their media contacts and media reporting in general. And last but not least, specific culturally dominant discourses such as technological progressivism, scientism and neoliberalism prevent critical sources from being accredited with legitimacy or critical stories to gain prominence. It is here, exactly, that important questions come forward concerning the social role of media in liberal democratic societies and the relationship between media, power and democracy. Since what is at stake is the creation of a discursive space for an ideological struggle between alternative futures, environmental issues should be approached from a framework of political conflict, entailing ideological discussion and collective debate and choice. Therefore, to avoid the post-political trap in sociological approaches, it is instructive to exceed a non-committal focus on meaning-making processes underlying media(ted) discourses, and develop a research agenda regarding the contribution of public discourse to democratic debate and citizenship. This requires an accommodation of research designs to function as spaces for conflict and dissent to be expressed and registered.

Media research on climate change

With the rise of climate change as an acknowledged and institutionalized global problem, there has been a similar rise of research in media and communication studies on the representation of climate change in public discourse. A recent review by Pepermans and myself revealed that much of the existing research literature primarily evaluates public discourse on the extent to which it either contributes to communicating a scientific consensus[5] or to achieving a social consensus (Pepermans and Maeseele 2014).

One important school within this literature starts from the assumption of a failed discursive translation between scientific and popular discourse (Antilla 2005, Boykoff and Boykoff 2004, Dispensa and Brulle 2003). Conceptually and empirically, most of these studies have in common that they distinguish between actors and demands informed by the scientific consensus, and others who are not, and blame the journalistic norm of objectivity that leads journalists to balance two contrasting positions, for effectively instilling bias (or: 'balance as bias', see Boykoff and Boykoff 2004). According to this school, this creates the undue perception that both sides present equally scientific claims, which confuses and misinforms the public, potentially delaying the necessary action to address climate change. Partly in response to this, a second school aims at 'saving' the consensus by focusing on how to overcome conflicting positions through reframing climate change in a specific way, which allows the scientific consensus to either be translated better or communicated successfully across ideological divides (Lakoff 2010, Nisbet 2009, Segnit and Ereaut 2007). The aim of a third school is not the translation of a scientific consensus, but how to achieve a social consensus and overcome conflicting values and interests in terms of creating the conditions for rational dialogue based on 'substantive' considerations (Jönsson 2012, Kumpu and Kunelius 2012, Malone 2009).

These three schools have in common that they start from exclusionary mechanisms discriminating between who/what is recognized as legitimate and who/what is recognized as illegitimate, thereby excluding those actors and/or demands that are not conforming to a (predefined) scientific or social consensus from democratic debate. However, considering their focus on an epistemic framing of climate change and their underlying desire for consensus, the respective analytical frameworks are not only incapable of recognizing or addressing these processes, but simultaneously contribute to them: in other words, their specific conceptual and empirical choices not only actively contribute to the *de-politicization* of climate change, but actually presume that (an 'unjustified') *politicization* is the problem to overcome.

Second, although most of these studies reject a transmission model of communication and take into account the wider social context and media logics, public discourse is nonetheless (explicitly or implicitly) evaluated on the extent to which it accurately represents a scientific consensus; therefore, the underlying relationship between science and society is characterized as a problem of communication and information. In other words, despite an often explicit epistemological and conceptual distance from the traditional model, there remains a clear normative affinity with it in academic discourse on environment and communication. On the other hand, the advocates of a consensus are always likely to prefer the traditional model, because of its politico-ideological functions (which, ironically, are to reject anything politico-ideological), its assumption of an instrumental role for media and communication and its equation of a lack of environmental progress to resistance towards specific scientific facts. So, although the research questions of these schools are inspired by the sociological approach, these schools somehow lament the fact that the conclusion of a relatively effective control of science's public image apparently does not sufficiently hold up in the case of climate change.

Two recent publications, however, succeed in exceeding these pitfalls by integrating insights from agonistic democratic theory. In their edited volume *Climate change politics: Communication and public engagement*, Carvalho and Peterson (2012) start from the observation that citizen engagement in climate change politics is remarkably low, therefore the various chapters investigate in what ways communication could be able to contribute to a transformation of politics. They identify three modes of public engagement, based on different underlying views of climate change communication and politics: (i) social marketing, (ii) public participation and (iii) agonistic pluralism. In their own words:

> Whereas social marketing and formal public participation are top-down managerial practices, citizen-led political participation is initiated from the bottom-up. Engagement starts with citizens who see faults in the ways formal political institutions deal with climate change and advance alternative forms of governance, whether through proposals for different governmental policies or through social and economic changes. This involves dissent over alternative political projects. The [agonistic pluralism] mode of engagement cultivates political conflict and rejects the viability of consensus between opposing viewpoints.
>
> (Carvalho and Peterson 2012: 12)

The impact of the first two approaches is seen to be limited in challenging the hegemonic technomanagerial approach to climate change: the social marketing approach individualizes responsibility and addresses people as consumers, thereby reducing the political realm to lifestyle choices, and the public participation approach often functions as an exclusionary legitimation tool and top-down approach to the production of consensus. The agonistic pluralist approach on the other hand is seen as conducive to helping to transform the socio-political status quo. The three chapters covering this approach in the volume subsequently focus on the role of art in questioning and subverting politically dominant discourses (Polli 2012), the role of alternative media in developing agonistic politics on climate change (Gunster 2012) and an analysis of the exclusion of voices and views of large parts of society from climate change politics and how to include them (Scandrett *et al.* 2012).

Lastly, Berglez and Olausson (2014) recently published an empirical exploration of how the post-political condition of climate change was discursively established in a 2009 focus-group study with Swedish citizens, by focusing on the ideological nature of their discourses. The authors identify three features counteracting radical political discourse capable of challenging the socio-political status quo: (i) emotional indifference when it comes to personal experiences of a changing climate, (ii) the fragmentation into various particular causes of climate change underlying the 'belief' in climate science (and not market capitalism as such) and (iii) individual responsibility and behaviour change. The authors conclude that only the recognition of market capitalism as the 'singular Cause' of climate change, will enable the articulation of alternative socio-environmental futures.

However, neither of these publications put forward an analytical framework on how to systematically identify processes of politicization and de-politicization in public discourse. Berglez and Olausson (2014: 2) recognize this deficit: 'This process of post-politicalization is rather well theorized but has seldom been substantiated with empirical evidence; thus, there is a need for discourse analyses that are able to empirically explore the discursive elements that function as post-political building blocks.' In the following section, an analytical framework is proposed that should answer their call.

The risk conflicts-perspective

In this section, the risk conflicts-perspective is put forward, which allows us to draw conclusions on the contribution of public discourse(s) to facilitating democratic debate and citizenship. It will be discussed in terms of its respective conceptual, methodological and empirical choices.

Conceptual choices

Since the primary condition to avoid the post-political trap is the recognition of an underlying politico-ideological conflict, this requires (i) to put forward environmental issues as a new type of *social conflict* in late modern societies, and (ii) to distinguish between (conflicting) *scientific rationality claims* on the one hand and (conflicting) *values* and *interests* on the other. These 'risk conflicts' then involve contestation between various social actors over conflicting risk definitions that are based on the confluence of conflicting (i) scientific rationality claims, (ii) values and (iii) interests. Depending on their respective material and/or ideological interests towards a given status quo, these social actors aim at either the politicization or de-politicization of an environmental issue. The *risk conflicts-perspective* then entails a Copernican revolution: instead of approaching controversies in terms of identifying the relevant science and its spokespersons versus those who distort 'the science' in promotion of 'special interests', it proposes to start from an identification of the competing values and interests at work, to arrive at the heart of what is at stake: an *ideological* struggle between *alternative* (technological or socio-ecological) *futures* that are based on competing analyses of the current and ideal state of affairs.

In elevating the struggle between processes of politicization and de-politicization as the main analytical concern, the legitimacy of the participants and the debate as such become a central issue: processes of de-politicization refer to discursive strategies in which legitimate, responsible actors and demands are distinguished from illegitimate, irresponsible actors and demands, based on the assumption of a (predefined) moral or rational consensus. In these processes, *moral* or *rational* demands of responsible actors are distinguished from *radical* epistemically vacuous concerns of irresponsible actors, thereby stigmatizing adversaries as enemies of an existing or potential consensus. These moral or rational imperatives effectively shift the *site of struggle* from politico-ideological conflict between alternative futures to a struggle between 'good' and 'evil' or 'rational' and 'irrational' (or 'scientific' and 'unscientific', or 'economic' and 'uneconomic'), and in doing so, act in the service of concealing rather than revealing what is at stake. Eventually, these processes of de-politicization aim to preclude an adversarial democratic debate in favour of consensual technocratic decision making and/or market forces. Processes of politicization, on the other hand, refer to discursive strategies that, instead of amplifying a moral or rational consensus, aim at revealing competing sets of rationality claims, values and interests underlying competing responses to uncertainty, and relate these to underlying alternative visions of society, which are subsequently made subject of public debate. This cultivates a discursive space for a democratic debate between conflicting, yet legitimate, demands of political adversaries, and a democratic struggle between alternative futures beyond the existing status quo.

Eventually, democratic debate is found to be facilitated when an environmental issue is framed as an (*ideological*) debate involving key *political* choices between *alternative* futures. On the other hand, democratic debate is found to be impeded when an issue is framed as a (*predefined consensual*) matter best left to *technocratic* decision making and/or *market* forces. In other words, when public discourse(s) contribute(s) to processes of *politicization* or *de-politicization*, respectively.

Methodological choices

To allow for the combination of an in-depth examination of discursive strategies and ideological preferences, a strong focus on language use and the relationship between discourses on the one hand and specific social, political and cultural contexts on the other is required. This implies that qualitative content analytic methods are preferred over quantitative content analytic methods, which generally aim at identifying the frequency of predefined thematic categorizations (often referred to as 'frames') and positive/negative evaluations of actors, technologies or demands. More specifically, critical discourse analysis allows us to reveal the role of discursive strategies and practices in the creation and reproduction of (unequal) relations of power (Carvalho 2008, Jørgensen and Phillips 2002, Maeseele 2013b, Raeijmaekers and Maeseele 2014). On the other hand, specific (critical) framing analysis approaches that start from a dynamic framing concept, could also be used in this respect (Maeseele 2010, 2011).

With regard to the role of ideology in public discourse, and media(ted) discourses in specific, we draw on the work of Carvalho (2007, 2008). She defines ideology as 'a system of values, norms and political preferences, linked to a program of action vis-à-vis a given social and political order' (Carvalho 2007: 225). Furthermore, she puts forward a concept of ideological cultures that emphasizes their socially constructed nature and allows a certain level of diversity and pluralism. Ideological cultures then refer to 'communities of ideas, values and preferences inside media organizations and in their particular audiences' (Carvalho 2007: 239–240). To be able to reveal the respective ideological cultures, four subsequent steps are needed: in a first step, the relevant ideological fault lines are identified. In the case of environment and communication, the most relevant fault lines as identified in research on climate change and GM food have been the techno-environmental and socio-economic fault lines (Carvalho 2007; Maeseele 2010). In the former, a Promethean discourse in which there are no limits to the mastery and exploitation of nature for material development or economic growth is distinguished from a discourse of harmony with and respect for nature and other species. In the latter, a non-regulatory approach driven by values of market liberalism, individual freedom and profitability is distinguished from a public accountability approach, in which the precautionary principle (i.e. technological innovations should not be commercialized before risks and dangers are sufficiently known), social responsibility and equity warrant political action. In a second step, the relevant discursive strategies are identified, such as framing (selection and composition), positioning, homogenization, universalization, de/legitimation, etc., which allow us to decide on processes of de/politicization. A third step identifies which ideological preferences and discursive strategies relate to each other and groups these as the respective ideological cultures. In a potential fourth step, the dominant problem definition of an environmental issue is induced from the ideological culture characterized by de-politicizing discursive strategies, indicating the naturalized ideological preferences.

Empirical choices

In terms of empirical framework, we acknowledge the urgent need recently formulated by Hansen (2011: 8): 'for reconnecting and reintegrating the traditional, but traditionally also relative distinct, three major foci of communication research on media and environmental issues.' This means that the traditional reflexive circuit between social actors, media(ted) discourses and audience discourses not only remains as relevant as ever, but in fact appears as most promising to reveal the material needed to draw conclusions on the contribution of media(ted) discourses to facilitating or impeding democratic debate and citizenship. Regarding the level of social actors, it is important to identify the respective strategic actors (those individuals or groups with interests in an issue's representation) and to reveal from their strategic communication documents not only whether they sponsor politicizing or de-politicizing discourses, but more importantly, the ideological nature of these discourses. In addition, their communication activities and media strategies should be mapped so as to gain insights into the nature and success of their claims-making efforts. Regarding the media level, it is important to examine which social actors and discourses influence or fail to influence media(ted) discourses and to reveal the nature and extent of ideological discussion. Here it is important to include a wide range of media, such as new alternative, online media, since these are most likely to produce alternative, i.e. politicizing, discourses. Preferably, the analysis is restricted to 'critical discourse moments', selected cases marked by topical (scientific, political, etc.) events that have the potential to transform the configuration of meaning. With respect to timescales, these cases should be analysed along two axes: a comparative-synchronic axis (simultaneous coverage in different media) and a historical-diachronic axis (consecutive coverage in one media outlet). Lastly, to seize the significance of these discourses, audience discourses should be studied using reception analyses, with a focus on the extent to which media users succeed in making media(ted) discourses relevant on a personal and/or political-societal level. Eventually, once we are able to draw conclusions on the contribution of public discourse to democratic debate and citizenship regarding an environmental issue, it is fairly straightforward to conclude on how to communicate this issue more effectively from the perspective of democratic politics.

Conclusion

The literature on agonistic democratic theory allows us to identify three conditions required for a research agenda on the evaluation of public discourse with regard to its contribution to facilitating democratic debate and citizenship. Since the articulation of alternative futures depends on conflict about the naming and trajectories of competing futures, the first condition, the recognition of an underlying politico-ideological conflict, is found to be the primary factor in shifting the site of struggle from the epistemic level to the political level, academically as well as socially. Furthermore, the evaluation of public discourse in this respect is found to be dependent on a corresponding *politicization* of academic discourse, since this research agenda requires fundamental conceptual and methodological shifts: only when nature and the environment are recognized as political categories, resisting singular views or framings exclusively in terms of scientific rationality claims, a real confrontation between alternative futures becomes possible.

Notes

1 In this chapter, the concept of academic discourse is used to refer to scholarly discourses, while public discourse is used to refer to discourses circulating in the public sphere, such as media(ted) discourses, citizen discourses, strategic communication by organizations, etc.

2 While the IPCC reports have played a vital role in framing climate change in terms of a scientific battle against CO_2 emissions (IPCC reports), the annual climate summits have narrowed the politics of climate change to debates on particular technologies and market mechanisms.

3 Geo-engineering, carbon capture and storage, nuclear power, biofuels, etc.

4 Interestingly, the inflation and tenacity of the traditional model in the guise of the Public Understanding of Science movement since the mid 1980s has been explained in terms of a response of the scientific establishment to a widely perceived legitimation vacuum and crisis in public trust in a period in which the commercialization of science took off in leading areas of biotechnology (Durant 1999, Wynne 1995). This is ironic, since its politico-ideological function is exactly to naturalize the existing institutionalized culture of science in terms of its representation, organization, patronage, control and social relations, while systematically deleting the institutional and epistemic characteristics of dominant forms of science (Wynne 1995).

5 Presuming that public discourse accurately represents what is commonly put forward as the established consensus within climate science regarding the anthropogenic nature of climate change.

References

Anderson, A., Allan, S., Petersen, A. and Wilkinson, C. (2005) 'The framing of nanotechnologies in the British newspaper press', *Science Communication*, 27: 200–220.

Antilla, L. (2005) 'Climate of scepticism: US newspaper coverage of the science of climate change', *Global Environmental Change Part A*, 15: 338–352.

Berglez, P. and Olausson, U. (2014) 'The post-political condition of climate change: An ideology approach', *Capitalism Nature Socialism*, 25 (1): 54–71.

Boykoff, M. and Boykoff, J.M. (2004) 'Balance as bias: global warming and the US prestige press', *Global Environmental Change*, 14: 125–136.

Bucchi, M. (1996) 'When scientists turn to the public: Alternative routes in science communication', *Public Understanding of Science*, 5: 375–394.

Carvalho, A. (2007) 'Ideological cultures and media discourses on scientific knowledge: Re-reading news on climate change', *Public Understanding of Science*, 16: 223–243.

Carvalho, A. (2008) 'Media(ted) discourse and society: Rethinking the framework of Critical Discourse Analysis', *Journalism Studies*, 9: 161–177.

Carvalho, A. and Peterson, T.R. (eds) (2012) *Climate change politics: Communication and public engagement*, Amherst, NY: Cambria Press.

Dispensa, J.M. and Brulle, R. (2003) 'Media's social construction of environmental issues: Focus on global warming – A comparative study', *International Journal of Sociology and Social Policy*, 23 (10): 74–105.

Durant, J. (1999) 'Participatory technology assessment and the democratic model of the public understanding of science', *Science and Public Policy*, 26: 313–319.

Farago, P. (1976) *Science and the media*, Oxford: Oxford University Press.

Fukuyama, F. (1992) *The end of history and the last man*, New York: Avon Books.

Giddens, A. (1994) *Beyond left and right: The future of radical politics*, Redwood City, CA: Stanford University Press.

Goeminne, G. (2010) 'Climate policy is dead, long live climate politics!', *Ethics, Place and Environment*, 13 (2): 207–214.

Goeminne, G. (2012) 'Lost in translation: Climate denial and the return of the political', *Global Environmental Politics*, 12 (2): 1–8.

Gunster, S. (2012) 'Visions of climate politics in alternative media', in A. Carvalho and T.R. Peterson (eds) *Climate change politics: Communication and public engagement*, Amherst: Cambria Press, pp. 239–267.

Hansen, A. (2011) 'Communication, media and environment: Towards reconnecting research on the production, content and social implications of environmental communication', *The International Communication Gazette*, 73: 7–25.

Hartz, J. and Chapell, R. (1997) *Worlds apart: How the distance between science and journalism threatens America's future*, Nashville, TN: First Amendment Center.

Jönsson, A.M. (2012) 'Climate governance and virtual public spheres', in A. Carvalho and T.R. Peterson (eds) *Climate change politics: Communication and public engagement*, Amherst: Cambria Press, pp. 163–192.

Jørgensen, M.W. and Phillips, L.J. (2002) *Discourse analysis as theory and method*, London: Sage.

Kenis, A. and Lievens, M. (2015, forthcoming) 'Greening the economy or economizing the green project? When environmental concerns are turned into a means to save the market', *Review of Radical Political Economics*.

Krieghbaum, H. (1967) *Science and the mass media*. New York: New York University Press.

Kumpu, V. and Kunelius, R. (2012) 'Attention, access and dialogue in the global newspaper sample: Notes on the dependency, complexity and contingency of climate summit journalism', in E. Eide and R. Kunelius (eds), *Media Meets Climate. The Global Challenge for Journalism*, Götheborg: Nordicom, pp. 313–330.

Lakoff, G. (2010) 'Why it matters how we frame the environment?', *Environmental Communication: A Journal of Nature and Culture*, 4 (1): 70–81.

Machin, A. (2013) *Negotiating climate change: Radical democracy and the illusion of consensus*, London: Zed Books.

Maeseele, P. (2010) 'On neo-luddites led by ayatollahs: The frame matrix of the GM good debate in Northern Belgium', *Journal of Environmental Communication: A Journal of Nature and Culture*, 4: 277–300.

Maeseele, P. (2011) 'On news media and democratic debate: Framing agricultural biotechnology in Northern Belgium', *International Communication Gazette*, 73: 83–105.

Maeseele, P. (2013a) 'On media and science in late modern societies', in E. Cohen (ed.), *Communication Yearbook 37*, New York: Routledge, pp. 154–181.

Maeseele, P. (2013b) 'Risk conflicts, critical discourse analysis, and media discourses on GM crops and food'. Available online at http://jou.sagepub.com/content/early/2013/12/09/1464884913511568.abstract (accessed 11 November 2013).

Malone, E. (2009) *Debating climate change: Pathways through argument to agreement*, London: Earthscan.

Meyer, G. (2006) 'Journalism and science: How to erode the idea of knowledge', *Journal of Agricultural and Environmental Ethics*, 19 (3): 239–252.

Mouffe, C. (2005) *On the political*, London: Routledge.

Murdock, G. (2004) 'Past the posts: Rethinking change, retrieving critique', *European Journal of Communication*, 19 (1): 19–38.

Nisbet, M. (2009) 'Communicating climate change: Why frames matter for public engagement', *Environment magazine*, 51 (2): 12-23.

Nisbet, M. and Huge, M. (2006) 'Attention cycles and frames in the plant biotechnology debate: Managing power and participation through the press/policy connection', *Harvard International Journal of Press/Politics*, 11 (2): 3–40.

Pepermans, Y. and Maeseele, P. (2014) 'Democratic debate and mediated discourses on climate change: from consensus to de/politicization', *Environmental Communication: A Journal of Nature and Culture*, 8 (2): 216–232.

Polli, A. (2012) 'Art and political contestation in climate issues : Who owns the air', in A. Carvalho and T.R. Peterson (eds) *Climate change politics: Communication and public engagement*, Amherst: Cambria Press, pp. 212–246.

Raeijmaekers, D. and Maeseele, P. (2014) 'Ideologisch pluralisme in Vlaamse nieuwsmedia: de Belgische regeringsformatie 2010–2011' [Ideological pluralism in Flemish news media: the 2010–2011 government formation]. *Tijdschrift voor Communicatiewetenschap*, 42 (2): 145–164.

Rancière, J. (1998) *La chair des mots: Politiques de l'écriture*, Paris: Galilée.

Royal Society (2006) *Science communication: Survey of factors affecting science communication by scientists and engineers*, London: Royal Society.

Scandrett, E., Crowther, J. and McGregor, C. (2012) 'Poverty, protest, and popular education: Class interest in discourses of climate change', in A. Carvalho and T.R. Peterson (eds) *Climate change politics: Communication and public engagement*, Amherst, NY: Cambria Press, pp. 277–306.

Segnit, N. and Ereaut, G. (2007) *Warm Words II: How the climate story is evolving and the stories we can learn for encouraging public action*, London: Energy Saving Trust.

Swyngedouw, E. (2007) 'Impossible "sustainability" and the post-political condition', in D. Gibbs and R. Kreuger (eds) *The sustainable development paradox: Urban political economy in the United States and Europe*, New York: Guilford Press.

Swyngedouw, E. (2010) 'Apocalypse forever? Post-political populism and the spectre of climate change', *Theory, Culture and Society*, 27: 213–232.

Willems, J. (2003) 'Bringing down the barriers', *Nature*, 422 (6931): 470–470.

Wynne, B. (1995) 'Public understanding of science', in S. Jasanoff, G.E. Markle, J.C. Petersen and T. Pinch (eds) *Handbook of science and technology studies*, Thousand Oaks, CA/London/New Delhi: Sage, pp. 361–388.

Žižek, S. (1999) *The ticklish subject: The absent centre of political ontology*, London: Verso.

33

WHITHER THE HEART(-TO-HEART)?

Prospects for a humanistic turn in environmental communication as the world changes darkly

Susanne C. Moser

Introduction

In decades hence, we will look back to the first years of the twenty-first century as the years when environmental crises accelerated, when the impacts of global environmental changes such as climate change shifted from being pervasive if intangible problems, to lived-and-felt, everyday experiences. We may well view these as the years when disasters turned from being horrific but rare exceptions to the even more heart-wrenching condition of "normal" life in a climate-altered world. Cox (2007) placed the rise of the professional field of environmental communication since the early 1980s into the context of environmental risks and degradation, and—in the face of currently accelerating environmental challenges—charged the field to serve as an ethically motivated "crisis discipline."

While this notion of a "crisis discipline" was welcomed by some and sincerely debated or even contested by others (e.g., Heath et al. 2007; Killingsworth 2007; Schwarze 2007; Senecah 2007), it would be hard to deny that much of what has been written under the flag of "environmental communication" in the late twentieth and early twenty-first century is not somehow motivated by or linked to an unease about environmental events, trends, problems, or dangers—however perceived. The rise of the subfield of climate change communication can certainly serve as "Exhibit A" for this claim (e.g., Boykoff 2011; Carvalho 2008; Carvalho 2010; Moser and Dilling 2004; Moser and Dilling 2007). From this perspective, the practice of environmental communication for many is "instrumental": it aims to inform or help mobilize a more effective societal response to these growing dangers. In turn, much of environmental communication research has aimed and become more adept at untangling the various aspects of the communication process in an effort to make it more effective. We have tracked changing perceptions and attitudes to better address our various audiences; we have identified and tested different framings, channels, messages and messengers to reach those who might influence public and policy debates; and we have unearthed a range of influences on the communication process to render it more helpful, timely, and influential. Even the more "constitutive" approach to environmental communication, which looks at communication as a symbolic act that helps humans place themselves vis-à-vis the other-than-human life world, can be read as an attempt to reckon with the human footprint on Earth. Over the 30 years since the field's inception, environmental communication has indeed matured significantly in doing all of this. With a well established technical vernacular in place, a strong set of methodologies to examine communication efforts, and growing geographic coverage of investigations that enables comparative insights into the importance of culture, context, and communication practices, the field of environmental communication has become increasingly sophisticated (progress and achievements to which the contributions in this *Handbook* pay tribute).

At the same time that environmental crises are becoming commonplace, and environmental communication has come into its own, a third trend is inescapable in our field and in our lives: the rise of the internet, the near-saturation of social media in public use, and profound technological and political-economic changes in the media industry (Brenner and Smith 2013; Rainie 2013; The Pew Research Center 2013). Communication has become faster, more distributed, more fragmented, and yet also more media-ted as a result. Dominick (2010) has well delineated the social implications of these developments, including the growing speed of "news," the lack of gatekeepers sorting through the abundance, yes, overload of information, growing privacy concerns, the emergence of media use as escapism, and, disconcertingly, the growing social isolation despite virtual connectedness (for a visual commentary on just this effect of social media, see Cohen 2013). Environmental communication practice, without critically questioning this trend, has instead fully embraced it. These developments in technology, research, and practice entail a certain degree of reification, of distancing from that which we study and do: humans trying—sometimes desperately—to connect with each other by way of words, images, gestures, and touch.

It is not unreasonable then to ask whether we in the environmental communication field may be losing touch with the very heart of communication at a crucial time. Despite all our communication options and opportunities, despite our skill and sophistication, are we still serving the deepest purpose of all communication, namely to exchange ideas and information, to hear and be heard, to create understanding and foster connection among us (some would extend the circle beyond humans (Peterson et al. 2007)), and, ultimately, to ensure survival? This question becomes ever more important to ask of the kind of communication needed most as environmental changes, disasters, and continual degradation of our life world take on a global scale. In such a time, I would submit, what is called for first and foremost is not persuasion, education, and deliberation (though none of these will lose in importance), but kind and compassionate human support. Not conversion but respect and dignity. Not a battle of the minds, but a meeting of the hearts.

In what follows, I will argue that the two major trends introduced above—the increasing frequency of environmental crises and the pervasiveness of technology-based communication—open up a gap, a profound need, that an environmental communication oriented toward human welfare and connection may be able to meet. I call such an environmental communication "humanistic" and offer it here as a promising future direction for our field. In the section below, I begin by making the case for how environmental crises are beginning to emerge in our collective experience. Next, I define and sketch the outlines of such a "humanistic" environmental communication, and then focus in on how it may serve a society increasingly in dire environmental straits. I will close with an appeal to both environmental communication researchers and practitioners to issue not just warnings and clarion calls to action but to partake in the restoration of our relationships to each other and between ourselves and the more-than-human world.

A world changing darkly

At the time of this writing, an American icon—Yosemite, the nation's oldest national park—is engulfed in flames. One of the largest wildfires in California history, in one of the worst fire seasons ever, in the context of a series of critical drought years in the state, is yet another "natural" disaster impinging on the national psyche. Going back in time through the last few years, the litany is almost biblical: extreme drought followed by floods across the mid section of the country, Superstorm Sandy in Fall 2012 unfurling its fury on the Atlantic seaboard—by many viewed as a "game changing" extreme event with lasting impacts not only on those directly affected, but on public and policy conversations in the US—and devastating tornadoes before and after that. The list goes on with record numbers of multibillion dollar events in 2011, and before that in 2010 (Smith and Katz 2013) (for ongoing updates see: www.ncdc.noaa.gov/

billions). The previous "watershed event," Hurricane Katrina in 2005, which caused more deaths than any other natural disaster in the US since the Galveston Hurricane in 1906, still echoes darkly through the media, policy debates, and environmental communication.

And that is just the picture within the US. Elsewhere, the story is equally heart-breaking and disconcerting: devastating floods in Canada and Europe—yet again—in 2013; record-breaking heat waves during the "angry" Austral summer of 2013 and long-lasting droughts with related agricultural and urban water shortages and bush fires in Australia in the years before that; in several recent years in Europe and Russia deadly cold winters; and extensive flooding from typhoons and monsoon rains, if they came at all, in Asia. How exactly all these events are related to climate change remains a hotly debated topic in science at this time, yet they all coalesce into a picture of a world increasingly perturbed, a world increasingly out of control.

To be clear, this description of havoc is not to be read as a sloppy statement about the causal links between every extreme event and climate change. Contemporary climate change did not "invent" extreme events, yet it is systemically changing the conditions for them. Careful work is being done by physical scientists to detect trends among rare events, to discern how the systemic changes relate to specific weather events, and which of these events can be attributed to human-caused global warming (Seneviratne et al. 2012). What we know is that this work is made difficult not just by the challenge of detecting trends among rare events in the midst of an always "noisy" climate, but by the fact that extreme events have more devastating impacts now because more people and more human-built structures are located in harm's way. In addition, our observation systems vary in spatial coverage and quality but many have become better over time, thus providing simply more data for recent times than for the past. And media attention to catastrophes is always a headline-grabbing business opportunity. All of these factors play into why it "seems" like there are more "catastrophes" now than there used to be. In some instances that has indeed been shown, in others uncertainty prevails—for now (Coumou and Rahmstorf 2012; IPCC 2012; MunichRe 2013; Peterson et al. 2012; Showstack 2013).

The point of this litany of extremes, instead, is precisely in the fact that "it seems" to us that change—undesirable change—is afoot. Both people's own, direct experience and the mediated communication about catastrophic events create a collective sense that something "weird" is going on in the world. Apocalypse, as Frederick Buell (2003) put it, is becoming a way of life.

And in fact, there is empirical evidence for a growing sense of unease, maybe even of doom, in the public; that people are "connecting the dots" between all these extreme events—almost in spite of scientists' ever-so-careful, scientifically correct, and yet sometimes evasive-sounding, attempts to not link any single disaster directly to climate change. For example, researchers in Europe, the US, India, the Arctic, and Australia have found that people increasingly perceive changes in their local environment (e.g., Akerlof et al. 2012, 2013; Krupnik and Jolly 2002; Leiserowitz et al. 2013b, c; Leiserowitz and Thaker 2012; Maibach et al. 2013), that extreme events heighten people's awareness and worry about climate change, and that the reverse is true as well (e.g., Capstick et al. 2013; Leiserowitz et al. 2013b; Reser et al. 2012). Researchers are also finding that in some, but not all instances, awareness or experience of extreme events heighten people's willingness to engage in preparedness measures and/or support adaptation and mitigation policies (e.g., Reser et al. 2012; Spence et al. 2011; Whitmarsh 2008; Zaalberg et al. 2009), and that direct experiences increase people's psychological distress (e.g., Albrecht et al. 2010; Coyle and Van Susteren 2011; Moser 2013b; Reser et al. 2012; The Climate Institute 2011). At the same time, there is still much observed "psychological distancing" from climate change observed among studied publics across the world, mechanisms that are both intra-psychically and socially reinforced (e.g., Leiserowitz et al. 2013a; Lertzman 2008; Norgaard 2011; Spence et al. 2012). These seemingly contradictory findings might suggest that people are caught in a tense dilemma between, on the one hand, a desire to avoid news of climate change (both current conditions and even more so a future projected to be worse) and the dawning realization of a climate reality that is pressing upon them in real time, on the other (Cramer 2008; Dickinson 2009; Pienaar 2011).

This emerging sense of climate change being here and maybe already worse than feared is significant as a social and psychological phenomenon, and it is at the heart of the question of what kind of environmental communication is called for in this and the coming crisis time.

The humanistic imperative of environmental communication in a world of crisis

What is meant by a "humanistic" environmental communication?

To begin to answer this question, it helps to place some definitional boundaries around the word "humanistic." What stance, perspective, or approach does the adjective describe? Among the most basic definitions I list here, the first two are of greatest interest for the purposes of this chapter, but the third is quite relevant to the topic of communication, too. Humanism is (1) a system of thought that rejects religious beliefs and centers on humans and their values, capacities, and worth;[1] (2) a deep concern with the interests, needs, values, as well as the dignity and welfare of humans; and (3) the study of the humanities, learning in the liberal arts (The Free Dictionary 2013). In short, a humanistic science—and practice—is centrally concerned with human experience, the whole of human subjectivity, and with the possibilities of fulfillment of the human potential in whatever circumstance—social, economic, cultural, ecological, and even cosmological—people find themselves in (Diaz-Laplante 2007; Kuhn 2001).

The spirit of humanism that I wish to invoke here can be further specified by elements of what humanistic communications research has to offer. According to the Humanistic Communication Research Institute, research in this field aims to understand the *substance* (its weightiness and meaning), not just the *contents* of communication (*Gehalt, nicht Inhalt*) (http://hcri.de/about). It focuses on cultural values, paradigms, and belief systems, on ethics and on how individuals and groups construct their realities and thus meaning and purpose, as well as on how communication functions within social systems and can be used responsibly within them.

Even greater inspiration, however, for a relevant environmental communication in times of crisis comes from humanistic psychology. Humanistic psychology is fundamentally interested in the subjective human experience and normatively aims for human welfare. It seeks to support individuals in a process of "self-actualization," i.e. maturing into a conscious and empowered place of self-determination, in which people creatively realize their full potential. The humanistic approach in psychology emphasizes wholeness, free will, and empathy, and stresses the good in human beings. (Even so, many humanists fully embrace both the light and dark side of being human, the cruelty and love of which humans are capable.) Over the past half century, humanistic psychology has moved from being narrowly client-centered and focused on the individual to increasingly engage the question of how individual psychology is holistically embedded in, and mutually constitutive of, social, political, and environmental contexts and challenges (e.g., Diaz-Laplante 2007; Kuhn 2001; Michael 2000). There is a deeply emancipatory impulse at the root of humanistic psychology, and as such a radical desire for human liberation from both inner and outer bondage—a normative stance that enjoys good company with other empowerment-oriented "liberation" disciplines (Moser 2013a).

This focus on understanding human experience together with the desire to support human unfolding may at first seem counter-intuitive as a crucial focus for environmental communication. Yet is that human experience not at the heart of living through a time that will be increasingly disrupted by environmental crises, surprises, and profound change? Is it not a task of environmental communication to help understand such a time and make sense of it? Are the questions of who we are as humans and how we are to behave in the Anthropocene not at the core of the question of how we relate to "nature"? Is our human unfolding from now on not deeply linked to the fate of the Earth (Barnosky et al. 2012; Chakrabarty 2009; Palsson et al. 2013; Rockström et al. 2009)?

It is to these questions that a third understanding of the "humanistic" speaks, namely the essential contributions from the humanities. Like them, a humanistic environmental communication must reflect back to us our past, present, and future, our actions and desires, our beliefs and illusions, our truths and deepest needs, our destructiveness and creativity, our brightest and darkest natures. Environmental communication and the humanities share an interest in culture, values, worldviews, and frames, in stories and other forms of artistic or culturally resonant expressions of the human–environment relationship. Like for the humanities, the *raison d'être* of a humanistic environmental communication lies not merely in dissecting analysis but in curative synthesis. While it arguably remains difficult for social scientists and the humanities to draw more fully on each other's work, a humanistic environmental communication can help foster this exchange by embracing and seeking to more fully understand from both perspectives the human experience in a climate-altered world.

How can a humanistic environmental communication serve a world in crisis?

Now, as the world is beginning to see more frequent crises and disruptions, now is the time to ask what kind of environmental communication is needed. Elsewhere, I considered the tasks of environmental leaders in such a world, and asked which metaphor best described their future assignment: is it to "[b]e a steward, shepherd, arbiter, crisis manager, grief counselor, future builder?" (Moser 2012: 435). Maybe it is all of these or maybe something altogether different. In any event,

> the leaders of the future will face not just new, more difficult, and more pervasive environmental challenges than past and present leaders do, but they will need to be adept in a range of psychological, social, and political skills to navigate the inevitable human crises that will precede, trigger, and follow environmental ones … [They] will need to mentor, guide, and assist people in processing enormous losses, human distress, constant crises, and the seemingly endless need to remain engaged in the task of maintaining, restoring, and rebuilding—despite all setbacks—a viable planet, the only place the human species can call its home.
>
> (Moser 2012: 435)

Supporting those in crisis compassionately

A first answer then to the question of how a more humanistic environmental communication—both as a field of research and as a field of practice—might serve a world of crisis is an instrumental, and therefore normative, one: to attend to the people and other-than-human members of our life world to help them cope with and adjust to a crisis-stricken world. This clearly adds a layer of meaning to the notion of a "crisis discipline." It is not just about speaking to an *environment in crisis*, and that something ought to be done about it, but it is about communicating meaningfully and supportively to those *living through crisis*. If communication is indeed both symbolic and material, about mirroring our world back to ourselves and constructing meaning, about self-expression and mutual understanding, about exchange of information and, ultimately, survival, then kind and supportive communication is essential to the ability—human and otherwise—to cope and adapt to a changing climate.

Truth telling

Such support entails first and foremost helping others and ourselves to face the unfolding changes and crises. We have a long way to go to "getting real" about the legacy we have created for ourselves and all the co-inhabitants of this planet (Moser 2012). There is much "truth telling" to be done, and not simply in the way of cataloguing the unfolding catastrophes, or blaming others or ourselves, but in the way David Orr has so beautifully appealed to us:

> Telling the truth means that the people must be summoned to a level of extraordinary greatness appropriate to an extraordinarily dangerous time … Telling the truth means that we will have to speak clearly about the causes of our failures that have led us to the brink of disaster … Telling the truth means summoning people to a higher vision.
>
> (Orr 2011a: 330–331)

Communicating the truth will help us be clearer, analyze our situation more honestly, and bring us to a higher vision of ourselves, i.e. help us move up the humanistic ladder of self-actualization.

Expressing grief safely

There will also be the endless need to create communal spaces in which our emotional experience of this world can be expressed safely. While most Western cultures do not support much public display of grief, grieving our (recurring) losses we will all do. A humanistic environmental communication can serve a crucial social healing function in making space for it in how it portrays and supports this inevitable human experience. In this way it will help strengthen people's capacity to be in their own and with others' distress.

Visioning alternative futures

To be clear, a humanistic environmental communication—in supporting the much-needed capacity to cope and adapt to change and crisis—is not to be construed as a handmaiden to silencing discontent. Instead, it will insist on the emancipatory, empowering impulse of the humanistic tradition. In seeking to support human evolution toward our highest possible self, it must seek, communicate, and engage in exploring new cultural ideals and aspirations that will lessen the destructive human impact on the planet. A humanistic environmental communication would be committed to serving social transformations toward a more sustainable existence on Earth.

Fostering authentic hope

To this end, as is increasingly well understood in climate change communication, more than "narrators of doom" are needed. To overcome feelings of overwhelm, anxiety, fear, and helplessness in the face of ongoing crises and seemingly insurmountable challenges, people need true hope. Such hope can only be constructed from realistic goals, a clear or at least imaginable path, from doable tasks and a meaningful role in addressing the problems at hand (Bell 2009; Fritze et al. 2008; Hicks et al. 2005; Orr 2011b). Hope thrives where such arduous work is undertaken together (Bonanno et al. 2011; Moser 2007; Wheatley 2002). As the environmental journalist and blogger, David Roberts, so aptly put it, "When we ask for hope, then, I think we're […] asking for fellowship. The weight of climate change, like any weight, is easier to bear with others" (Roberts 2013).

Framing and urging on the transition

A humanistic environmental communication must play a critical role in helping to hold that unavoidable tension between pain and hope, out of which forward movement will come. It must help build people's ability to hold the paradoxes of crisis and normality, of immediacy and long time horizons, of destruction and beauty, of change and durability. For between repeated crises, there will be rebuilding; between death and destruction there will be birth and renewal; between fire, flood, and furious storms, there will be regrowth, rest, and recovery. But in the midst of crisis we will forget larger goals. In the face of setbacks, we will waver in our commitment. Thus, a humanistic environmental communication must help frame the transition from that "truth" of what is, that David Orr called for, toward a different, more sustainable future. It must assist the rise and ongoing renewal of authentic hope. And because the weight of the work we must do is heavy and long, it must connect, more than divide us—back with each other, to our own humanity and deepest human nature, and to the more-than-human world around us.

Mirroring who we become along the journey

A humanistic environmental communication must not be merely instrumental, however, in the sense of assisting us in alleviating immediate needs or mobilizing us to address the underlying causes of our distress. It must also be constitutive: helping us to understand this time, what is happening around and with us, and why; assisting us in reconsidering and reimagining who we are in the Anthropocene and how we fit into and belong in the larger Earth community.

As the Pulitzer Prize-winning poet, Jorie Graham, discovered in compiling an anthology of American poetry entitled *The Earth Took of Earth*, the Age of Discovery fundamentally changed seafaring people's perception of the geography of the known. Setting out from a familiar place to completely unknown lands manifested in a changed imagination, in remarkable shifts in language, turning *journeys* from searches within familiar territory into dreams "of finding an unknown *destination*" (Graham 1996). Curiously, as Sodré (2005: 71) explained, in the Arabic language, the word "Earth" shares the same linguistic root as the word "destiny." A humanistic environmental communication can help map and mirror this new human journey when not our rootedness in place and geography is at stake but our steady, or at best cyclical, experience of time, our deep sense of stability. Human destiny understood then not as a divinely, prede-termined end state, but as an actively and interactively created becoming, an unfolding that arises out of our inescapable interwovenness with an Earth on which everything is now in flux and unstable in ways unknown to the human species, this human journey into *terra quasi incognita* (Schellnhuber 2009) will be a truly worthy topic for charting in our field.

Closing: the restorative power of a humanistic environmental communication

In this chapter, I have tried to build on the notion of environmental communication as a crisis discipline introduced by Cox (2007) by outlining what a humanistic emphasis in our field may look like, and how it may be increasingly needed if the world enters a perpetual "crisis mode." What seems to be needed most is in direct tension with increasingly technologized forms of communication.

As I have tried to show, the notion of a humanistic environmental communication in a world of crisis is in some sense an extension of the field of crisis communication (e.g., Bonanno et al. 2011; Heath et al. 2007; Kasperson and Kasperson 2005; Seeger 2006). It clearly must—at minimum—convey information about the crisis at hand, manage confusion, logistics, and public relations, and otherwise mobilize action to address them. But it must do far more than that as crises become our daily bread. As Seeger (2006) argued, environmental communication must "enhance the quality of public discourse and, in turn, the quality of public judgment regarding environmental matters" (p. 96). Such judgment will be improved if we are not in fast-paced reactivity, but instead can find havens of calm in which to heal, recover, regroup, and recommit to the rebuilding and transforming that requires thoughtful attention. What I have suggested here is that such havens are made from compassion, truth telling, grieving, visioning, true hope, support-ing movement toward a better future, and mirroring back to us our journey.

What we know from crisis communication is that to restore ourselves and our environment from the trauma experienced, we need to make sense of what is happening, we need to regain a sense of control and confidence, and reclaim our capacities to manage our lives. We will do so more easily if we can draw on or re-establish our social connections, and if at all possible, as well our sense of socio-ecological belonging, our sense of place. These conditions have been found repeatedly to aid in individual and community resilience (Armitage et al. 2012; Bonanno et al. 2011; Dynes 2005; Fresque-Baxter and Armitage 2012; Morrow 2008). A humanistic environmental communication, fundamentally driven by a desire to provide solace in a time of difficulty, a hope to foster understanding and create meaning in a disrupted, disruptive environment, a wish to restore and sustain human welfare in the midst of rapid change, and a longing to support human emancipation and evolution in the Anthropocene toward our highest selves, such a discipline and practice is not just a crisis discipline, but a restorative one.

It is in this notion of environmental communication as restorative that I return to the two strands of argument from which I began. One spoke to a kind of disassociation resulting from living in the cyberworld of technology-based communication, namely the dangers of people being engulfed in endless streams of information about the world and being seemingly connected with each other through digital media in virtual worlds yet becoming increasingly absent from the world of tacit experience and true social relations (for searing critiques and discussions of how such technologized communication changes us socially, psychologically, and physiologically, see, e.g., Carr 2010; Slater 2008). As David Orr sharply observed:

> Our experience of an increasingly uniform and ugly world is being engineered and shrink-wrapped by recreation and software industries and pedaled back to us as "fun" and "information." We've become a nation of television watchers, googlers, face bookers, text messengers, and twitterers, and it shows in the way we talk and what we talk about. More and more we speak as if we are voyeurs furtively peeking in on life, not active participants, moral agents, neighbors, friends, or engaged citizens.
> (Orr 2011c: 8)

Far from placing humanistic environmental communication on the Luddite end of the disciplinary spectrum, the appeal here is simply not to forget that which matters most in crisis, and even more so as crises become commonplace: real connection. While a tweet may help locate a survivor, a story intimately told might help us make sense amidst chaos. While a message board may give crucial information, a poem might restore sanity. While a blog may serve to air frustration, in-depth conversation and dialogue will help chart the difficult path forward. Thus, environmental communication, in remembering its shared etymological roots with words such as making common, communion, sharing and fellowship, must help restore real human connection.

The other strand of argument took off from a description of the environmental conditions, particularly climate change and growing climate-related disruptions, that will confront us with ever increasing incidences of extreme events and crises. The charge to environmental communication laid out in these pages is one of helping humans not just weather these coming storms and making sense of them, but assisting humans to restore our human-to-human and human-to-life world connections, and in the inevitable tension between loss and hope find our way toward a more sustainable expression of homo sapiens living on Earth. In this way it can be a discipline that helps restore heart—*eros*, the innate desire for wholeness and connectedness—back to our everyday lives, each other, and our relationship to the more-than-human world.

Note

1 Note, however, that the issue of whether religious, transcendent experience and thought should be included in humanistic thinking is a continually debated topic and for many resolved toward an inclusive answer (Edwords

2008; Gordon 2003; Kuhn 2001). This will not be further discussed here, but is of relevance to the extent that, in crises, humans not only seek explanations that frequently invoke the supra-natural ("acts of god"), they also seek solace in one form or another of religious faith and in spiritual community. I thus include the consideration of the spiritual, religious, and transcendent in the humanistic approach suggested here.

References

Akerlof, K., Rowan, K. E., Fitzgerald, D., and Cedeno, A. Y. (2012). "Communication of climate projections in US media amid politicization of model science." *Nature Climate Change*, 2, 648–654.

Akerlof, K., Maibach, E. W., Fitzgerald, D., Cedeno, A. Y., and Neuman, A. (2013). "Do people 'personally experience' global warming, and if so how, and does it matter?" *Global Environmental Change*, 23(1), 81–91.

Albrecht, G., Sartore, G.-M., Connor, L., Higginbotham, N., Freeman, S., Kelly, B., Stain, H., Tonna, A., and Pollard, G. (2010). "Solastalgia: The distress caused by environmental change." *Australasian Psychiatry*, 15(Suppl.), S95–S98.

Armitage, D., Béné, C., Charles, A. T., Johnson, D., and Allison, E. H. (2012). "The interplay of well-being and resilience in applying a social-ecological perspective." *Ecology and Society*, 17(4), C7–15.

Barnosky, A. D., Hadly, E. A., Bascompte, J., Berlow, E. L., Brown, J. H., Fortelius, M., Getz, W. M., Harte, J., Hastings, A., Marquet, P. A., Martinez, N. D., Mooers, A., Roopnarine, P., Vermeij, G., Williams, J. W., Gillespie, R., Kitzes, J., Marshall, C., Matzke, N., Mindell, D. P., Revilla, E., and Smith, A. B. (2012). "Approaching a state shift in Earth's biosphere." *Nature*, 486(7401), 52–58.

Bell, M. M. (2009). "Can we? The audacity of environmental hope." *Nature and Culture*, 4(3), 316–323.

Bonanno, G. A., Brewin, C. R., Kaniasty, K., and LaGreca, A. M. (2011). "Weighing the costs of disaster: Consequences, risks, and resilience in individuals, families, and communities." *Psychological Science in the Public Interest*, 11(1), 1–49.

Boykoff, M. T. (2011). *Who Speaks for the Climate: Making Sense of Media Reporting on Climate Change*, Cambridge University Press, New York.

Brenner, J. and Smith, A. (2013). "72% of online adults are social networking site users," The Pew Research Center, Pew Research Center, Internet & American Life Project, Washington, DC.

Buell, F. (2003). *From Apocalypse to Way of Life: Environmental Crisis in the American Century*, Routledge, London.

Capstick, S., Pidgeon, N., and Whitehead, M. (2013). *Public Perceptions of Climate Change in Wales: Summary Findings of a Survey of the Welsh Public Conducted During November and December 2012*. Climate Change Consortium of Wales, Cardiff.

Carr, N. (2010). *The Shallows: What the Internet Is Doing to Our Brains*, W.W. Norton & Co., New York, London.

Carvalho, A. (2008). "Communicating climate change: Discourses, mediations and perceptions." Centro de Estudos de Comunicacao e Sociedade, Universidade do Minho, Braga.

Carvalho, A. (2010). "Media(ted) discourses and climate change: A focus on political subjectivity and (dis)engagement." *WIREs Climate Change*, 1(2), 172–179.

Chakrabarty, D. (2009). "The climate of history: Four theses." *Critical Inquiry*, 35(Winter), 197–222.

Cohen, S. (2013). "The Innovation of Loneliness." Available online at: www.upworthy.com/loneliness-illustrated-so-beautifully-you-will-need-to-tell-someone (last accessed 23 January 2015).

Coumou, D. and Rahmstorf, S. (2012). "A decade of weather extremes." *Nature Climate Change*, 2(7), 491–496.

Cox, R. (2007). "Nature's 'crisis discipline': Does environmental communication have an ethical duty?" *Environmental Communication: A Journal of Nature and Culture*, 1(1), 5–20.

Coyle, K. J. and Van Susteren, L. (2011). *The Psychological Effects of Global Warming on the United States: And Why the U.S. Mental Health Care System Is Not Adequately Prepared*. National Wildlife Federation Climate Education Program, Washington, DC.

Cramer, P. (2008). "Seven pillars of defense mechanism theory." *Social and Personality Psychology Compass*, 2, 1963–1981.

Diaz-Laplante, J. (2007). "Humanistic psychology and social transformation: Building the path toward a livable today and a just tomorrow." *Journal of Humanistic Psychology*, 47(1), 54–72.

Dickinson, J. L. (2009). "The people paradox: Self-esteem striving, immortality ideologies, and human response to climate change." *Ecology and Society*, 14(1), 34. Available online at: www.ecologyandsociety.org/vol14/iss1/art34/ (last accessed 23 January 2015).

Dominick, J. R. (2010). *The Dynamics of Mass Communication: The Media in Transition*, McGraw-Hill International Edition, Columbus, OH.

Dynes, R. R. (2005). "Community social capital as the primary basis for resilience." University of Delaware, Disaster Research Center.

Edwords, F. (2008). "What is humanism?", American Humanist Association, Washington, DC. Available online at: http://americanhumanist.org/humanism/what_is_humanism (last accessed 23 January 2015).

Fresque-Baxter, J. A. and Armitage, D. (2012). "Place identity and climate change adaptation: A synthesis and framework for understanding." *WIREs Climate Change*, 3(3), 251–266.

Fritze, J. G., Blashki, G. A., Burke, S., and Wiseman, J. (2008). "Hope, despair and transformation: Climate change and the promotion of mental health and wellbeing." *International Journal of Mental Health Systems*, 2(1), [art 13]. Available online at: www.ijmhs.com/content/2/1/13 (last accessed 23 January 2015).

Gordon, K. (2003). "The impermanence of being: Toward a psychology of uncertainty." *Journal of Humanistic Psychology*, 43(2), 96–117.

Graham, J. (1996). "The Earth Took of Earth: Introduction." *The Earth Took of Earth: A Golden Ecco Anthology*, J. Graham, ed., The Ecco Press/HarperCollins, New York, 282 pp.

Heath, R. L., Palenchar, M. J., Proutheau, S., and Hocke, T. M. (2007). "Nature, crisis, risk, science, and society: What is our ethical responsibility?" *Environmental Communication: A Journal of Nature and Culture*, 1(1), 34–48.

Hicks, M. A., Berger, J. G., and Generett, G. G. (2005). "From hope to action: Creating spaces to sustain transformative habits of mind and heart." *Journal of Transformative Education*, 3(1), 57–75.

Intergovernmental Panel on Climate Change (IPCC). (2012). "Managing the risks of extreme events and disasters to advance climate change adaptation. A Special Report of Working Groups I and II of the Intergovernmental Panel on Climate Change." C. B. Field, V. Barros, T. F. Stocker, D. Qin, D. J. Dokken, K. L. Ebi, M. D. Mastrandrea, K. J. Mach, G.-K. Plattner, S. K. Allen, M. Tignor, and P. M. Midgley, eds, Cambridge University Press, Cambridge, UK and New York, 582 pp.

Kasperson, J. X., and Kasperson, R. E. (2005). *Social Contours of Risk, Volume 1: Publics, Risk Communication and the Social Amplification of Risk*, Earthscan, London.

Killingsworth, M. J. (2007). "A phenomenological perspective on ethical duty in environmental communication." *Environmental Communication: A Journal of Nature and Culture*, 1(1), 58–63.

Krupnik, I. and Jolly, D. (2002). *The Earth is Faster Now: Indigenous Observations of Arctic Environmental Change*. Arctic Research Consortium of U.S., in cooperation with the Arctic Studies Center, Smithsonian Institution, Fairbanks, Alaska, 384 pp.

Kuhn, J. L. (2001). "Toward an ecological humanistic psychology." *Journal of Humanistic Psychology*, 41(2), 9–24.

Leiserowitz, A. and Thaker, J. (2012). *Climate Change in the Indian Mind*. Yale University, Yale Project on Climate Change Communication, New Haven, CT.

Leiserowitz, A., Maibach, E., Roser-Renouf, C., Feinberg, G., and Howe, P. (2013a). *Climate Change in the American Mind: Americans' Global Warming Beliefs and Attitudes in April, 2013*. Yale University and George Mason University, Yale Project on Climate Change Communication, New Haven, CT.

Leiserowitz, A., Maibach, E., Roser-Renouf, C., Feinberg, G., and Howe, P. (2013b). *Extreme Weather and Climate Change in the American Mind: April 2013*. Yale University and George Mason University, Yale Project on Climate Change Communication, New Haven, CT.

Leiserowitz, A., Thaker, J., Feinberg, G., and Cooper, D. (2013c). *Global Warming's Six Indias*. Yale University, Yale Project on Climate Change Communication, New Haven, CT.

Lertzman, R. (2008). "The myth of apathy." *The Ecologist* (June), 16–17.

Maibach, E., Bloodheart, B., and Zhao, X. (2013). *Perceptions of Extreme Weather and Climate Change in Virginia: A Statewide Survey of Adults*. Center for Climate Change Communication, George Mason University, Fairfax, VA.

Michael, D. N. (2000). "Some observations with regard to a missing elephant." *Journal of Humanistic Psychology*, 40(1), 8–18.

Morrow, B. H. (2008). *Community Resilience: A Social Justice Perspective*. Community and Regional Resilience Institute, Oak Ridge, TN.

Moser, S. C. (2007). "More bad news: The risk of neglecting emotional responses to climate change information." In: *Creating a Climate for Change: Communicating Climate Change and Facilitating Social Change*, S. C. Moser and L. Dilling, eds, Cambridge University Press, Cambridge, UK, 64–80.

Moser, S. C. (2012). "Getting real about it: Navigating the psychological and social demands of a world in distress." In: *Sage Handbook on Environmental Leadership*, D. Rigling Gallagher, R. N. L. Andrews, and N. L. Christensen, eds, Thousand Oaks, CA: Sage, 432–440.

Moser, S. C. (2013a). "Individual and community empowerment for human security." In: *A Changing Environment for Human Security: Transformative Approaches to Research, Policy and Action*, L. Sygna, K. O'Brien, and J. Wolf, eds, Earthscan/Routledge, London, New York, 279–293.

Moser, S. C. (2013b). "Navigating the political and emotional terrain of adaptation: Community engagement when climate change comes home." In: *Successful Adaptation to Climate Change: Linking Science and Policy in a Rapidly Changing World*, S. C. Moser and M. T. Boykoff, eds, Routledge, London, 289–305.

Moser, S. C. and Dilling, L. (2004). "Making climate hot: Communicating the urgency and challenge of global climate change." *Environment*, 46(10), 32–46.

Moser, S. C. and Dilling, L. (2007). *Creating a Climate for Change: Communicating Climate Change and Facilitating Social Change*. Cambridge University Press, Cambridge, UK.

MunichRe. (2013). *2012 Natural Catastrophe Year in Review*. Munich Reinsurance America, Princeton, NJ. Available online at: www.munichre.com/us/property-casualty/events/webinar/index.html (last accessed 23 January 2015).

Norgaard, K. M. (2011). *Living in Denial: Climate Change, Emotions, and Everyday Life*, MIT Press, Cambridge, MA.

Orr, D. W. (2011a). "Hope (in a hotter time)." In: *Hope is an Imperative: The Essential David Orr*, D. W. Orr, ed., Island Press, Washington, DC, 324–332.

Orr, D. W. (2011b). *Hope is an Imperative: The Essential Orr*. Island Press, Washington, DC.

Orr, D. W. (2011c). "Verbicide." In: *Hope is an Imperative: The Essential David Orr*, D. W. Orr, ed., Island Press, Washington, DC, 5–12.

Palsson, G., Szerszynski, B., Sörlin, S., Marks, J., Avril, B., Crumley, C., Hackmann, H., Holm, P., Ingram, J., Kirman, A., Buendía, M. P., and Weehuizen, R. (2013). "Reconceptualizing the "Anthropos" in the Anthropocene: Integrating the social sciences and humanities in global environmental change research." *Environmental Science and Policy* 28, 3–13.

Peterson, N. M., Peterson, M. J., and Rai Peterson, T. (2007). "Environmental communication: Why this crisis discipline should facilitate environmental democracy." *Environmental Communication: A Journal of Nature and Culture*, 1(1), 74–86.

Peterson, T. C., Stott, P. A., and Herring, S. (2012). "Explaining extreme events of 2011 from a climate perspective." *BAMS*, 93(7), 1041–1067.

Pienaar, M. (2011). "An eco-existential understanding of time and psychological defenses: Threats to the environment and implications for psychotherapy." *Ecopsychology*, 3(1), 25–39.

Rainie, L. (2013). *New Media Ecology: Personal. Portable. Participatory. Pervasive. The Digital Landscape in 2013 and its Impact on Communities*. Pew Research Center, Internet & American Life Project, Washington, DC.

Reser, J. P., Bradley, G. L., Glendon, A. I., Ellul, M. C., and Callaghan, R. (2012). "Public risk perceptions, understandings and responses to climate change and natural disasters in Australia and Great Britain: Final report." National Climate Change Adaptation Research Facility, Gold Coast, Australia.

Roberts, D. (2013). "Hope and fellowship." *Grist* Magazine, August 30. Available online at: http://grist.org/climate-energy/hope-and-fellowship/ (last accessed 23 January 2015).

Rockström, J., Steffen, W., Noone, K., Persson, A., Chapin, F. S., Lambin, E. F., Lenton, T. M., Scheffer, M., Folke, C., Schellnhuber, H. J., Nykvist, B., de Wit, C. A., Hughes, T., van der Leeuw, S., Rodhe, H., Sorlin, S., Snyder, P. K., Costanza, R., Svedin, U., Falkenmark, M., Karlberg, L., Corell, R. W., Fabry, V. J., Hansen, J., Walker, B., Liverman, D., Richardson, K., Crutzen, P., and Foley, J. A. (2009). "A safe operating space for humanity." *Nature*, 461(7263), 472–475.

Schellnhuber, H.-J. (2009). "Terra Quasi-Incognita: Beyond the 2°C Line." Keynote Address presented at the "4 Degrees & Beyond: International Climate Conference." Available online at: www.eci.ox.ac.uk/4degrees/keynote.php (last accessed 23 January 2015).

Schwarze, S. (2007). "Environmental communication as a discipline of crisis." *Environmental Communication: A Journal of Nature and Culture*, 1(1), 87–98.

Seeger, M. W. (2006). "Best practices in crisis communication: An expert panel process." *Journal of Applied Communication Research*, 34(3), 232–244.

Senecah, S. L. (2007). "Impetus, mission, and future of the Environmental Communication Commission/Division: Are we still on track? Were we ever?" *Environmental Communication: A Journal of Nature and Culture*, 1(1), 21–33.

Seneviratne, S. I., Nicholls, N., Easterling, D., Goodess, C. M., Kanae, S., Kossin, J., Luo, Y., Marengo, J., McInnes, K., Rahimi, M., Reichstein, M., Sorteberg, A., Vera, C., and Zhang, X. (2012). "Changes in climate extremes and their impacts on the natural physical environment." In: *Managing the Risks of Extreme Events and Disasters to Advance Climate Change Adaptation. A Special Report of Working Groups I and II of the Intergovernmental Panel on Climate Change*, C. B. Field, V. Barros, T. F. Stocker, D. Qin, D. J. Dokken, K. L. Ebi, M. D. Mastrandrea, K. J. Mach, G.-K. Plattner, S. K. Allen, M. Tignor, and P. M. Midgley, eds, Cambridge University Press, Cambridge, UK, 109–230.

Showstack, R. (2013). "Report ranks U.S. coastal state protection of marine areas." *Eos, Transactions of the American Geophysical Union*, 94(24), 216–216.

Slater, G. (2008). "Numb." In: *Archetypal Psychologies: Reflections in Honor of James Hillman*, S. Marlan, ed., Spring Journal, New Orleans, LA, 351–368.

Smith, A., and Katz, R. (2013). "U.S. billion-dollar weather and climate disasters: Data sources, trends, accuracy and biases." *Natural Hazards*, 67, 387–410.

Sodré, M. (2005). "For a feeling humanism: The political emergence of the emotions." *Diogenes*, 52(2), 71–78.

Spence, A., Poortinga, W., Butler, C., and Pidgeon, N. F. (2011). "Perceptions of climate change and willingness to save energy related to flood experience." *Nature Climate Change*, 1(1), 46–49.

Spence, A., Poortinga, W., and Pidgeon, N. (2012). "The psychological distance of climate change." *Risk Analysis*, 36(6), 957–972.

The Climate Institute. (2011). *A Climate of Suffering: The Real Costs of Living with Inaction on Climate Change*. The Climate Institute, Melbourne, Sydney.

The Free Dictionary. (2013). "Humanism (definition)." The Free Dictionary, available online at: www.thefreedictionary.com/humanism (last accessed 23 January 2015).

The Pew Research Center. (2013). *The State of the News Media in 2013: An Annual Report on American Journalism*. The Pew Research Center, Project for Excellence in Journalism, Washington, DC.

Wheatley, M. J. (2002). *Turning to One Another: Simple Conversations to Restore Hope to the Future*, Berrett-Koehler Publishers, Berkeley, CA.

Whitmarsh, L. (2008). "Are flood victims more concerned about climate change than other people? The role of direct experience in risk perception and behavioural response." *Journal of Risk Research*, 11(3), 351–374.

Zaalberg, R., Midden, C., Meijnders, A., and McCalley, T. (2009). "Prevention, adaptation, and threat denial: Flooding experiences in the Netherlands." *Risk Analysis*, 29(12), 1759–1778.

INDEX

For Product Safety Concerns and Information please contact our EU
representative GPSR@taylorandfrancis.com
Taylor & Francis Verlag GmbH, Kaufingerstraße 24, 80331 München, Germany